VOLUME FOUR TWENTY EIGHT

Methods in
ENZYMOLOGY

Osmosensing and Osmosignaling

METHODS IN ENZYMOLOGY

Editors-in-Chief

JOHN N. ABELSON AND MELVIN I. SIMON

Division of Biology
California Institute of Technology
Pasadena, California

Founding Editors

SIDNEY P. COLOWICK AND NATHAN O. KAPLAN

VOLUME FOUR TWENTY EIGHT

METHODS IN ENZYMOLOGY

Osmosensing and Osmosignaling

EDITED BY

DIETER HÄUSSINGER
Clinic for Gastroenterology, Hepatology, and Infectiology
University Hospital
Düsseldorf, Germany

HELMUT SIES
Heinrich-Heine-University Düsseldorf
Institute for Biochemistry and Molecular Biology I
Düsseldorf, Germany

AMSTERDAM • BOSTON • HEIDELBERG • LONDON
NEW YORK • OXFORD • PARIS • SAN DIEGO
SAN FRANCISCO • SINGAPORE • SYDNEY • TOKYO
Academic Press is an imprint of Elsevier

ELSEVIER

Academic Press is an imprint of Elsevier
525 B Street, Suite 1900, San Diego, California 92101-4495, USA
84 Theobald's Road, London WC1X 8RR, UK

This book is printed on acid-free paper. ∞

For information on all Elsevier Academic Press publications
visit our Web site at www.books.elsevier.com

ISBN: 978-0-12-373921-6

PRINTED IN THE UNITED STATES OF AMERICA
07 08 09 10   9 8 7 6 5 4 3 2 1

CONTENTS

CONTRIBUTORS

Darwin O. V. Alonso
Department of Medicinal Chemistry, University of Washington, Seattle, Washington

Matthew Auton
Department of Biochemistry and Molecular Biology, University of Texas Medical Branch, Galveston, Texas

Wendy Bartlett
School of Medical Sciences, University of Aberdeen, Institute of Medical Sciences, Foresterhill, Aberdeen, United Kingdom

David A. C. Beck
Department of Medicinal Chemistry, University of Washington, Seattle, Washington

Brian J. Bennion
Department of Medicinal Chemistry, University of Washington, Seattle, Washington

Adam Bertl
Institute of Botany, Applied Plant Sciences, Darmstadt University of Technology, Darmstadt, Germany

Susan Black
School of Medical Sciences, University of Aberdeen, Institute of Medical Sciences, Foresterhill, Aberdeen, United Kingdom

D. Wayne Bolen
Department of Biochemistry and Molecular Biology, University of Texas Medical Branch, Galveston, Texas

Ian R. Booth
School of Medical Sciences, University of Aberdeen, Institute of Medical Sciences, Foresterhill, Aberdeen, United Kingdom

Carl D. Bortner
Laboratory of Signal Transduction, National Institute of Environmental Health Sciences, Department of Health and Human Services, National Institutes of Health, Research Triangle Park, North Carolina

Alice G. M. Bot
Department of Biochemistry, Erasmus University Medical Center, Rotterdam, The Netherlands

Maurice B. Burg
Laboratory of Kidney and Electrolyte Metabolism, National Heart, Lung, and Blood Institute, Department of Health and Human Services, National Institutes of Health, Bethesda, Maryland

Ivan L. Cameron
Department of Cellular and Structural Biology, University of Texas HSC at San Antonio, San Antonio, Texas

John A. Cidlowski
Laboratory of Signal Transduction, National Institute of Environmental Health Sciences, Department of Health and Human Services, National Institutes of Health, Research Triangle Park, North Carolina

J. Clotet
Department of Molecular and Cellular Biology, Universitat Internacional de Catalunya (UIC), Sant Cugat del Vallès, Spain

Valerie Daggett
Department of Medicinal Chemistry, University of Washington, Seattle, Washington

Hugo R. de Jonge
Department of Biochemistry, Erasmus University Medical Center, Rotterdam, The Netherlands

Natalia I. Dmitrieva
Laboratory of Kidney and Electrolyte Metabolism, National Heart, Lung, and Blood Institute, National Institutes of Health, Bethesda, Maryland

Michelle D. Edwards
School of Medical Sciences, University of Aberdeen, Institute of Medical Sciences, Foresterhill, Aberdeen, United Kingdom

Joan D. Ferraris
Laboratory of Kidney and Electrolyte Metabolism, National Heart, Lung, and Blood Institute, Department of Health and Human Services, National Institutes of Health, Bethesda, Maryland

Michael Föller
Department of Physiology, University of Tübingen, Germany

Gary D. Fullerton
Department of Radiology, University of Texas HSC at San Antonio, San Antonio, Texas

Lila M. Gierasch

Departments of Biochemistry & Molecular Biology and Chemistry, University of Massachusetts, Amherst, Massachusetts

Teresa Gonzalez

INSERM U 568, University of Nice Sophia-Antipolis, Nice, France

Thierry Gremeaux

INSERM U 568, University of Nice Sophia-Antipolis, Nice, France

Philippe Gual

INSERM U 568, University of Nice Sophia-Antipolis, Nice, France

Erich Gulbins

Department of Molecular Biology, University Duisburg-Essen, Duisburg, Germany

Dieter Häussinger

Clinic of Gastroenterology, Hepatology and Infectiology, Heinrich-Heine-University Düsseldorf, Düsseldorf, Germany

Helmut Haas

Department of Neurophysiology, University of Düsseldorf, Düsseldorf, Germany

Birgit Heller-Stilb

Clinic for Gastroenterology, Hepatology and Infectiology, University of Düsseldorf, Germany

Thomas L. Hilder

Department of Pharmacology and Lineberger Comprehensive Cancer Center, University of North Carolina at Chapel Hill, Chapel Hill, North Carolina

Stefan Hohmann

Department of Cell and Molecular Biology, Göteborg University, Göteborg, Sweden

Zebo Huang

College of Pharmacy, Wuhan University, Wuhan, China, and Institute of Biotechnology, University of Cambridge, Cambridge, United Kingdom

Stephan M. Huber

Department of Physiology, University of Tübingen, Germany

Zoya Ignatova

Max Planck Institute for Biochemistry, Martinsried, Germany

Un Sil Jeon

Department of Medicine, University of Maryland, Baltimore, Maryland

Gary L. Johnson
Department of Pharmacology and Lineberger Comprehensive Cancer Center, University of North Carolina at Chapel Hill, Chapel Hill, North Carolina

Ralf Kaldenhoff
Institute of Botany, Applied Plant Sciences, Darmstadt University of Technology, Darmstadt, Germany

Jeong Ah Kim
Department of Medicine, University of Maryland, Baltimore, Maryland

Michael C. Konopka
Department of Chemistry, University of Wisconsin–Madison, Madison, Wisconsin

Marcus Krantz
The Systems Biology Institute, Keio University School of Medicine, Shinjuku-ku, Tokyo, Japan

Ralf Kubitz
Clinic of Gastroenterology, Hepatology, and Infectiology, Heinrich-Heine-University Düsseldorf, Düsseldorf, Germany

H. Moo Kwon
Department of Medicine, University of Maryland, Baltimore, Maryland

Min Seong Kwon
Department of Medicine, University of Maryland, Baltimore, Maryland

Florian Lang
Department of Physiology, University of Tübingen, Germany

Karl Lang
Department of Physiology, University of Tübingen, Germany

Philipp Lang
Department of Physiology, University of Tübingen, Germany

Christina H. Lim
Department of Biochemistry, Erasmus University Medical Center, Rotterdam, The Netherlands

Sun Woo Lim
Department of Medicine, University of Maryland, Baltimore, Maryland

Michael H. Malone
Department of Pharmacology and Lineberger Comprehensive Cancer Center, University of North Carolina at Chapel Hill, Chapel Hill, North Carolina

Yannick Le Marchand-Brustel
INSERM U 568, University of Nice Sophia-Antipolis, Nice, France

Samantha Miller
School of Medical Sciences, University of Aberdeen, Institute of Medical Sciences, Foresterhill, Aberdeen, United Kingdom

Menachem Moshelion
The Robert H. Smith Institute of Plant Sciences and Genetics in Agriculture, Faculty of Agricultural, Food & Environmental Quality Science, The Hebrew University of Jerusalem, Rehovot, Israel

Kazuo Nakashima
Biological Resources Division, Japan International Research Center for Agricultural Sciences, Tsukuba, Ibaraki, Japan

Bernd Nilius
Department of Molecular Cell Biology, Division of Physiology, KU Leuven, Campus Gasthuisberg, Leuven, Belgium

Bodil Nordlander
Department of Cell and Molecular Biology, Göteborg University, Göteborg, Sweden

Evelyn Oermann
C.&O. Vogt Institute for Brain Research, University of Düsseldorf, Germany

Beate Otto
Institute of Botany, Applied Plant Sciences, Darmstadt University of Technology, Darmstadt, Germany,

Evangelia A. Papakonstanti
Department of Biochemistry, University of Crete Medical School, Heraklion, Greece

Stine Falsig Pedersen
Department of Molecular Biology, University of Copenhagen, Copenhagen, Denmark

F. Posas
Cell Signaling Unit, Departament de Ciències Experimentals i de la Salut, Universitat Pompeu Fabra (UPF), Barcelona, Spain

Akiko Rasmussen
School of Medical Sciences, University of Aberdeen, Institute of Medical Sciences, Foresterhill, Aberdeen, United Kingdom

Tim Rasmussen
School of Medical Sciences, University of Aberdeen, Institute of Medical Sciences, Foresterhill, Aberdeen, United Kingdom

M. Thomas Record, Jr.
Department of Chemistry and Biochemistry, University of Wisconsin–Madison, Madison, Wisconsin

Roland Reinehr
Clinic for Gastroenterology, Hepatology, and Infectiology, Heinrich–Heine–University Düsseldorf, Düsseldorf, Germany

Markus Ritter
Institute of Physiology and Pathophysiology, Paracelsus Medical University, Salzburg, Austria

Jörg Rösgen
Department of Biochemistry and Molecular Biology, University of Texas Medical Branch, Galveston, Texas

Freimut Schliess
Clinic for Gastroenterology, Hepatology, and Infectiology, Heinrich-Heine-University Düsseldorf, Düsseldorf, Germany

Ulrike Schumann
School of Medical Sciences, University of Aberdeen, Institute of Medical Sciences, Foresterhill, Aberdeen, United Kingdom

Kazuo Shinozaki
RIKEN Plant Science Center, Tsurumi-ku, Yokohama, and Core Research for Evolution Science and Technology, Japan Science and Technology, Kawaguchi, Saitama, Japan

Maria I. Sifre
Laboratory of Signal Transduction, National Institute of Environmental Health Sciences, Department of Health and Human Services, National Institutes of Health, Research Triangle Park, North Carolina

Christos Stournaras
Department of Biochemistry, University of Crete Medical School, Heraklion, Greece

Ildiko Szabo
Department of Biology, University of Padova, Padova, Italy

Jean-François Tanti
INSERM U 568, University of Nice Sophia-Antipolis, Nice, France

Ben C. Tilly
Department of Biochemistry, Erasmus University Medical Center, Rotterdam, The Netherlands

Lam-Son Phan Tran
Biological Resources Division, Japan International Research Center for Agricultural Sciences, Tsukuba, Ibaraki, Japan

Alan Tunnacliffe
Institute of Biotechnology, University of Cambridge, Cambridge, United Kingdom

Narendra Tuteja
Plant Molecular Biology, International Centre for Genetic Engineering and Biotechnology, Aruna Asaf Ali Marg, New Delhi, India

Norbert Uehlein
Institute of Botany, Applied Plant Sciences, Darmstadt University of Technology, Darmstadt, Germany

Alexey Vereninov
Institute of Cytology, Russian Academy of Sciences, St. Petersburg, Russia

Ulrich Warskulat
Clinic for Gastroenterology, Hepatology and Infectiology, University of Düsseldorf, Germany

James C. Weisshaar
Department of Chemistry, University of Wisconsin–Madison, Madison, Wisconsin

Janet M. Wood
Department of Molecular and Cellular Biology, University of Guelph, Guelph, Ontario, Canada

Kazuko Yamaguchi-Shinozaki
Biological Resources Division, Japan International Research Center for Agricultural Sciences, Tsukuba, Ibaraki, and Core Research for Evolution Science and Technology, Japan Science and Technology, Kawaguchi, Saitama, and Laboratory of Plant Molecular Physiology, Graduate School of Agricultural and Life Sciences, The University of Tokyo, Bunkyo-ku, Tokyo, Japan

Karl Zilles
C.&O. Vogt Institute for Brain Research, University of Düsseldorf, and Institute of Medicine, Research Center Jülich, Germany

Preface

One prerequisite for life is the ability of cells to maintain their volume, which may change in response to alterations of ambient osmolarity and nutrient supply. Thus, cells have evolved powerful mechanisms for stabilization of cell volume. These mechanisms allow fluctuations in cell hydration, which have been recognized as important signals. By these means, ambient osmolarity and activity of transport systems in the plasma membrane affect not only cell water content but also cell structure, function, and gene expression. Osmosensors and transducers exist in many cell types that couple to osmosignaling pathways, ultimately triggering functional adaptation.

Whereas in yeast and bacteria such adaptations primarily aim to stabilize cell volume, in higher organisms with stable extracellular osmolarity, cell water changes are used for metabolic regulation. Mechanotransduction plays a major role in osmosensing. Osmosensing and osmosignaling may not only maintain cell shape, water, and ionic composition of the cell, but may also be activated to adapt metabolic cell function and gene expression. Organic osmolytes are involved in counteracting osmotic stress in part by stabilizing protein structure. Disturbance of water homeostasis and osmolyte handling is pathophysiologically relevant in a variety of diseases.

We are pleased to present current knowledge in this rapidly expanding field of research. The leading experts in the areas of osmosensing and osmosignaling in bacteria, yeast, and higher organisms contributed to this volume, to all of whom we are grateful. We also would like to express our thanks to Marlies Scholtes and Cindy Minor for their excellent help.

Dieter Häussinger and Helmut Sies

METHODS IN ENZYMOLOGY

CHAPTER ONE

WATER COMPARTMENTS IN CELLS

Gary D. Fullerton* *and* Ivan L. Cameron[†]

Contents

* Department of Radiology, University of Texas HSC at San Antonio, San Antonio, Texas
[†] Department of Cellular and Structural Biology, University of Texas HSC at San Antonio, San Antonio, Texas

Methods in Enzymology, Volume 428
ISSN 0076-6879, DOI: 10.1016/S0076-6879(07)28001-2

Abstract

Human experience in the macrobiological world leads scientists to visualize water compartments in cells analogous to the bladder in the human pelvis or ventricles in the brain. While such water-filled cellular compartments likely exist, the volume contributions are insignificant relative to those of biomolecular hydration compartments. The purpose of this chapter is to identify and categorize the molecular water compartments caused by proteins, the primary macromolecular components of cells. The categorical changes in free energy of water molecules on proteins cause these compartments to play dominant roles in osmoregulation and provide important adjuncts to fundamental understanding of osmosensing and osmosignaling mechanisms. Water compartments possess differences in molecular motion, enthalpy, entropy, freezing point depression, and other properties because of electrostatic interaction of polar water molecules with electric fields generated by covalently bound pairs of opposite charge caused by electronegative oxygen and nitrogen atoms of the protein. Macromolecules, including polypeptides, polynucleotides, and polysaccharides, are stiff molecular chains with restricted folding capacities due to inclusion of rigid ring structures or double amide bonds in the backbone sequence. This creates "irreducible spatial charge separation" between positive and negative partial charges, causing elevated electrostatic energy. In the fully hydrated *in vivo* state of living cells the high dielectric coefficient of water reduces protein electrostatic free energy by providing polar "water bridge networks" between charges, thereby creating four measurably different compartments of bound water with distinct free energy differences.

1. INTRODUCTION

This chapter describes general considerations regarding experiments used to detect and measure compartmental hydration capacities of proteins, the primary macromolecular components of cells, and to predict the influence of hydration compartments on cellular osmoregulation using the red blood cell (RBC) as a model cellular system. The discussion embarks with a brief discussion of the electrostatic energy source for directly binding different categories of water molecules as single water bridges, double water bridges, dielectric water clusters (DWC) around the bridges, and finally as encapsulated bulk water sequestered inside the protein volume. The description leads to presentation of molecular calculation methods to predict compartmental water capacities and confirms predictions by comparison to biophysical measurements on fibrillar collagen and two globular proteins, serum albumin and lysozyme. Model predictions are validated by comparison to a wide range of biophysical measurements from the literature to confirm the wide extent of observations attributable to compartmental hydration effects. Four methods used to measure compartmental capacities of single proteins and/or cells accurately are described to assist experimental confirmation of predictions by the reader. The discussion includes a brief review of enzyme behavior as a function of compartmental hydration to facilitate the translation of cellular water and protein hydration compartment concepts to commercial biotechnology applications and concludes by relating molecular hydration mechanisms as source causes of osmosensing, osmosignaling, and osmoregulation effects.

2. THE STOICHIOMETRIC HYDRATION MODEL (SHM)

2.1. Water bridges

As shown in Figs. 1.1 and 1.2, water bridges (Ramachandran and Chandrasekharan, 1968; Bella et al., 1995, 1996; Fullerton and Rahal, 2007) are bound by the change in electrostatic free energy caused when stiff macromolecules with hydrophilic sites on the molecular backbone are immersed in dielectric water. The free energy $\Delta G = \Delta H - T\Delta S$ of water binding is the sum of negative electrostatic enthalpy ΔH partially compensated by the positive contributions from negative entropic change ΔS caused by the increased order of dielectrically aligned water molecules. Two categories of water bridges with the deepest ΔG free energy wells are found on the backbone due to the distance of separation between charges: the single water bridge (Ramachandran bridge, zone 0), requiring only one molecule for small separations, and double water bridge, requiring three water molecules (Fullerton and Rahal, 2007) (zone I) for larger separations, as shown in Fig. 1.2. Two additional water compartments with much lower binding energies are primary water molecules bound directly to side chain hydrophilic sites (zone II) and secondary molecules with the lowest binding energies bridging between directly bound water to bridge over exposed hydrophobic regions (zone III) (for details, see Fullerton and Rahal, 2007).

2.2. Methods used to calculate compartmental capacities

Methods used for calculating hydration fractions on collagen (Fullerton and Amurao, 2006; Fullerton and Rahal, 2007; Fullerton et al., 2006c) are summarized in Table 1.1. and briefly described here. Molecular weights

Work W_V to separate NH^+ and CO^-
in vacuum is reduced in dielectric water

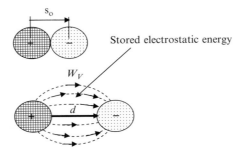

In vacuum: $W_V = q_1 q_2/(4\pi\varepsilon_o)/(1/r_{12} - 1/s_o)$
              $r_{12} = s_o + d$

In water: $W_W = W_V/\kappa = W_V/80$
              where κ = water dielectric constant ~80

Figure 1.1 Work is required to separate a charge pair. Thus the irreducible separation of partial charges on the backbone of macromolecular polymers such as proteins creates an electrostatic field that stores this energy. Electrostatic energy is reduced by insertion of a dielectric between the charge pair. The water molecules forming the dielectric are both aligned in the electric field and bound to the protein by the energy differential between vacuum and water. Ramachandran coined the term "water bridge" to describe this effect for a single water bridge, which we refer to as "Ramachandran bridges." Reproduced with permission from Fullerton and Rahal (2007).

and average residue molecular weights are calculated from the number of constituent amino acids downloaded from the Protein Data Bank (http://www.rcsb.org/pdb/home/home.do) and known molecular weights of the constituent amino acids using the Excel spreadsheet displayed in Table 1.1. Hydration fractions are calculated using stoichiometric hydration rules derived for the collagen molecule structure relative to the fundamental tripeptide unit. The hydration capacities h equal the molecular weight of water $M_w = 18.015$ Da divided by three times the mean amino acid residue molecular weight of the protein multiplied by stoichiometric hydration numbers; one single water bridge per tripeptide for h_{Ra}, four waters for h_B, and 12.5 water molecules per tripeptide h_{Pr} to complete coverage of hydrophilic sites (Fullerton and Rahal, 2007). Thus $h_{Ra} = 18.015/(3 \times$ average amino acid molecular weight $= 3 \times 91.2$ Da$) = 0.065$ g/g, and $h_B = 4 \times h_{Ra} = 0.262$ g/g and $h_{Pr} = 12.5 \times h_{Ra} = 0.81$ g/g for collagen as calculated at the bottom of Table 1.1. The calculation procedures were validated for collagen by comparison of model predicted hydration capacities with measurements using six biophysical measurement methods (Fullerton and Rahal, 2007). The calculation of h_M in the same spreadsheet uses *solvent-accessible surface areas* (SASA)

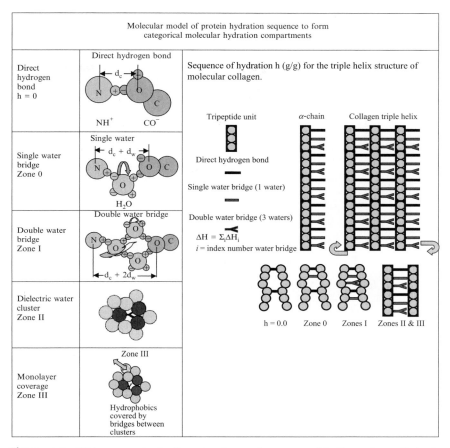

Figure 1.2 This conceptual diagram shows the sequential formation of single water bridges (zone 0), double water bridges (zone I), dielectric water clusters (zone II), and monolayer coverage (zone III) with completed fourfold hydrogen bonding of all bound waters in the dielectric clusters. Water molecules in each zone have differences in free energy, melting temperature, osmosis, and other properties of biological significance. (See color insert.)

as described in the next section. The collagen hydration capacities are summarized in Fig. 1.3.

2.3. Solvent-accessible surface area calculation method

The SASA calculation method developed by Miller and colleagues (1987a,b) is used to calculate SASA for polar, nonpolar, and total molecular surfaces. The area occupied by a water molecule is calculated with Eq. (1.1) using an effective radius of separation at room temperature for water neighbors

Table 1.1 Protein molecular calculation method based on type 1 collagen

| Amino acid | | Backbone SASA | | | Side chain SASA | | | AA | Protein composition | | Protein Max SASA | | |
Name symbol	AA MW (Da)	Nonpolar (Å²)	Polar (Å²)	Total (Å²)	Nonpolar (Å²)	Polar (Å²)	Total (Å²)	Total (Å²)	Number residue	Residue MW	Nonpolar (Å²)	Polar (Å²)	Total (Å²)
Ala	89.09	20.23	25.77	46	67	0	67	113	119	8458	10380	3067	13447
Arg	174.21	19.23	25.77	45	89	107	196	241	64	9996	6927	8497	15424
Asn	132.12	19.23	25.77	45	44	69	113	158	24	2739	1518	2274	3792
Asp	133.1	19.23	25.77	45	48	58	106	151	23	2647	1546	1927	3473
Cys	121.15	10.23	25.77	36	35	69	104	140	0	0	0	0	0
Gln	146.15	19.23	25.77	45	53	91	144	189	24	3075	1734	2802	4536
Glu	147.13	19.23	25.77	45	61	77	138	183	53	6843	4252	5447	9699
Gly	75.07	59.23	25.77	85	0	0	0	85	394	22480	23337	10153	33490
His	155.16	17.23	25.77	43	102	49	151	194	5	686	596	374	970
Ile	131.18	16.23	25.77	42	140	0	140	182	15	1697	2343	387	2730
Leu	131.18	17.23	25.77	43	137	0	137	180	40	4527	6169	1031	7200
Lys	146.19	18.23	25.77	44	119	48	167	211	24	3076	3294	1770	5064
Hylys	162.19	18.23	25.77	44	78	43	169	213	13	1874	1251	894	2145
Met	149.22	18.23	25.77	44	117	43	160	204	12	1574	1623	825	2448
Phe	165.19	17.23	25.77	43	175	0	175	218	15	2208	2883	387	3270
Pro	115.13	12.23	25.77	38	105	0	105	143	124	12042	14537	3195	17732
Hypro	131.13	12.23	25.77	38	64	43	107	145	135	15271	10291	9284	19575
Ser	105.09	16.23	25.77	42	44	36	80	122	42	3657	2530	2594	5124
Thr	119.12	18.23	25.77	44	74	28	102	146	19	1921	1752	1022	2774
Trp	204.22	16.23	25.77	42	190	27	217	259	0	0	0	0	0
Tyr	181.19	16.23	25.77	42	144	43	187	229	0	0	0	0	0
Val	117.15	17.23	25.77	43	117	0	117	160	34	3371	4564	876	5440
							SUM =		1179	108142	101526	56807	158333
									Avg. res. MW =	91.7			

See section SASA discussion of variable native monolayer coverage

$h_{Ra} =$ 0.065 $h_{B} =$ 0.262

$h_{M}(full) =$ 3.07
$h_{Pr} =$ 0.84
$h_{M}(nat) =$ 1.63

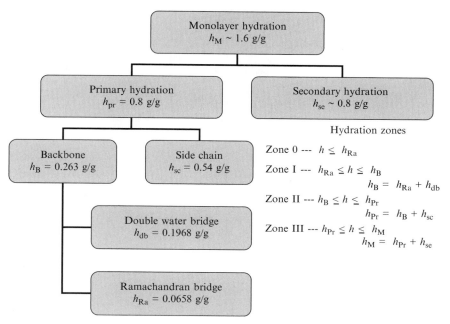

Figure 1.3 There are a total of seven water hydration compartments predicted by the molecular hydration model as shown here for collagen. Of the seven compartments, four capacities for zones 0, I, II, and III, h_{Ra}, h_B, h_{Pr}, and h_M, can be measured directly, whereas the other three are calculated from the measured results. Reproduced with permission from Fullerton and Rahal (2007). (See color insert.)

$r_e = 2.37$ Å as calculated and verified by Berendsen (1962) and the assumption of hexagonal close packing to calculate hydration fractions in Table 1.2.

$$A_{ew} = \pi r_w^2 \frac{2\sqrt{3}}{\pi} = r_w^2 2\sqrt{3} \tag{1.1}$$

$$h_{max} = \frac{18 \times SAS/A_{ew}}{MW_{coll}} \tag{1.2}$$

Thus the maximum hydration fraction is calculated with Eq. (1.1) by dividing the SASA by $A_{ew} = 8.59$ Å^2 per water molecule to calculate the number of surface water molecules and multiplying by the molar mass of water divided by the molar mass of collagen to give the monolayer maximum h_{max}(full) = h_{max}(polar-full) + h_{max}(nonpolar-full) = 1.06 g/g + 2.07 g/g = 3.13 g/g as summarized in Table 1.2. The SASA maximum possible hydration for the fully extended form of globular proteins as a function of molecular weight using the SASA empirical equation of Miller and colleagues (1987a) SASA (Å^2) = 1.48 × MW + 21 to calculate the area

Table 1.2 Calculation of collagen hydration from solvent-accessible surface areas (SASA)

Surface	$SASA_{max}(Å^2)$	$h_{max}(g/g)$	$h_{dis}(g/g)$	$h_{nat}(g/g)$
Polar	205218	1.06	0.26 (17%)	0.80 (50%)
Nonpolar	402197	2.07	1.25 (83%)	0.80 (50%)
Total	607416	3.13	1.51 (100%)	1.60 (100%)

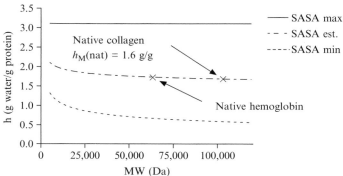

Figure 1.4 Plots of the calculated monolayer hydration of globular proteins using the method of Miller and colleagues (1987a,b) to calculate the surface area of a theoretical fully hydrated protein (SASA max), a compact protein with surface-only hydration (SASA min), and an estimated native monolayer hydration with internal water molecules based on native collagen in tendon (Fullerton and Amurao, 2006). The hydration of hemoglobin measured by osmotic compression as shown in Fig. 1.9 agrees with this estimate.

$(Å^2)$ and then converting to hydration fraction $h_{max} = (1.48MW + 21)/8.59 \times (18/MW) = 3.10$ g/g independent of molecular weight as shown in Fig. 1.4. Thus the monolayer hydration of fully exposed collagen is identical to the average monolayer hydration of fully extended globular proteins over the entire range of molecular weights. Monolayer surface hydration of globular proteins is reduced, however, by folding to achieve a minimum free energy by removing as much hydrophobic surface as possible from exposure to water.

2.4. SASA compensation assuming compact folding of globular proteins

The minimum possible surface hydration $h_{min} = (18 \times 6.3 \times MW^{0.73}/8.59\ Å^2)/MW$ is calculated using Miller's empirical equation for the SASA of compactly folded globular proteins with no internal water (Miller *et al.*, 1987a,b).

The SASA minimum hydration fraction, h_{min}, decreases slowly from approximately 1.4 g water/g protein as a function of molecular weight for small proteins to approximately 0.5 g/g for large proteins, as shown by the dotted line in Fig. 1.4. The molecular hydration model, however, contradicts the possibility of such compact folding due to the need for water bridges on the backbone and adjacent DWC. Thus the stoichiometric hydration model predicts that the hydration of native globular proteins must lie between the curves for h_{max} and h_{min}. As a first estimate we assume that the fraction of internal surface exposed to water for globular proteins is identical to collagen.

2.5. SASA estimate for globular proteins based on the hydration model

Using the measured monolayer hydration for native collagen or tendon $h = 1.62$ g/g (indicated by the X) as a starting point, one estimates monolayer hydration of globular proteins with the equation $h_M = h_{min} + (h_{max} - h_{min}) \times (1.62 - 0.4)/(3.1 - 0.4)$ as shown by the dot-dash line in Fig. 1.4. The native hydration of hemoglobin $h_M(native) = 1.7$ g/g extracted from the osmotic compression data of Adair (1928) using the analysis methods of Fullerton and colleagues (1992, 2006a,b) is plotted for comparison. The good agreement between the measured value and the predicted hydration suggests that the procedure gives reasonable estimates of native protein hydration. Let us emphasize, however, that native monolayer hydration is easily perturbed from the estimated monolayer hydration value by nonnative cosolute concentration, cosolute character, and pH, as well as protein surface interactions with ligands that displace additional surface water molecules (see discussion of osmotic compression measurements).

 ## 3. PREDICTIONS USING THE SHM

3.1. Prediction of hydration capacities h_{Ra}, h_B, h_{Pr}, and h_M(native)

Protein backbone stoichiometric relationships were used to calculate h_{Ra}, h_B, and h_{Pr} for collagen as shown in Table 1.1, whereas surface areas were used to calculate and estimate h_M(nat). The close agreement of collagen hydration values with previously measured compartmental capacities for globular proteins leads to the hypothesis extending the SHM to globular proteins. The collagen stoichiometric hydration rules defined in the spreadsheet were used to calculate the values for h_{Ra}, h_B, h_{Pr}, and h_M as summarized for two types of fibrillar collagen, six globular proteins, and three homopolypeptides in Table 1.3 by substituting the number of amino acids for each protein or

Table 1.3 Comparison of predicted and measured hydration values

Protein structural data input			Calculated hydration h(g water/g dry protein)					Measured hydration h(g/g) [a]			
Protein	MW (Da)	# Residues	h_{Ra}	h_B	ce: italic>h_{Pr}	h_M (full)	h_M (native)	h_{Ra}	h_B	h_{Pr}	$h_M(nat)$
Collagen type 1	324426	3537	0.065	0.262	0.818	3.071	1.627	0.06	0.26	0.80	1.62
Collagen type 4 [b]	488340	5044	0.0620	0.248	0.762	3.137	1.613	—	—	—	—
Human Hb	61915	574	0.056	0.223	0.696	3.152	1.630	—	—	—	1.72
Lysozyme	14296	129	0.054	0.217	0.677	3.099	1.737	0.05	0.22	—	—
BSA	66336	582	0.053	0.211	0.658	3.127	1.727	0.05	0.22	—	—
Myglobin	17181	153	0.053	0.214	0.668	3.174	1.679	—	—	—	—
Ovalbumin	43169	386	0.054	0.215	0.671	3.131	1.685	—	—	—	—
Insulin	5778	5778	0.053	0.212	0.663	3.088	1.706	—	—	—	—
Polyasparagine	10473	91	0.052	0.212	0.652	3.088	1.706	0.05	0.22	—	—
Polyglutamate	17431	135	0.047	0.186	0.581	2.972	2.005	0.05	0.19	—	—
Polylysine	12433	97	0.047	0.187	0.586	3.452	1.918	0.05	0.19	—	—

[a] Measurement methods used to confirm hydration fractions include (1) DSC of protein with variable hydration, (2) isotherm rehydration from the vapor phase, (3) NMR water titration, (4) NMR freezing point depression, (5) high G-force dehydration, and (6) hydration force (osmotic compression). All methods confirmed on type 1 collagen. Calculations for most common $2 \times \alpha1$ and $1 \times \alpha2$ but calculations for $\alpha1$, $\alpha2$, $\alpha3$, $\alpha4$, $\alpha5$, and $\alpha6$ show negligible differences in hydration fractions.

[b] Calculations for most common $2 \times \alpha1$ and $1 \times \alpha2$ but calculations for $\alpha1$, $\alpha2$, $\alpha3$, $\alpha4$, $\alpha5$, and $\alpha6$ show negligible differences in hydration fractions.

homopolypeptides. These calculations depend only on the amino acid composition of each specific protein. Substitution of amino acid compositions of other proteins and homopolypeptides in the spreadsheet predicts four hydration fractions shown in Table 1.3 for two fibrillar proteins, six globular proteins, and three homopolypeptides. It is noteworthy how little variation occurs in calculated hydration capacities due to the relatively narrow range of amino acid masses and statistical averaging of compositions. Only the homopolypeptides that do not exist in nature deviate significantly. Comparison of biophysical measurements of the hydration capacities for all 11 polypeptides in the authors' laboratory confirm SHM predictions with no exception yet found.

3.2. Comparison of hydration model predictions to measured capacities

Figure 1.5 summarizes comparison of $n = 45$ measured hydration capacities h_{Ra}, h_B, h_{Pr}, and h_M for lysozyme, bovine serum albumin (BSA), and collagen using 18 different biophysical measurement techniques against calculated SHM capacities from Table 1.3. The highly significant slope $S = 1.019 \pm 0.049$ and constant $C = -0.000758 \pm 0.036$ of the regression equation validate the capacity of SHM to predict the hydration compartment on globular proteins as well as collagen. It simultaneously confirms that molecular hydration compartments on the backbone of the protein are responsible for effects ranging from proton nuclear magnetic resonance (NMR) line width, proton NMR spin–lattice, heat capacity changes, and all the other 18 different types of measurements included in the plot (see the legend to Fig. 1.5 for details). The 95% confidence intervals show a discrepancy of only 0.07 g/g such that one can predict measured compartmental capacities quite accurately but even here the primary source of measurement error comes from measures of h_M. As shown in the next section, methods used to measure h_M are, in reality, measures of the total protein encapsulated water that can include a bulk water fraction when environmental factors are not controlled carefully and cause large deviation from the monolayer hydration capacity.

4. BIOPHYSICAL MEASUREMENTS OF COMPARTMENTAL HYDRATION CAPACITIES

4.1. Selection of methods

As shown in the last section there are a large number of biophysical measurement methods available to measure protein hydration compartment capacities of which four are selected for both simplicity and accuracy of results. (1) The

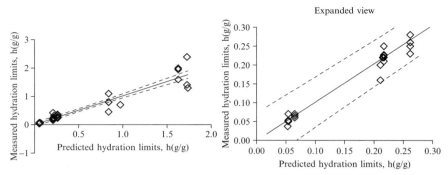

Figure 1.5 Plots of measured hydration capacities from an extensive literature versus hydration limits h_{Ra}, h_B, h_{Pr}, and h_M as calculated for lysozyme, BSA, and collagen in Table 1.3 using the stoichiometric hydration model. Capacity measurements of hydration compartments ($n = 45$) were made with multiple methods and estimated empirically by the authors of each report as follows: (1) NMR line width (Fuller and Brey, 1968), (2) isotherm (Careri and Giansanti, 1979; Ling, 1972), (3) optical absorbance (Careri and Giansanti, 1979; Careri et al., 1980), (4) heat capacity (Careri et al., 1980; Haly and Snaith, 1971), (5) ESR TEMPONE probe (Rupley et al., 1980), (6) dynamic mechanical response (Nomura et al., 1977), (7) differential scanning calorimetry (Careri et al., 1980; Haly and Snaith, 1971; Luescher et al. 1974; Miles and Ghelashvili, 1999; Zhang et al., 1985), (8) X-ray diffraction (Fullerton and Amurao, 2006; Haly and Snaith, 1971; Sasaki et al., 1983) (9) NMR water titration (Fullerton et al., 1986, 2006c), (10) rehydration rate (Cameron et al., 2007), (11) self-diffusion (Klotz, 1958), (12) dielectric dispersion (Oncley, 1941), (13) diamagnetic susceptibility (Careri et al., 1980), (14) enzyme activity (Careri et al., 1980; Rupley and Careri, 1991), (14) centrifugation dehydration rate (Cameron et al., 2007), (15) hydration force (Fullerton and Rahal, 2007; Leikin et al., 1997), (16) Raman spectroscopy (Fullerton and Rahal, 2007; Leikin et al., 1997), (17) computed tomography dilatometry (Fullerton and Amurao, 2006), and (18) osmotic compression (Fullerton et al., 1992; Zimmerman et al., 1995). The linear regression of the measured hydration capacities versus prediction from the molecular hydration model is plotted as a solid line with 95% CI as dotted lines. The slope 1.019 ± 0.049 is not significantly different from one and shows that the hydration model predicts hydration capacities measured with 18 biophysical effects of hydration with a p value <0.0001.

rehydration method is simple and gives accurate measures of h_{Ra}. (2) The proton NMR water titration relaxation method gives measurements of h_{Ra}, h_B, h_{Pr}, and h_M(nat) but requires more specialized equipment. (3) Differential scanning calorimetry gives measures h_{Ra}, h_B, and h_{Pr} but is more time-consuming and expensive. (4) The osmotic compression technique measures the encapsulated water fraction h_{enc} accurately and reproducibly, which is equivalent to h_M(nat) *only* under limited environmental conditions.

4.2. Gravimetric measurement of hydration

All four measurement techniques require careful, accurate measurement of the hydration fraction h at the time of biophysical measurement. Wet sample mass is calculated by subtracting the weighing boat mass from the boat + sample mass. Samples are then dried to equilibrium in a vacuum oven at room temperature to avoid denaturing damage and possible loss or gain of mass from chemical reactions with the atmosphere. The temperature is then raised to 50° in a vacuum for 1 day and then to 90° for 4 additional days to achieve removal of the most tightly bound water fraction. The vacuum oven is then allowed to cool to room temperature and the chamber is back filled with dry nitrogen. The mass of the dry sample + boat is measured rapidly to minimize exposure to reabsorption of atmospheric water on the sample. The mass of the water is calculated from the mass (boat + wet sample) − mass(boat + dry sample), whereas the mass of dry protein is calculated from the mass(boat + dry sample) − mass(boat). The hydration is h = (mass water)/(mass protein). Careful adherence to a systematic gravimetric drying protocol is crucial to avoid loss or gain of sample mass as a consequence of chemical interaction with the atmosphere and/or unexpected regain of water from the laboratory atmosphere.

4.3. Protein rehydration rate method

The rehydration rate for collagen from the absolute dry state (Cameron *et al.*, 2007) using measurements of wet protein mass as a function of time is a simple but accurate method used to measure the single water bridge hydration h_{Ra}, as shown in Fig. 1.6. The basis for analysis uses well-known ligand to protein formalism to describe and evaluate drug-binding constants where the mass action relationship defines chemical interaction of water ligand with the protein k_{in} [water] [collagen] $\leftrightarrow$ k_{out} [collagen·water]. The general

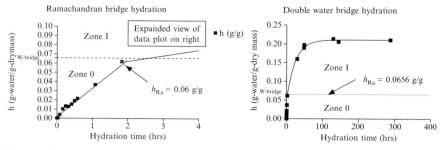

Figure 1.6 The rehydration of collagen from the "absolute" dry state in a wet atmosphere of 35 % relative humidity at 22° is described by two zones with boundary between zone 0 (Ramachandran bridges) and zone I (double water bridges) as defined in Fig. 1.3. Rehydration of the single water bridges occurs in less than 2 h as shown on the left, whereas accumulation of water with a rehydration time of several days to equilibrium is shown on the right. The nonlinear best fit to the two-zone data set governing gives h_{Ra} = 0.06 g/g at the intersection shown by the "X" at the intersection that is compared with the theoretical value h_{Ra} = 0.0656 shown by the dotted line or a measurement error of approximately 2 %. Reproduced with permission from Cameron *et al.* (2007). (See color insert.)

differential equation Eq. (1.3) describing the rehydration process has two terms; one describes the water input from the

$$\frac{dh}{dt} = k_{in}\rho_w\rho_c - k_{out}(1+h)\rho_c \tag{1.3}$$

atmosphere using the input constant k_{in} times the product of the water concentration ρ_w in the atmosphere times the concentration of the collagen ρ_c in the sample minus the output or return of water to the atmosphere from the hydrated protein expressed by k_{out} times the concentration of hydrated protein $(1+h)\,\rho_c$, where h is the hydration fraction such that application of the molecular model of collagen hydration allows us to predict two different solutions to this equation. The solution [Eq. (1.4)] is linear in zone 0 because the output constant for the Ramachandran water bridge is extremely small with exchange times on the order of hours or weeks for water bridges. In zone 0 the constant $k_{out} \rightarrow 0$ such that the water out term can be neglected and the solution is a linear equation as a function of time with a slope $K = k_{in}\rho_{abs}r_h\rho_c$ that depends on the absolute concentration of water in air ρ_{abs} (as available in Lide, 2004) at a temperature that must be controlled and the relative humidity r_h that must be either measured or controlled. This is the relationship responsible for the linear region shown for zone 0 in Fig. 1.6.

Zone 0 rehydration solution:

$$\begin{aligned}\frac{dh}{dt} &= k_{in}\rho_w\rho_c = k_{in}\rho_{abs}r_h\rho_c \\ \rho_w &= \rho_{abs}r_h \\ h &= k_{in}\rho_{abs}r_h\rho_c t = kt\end{aligned} \tag{1.4}$$

In zone I where k_{out} is significant there is a more typical exponential recovery described by Eq. (1.5) in zone I where water molecules in the double water bridge are more labile, the differential equation describing the hydration rate has both in and out terms. The solution to this equation in the coordinate system beginning at $h = 0$ at $t = 0$ is given here for zone I.

Zone I rehydration solution:

$$\frac{dh}{dt} = k_{in}\rho_w\rho_c - k_{out}(1+h)\rho_c$$

$$h = h_{Ra} + \left(\frac{k'_{in}}{k'_{out}}\rho_{abs}r_h - 1\right)\left(1 - e^{-k_{out}\rho_c(h_{Ra}+1)(t-h_{Ra}/k)}\right) \tag{1.5}$$

This equation was used to calculate the nonlinear best-fit line shown in Fig. 1.6 and shows that water binding to collagen during atmospheric rehydration can be simply described by the dynamic mass action relationship. The value of h_{R_a} is accurately extracted ($\approx 2\%$) from these measurements as the hydration level where the zone 0 and zone I relationships intersect, as shown in Fig. 1.6 by the large X.

4.4. Differential scanning calorimetry method

The DSC method relies directly on measurement of the energy associated with the removal of water bridges from the protein surface (Rahal and Fullerton, 2007). Repeated measures of ΔH and peak temperature T_m versus hydration h to determine the intercept of linear region zone I with the constant region zone II provide a biophysical direct measure of backbone hydration h_B, as shown in Fig. 1.7 (Miles and Ghelashvili, 1999; Rahal and Fullerton, 2007). Comparison of the intersection at $h = 0.26$ with the model prediction $h_B = 0.265$ indicates a measurement error approximately 2% using the DSC experimental method.

4.5. Proton NMR titration method

Repeated measures of spin–lattice relaxation rate $1/T1$ and spin–spin relaxation rate $1/T2$ provide data to measure h_B the hydration of limit of all water bound to the backbone of protein and also give a good measure of h_{R_a}. The method was first applied in 1986 to the globular protein lysozyme (Fullerton et al., 1986c), as shown in Fig. 1.8. Results were nearly identical in general form to the more recent results measured for collagen in 2006 (Fullerton et al., 2006c). The correlation of NMR titration results with the structural information provided by collagen provides the framework to correlate specific types of hydration compartments with slowed water motions. The plots of spin echo relaxation times and amplitudes in Fig. 1.9 show sharp changes at hydration capacities near the monolayer hydration capacity h_M as T2 relaxation is dominated by slowed motion of protein tumbling when surface water is lost.

4.6. Osmotic compression method

Studies of protein hydration compartments suggested that water bound to the protein must directly affect the measured osmotic pressure. A simple method to correct $PV = nRT$ for bound water was proposed as summarized in Fig. 1.10 to create the osmotic compression method. The slope of plots of $1/h = M_w/M_p$ as a function of $1/P$ shown in Fig. 1.11 gives accurate measures of protein molecular weight (Cameron and Fullerton, 1990;

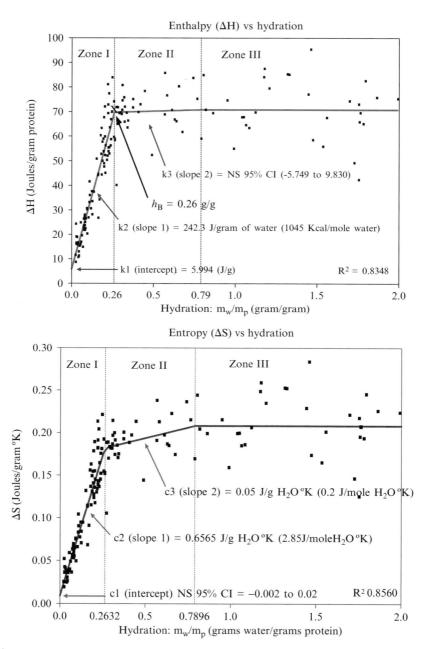

Figure 1.7 DSC measurements of the enthalpy ΔH of melting of rat tail tendon and calculated values of $\Delta S = \Delta H / T_m$ as functions of hydration h (g/g) are shown. Hydration zone limits are defined by the molecular hydration model zone I maximum $h_B = 0.26$ g/g and zone II maximum $h_{pr} = 0.79$ g/g. Zone III extends above native hydration $h(\text{native}) = 1.6$ g/g.

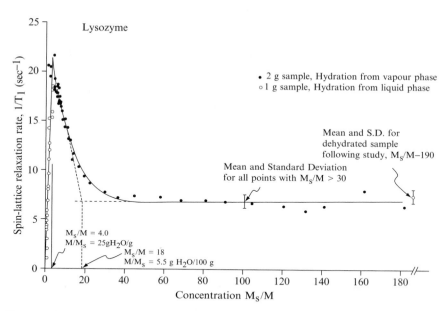

Figure 1.8 NMR measures of the spin–lattice relaxation rate $1/T1$ for lysozyme as a function of hydration level expressed as the inverse of hydration $1/h$ demonstrate linear regions indicative of fast exchange of protons between water compartments (Fullerton *et al.*, 1986). The intersection of the linear section gives capacities of $h_B = 0.25$ g/g and $h_{Ra} = 0.055$ g/g in good agreement with predictions from Table 1.3. Reproduced with permission from Fullerton *et al.* (1986).

Cameron *et al.*, 1990, 1997; Fullerton *et al.*, 1992, 2006a,b; Kanal *et al.*, 1994; Zimmerman *et al.*, 1994, 1995) while the constant in the linear equation accurately measures h_{enc}, the total hydration encapsulated by the protein. Studies of BSA under nonphysiologic conditions showed that h_{enc} can be $> h_M$ when the protein expands from the compact native form into a molten globule (Kanal *et al.*, 1994; Zimmerman *et al.*, 1994). Thus estimates of h_M from measurements of h_{enc} can be highly dependent on solution cofactors such as pH and salt as shown in Fig. 1.12 for BSA unless one selects the unique minimum value achieved by adjusting over a range of pH values (Kanal *et al.*, 1994). The SHM and DSC measurements on collagen indicate that most encapsulated water is thermodynamically identical to bulk water. However, encapsulated water differs from bulk external water in that it never has an opportunity to participate in the bombardment of adjacent semipermeable osmotic membranes. More recently, the osmotic compression method was used to show that the water content of mammalian cells responds to cosolute factors that influence protein conformation (Cameron *et al.*, 2006; Fullerton *et al.*, 2006a,b). Comparison of the I value measured for hemoglobin $h_m = 1.7$ g/g as shown in Fig. 1.12 agrees with the value for

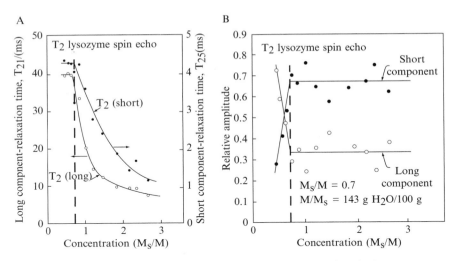

Figure 1.9 NMR measures of the spin–echo relaxation rate 1/T2 for lysozyme as a function of hydration level expressed as the inverse of hydration 1/h demonstrate linear regions have biphasic decay with two proton compartments associated with the protein that sharp changes in relaxation time and amplitude at $h_M = 1.43$ g/g (Fullerton *et al.*, 1986) due to increased contact when less than monolayer water coverage remains. Measured value is in reasonable agreement with predicted $h_M = 1.74$ g/g from Table 1.3. Reproduced with permission from Fullerton *et al.* (1986).

the RBC in the native isotonic region. RBC hydration depends on the $h_{enc} = 1.7$ g/g of hemoglobin plus the water content allowed by response to the osmotic pressure of the nonpenetrating osmolytes outside the cell $h_{os} \approx 0.4$ g/g or a total RBC water content $h_{RBC} \sim 2.1$ g/g or 68% water. Thus $1.7/2.1 = 81\%$ of the water in the RBC is water encapsulated by the hemoglobin, compelling evidence that protein hydration must be considered when evaluating tissue osmoregulation.

5. RELATIONSHIP OF THE HYDRATION MODEL TO OSMOSENSING AND OSMOSIGNALING

The measurement of the encapsulated water fraction discussed earlier and hydration force measurements (Fullerton *et al.*, 2006a,b; Leikin *et al.*, 1994, 1995) indicate that hydrations on proteins under native condition range from 1.4 to 4 g-water per g-protein, which implies a water content of 58 to 80%. This range of hydration brackets the range of most mammalian cells and suggests that most of the water in mammalian cells is sequestered by

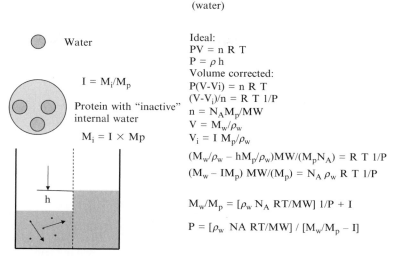

Osmotically inactive volume
(water)

Water

$I = M_i/M_p$

Protein with "inactive"
internal water

$M_i = I \times M_p$

Ideal:
$PV = n R T$
$P = \rho h$
Volume corrected:
$P(V-V_i) = n R T$
$(V-V_i)/n = R T \, 1/P$
$n = N_A M_p/MW$
$V = M_w/\rho_w$
$V_i = I \, M_p/\rho_w$

$(M_w/\rho_w - h M_p/\rho_w)MW/(M_p N_A) = R T \, 1/P$
$(M_w - I M_p) \, MW/(M_p) = N_A \rho_w \, R \, T \, 1/P$

$M_w/M_p = [\rho_w N_A \, RT/MW] \, 1/P + I$

$P = [\rho_w \, NA \, RT/MW] \, / \, [M_w/M_p - I]$

Figure 1.10 Derivation of the volume–corrected expression for osmotic pressure that accounts for encapsulated water fraction $h_{enc} = I$, including water bridges, dielectric water clusters, and internal bulk water. As shown in Figs. 1.11 and 1.12, the encapsulated water fraction is variable but generally in the range from 1.4 to 4.0 g/g for most proteins under native mammalian conditions (Fullerton *et al.*, 1993). (See color insert.)

macromolecular components of the cell. The water content of tissues responds to the osmotic status of the plasma and causes variation in tissue water content. For example, the hydration of native tendon increases from 1.6 to 2.0 g/g in buffered saline and even higher in pure water where $h_M(neat) = 2.39$ g/g. Exposure of rat tail tendon fibers to changing salt, pH, and other cosolute concentrations causes changes in the collagen hydration and related mechanical changes in length. Similar expansion and contractions are predicted for globular proteins as a function of cosolute concentrations and suggest that hydration compartments play key roles in osmosensing, osmosignaling, and osmoregulation.

6. RELATIONSHIP OF THE SH MODEL TO ENZYME FUNCTION

6.1. Enzyme hydration and activity

Seminal work relating enzyme catalytic activity to hydration level reported in 1991 (Rupley and Careri, 1991) led to the general view that water is required for enzyme function. This early work evaluated the enzyme activity of hen egg lysozyme plus substrate using rapid freezing, lyophilizing

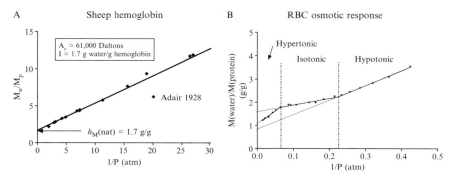

Figure 1.11 (A) Plots of h(g-water/g-protein) versus the inverse of osmotic pressure 1/P gives linear plots for which the slope varies as the inverse of the effective molecular weight and the constant is the solute solvent interaction parameter expressing hydration encapsulated by the protein h_{enc} (Fullerton *et al.*, 1993). The value h_{enc} includes water bridges, dielectric water clusters, and bulk water fraction that are encapsulated inside the protein such that they are excluded from thermodynamic solution expressions for osmotic pressure and freezing depression. Encapsulated water molecules never interact with the osmotic membrane but can change dramatically, as shown in Figure 1.12 with solution parameters and most especially with pH. Under native condition, however, the measured $h_{enc} \rightarrow h_M(nat)$ as it does here for hemoglobin. (B) Reevaluation of RBC volume measurements as a function of osmotic pressure show that the $h_{enc} = h_M(nat) = 1.7$ g/g in the isotonic region. There are changes in the h_{enc} in the hypotonic and hypertonic regions. Reproduced with permission from Fullerton *et al.* (2006a).

to a dry powder, and then subjecting to different levels of rehydration to demonstrate that the onset of enzyme catalysis occurs at a hydration level of ≈ 0.2 g H_2O/g dry lysozyme. This level of hydration was said to coincide with the onset of intermolecular motions, a greater conformational freedom of water arrangements at the protein surface, and a transition onset of more weakly interactive regions at the surface of the enzyme. The SH model prediction $h_B = 0.217$ g/g agrees with these observations and, furthermore, predicts water in the zone II between $h = 0.217$ and 0.677 will form dielectric water clusters (Rahal and Fullerton, 2007) around the bridges, as shown in Fig. 1.2. These DWC water molecules bridge to remaining hydrophilic sites on the protein side chains. As shown in Fig. 1.1 there is small energy difference between polar sites in direct contact (nonhydrated) and those separated by a distance d when hydrated (Rahal and Fullerton, 2007). Because side chains have mobility, they are free to move and touch when dehydrated. When the protein is hydrated above $h = 0.217$ g/g the charge pairs on the side chains immobilized by strong electrostatic attraction are liberated, as the dielectric coefficient of water is so high that the electrostatic energy of the water-separated pair is nearly equivalent to the dry intimate-contact position. Thus the protein tertiary structure can now oscillate back and forth between two energetically equivalent configurations.

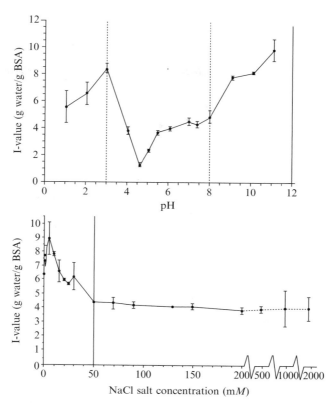

Figure 1.12 Plots of the solute–solvent interaction parameter I value (total encapsulated hydration h_{enc}) for bovine serum albumin as a function of pH and salt concentration show that the osmotically unresponsive water can greatly exceed monolayer hydration for a labile protein-like BSA under nonnative conditions. It is worth noting that the $h_{enc} \approx 4$ in the physiological range near pH 7 and salt concentration 100/200 mM, which implies water content $\approx$80%. Reproduced with permission from Zimmerman *et al.* (1995).

With less than $h = 0.217$ g/g, polar amino acid side chains are locked in a dry charge-touching position. Only when water is added can the side chains move away to more distant positions, thereby allowing an internal vibration.

6.2. Neutron scattering assessment of enzyme motion and activity

To address the influence of hydration on the internal dynamics (flexibility) of hen egg lysozyme, Roh and colleagues (2006) turned to quasielastic neutron and light scattering, as well as molecular dynamic simulations, as summarized in Fig. 1.13. They found three main relaxation processes in the pico-to-nanosecond time range: methyl group fluctuations, fast picosecond

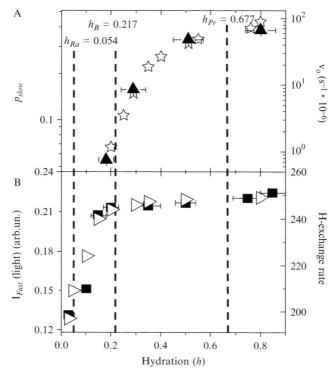

Figure 1.13 Comparison of critical molecular hydration levels for lysozyme with measured biophysical parameters from multiple studies as summarized by Roh *et al.* (2006). (A) Parallel comparison of the mobile fraction of hydrogen atoms involved in a slow relaxation process (▲▲) identified by neutron scattering and enzymatic reaction rate, v_o, of the lysozyme to hexasaccharide of N-acetylglucosamine (☆) as estimated by Rupley and colleagues (1980) shows good correlation of these parameters. The molecular hydration model attributes change in the hydration range $h_B = 0.217$ to $h_{Pr} = 0.677$ to completion of the hydration of all polar sites on the protein. This would eliminate all direct hydrogen–bonding anchors to sites on the side chains to release the slower undulating vibration of tertiary protein structures. (B) Parallel comparison of hydration dependences of the integrated QES intensity of the fast vibration process (■) and hydrogen exchange rate (▷) in units of moles of exchanged H atoms/mole of lysozyme/ 24 h as estimated by Schinkel and colleagues (1985) is well correlated. The SH model attributes hydration in the range from zero to $h_B = 0.217$ to completion of tightly bound water bridges to the backbone of the protein. Hydration of amide sites by water bridges liberates high-frequency vibration of the secondary structures of the protein (i.e., α helices and β sheets) and also increases the rate of exchange of hydrogen atoms from amide groups that anchor the water bridges. Reproduced with permission from Roh *et al.* (2006).

relaxation, and a slow relaxation process. The methyl group vibration was not sensitive to the hydration level. For hydrations in zone II ($h_B = 0.217$ g/g $< h < h_{Pr} = 0.677$ g/g), there is a sharp increase in the slow relaxation fraction that correlates with the enzymatic activity results from Rupley $et\ al.$ (1980), as shown in Fig. 1.13A. Roh and co-workers (2006) concluded that the slow relaxation process might be necessary for lysozyme function, perhaps related to molecular motions of secondary structures of the lysozyme. Enzyme activity in the lysozyme system continued to increase until asymptotically approaching the level of activity of the enzyme in dilute solution (T = 295°) at hydration near $h_{Pr} = 0.677$. At low hydration levels (zones 0 and I), $0 < h < h_B = 0.217$ g/g, Roh $et\ al.$ (2006) reported that both light- and neutron-scattering data demonstrate a similar sharp hydration dependence of the fast relaxation process, as shown in Fig. 1.13B. Fast relaxation dependence on hydration correlates with the hydrogen exchange rate of lysozyme reported by Shinkel and colleagues (1985). The fast relaxation increases more slowly in zone II above h_B.

6.3. The SH model predicts both vibrational dynamics and enzyme activity changes

Enzyme activity, vibration, and hydrogen exchange all show compartmental capacities that correlate with predictions of the SHM. Furthermore, the SHM relates fast conformational fluctuations in the picosecond range to hydration of the protein backbone. As globular lysozyme consists of two domains with both α-helix and β-sheet components, it leads to the hypothesis that fast fluctuations are vibrations of tightly bound protein secondary structures. Additionally, backbone water bridges participate in tight hydrogen bonding between amide and carbonyl groups, which predicts a monotonic increase in the proton exchange rate as the number of backbone amide protons associated with water increases with each bridge until all sites are fully occupied at $h_B = 0.217$ g/g. The slow relaxation process starts to increase above $h_B = 0.217$ g/g where the SH model predicts DWC form around the water bridges. DWC cover charge sites on the immobilized water bridges and remaining polar sites on the side chains of the protein. The dielectric water allows a range of spatial configurations with nearly the same free energy that promotes vibrational motion of the protein tertiary structure. No further increase in slow vibration or enzymatic activity occurs above $h_{Pr} = 0.677$. Thus the correlations of enzymatic activity, proton exchange, and vibrational motion of the molecule are consistent with the SH model water compartments. The conclusion from these enzyme dynamic flexibility studies is that hydration changes the shape of the protein and liberates large-scale motion necessary for functional enzymes. Dry enzymes are not functional, as water is required to perfect the correct lock/key shape, provide enzyme motional freedom, and release the completed product from the enzyme in a timely manner.

7. Relationship of the SH Model to Cellular Function

7.1. Cellular function depends on water

Most scientists accept that life as we know it requires water and have gone so far as to say there can be no life without water. Why is water necessary for life and what are the functions of water in biomolecular processes? The focus to this point in this chapter has been on the SHM of protein hydration and how it relates to enzyme activity, starting from the primary amino acid sequence extending to macromolecular structure. As cellular functions depend on enzyme actions it seems logical to expect that the interfacial interactions of water with the primary structure of proteins such as lysozyme will extend to cells and the functional activity of cells. Cellular systems should display changes occurring at different levels of hydration as a consequence of the motional and thermodynamic properties of water. The different fractions of water involved in these biological systems surely play a role in regulation of function. The most carefully studied area of research on the role of hydration levels at a cellular level relies on desiccation-tolerant biological systems that survive desiccation and resume enzyme activity and life upon rehydration (Clegg, 1986; Sun, 2000).

7.2. Hydration compartments and cellular systems correlate with SH model predictions

Isothermal sorption measures on desiccation-tolerant biological systems give evidence of three hydration fractions: at low water content, a strong water-binding fraction; at intermediate water content, a weak water-binding fraction; and at high water content, a loose water-binding fraction (Clegg, 1978). More recent and sensitive methods identify at least four fractions (Sun, 2000). Some of the biological systems used in rehydration studies rely on desiccation-tolerant seeds (soybean and cowpea), pollen, and brine shrimp cysts, as well as dehydration-intolerant seeds (red oak acorn) (Clegg, 1978, 1986; Clegg et al., 1978; Pagnotta and Bruni, 2006; Sun, 2000). Onset of biological function for dehydration-tolerant species relates to an increased level of hydration to liberate the organism from a dormant state (Bruni et al., 1989a,b; Clegg, 1982; Clegg et al., 1982; Vertucci and Roos, 1990). Survival but no metabolic activity was observed below 0.10 g/g in desiccation-tolerant species, whereas desiccation-intolerant species do not survive such dehydration. This correlates with concepts that the

SH model at hydration below $h_B \approx 0.22$ g/g removes structural water bridges and exposes protein molecule to high mechanical stress (Fullerton and Amurao, 2006). Enzyme activity was first observed at 0.20 to 0.30 g/g in brine shrimp cysts, which is slightly greater than $h_B \approx 0.22$ g/g and correlates with the earlier discussion of the hydration dependence of lysozyme enzymatic activity. Higher levels of metabolic activity and respiration occurred at hydration levels of greater than 0.40 g/g for seeds and at greater than 0.6 g/g for brine shrimp; this correlates with the hydration behavior of the lysozyme shown in Fig. 1.13.

8. SUMMARY AND CONCLUSIONS

This chapter reviewed physical evidence for the formation of categorical water compartments on proteins due to "irreducible separation" of electrical charges on the protein backbone. Investigations of collagen and correlation of results on globular proteins led to the hypothesis of a stoichiometric hydration model applicable to all proteins. Comparison of predicted hydration values with measurement of hydration capacities for multiple proteins confirmed the utility of the SH model. Four methods were selected and presented to make accurate measurements of the hydration fractions on other proteins and tissues. The relatively uniform size of hydration fractions is directly due to the uniform character of protein backbones. Correlation of SH predictions with enzyme mobility and activity measurements as a function of hydration are in close agreement. Finally, measurements of biological activity of cells also show values consistent with SH predictions.

REFERENCES

Adair, G. S. (1928). A theory of partial osmotic pressures and membrane equilibia, with special reference to the application of Dalton's law to hemoglobin solution in presence of salts. *Proc. Roy. Soc. Lond. Ser. A* **120**, 573–603.

Bella, J., Brodsky, B., and Berman, H. M. (1995). Hydration structure of a collagen peptide. *Structure* **3**, 893–906.

Bella, J., Brodsky, B., and Berman, H. M. (1996). Disrupted collagen architecture in the crystal structure of a triple-helical peptide with a Gly→Ala substitution. *Connect. Tissue Res.* **35**, 401–406.

Berendsen, H. J. (1962). Nuclear magnetic resonance study of collagen hydration. *J. Chem. Phys.* **16**, 3297–3305.

Bruni, F., Careri, G., and Clegg, J. S. (1989a). Dielectric properties of Artemia cysts at low water contents: Evidence for a percolative transition. *Biophys. J.* **55**, 331–338.

Bruni, F., Careri, G., and Leopold, A. C. (1989b). Critical exponents of protonic percolation in maize seeds. *Physical. Rev. A* **40**, 2803–2805.

Cameron, I., Short, N. J., and GD, F (2007). Verification of simple hydration/dehydration methods to characterize multiple water compartments in tendon type 1 collagen. *Cell. Biol. Int.* **31**(6), 531–539.

Cameron, I. L., and Fullerton, G. D. (1990). A model to explain the osmotic pressure behavior of hemoglobin and serum albumin. *Biochem. Cell. Biol.* **68**, 894–898.

Cameron, I. L., Kanal, K. M., and Fullerton, G. D. (2006). Role of protein conformation and aggregation in pumping water into and out of a cell. *Cell. Biol. Int.* **30**, 78–85.

Cameron, I. L., Kanal, K. M., Keener, C. R., and Fullerton, G. D. (1997). A mechanistic view of the non-ideal osmotic and motional behavior of intracellular water. *Cell. Biol. Int.* **21**, 99–113.

Cameron, I. L., Merta, P., and Fullerton, G. D. (1990). Osmotic and motional properties of intracellular water as influenced by osmotic swelling and shrinkage of *Xenopus* oocytes. *J. Cell. Physiol.* **142**, 592–602.

Careri, G., and Giansanti, A. (1979). Lysozyme film hydration events: An IR and gravimetric study. *Biopolymers* **18**, 1187–1203.

Careri, G., Gratton, E., Yang, P. H., and Rupley, J. A. (1980). Correlation of IR spectroscopic, heat capacity, diamagnetic susceptibility and enzymatic measurements on lysozyme powder. *Nature* **284**, 572–573.

Clegg, J. S. (1978). Interrelationships between water and cellular metabolism in Artemia cysts. VIII. Sorption isotherms and derived thermodynamic quantities. *J. Cell. Physiol.* **94**, 123–137.

Clegg, J. S. (1982). Interrelationships between water and cell metabolism in Artemia cysts. IX. Evidence for organization of soluble cytoplasmic enzymes. *Cold Spring Harb. Symp. Quant. Biol.* **46**(Pt. 1), 23–37.

Clegg, J. S. (1986). Artemia cysts as a model for the study of water in biological systems. *Methods Enzymol.* **127**, 230–239.

Clegg, J. S., Seitz, P., Seitz, W., and Hazlewood, C. F. (1982). Cellular responses to extreme water loss: The water-replacement hypothesis. *Cryobiology* **19**, 306–316.

Clegg, J. S., Zettlemoyer, A. C., and Hsing, H. H. (1978). On the residual water content of dried but viable cells. *Experientia* **34**, 734.

Fuller, M. E., 2nd, and Brey, W. S., Jr. (1968). Nuclear magnetic resonance study of water sorbed on serum albumin. *J. Biol. Chem.* **243**, 274–280.

Fullerton, G. D., and Amurao, M. R. (2006). Evidence that collagen and tendon have monolayer water coverage in the native state. *Cell. Biol. Int.* **30**, 56–65.

Fullerton, G. D., Kanal, K. M., and Cameron, I. L. (2006a). On the osmotically unresponsive water compartment in cells. *Cell. Biol. Int.* **30**, 74–77.

Fullerton, G. D., Kanal, K. M., and Cameron, I. L. (2006b). Osmotically unresponsive water fraction on proteins: Non-ideal osmotic pressure of bovine serum albumin as a function of pH and salt concentration. *Cell. Biol. Int.* **30**, 86–92.

Fullerton, G. D., Nes, E., Amurao, M., Rahal, A., Krasnosselskaia, L., and Cameron, I. (2006c). An NMR method to characterize multiple water compartments on mammalian collagen. *Cell. Biol. Int.* **30**, 66–73.

Fullerton, G. D., Ord, V. A., and Cameron, I. L. (1986). An evaluation of the hydration of lysozyme by an NMR titration method. *Biochem. Biophys. Acta* **869**, 230–246.

Fullerton, G. D., and Rahal, A. (2007). Collagen structure: The molecular source of the tendon magic angle effect. *J. Magn. Reson. Imag.* **25**, 345–361.

Fullerton, G. D., Zimmerman, R. J., Cantu, C., 3rd, and Cameron, I. L. (1992). New expressions to describe solution nonideal osmotic pressure, freezing point depression, and vapor pressure. *Biochem. Cell. Biol.* **70**, 1325–1331.

Fullerton, G. D., Zimmerman, R. J., Kanal, K. M., Floyd, J., and Cameron, I. L. (1993). Methods to improve the accuracy of membrane osmometry measures of protein molecular weight. *J. Biochem. Biophys. Methods* **20**, 299–307.

Haly, A. R., and Snaith, J. W. (1971). Calorimetry of rat tail tendon collagen before and after denaturation: The heat of fusion of its absorbed water. *Biopolymers* **10**, 1681–1699.

Kanal, K. M., Fullerton, G. D., and Cameron, I. L. (1994). A study of the molecular sources of nonideal osmotic pressure of bovine serum albumin solutions as a function of pH. *Biophys. J.* **66**, 153–160.

Klotz, I. M. (1958). Protein hydration and behavior; many aspects of protein behavior can be interpreted in terms of frozen water of hydration. *Science* **128**, 815–822.

Leikin, S., Parsegian, V. A., Yang, W., and Walrafen, G. E. (1997). Raman spectral evidence for hydration forces between collagen triple helices. *Proc. Natl. Acad. Sci. USA* **94**, 11312–11317.

Leikin, S., Rau, D. C., and Parsegian, V. A. (1994). Direct measurement of forces between self-assembled proteins: Temperature-dependent exponential forces between collagen triple helices. *Proc. Natl. Acad. Sci. USA* **91**, 276–280.

Leikin, S., Rau, D. C., and Parsegian, V. A. (1995). Temperature-favoured assembly of collagen is driven by hydrophilic not hydrophobic interactions. *Nat. Struct. Biol.* **2**, 205–210.

Lide, D. R. (2004). "Handbook of chemistry and physics," 85th ed. CRC press, Beca Raton. pp. 6–10.

Ling, G. N. (1972). Hydration of macromolecules. *In* "Water and Aqueous Solutions: Structure, Thermodynamics and Transport Processes" (R. A. Horne, ed.), Vol. 1, pp. 663–700. Wiley Interscience, New York.

Luescher, M., Ruegg, M., and Schindler, P. (1974). Effect of hydration upon the thermal stability of tropocollagen and its dependence on the presence of neutral salts. *Biopolymers* **13**, 2489–2503.

Miles, C. A., and Ghelashvili, M. (1999). Polymer-in-a-box mechanism for the thermal stabilization of collagen molecules in fibers. *Biophys. J.* **76**, 3243–3252.

Miller, S., Janin, J., Lesk, A. M., and Chothia, C. (1987a). The accessible surface area and stability of oligomeric proteins. *Nature* **328**, 834–836.

Miller, S., Janin, J., Lesk, A. M., and Chothia, C. (1987b). Interior and surface of monomeric proteins. *J. Mol. Biol.* **196**, 641–656.

Nomura, S., Hiltner, A., Lando, J. B., and Baer, E. (1977). Interaction of water with native collagen. *Biopolymers* **16**, 231–246.

Oncley, J. L. (1941). Evidence from physical chemistry regarding the size and shape of protein molecules from ultra-centrifugation, diffusion, viscosity, dielectric dispersion and double refraction of flow. *Ann. N.Y. Acad. Sci.* **41**, 121–150.

Pagnotta, S. E., and Bruni, F. (2006). The glassy state of water: A 'stop and go' device for biological processes. *In* "Water and the Cell" (G. H. Pollack, I. L. Cameron, and D. N. Wheatley, eds.), pp. 93–112. Springer, Dordrecht.

Rahal, A., and Fullerton, G. (2007). Cause of collagen melting: Thermal excitation of single and double water bridges on the protein backbone. Submitted for publication.

Ramachandran, G. N., and Chandrasekharan, R. (1968). Interchain hydrogen bonds via bound water molecules in the collagen triple helix. *Biopolymers* **6**, 1649–1658.

Roh, J. H., Curtis, J. E., Azzam, S., Novikov, V. N., Peral, I., Chowdhuri, Z., Gregory, R. B., and Sokolov, A. P. (2006). Influence of hydration on the dynamics of lysozyme. *Biophys. J.* **91**, 2573–2588.

Rupley, J. A., and Careri, G. (1991). Protein hydration and function. *Adv. Prot. Chem.* **41**, 37–172.

Rupley, J. P., Yang, P. H., and Tollin, G. (1980). Thermodynamic and related studies of water interacting with proteins. *In* "Water in Polymers" (S. Rowland, ed.), Vol. 127, pp. 111–132. American Chemical Society, Washington, DC.

Sasaki, N., Shiwa, S., Yagihara, S., and Hikichi, K. (1983). X-ray diffraction studies on the structure of hydrated collagen. *Biopolymers* **22**, 2539–2547.

Schinkel, J. E., Downer, N. W., and Rupley, J. A. (1985). Hydrogen exchange of lysozyme powders: Hydration dependence of internal motions. *Biochemistry* **24**, 352–366.

Sun, W. Q. (2000). Dielectric relaxation of water and water-plasticized biomolecules in relation to cellular water organization, cytoplasmic viscosity, and desiccation tolerance in recalcitrant seed tissues. *Plant Physiol.* **124**, 1203–1216.

Vertucci, C. W., and Roos, E. E. (1990). Theoretical basis of protocols for seed storage. *Plant Physiol.* **94**, 1019–1023.

Zhang, J., Zhang, Z., Zhang, W., Fu, Y., Ye, G., and Nin, Z. (1985). Hydration and thermotransition of collagen fibre. *In* "Water and Ions in Biological Systems" (A. Pullman, V. Vasilescu, and L. Packer, eds.), pp. 197–214. Plenum, New York.

Zimmerman, R. J., Kanal, K. M., Sanders, J., Cameron, I. L., and Fullerton, G. D. (1995). Osmotic pressure method to measure salt induced folding/unfolding of bovine serum albumin. *J. Biochem. Biophys. Methods* **30**, 113–131.

CHAPTER TWO

YEAST OSMOREGULATION

Stefan Hohmann,* Marcus Krantz,† *and* Bodil Nordlander*

Contents

Abstract

Osmoregulation is the active control of the cellular water balance and encompasses homeostatic mechanisms crucial for life. The osmoregulatory system in the yeast *Saccharomyces cerevisiae* is particularly well understood. Key to yeast osmoregulation is the production and accumulation of the compatible solute glycerol, which is partly controlled by the high osmolarity glycerol (HOG) signaling system. Genetic analyses combined with studies on protein–protein interactions have revealed the wiring scheme of the HOG signaling network, a branched mitogen-activated protein (MAP) kinase (MAPK) pathway that eventually converges on the MAPK Hog1. Hog1 is activated following cell shrinking and controls posttranscriptional processes in the cytosol as well as gene expression in the nucleus. HOG pathway activity can easily and rapidly be controlled experimentally by extracellular stimuli, and signaling and adaptation can be separated by a system of forced adaptation. This makes yeast osmoregulation suitable for studies on system properties of signaling and cellular adaptation via mathematical modeling. Computational simulations and parallel quantitative time course experimentation on different levels of the regulatory system have provided a stepping stone toward a holistic understanding, revealing how the HOG pathway can combine rigorous feedback control with maintenance of signaling competence. The abundant tools make yeast a suitable model for an integrated analysis of cellular osmoregulation.

* Department of Cell and Molecular Biology, Göteborg University, Göteborg, Sweden
† The Systems Biology Institute, Keio University School of Medicine, Shinjuku-ku, Tokyo, Japan

Methods in Enzymology, Volume 428

ISSN 0076-6879, DOI: 10.1016/S0076-6879(07)28002-4

Maintenance of the cellular water balance is fundamental for life. All cells, even those in multicellular organisms with an organism-wide osmoregulation, have the ability to actively control their water balance. Osmoregulation encompasses homeostatic processes that maintain an appropriate intracellular environment for biochemical processes as well as turgor of cells and organism. In the laboratory, the osmoregulatory system is studied most conveniently as a response to osmotic shock, causing rapid and dramatic changes in the extracellular water activity. Those rapid changes mediate either water efflux (hyperosmotic shock), and hence cell shrinkage, or influx (hypoosmotic shock), causing cell swelling. The yeast *S. cerevisiae*, as a free-living organism experiencing both slow and rapid changes in extracellular water activity, has proven a suitable and genetically tractable experimental system in studying the underlying signaling pathways and regulatory processes governing osmoregulation. Although far from complete, the present picture of yeast osmoregulation is both extensive and detailed (de Nadal *et al.*, 2002; Hohmann, 2002; Klipp *et al.*, 2005). Simulations using mathematical models combined with time course measurements of different molecular processes in signaling and adaptation have allowed elucidation of the first system properties on the yeast osmoregulatory network.

1. Time Line: Response to Osmotic Shock

Following a hyperosmotic shock, yeast cells shrink within seconds. Within less than 1 min the glycerol export channel Fps1 closes rapidly to prevent glycerol leakage and enables the accumulation of this compatible solute (Tamás *et al.*, 1999, 2003; Thorsen *et al.*, 2006). Within the same time frame, the high osmolarity glycerol (HOG) signal transduction system is activated (de Nadal *et al.*, 2002; Hohmann, 2002) (Fig. 2.1). The initial responses mediated by the active Hog1 protein kinase appear to encompass the stimulation of ion export (Proft and Struhl, 2004), arrest of the cell cycle (Clotet *et al.*, 2006; Escote *et al.*, 2004), diminishing of translational capacity (Bilsland *et al.*, 2004; Bilsland–Marchesan *et al.*, 2000; Teige *et al.*, 2001), and stimulation of glycolysis to enhance production of the osmolyte glycerol (Dihazi *et al.*, 2004). A significant portion of active Hog1 accumulates in the nucleus where it is recruited to target promoters to control gene expression (O'Rourke and Herskowitz, 2004; Pokholok *et al.*, 2006; Posas *et al.*, 2000; Rep *et al.*, 2000). Genes encoding enzymes in glycerol production and active glycerol uptake are among those upregulated in a Hog1–dependent manner, but the bulk of the transcriptional change seems to revolve around metabolism and is similar to that caused by other stresses (Causton *et al.*, 2001; Gasch *et al.*, 2000). Adaptation depends on glycerol accumulation, and once the intracellular glycerol level has reached about one-third of its maximum, Hog1 is dephosphorylated and gene expression responses attenuate (Hohmann, 2002; Klipp *et al.*, 2005).

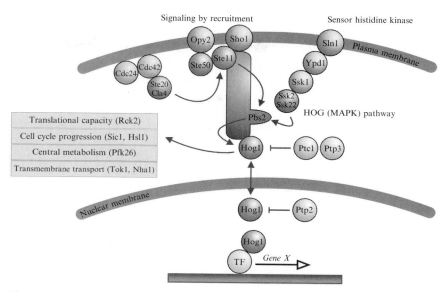

Figure 2.1 The yeast HOG pathway signaling system and overview of response mechanisms. The pathway consists of two branches: the Sho1 branch (Cdc42/24: G protein, Ste20: PAK protein kinase, Opy2, Sho1: membrane protein scaffold, Ste11/Ste50: MAPKKK, Pbs2: MAPKK) and the Sln1 branch (with Sln1–Ypd1–Ssk1: sensing histidine kinase phosphorelay system, Ssk2/22: MAPKKK), Hog1: MAPK, Ptc1, Ptp2/3: protein phosphatases, TF: transcription factor. (See color insert.)

Less is known about the time line of the response to a hypoosmotic shock. Cell volume increases rapidly following such treatment. The Fps1 glycerol export channel opens to allow glycerol release; this process is essential for survival (Luyten *et al.*, 1995; Tamás *et al.*, 1999). In addition, the cell integrity pathway with its mitogen–activated protein kinase (MAPK) Slt2 is activated rapidly and transiently (Davenport *et al.*, 1995). Active Slt2 kinase stimulates the transcription of genes encoding proteins required for cell wall assembly, which may be important to adapt to the new conditions (Jung and Levin, 1999).

2. GLYCEROL METABOLISM AND THE AQUAGLYCEROPORIN FPS1

During growth on glucose as a carbon source, glycerol for osmoadaptation is produced via a short branch of glycolysis consisting of two enzymatic steps (Ansell *et al.*, 1997; Norbeck *et al.*, 1996). The yeast genome encodes two isoforms for both steps: Gpd1 and Gpd2 for glycerol-3-phosphate dehydrogenase and Gpp1 and Gpp2 for glycerol-3-phosphatase.

Elimination of glycerol production by gene knockout causes an inability of yeast cells to grow in growth medium with more than about 0.3 *M* NaCl. The expression of *GPD1*, *GPP2*, and, to a lesser extent, *GPP1* is stimulated by hyperosmotic shock (Albertyn *et al.*, 1994; Rep *et al.*, 2000) (Fig. 2.2). However, the observed glycerol accumulation immediately after osmotic shock is probably mainly because of stimulated metabolic flux and prevention of glycerol efflux. The yeast plasma membrane is rather impermeable for glycerol (Karlgren *et al.*, 2005; Oliveira *et al.*, 2003) and hence a gated glycerol export channel allows yeast cells to rapidly control and fine-tune their glycerol content. This role is taken by the aquaglyceroporin Fps1. These type of proteins are characterized by six transmembrane domains and two loops, B and E, that dip into the membrane from both sides, essentially forming a seventh transmembrane domain. Mutants lacking Fps1 display a higher intracellular glycerol level (Tao *et al.*, 1999) but accumulate glycerol with a similar profile as wild-type cells under hyperosmotic conditions (Tamás *et al.*, 1999). However, such *fps1Δ* mutants poorly survive a hypoosmotic shock, that is, when yeast cells are transferred into distilled water.

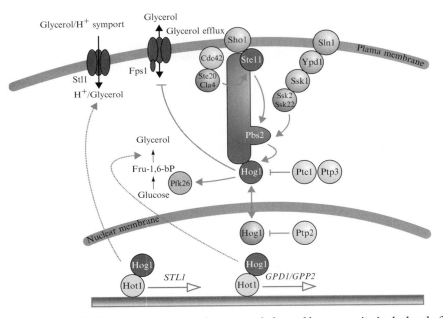

Figure 2.2   The HOG pathway controls water and glycerol homeostasis. At the level of gene expression the capacity for glycerol uptake and production are increased by Hog1. Hog1 also appears to stimulate glycolytic flux by activating Pfk26 to produce fructose-2,6-bisphosphate. Finally, Hog1 may participate in the control of the glycerol efflux channel Fps1. (See color insert.)

Under such conditions most of the intracellular glycerol is released to the surrounding medium within less than 3 min in an Fps1-dependent manner (Tamás et al., 1999). Only a fraction of *fps1Δ* mutants cells survive such a treatment (Tamás et al., 1999), illustrating the role of Fps1 as an osmotic safety valve.

Constitutively open Fps1 mutants exist, which mediate glycerol export even under hyperosmotic conditions. As a consequence, yeast cells that express them are unable to accumulate glycerol, which they try to compensate by enhanced glycerol production (Hedfalk et al., 2004; Karlgren et al., 2004; Tamás et al., 1999, 2003). Mutants expressing constitutively open Fps1 grow more poorly under hyperosmotic conditions. Hence, mutants lacking Fps1 are hypoosmosensitive, whereas mutants with overactive Fps1 are hyperosmosensitive, again illustrating the importance of Fps1 for yeast osmoregulation. While proteins with similar functions in regulated osmolyte export are likely to exist in most organisms, the precise type of aquaglyceroporin represented by *S. cerevisiae* Fps1 appears to be specific to yeasts (Pettersson et al., 2005).

The control mechanism of Fps1 has been investigated by targeted and random mutagenesis. It appears that three regions play a role in Fps1 gating: the B loop (Karlgren et al., 2004), the region of about 40 amino acids proximal to the first transmembrane domain (Karlgren et al., 2004; Tamás et al., 1999, 2003), and the 10 amino acids immediately distal to the sixth transmembrane domain (Hedfalk et al., 2004; Karlgren et al., 2004). All these elements are highly conserved in Fps1 orthologues from other yeasts.

3. THE HOG SIGNALING SYSTEM

The HOG pathway (Fig. 2.1) is one of the best understood and most intensively studied MAPK systems. First, components (Hog1 and Pbs2) were identified in a genetic screen for osmosensitive mutants deficient in glycerol accumulation (Brewster et al., 1993). In parallel, inactivation of SLN1, encoding the single yeast sensor histidine kinase, was found to be lethal (Ota and Varshavsky, 1993). This lethality, which was later shown to be because of inappropriate overactivation of the Hog1 kinase, was suppressed by knockout of any of the genes SSK1, SSK2, PBS2, and HOG1, thereby defining a linear pathway from Sln1 to Hog1. In addition, overexpression of PTC1, PTP2, or PTP3 suppressed lethality of the sln1Δ mutant, defining those as negative elements of the pathway (Maeda et al., 1994; Posas and Saito, 1997; Posas et al., 1996). Finally, the observation that ssk1Δ as well as ssk2Δ ssk22Δ mutants were osmotolerant while deletion of PBS2 and HOG1 caused osmosensitivity prompted genetic screens employing synthetic enhancement that identified SHO1, STE20, and STE11 as encoding components of the Sho1 branch (Maeda *et al.*, 1995). Identification of the pathway components and characterization of their order of function represent textbook examples of the power of both targeted and global yeast genetics approaches. In fact, forward genetics, suppressor mutation, multicopy suppression, synthetic enhancement, epistasis analysis, and yeast two hybrid screens were all employed in this context. Particular powerful genetic tools are mutations that activate signaling constitutively. Significant knowledge has emerged on the flow of information through the pathway and hence the mechanisms of signal transduction by combining the genetic tools with *in vitro* and *in vivo* protein interaction assays, as well as *in vitro* protein kinase assays (de Nadal *et al.*, 2002; Hohmann, 2002; O'Rourke *et al.*, 2002; Saito and Tatebayashi, 2004; Tatebayashi *et al.*, 2006).

The HOG signaling system consists of two branches that converge on the MAPKK Pbs2, the Sln1, and the Sho1. Components of the Sho1 branch also take part in pseudohyphal development and mating in *S. cerevisiae* (O'Rourke and Herskowitz, 1998). In many fungi, it appears that the Sho1 module is not connected to Pbs2 and hence is not involved in osmotic responses (Furukawa *et al.*, 2005; Krantz *et al.*, 2006). This indicates that the Sho1 module might not primarily have a role in osmosensing but rather perceives signals related to cell shape and/or cell surface conditions, in accordance with the role in activation played by the cell polarity machinery. Sho1 is specifically located at sites of cell growth and does not appear to sense turgor changes (Reiser *et al.*, 2000, 2003).

The Sho1 branch consists almost exclusively of proteins shared with the pseudohyphal development pathway and the pheromone response pathway. Signaling specificity seems to be assured by recruitment to scaffold proteins (Sho1, Opy2, Pbs2) and requires the Hog1 kinase. In *hog1Δ* mutants, exposure to osmotic stress causes activation of the pseudohyphal and pheromone response pathways and morphological aberrations (Davenport *et al.*, 1999; O'Rourke and Herskowitz, 1998; Rep *et al.*, 2000). The mechanism by which Hog1 prevents such cross talk has not yet been elucidated. Mechanisms involved in activation of the Sho1 branch following osmotic shock have been described in detail using constitutively active Ste11 and Sho1 mutants as well as protein interaction studies (Tatebayashi *et al.*, 2006). As indicated earlier, the sensing mechanism of osmotic changes in the Sho1 branch is not understood at this point but must be closely related to Sho1 (Tatebayashi *et al.*, 2006). The observation that Sho1 can be replaced by engineered proteins that recruit Pbs2 to the plasma membrane suggests that Sho1 does not function as a sensor itself (Raitt *et al.*, 2000). Sho1 shows much less variation in size than in primary sequence (Krantz and Hohmann, 2006), indicating a structural rather than an enzymatic function.

Sln1 is a sensor histidine kinase related to bacterial two-component systems. Such proteins are widespread in fungi and plants (Catlett *et al.*, 2003).

Sln1 has a similar domain organization as the bacterial osmosensing histidine kinase EnvZ. Both proteins have two transmembrane domains at their N terminus, which are connected by a large extracellular loop, about 300 amino acids in yeasts. It is believed that the extracellular loop and the transmembrane domains sense turgor changes (Reiser et al., 2003), perhaps by responding to movements of the plasma membrane relative to the cell wall. The homodimer is likely regulated by a structural change, which is propagated from the extracellular sensing domain to the intracellular histidine kinase domain of Sln1 (Posas et al., 1996; Reiser et al., 2003). In S. cerevisiae the Sln1 histidine kinase is a negative regulator of the downstream MAPK cascade; deletion of SLN1 or inactivation of the kinase results in lethal Hog1 overactivation (Maeda et al., 1994). When active (i.e., under ambient conditions), the Sln1 histidine kinase cross–phosphorylates within a dimer (Posas et al., 1996), and the phosphate group is transferred via the Sln1 receiver and response regulator domains as well as the Ypd1 phosphotransfer protein to the Ssk1 response regulator protein. Hyperosmotic shock causes inactivation of Sln1 kinase activity and dephosphorylation of Ssk1. This scenario is well supported by mutational analysis of all steps in the phosphorelay system (Posas et al., 1996). Unphosphorylated Ssk1 mediates activation of the redundant MAPKKKs Ssk2 and Ssk22, which in turn activate Pbs2.

The activity and the relative contribution of the two pathway branches to Hog1 kinase activity are usually measured in mutants that are blocked in either branch (Maeda et al., 1995; O'Rourke and Herskowitz, 2004). Whether such experiments reflect activity of the two branches in wildtype cells is presently unknown. It appears that the Sho1 branch has a higher stress threshold for activation (Maeda et al., 1995; O'Rourke and Herskowitz, 2004) and that it is insufficient to mediate maximal pathway activation alone (unpublished data).

4. Transcriptional Responses

Several studies have reported global gene expression analyses employing different types of microarrays following a hyperosmotic shock of different intensity (Causton et al., 2001; Gasch et al., 2000; Posas et al., 2000; Rep et al., 2000; Yale and Bohnert, 2001). The reported number of genes up- or downregulated following a hyperosmotic shock varies between studies because of different growth conditions and thresholds applied in the analyses. It appears that about 200 to 400 genes are upregulated and that some 150 to 250 genes are downregulated. Some important conclusions from these analyses are as follows (Hohmann, 2002).

1. The response is transient. Most genes are up- or downregulated only for a certain period of time, which depends on the strength of the osmotic treatment from a few minutes to more than 1 h.
2. The response is largely general. Essentially all genes showing robust up- or downregulation following osmotic shock are affected in a similar way by a range of different stress conditions. This general stress response is regulated negatively by protein kinase A.
3. Downregulated genes encode proteins related to growth and biomass formation. In fact, most of those genes encode ribosomal proteins, translation factors, and glycolytic enzymes. Those genes are regulated positively by protein kinase A.
4. Upregulated genes are not necessary for osmotolerance. There is essentially no overlap between the set genes whose expression is upregulated by osmotic shock and those whose knockout affects growth in high osmolarity conditions (Warringer *et al.*, 2003). Probably, genes encoding functions required to sense and respond to stress need to be present at all times and hence may not be stress regulated. Many proteins whose production is upregulated may rather be relevant for sustaining an osmotolerant state and their deletion may not cause a strong phenotype. In addition, several upregulated genes, such as those encoding enzymes in glycerol production, are duplicated and only one copy is stress controlled. Its deletion is, at least partly, compensated by the presence of the second isoform.
5. Hog1 is required only for a fraction of the observed transcriptional response. The osmostress–induced upregulation of only about 20% of the genes is fully or largely Hog1 dependent. However, those typically include the most strongly osmoregulated genes.

Several transcription factors seem to be involved in Hog1-dependent responses: Hot1, Sko1, Msn2/Msn4, Msn1, and Smp1 (Fig. 2.3). The first four of those are best characterized and also seem to play the most significant roles.

Hot1 (Alepuz *et al.*, 2003; Rep *et al.*, 1999, 2000) is a nuclear protein that seems to control a set of less than 10 genes, including those that encode proteins in glycerol metabolism and uptake. Hot1 recruits Hog1 to target promoters.

Sko1 binds to cAMP response element sites in targets promoters, although not all of the about 40 target genes seem to contain such a site (Proft and Struhl, 2002; Proft *et al.*, 2001, 2005; Rep *et al.*, 2001). Active Hog1 converts Sko1 from a repressor to an activator (Proft and Struhl, 2002; Rep *et al.*, 2001). It appears that Sko1 controls the expression of several regulators of the osmoresponse systems, such as the Msn2 transcription factor and the Ptp3 protein phosphatase (Proft *et al.*, 2005). Hence, Sko1 may be central to a transcriptional network important for establishing the adapted state in osmoregulation.

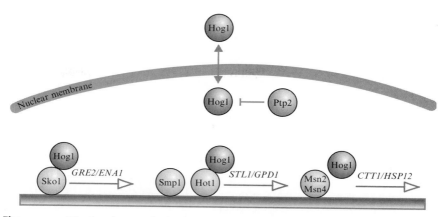

Figure 2.3 Hog1 and control of gene expression. Hog1 interacts with different transcription factors in the nucleus, such as Sko1, Hot1, and probably also Smp1, to control different sets of genes (some typical examples are shown). It appears that Hog1 thereby itself is the activating factor. (See color insert.)

Msn2 and Msn4 are two redundant proteins mediating a general stress response (Ruis and Schüller, 1995). Msn2/Msn4 nuclear localization is negatively controlled by protein kinase A. Although Msn2/Msn4 is required for transcriptional responses to a range of stress conditions, stimulation of its targets under osmotic stress requires the Hog1 kinase (Rep et al., 2000).

It appears that the Hog1 kinase itself is the activating transcription factor for its targets. Hog1 is recruited to target promoters by Hot1, Sko1, and Msn2/Msn4 (Alepuz et al., 2001, 2003; De Nadal et al., 2004; Pokholok et al., 2006; Proft and Struhl, 2002). Hog1 is not only required for initiating transcription of target genes, but also for the transcriptional elongation step (Pokholok et al., 2006; Proft et al., 2006). Hence, the perspective of how a protein kinase controls gene expression has changed substantially through studies on the Hog1 MAPK. Rather than phosphorylating and thereby (in) activating a DNA-binding transcription factor, Hog1 (and also other kinases [Pokholok et al., 2006]) associates with DNA-binding proteins to mediate gene-specific control of transcription via general components of the gene regulation machinery.

5. STUDYING THE CONTROL OF SIGNALING

The response of yeast cells to a hyperosmotic shock is transient (Fig. 2.4). Hence, HOG pathway activation is controlled by feedback mechanisms that ensure that the pathway is inactivated during adaptation.

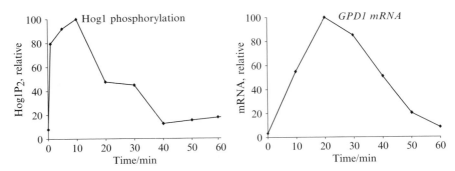

Figure 2.4 Osmotic responses are transient. An example of quantification of the level of phospho–Hog1 and the mRNA of the Hog1-target gene *GPD1* is shown. Cells were treated with 0.5 *M* NaCl at time 0. (See color insert.)

This means that the observed, strong HOG pathway activation following a hyperosmotic shock mainly serves to trigger the adaptation process but that maintenance of the adapted state is supported either by a much smaller level of active Hog1 or by entirely different mechanisms.

The yeast osmosensing HOG system controls a homeostatic process and hence needs to remain in a signaling-competent state at all times. Indeed, the HOG pathway, and HOG-dependent gene expression, can be reactivated readily with essentially the same time course and amplitude by a second osmotic shock at different times following a first shock (Klipp *et al.*, 2005). Hence, mechanisms that control signaling through the HOG pathway combine feedback with maintenance of signaling competence. This property is shared by both branches of the HOG pathway when they control Hog1 activation individually (unpublished data).

In order to better understand the dynamics of osmoregulation and signaling we study how several biological processes can be integrated into one system. To do quantitative time-resolving measurements of biomolecules, such as phosphorylated Hog1, mRNAs, proteins, and metabolites provide insight into the causal and temporal relationships between different biological events. Based on biochemical knowledge of the system and quantitative biological data, a mathematical reconstruction of the system was undertaken in order to be able to simulate and address specific biological questions, such as the mechanism underlying feedback control and maintenance of signaling competence (Klipp *et al.*, 2005).

In its present state, the model encompasses sensing through the Sln1 phosphorelay system, signaling through the MAPK cascade, gene and protein expression for glycerol production, and glycolytic metabolism, including glycerol production and regulated export through Fps1, as well as changes in cell volume and turgor. These items were initially set up as modules on the basis of literature and large-scale genomics data and were tested individually before they were connected to an overall model of 35 differential equations and 70 parameters. Some of the parameters were calculated on the basis of standard experiments (osmotic shock with 0.5 M NaCl) and time course data on relative levels of phosphorylated Hog1, levels of mRNA of HOG target genes, and specific activity of an enzyme in glycerol production, as well as the levels of intracellular and total glycerol. Other parameters were calculated and optimized to best fit experimental data. The model was then tested and improved iteratively using additional experimental data from perturbations such as different types of stress treatments or genetic changes, including inability to accumulate glycerol. Furthermore, the model will be used to test specific hypothesis and will serve as a guideline of what types of experimental data should be generated in order to increase our knowledge of osmoregulation and signaling in yeast.

Like any other MAPK signaling pathway, the HOG pathway is controlled negatively by protein phosphatases (Fig. 2.5). Three phosphatases are physiologically relevant: Ptp2, a nuclear phosphotyrosine–specific phosphatase; Ptp3, a cytosolic phosphotyrosine–specific phosphatase; and Ptc1, a type 2C protein phosphatase (PP2C) located in both the cytosol and the nucleus. Simultaneous knockout of *PTP2* and *PTC1* is lethal because of overactive Hog1 (Maeda *et al.*, 1993), demonstrating that the phosphatases are needed to control noise and thresholds of signaling.

Simulations (Klipp *et al.*, 2005) were used to test if Hog1–mediated upregulation of protein phosphatases could account for the observed feedback control on the Hog1 phosphorylation status. Although it could possibly account for the observed downregulation, it is incompatible with the ability of the HOG pathway to maintain signaling competence. In osmostressed yeast

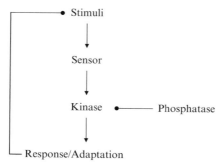

Figure 2.5 Feedback of the osmotic response. Osmotic adaptation, i.e., turgor recovery, seems to determine the time point when HOG signaling is downregulated. The onset of cell reswelling probably terminates initial stimuli. Then, protein phosphatases deactivate the pathway. Upregulation of those phosphatases, indicated here as a transcriptional feedback loop, may play a role in adaptation or, rather, contribute to establish a new steady state in adapted cells.

cells, it appears that the timing of downregulation is determined by adaptation, that is, recovery of volume and turgor, which in turn is mediated by glycerol accumulation (Fig. 2.5). This scenario directly implies that the sensors of the HOG pathway detect an osmotic change rather than an osmotic status. In this picture, the phosphatases act constitutively and keep the phosphorylation state of Hog1 dependent on sensor activity. The roles of the phosphatases are to adjust the dynamic range, control noise, and determine signaling thresholds of the entire signaling system depending on environmental requirements.

Several experimental scenarios support the idea that the Hog1 phosphorylation state is mainly controlled by the osmotic changes caused by external stimuli or cellular adaptation rather than intrinsic feedback mechanisms. First, mutations that prevent glycerol production ($gpd1\Delta$ $gpd2\Delta$) or that delay glycerol accumulation despite glycerol overproduction (constitutively open Fps1) cause sustained Hog1 phosphorylation (Klipp *et al.*, 2005; Siderius *et al.*, 2000). Such cells do not adapt to increased osmolarity or only with a significant delay. Second, cells that accumulate glycerol faster display a shorter period of Hog1 phosphorylation (Krantz *et al.*, 2004). Such faster glycerol accumulation can be achieved by treating cells growing under anaerobic conditions (the rate of glycerol production is higher in such cells) or by overexpressing *GPD1* from a multicopy plasmid. Third, it is well known that yeast cells treated with an osmotic shock of higher intensity display a longer period of Hog1 phosphorylation than cells treated with lower stress levels (Van Wuytswinkel *et al.*, 2000). Finally, there is a clear correlation between the apparent period of osmotic adaptation and Hog1 phosphorylation (Karlgren *et al.*, 2005) in an experimental setup of forced adaptation (or conditional osmotic stress). This experimental setup makes use of yeast cells unable to produce any glycerol ($gpd1\Delta$ $gpd2\Delta$) and hence unable to adapt to increased osmolarity. Even treatments with glycerol, xylitol, and sorbitol prevent cell growth and sustained activation of the HOG pathway because of rapid closing of the Fps1 glycerol channel, which is independent of the osmoticum employed. Under these conditions, such mutant cells can be rescued when transformed with a plasmid that expresses a constitutively open Fps1. Initially these cells shrink, but the polyols equilibrate over the plasma membrane via the aquaglyceroporin, with a rate decreasing from glycerol to sorbitol, such that they relieve the osmotic stress at different rates. Consequently, the period of Hog1 phosphorylation is shortest with glycerol, longer with xylitol, and longest with sorbitol (Karlgren *et al.*, 2005), demonstrating a clear link between osmotic adaptation and the period of Hog1 phosphorylation/activity. This experimental setup employing constitutively open Fps1 and heterologous aquaglyceroporin (such as rat AQP9), as well as different polyols, allows different periods of osmotic adaptation independent of the initial stress level, providing an experimental tool to separate the two. An additional advantage of the system is that cells adapt independently of their own activity. Hence, this setup allows addressing a range of questions on osmosensing feedback control and the consequences of different periods of Hog1 activity, for instance on the global gene expression profile.

By controlling the osmotic response and the timing of its downregulation mainly via actual osmotic adaptation, probably via the activity of the sensors, the cell combines effective feedback control with signaling competence, as required for a homeostatic process. However, there are clearly additional mechanisms that impinge on HOG pathway activity and feedback control. Ssk1 in the Sln1 pathway is destabilized in osmostressed cells and this mechanism may contribute to pathway downregulation, although it has to take into account maintenance of signaling competence (Sato *et al.*, 2003). Moreover, Ste11 in the Sho1 pathway is an unstable protein and is degraded at least during the pheromone response (Esch and Errede, 2002).

6. FUTURE DIRECTIONS

Quantitative analyses and combination of modeling and experimentation will provide new insights into the mechanisms of osmotic homeostasis. For instance, the contribution of the two HOG pathway branches to signaling has so far only been studied in mutants blocked in one branch, which may give an incorrect impression of their roles in wild-type cells. Monitoring the activity of different proteins in the two branches by more sophisticated experimental setups in combination with modeling may help elucidate the precise role of the Sln1 and Sho1 branches in the overall response. Another interesting question concerns the quantitative and temporal contributions of different control mechanisms in glycerol accumulation: Fps1 closure and control of glycolytic flux, as well as upregulation of enzymes in glycerol production and uptake (Fig. 2.2). Finally, most measurements done so far monitor osmotic responses in cell populations. Such response profiles may give a misleading impression, as they could be composed of single cell responses that follow a very different profile. Such single cell measurements are feasible using specific fluorescent reporters combined with flow cytometry or time-lapse microscopy combined with microfluidics and optical manipulations tools. Hence, systems level understanding of cell responses, such as analysis of thresholds, noise regulation, and feedback control, requires acquisition of a new type of data and hence encompasses both computational and experimental challenges.

ACKNOWLEDGMENTS

Relevant work in the Hohmann laboratory is supported by grants from the Swedish Research Council and the European Commission (QUASI: LSHG-CT2003-503230; SystemsBiology: MEST2-CT2004-514169). VINNOVA supports the collaboration between the Hohmann laboratory and the Systems Biology Institute (Interdisciplinary BIO Sweden–Japan). MK is a postdoctoral fellow supported by the Japan Society for the Promotion of Science.

REFERENCES

Albertyn, J., Hohmann, S., Thevelein, J. M., and Prior, B. A. (1994). *GPD1*, which encodes glycerol-3-phosphate dehydrogenase is essential for growth under osmotic stress in *Saccharomyces cerevisiae* and its expression is regulated by the high-osmolarity glycerol response pathway. *Mol. Cell. Biol.* **14**, 4135–4144.

Alepuz, P. M., de Nadal, E., Zapater, M., Ammerer, G., and Posas, F. (2003). Osmostress-induced transcription by Hot1 depends on a Hog1-mediated recruitment of the RNA Pol II. *EMBO J.* **22**, 2433–2442.

Alepuz, P. M., Jovanovic, A., Reiser, V., and Ammerer, G. (2001). Stress-induced MAP kinase Hog1 is part of transcription activation complexes. *Mol. Cell* **7**, 767–777.

Ansell, R., Granath, K., Hohmann, S., Thevelein, J. M., and Adler, L. (1997). The two isoenzymes for yeast NAD⁺-dependent glycerol 3-phosphate dehydrogenase encoded by *GPD1* and *GPD2* have distinct roles in osmoadaptation and redox regulation. *EMBO J.* **16**, 2179–2187.

Bilsland-Marchesan, E., Arino, J., Saito, H., Sunnerhagen, P., and Posas, F. (2000). Rck2 kinase is a substrate for the osmotic stress-activated mitogen-activated protein kinase Hog1. *Mol. Cell. Biol.* **20**, 3887–3895.

Bilsland, E., Molin, C., Swaminathan, S., Ramne, A., and Sunnerhagen, P. (2004). Rck1 and Rck2 MAPKAP kinases and the HOG pathway are required for oxidative stress resistance. *Mol. Microbiol.* **53**, 1743–1756.

Brewster, J. L., de Valoir, T., Dwyer, N. D., Winter, E., and Gustin, M. C. (1993). An osmosensing signal transduction pathway in yeast. *Science* **259**, 1760–1763.

Catlett, N. L., Yoder, O. C., and Turgeon, B. G. (2003). Whole-genome analysis of two-component signal transduction genes in fungal pathogens. *Eukaryot. Cell* **2**, 1151–1161.

Causton, H. C., Ren, B., Koh, S. S., Harbison, C. T., Kanin, E., Jennings, E. G., Lee, T. I., True, H. L., Lander, E. S., and Young, R. A. (2001). Remodeling of yeast genome expression in response to environmental changes. *Mol. Biol. Cell* **12**, 323–337.

Clotet, J., Escote, X., Adrover, M. A., Yaakov, G., Gari, E., Aldea, M., de Nadal, E., and Posas, F. (2006). Phosphorylation of Hsl1 by Hog1 leads to a G2 arrest essential for cell survival at high osmolarity. *EMBO J.* **25**, 2338–2346.

Davenport, K. D., Williams, K. E., Ullmann, B. D., and Gustin, M. C. (1999). Activation of the *Saccharomyces cerevisiae* filamentation/invasion pathway by osmotic stress in high-osmolarity glycogen pathway mutants. *Genetics* **153**, 1091–1103.

Davenport, K. R., Sohaskey, M., Kamada, Y., Levin, D. E., and Gustin, M. C. (1995). A second osmosensing signal transduction pathway in yeast: Hypotonic shock activates the PKC1 protein kinase-regulated cell integrity pathway. *J. Biol. Chem.* **270**, 30157–30161.

de Nadal, E., Alepuz, P. M., and Posas, F. (2002). Dealing with osmostress through MAP kinase activation. *EMBO Rep.* **3**, 735–740.

De Nadal, E., Zapater, M., Alepuz, P. M., Sumoy, L., Mas, G., and Posas, F. (2004). The MAPK Hog1 recruits Rpd3 histone deacetylase to activate osmoresponsive genes. *Nature* **427**, 370–374.

Dihazi, H., Kessler, R., and Eschrich, K. (2004). High osmolarity glycerol (HOG) pathway-induced phosphorylation and activation of 6-phosphofructo-2-kinase are essential for glycerol accumulation and yeast cell proliferation under hyperosmotic stress. *J. Biol. Chem.* **279**, 23961–23968.

Esch, R. K., and Errede, B. (2002). Pheromone induction promotes Ste11 degradation through a MAPK feedback and ubiquitin-dependent mechanism. *Proc. Natl. Acad. Sci. USA* **99**, 9160–9165.

Escote, X., Zapater, M., Clotet, J., and Posas, F. (2004). Hog1 mediates cell-cycle arrest in G1 phase by the dual targeting of Sic1. *Nat. Cell Biol.* **6**, 997–1002.

Furukawa, K., Hoshi, Y., Maeda, T., Nakajima, T., and Abe, K. (2005). *Aspergillus nidulans* HOG pathway is activated only by two-component signalling pathway in response to osmotic stress. *Mol. Microbiol.* **56**, 1246–1261.

Gasch, A. P., Spellman, P. T., Kao, C. M., Carmel-Harel, O., Eisen, M. B., Storz, G., Botstein, D., and Brown, P. O. (2000). Genomic expression programs in the response of yeast cells to environmental changes. *Mol. Biol. Cell* **11**, 4241–4257.

Hedfalk, K., Bill, R. M., Mullins, J. G., Karlgren, S., Filipsson, C., Bergstrom, J., Tamas, M. J., Rydstrom, J., and Hohmann, S. (2004). A regulatory domain in the C-terminal extension of the yeast glycerol channel Fps1p. *J. Biol. Chem.* **279**, 14954–14960.

Hohmann, S. (2002). Osmotic stress signaling and osmoadaptation in yeasts. *Microbiol. Mol. Biol. Rev.* **66**, 300–372.

Jung, U. S., and Levin, D. E. (1999). Genome-wide analysis of gene expression regulated by the yeast cell wall integrity signalling pathway. *Mol. Microbiol.* **34**, 1049–1057.

Karlgren, S., Filipsson, C., Mullins, J. G., Bill, R. M., Tamas, M. J., and Hohmann, S. (2004). Identification of residues controlling transport through the yeast aquaglyceroporin Fps1 using a genetic screen. *Eur. J. Biochem.* **271**, 771–779.

Karlgren, S., Pettersson, N., Nordlander, B., Mathai, J. C., Brodsky, J. L., Zeidel, M. L., Bill, R. M., and Hohmann, S. (2005). Conditional osmotic stress in yeast: A system to study transport through aquaglyceroporins and osmostress signaling. *J. Biol. Chem.* **280**, 7186–7193.

Klipp, E., Nordlander, B., Kruger, R., Gennemark, P., and Hohmann, S. (2005). Integrative model of the response of yeast to osmotic shock. *Nat. Biotechnol.* **23**, 975–982.

Krantz, M., Becit, E., and Hohmann, S. (2006). Comparative genomics of the HOG-signalling system in fungi. *Curr. Genet.* **49**, 137–151.

Krantz, M., and Hohmann, S. (2006). Employing protein size in the functional analysis of orthologous proteins, as illustrated with the yeast HOG pathway. *In* "Comparative Genomics: Using Fungi as Models" (P. Sunnerhagen and J. Piskur, eds.), Vol. 15, pp. 131–143. Springer-Verlag, Heidelberg.

Krantz, M., Nordlander, B., Valadi, H., Johansson, M., Gustafsson, L., and Hohmann, S. (2004). Anaerobicity prepares *Saccharomyces cerevisiae* cells for faster adaptation to osmotic shock. *Eukaryot. Cell* **3**, 1381–1390.

Luyten, K., Albertyn, J., Skibbe, W. F., Prior, B. A., Ramos, J., Thevelein, J. M., and Hohmann, S. (1995). Fps1, a yeast member of the MIP family of channel proteins, is a facilitator for glycerol uptake and efflux and is inactive under osmotic stress. *EMBO J.* **14**, 1360–1371.

Maeda, T., Takekawa, M., and Saito, H. (1995). Activation of yeast PBS2 MAPKK by MAPKKKs or by binding of an SH3-containing osmosensor. *Science* **269**, 554–558.

Maeda, T., Tsai, A. Y. M., and Saito, H. (1993). Mutations in a protein tyrosin phosphatase gene (*PTP2*) and a protein serine/threonine phosphatase gene (*PTC1*) cause a synthetic growth defect in *Saccharomyces cerevisiae*. *Mol. Cell. Biol.* **13**, 5408–5417.

Maeda, T., Wurgler-Murphy, S. M., and Saito, H. (1994). A two-component system that regulates an osmosensing MAP kinase cascade in yeast. *Nature* **369**, 242–245.

Norbeck, J., Påhlman, A. K., Akhtar, N., Blomberg, A., and Adler, L. (1996). Purification and characterization of two isoenzymes of DL-glycerol-3-phosphatase from *Saccharomyces cerevisiae*: Identification of the corresponding *GPP1* and *GPP2* genes and evidence for osmotic regulation of Gpp2p expression by the osmosensing mitogen-activated protein kinase signal transduction pathway. *J. Biol. Chem.* **271**, 13875–13881.

Oliveira, R., Lages, F., Silva-Graca, M., and Lucas, C. (2003). Fps1p channel is the mediator of the major part of glycerol passive diffusion in *Saccharomyces cerevisiae:* Artefacts and re-definitions. *Biochim. Biophys. Acta* **1613,** 57–71.

O'Rourke, S. M., and Herskowitz, I. (1998). The Hog1 MAPK prevents cross talk between the HOG and pheromone response MAPK pathways in *Saccharomyces cerevisiae. Genes Dev.* **12,** 2874–2886.

O'Rourke, S. M., and Herskowitz, I. (2004). Unique and redundant roles for HOG MAPK pathway components as revealed by whole-genome expression analysis. *Mol. Biol. Cell* **15,** 532–542.

O'Rourke, S. M., Herskowitz, I., and O'Shea, E. K. (2002). Yeast go the whole HOG for the hyperosmotic response. *Trends Genet.* **18,** 405–412.

Ota, I. M., and Varshavsky, A. (1993). A yeast protein similar to bacterial two-component regulators. *Science* **262,** 566–569.

Pettersson, N., Filipsson, C., Becit, E., Brive, L., and Hohmann, S. (2005). Aquaporins in yeasts and filamentous fungi. *Biol. Cell* **97,** 487–500.

Pokholok, D. K., Zeitlinger, J., Hannett, N. M., Reynolds, D. B., and Young, R. A. (2006). Activated signal transduction kinases frequently occupy target genes. *Science* **313,** 533–536.

Posas, F., Chambers, J. R., Heyman, J. A., Hoeffler, J. P., de Nadal, E., and Arino, J. (2000). The transcriptional response of yeast to saline stress. *J. Biol. Chem.* **275,** 17249–17255.

Posas, F., and Saito, H. (1997). Osmotic activation of the HOG MAPK pathway via Ste11p MAPKKK: Scaffold role of Pbs2p MAPKK. *Science* **276,** 1702–1705.

Posas, F., Wurgler-Murphy, S. M., Maeda, T., Witten, E. A., Thai, T. C., and Saito, H. (1996). Yeast HOG1 MAP kinase cascade is regulated by a multistep phosphorelay mechanism in the SLN1-YPD1-SSK1 "two-component" osmosensor. *Cell* **86,** 865–875.

Proft, M., Gibbons, F. D., Copeland, M., Roth, F. P., and Struhl, K. (2005). Genomewide identification of Sko1 target promoters reveals a regulatory network that operates in response to osmotic stress in *Saccharomyces cerevisiae. Eukaryot. Cell* **4,** 1343–1352.

Proft, M., Mas, G., de Nadal, E., Vendrell, A., Noriega, N., Struhl, K., and Posas, F. (2006). The stress-activated Hog1 kinase is a selective transcriptional elongation factor for genes responding to osmotic stress. *Mol. Cell* **23,** 241–250.

Proft, M., Pascual-Ahuir, A., de Nadal, E., Arino, J., Serrano, R., and Posas, F. (2001). Regulation of the Sko1 transcriptional repressor by the Hog1 MAP kinase in response to osmotic stress. *EMBO J.* **20,** 1123–1133.

Proft, M., and Struhl, K. (2002). Hog1 kinase converts the Sko1-Cyc8-Tup1 repressor complex into an activator that recruits SAGA and SWI/SNF in response to osmotic stress. *Mol. Cell* **9,** 1307–1317.

Proft, M., and Struhl, K. (2004). MAP kinase-mediated stress relief that precedes and regulates the timing of transcriptional induction. *Cell* **118,** 351–361.

Raitt, D. C., Posas, F., and Saito, H. (2000). Yeast Cdc42 GTPase and Ste20 PAK-like kinase regulate Sho1-dependent activation of the Hog1 MAPK pathway. *EMBO J.* **19,** 4623–4631.

Reiser, V., Raitt, D. C., and Saito, H. (2003). Yeast osmosensor Sln1 and plant cytokinin receptor Cre1 respond to changes in turgor pressure. *J. Cell Biol.* **161,** 1035–1040.

Reiser, V., Salah, S. M., and Ammerer, G. (2000). Polarized localization of yeast Pbs2 depends on osmostress, the membrane protein Sho1 and Cdc42. *Nat. Cell Biol.* **2,** 620–627.

Rep, M., Krantz, M., Thevelein, J. M., and Hohmann, S. (2000). The transcriptional response of *Saccharomyces cerevisiae* to osmotic shock: Hot1p and Msn2p/Msn4p are required for the induction of subsets of high osmolarity glycerol pathway-dependent genes. *J. Biol. Chem.* **275,** 8290–8300.

Rep, M., Proft, M., Remize, F., Tamas, M., Serrano, R., Thevelein, J. M., and Hohmann, S. (2001). The *Saccharomyces cerevisiae* Sko1p transcription factor mediates HOG pathway-dependent osmotic regulation of a set of genes encoding enzymes implicated in protection from oxidative damage. *Mol. Microbiol.* **40**, 1067–1083.

Rep, M., Reiser, V., Holzmüller, U., Thevelein, J. M., Hohmann, S., Ammerer, G., and Ruis, H. (1999). Osmotic stress-induced gene expression in *Saccharomyces cerevisiae* requires Msn1p and the novel nuclear factor Hot1p. *Mol. Cell. Biol.* **19**, 5474–5485.

Ruis, H., and Schüller, C. (1995). Stress signaling in yeast. *Bioessays* **17**, 959–965.

Saito, H., and Tatebayashi, K. (2004). Regulation of the osmoregulatory HOG MAPK cascade in yeast. *J. Biochem. (Tokyo)* **136**, 267–272.

Sato, N., Kawahara, H., Toh-e, A., and Maeda, T. (2003). Phosphorelay-regulated degradation of the yeast Ssk1p response regulator by the ubiquitin-proteasome system. *Mol. Cell. Biol.* **23**, 6662–6671.

Siderius, M., Van Wuytswinkel, O., Reijenga, K. A., Kelders, M., and Mager, W. H. (2000). The control of intracellular glycerol in *Saccharomyces cerevisiae* influences osmotic stress response and resistance to increased temperature. *Mol. Microbiol.* **36**, 1381–1390.

Tamás, M. J., Karlgren, S., Bill, R. M., Hedfalk, K., Allegri, L., Ferreira, M., Thevelein, J. M., Rydstrom, J., Mullins, J. G., and Hohmann, S. (2003). A short regulatory domain restricts glycerol transport through yeast Fps1p. *J. Biol. Chem.* **278**, 6337–6345.

Tamás, M. J., Luyten, K., Sutherland, F. C. W., Hernandez, A., Albertyn, J., Valadi, H., Li, H., Prior, B. A., Kilian, S. G., Ramos, J., Gustafsson, L., Thevelein, J. M., and Hohmann, S. (1999). Fps1p controls the accumulation and release of the compatible solute glycerol in yeast osmoregulation. *Mol. Microbiol.* **31**, 1087–1104.

Tao, W., Deschenes, R. J., and Fassler, J. S. (1999). Intracellular glycerol levels modulate the activity of Sln1p, a *Saccharomyces cerevisiae* two-component regulator. *J. Biol. Chem.* **274**, 360–367.

Tatebayashi, K., Yamamoto, K., Tanaka, K., Tomida, T., Maruoka, T., Kasukawa, E., and Saito, H. (2006). Adaptor functions of Cdc42, Ste50, and Sho1 in the yeast osmoregulatory HOG MAPK pathway. *EMBO J.* **25**, 3033–3044.

Teige, M., Scheikl, E., Reiser, V., Ruis, H., and Ammerer, G. (2001). Rck2, a member of the calmodulin-protein kinase family, links protein synthesis to high osmolarity MAP kinase signaling in budding yeast. *Proc. Natl. Acad. Sci. USA* **98**, 5625–5630.

Thorsen, M., Di, Y., Tangemo, C., Morillas, M., Ahmadpour, D., Van der Does, C., Wagner, A., Johansson, E., Boman, J., Posas, F., Wysocki, R., and Tamas, M. J. (2006). The MAPK Hog1p modulates Fps1p-dependent arsenite uptake and tolerance in yeast. *Mol. Biol. Cell.* **17**, 4400–4410.

Van Wuytswinkel, O., Reiser, V., Siderius, M., Kelders, M. C., Ammerer, G., Ruis, H., and Mager, W. H. (2000). Response of *Saccharomyces cerevisiae* to severe osmotic stress: Evidence for a novel activation mechanism of the HOG MAP kinase pathway. *Mol. Microbiol.* **37**, 382–397.

Warringer, J., Ericson, E., Fernandez, L., Nerman, O., and Blomberg, A. (2003). High-resolution yeast phenomics resolves different physiological features in the saline response. *Proc. Natl. Acad. Sci. USA* **100**, 15724–15729.

Yale, J., and Bohnert, H. J. (2001). Transcript expression in *Saccharomyces cerevisiae* at high salinity. *J. Biol. Chem.* **276**, 15996–16007.

CHAPTER THREE

PHYSIOLOGICAL ANALYSIS OF BACTERIAL MECHANOSENSITIVE CHANNELS

Ian R. Booth, Michelle D. Edwards, Susan Black,
Ulrike Schumann, Wendy Bartlett, Tim Rasmussen,
Akiko Rasmussen, *and* Samantha Miller

Contents

Abstract

Bacterial mechanosensitive (MS) channels play a significant role in protecting cells against hypoosmotic shock. Bacteria that have been diluted from high osmolarity medium into dilute solution are required to cope with sudden water influx associated with an osmotic imbalance equivalent to 10 to 14 atm. The cell wall is only poorly expansive and the cytoplasmic membrane even less so. Thus, swelling is not an option and the cell must rapidly eject solutes to diminish the osmotic gradient and thereby preserve structural integrity. This chapter describes cellular assays of MS channel function and their interpretation.

1. INTRODUCTION

Bacterial mechanosensitive (MS) channels were discovered by application of patch clamp electrophysiology in the late 1980s (Martinac *et al.*, 1987), but the analysis of their physiological roles was delayed a further 10 years because of the lack of comprehensive sets of mutants (Levina *et al.*, 1999; Sukharev *et al.*, 1994). Multiple MS channels in bacterial cells, and in

School of Medical Sciences, University of Aberdeen, Institute of Medical Sciences, Foresterhill, Aberdeen, United Kingdom

Methods in Enzymology, Volume 428
ISSN 0076-6879, DOI: 10.1016/S0076-6879(07)28003-6

some eukaryotic organelles, is an almost universal observation (Haswell and Meyerowitz, 2006; Pivetti *et al.*, 2003). Since the channels usually have overlapping functions, a complete analysis of their physiological roles is dependent on mutants lacking multiple channels (Levina *et al.*, 1999). Genome sequencing has made it possible to identify the distribution of the three channels for which genes have been described: MscL (*mscL*), MscS (*yggB*), and MscK (*kefA* or *aefA*).[1] However, such a simplistic view is confounded by several factors. First, most organisms possess multiple homologues of the MscS and MscK channel types but more rarely have multiple *mscL* genes (Pivetti *et al.*, 2003). A truly complete understanding of the physiology might depend on the creation of strains lacking all functional MS channel homologues. A short cut can be taken by eliminating those gene products that bear the strongest resemblance to the characterized *Escherichia coli* genes, but this is by no means a secure pathway (Touze *et al.*, 2001). Second, there is a class of MS channel activity evident in *E. coli* and, by inference, other bacteria, for which no gene has been identified. The so-called miniconductance channel MscM is not even universally accepted by all electrophysiologists performing measurements on *E. coli* and no structural gene has yet been assigned to this activity. MscM often appears as a low conductance (≈ 0.3 nS) compared to MscS (≈ 1.0 nS) and MscL (≈ 2.5 nS) and is not present in all membrane patches or under all conditions, hence the reason for doubt about whether it is a true activity. Third, although MS channels have now been proved to be a major pathway for solute movement when *E. coli* cells are challenged with an extreme hypoosmotic shock, this is by no means the only pathway operating (Bakker *et al.*, 1987; Koo *et al.*, 1991). Specific exit routes for compatible solutes and K^+ exist in a range of mutants that lack the major demonstrable MS channels (Koo *et al.*, 1991; Nottebrock *et al.*, 2003). These systems have been evident from a range of different physiological assays, but are detected most readily as rapid solute efflux when the major channels have been eliminated by mutation (Folgering *et al.*, 2005; W. Bartlett, S. Miller, and I. R. Booth, unpublished data). For these reasons the outcome of physiological analyses of MS channel activity needs to be interpreted with care.

The MS channel field is now well developed and a wide range of approaches has been brought to bear on their analysis. Methods for analysis of their activity, particularly using electrophysiology approaches, were last reviewed systematically in 1999 (Blount *et al.*, 1999). Little has changed that significantly affects the research methodology of patch clamp in this time, although some automated methods have started to be introduced but not yet applied to bacteria. The biotechnological applications of MS channels have started to develop, frequently using the purified and reconstituted

---

[1] Using the *E. coli* gene nomenclature.

channel proteins, but essentially with methods similar to those described previously (Folgering *et al.*, 2004; van den Bogaart *et al.*, 2006). The major significant changes since the previous review in this series have been the description of the structure for the MscS channel (Bass *et al.*, 2002; Steinbacher *et al.*, 2007) and the subsequent recognition of the greater prevalence of this channel type in bacteria, archaea, and plants when compared with the species distribution of MscL (Pivetti *et al.*, 2003). The discovery of the gene for MscS, which was a fortunate by-product of the discovery of the *kefA* gene (Levina *et al.*, 1999; McLaggan *et al.*, 2002), opened the way to physiological analysis of MS channel function. This chapter has little to add to the description of electrophysiological methodology but will focus on physiological assays of MS channels and the interpretation of data arising from these assays.

Analytical methods can be broadly broken down into two categories: those that are specifically applied to the native channel and those that apply most directly to the analysis of mutants. Some techniques are equally applicable to both.

2. ANALYSIS OF NATIVE CHANNEL FUNCTION

The standard assay of MS channel function is through hypoosmotic challenge (Levina *et al.*, 1999). The original method works well for K-12 strains of *E. coli*, but has not proven as effective in other organisms (Folgering *et al.*, 2005; Nottebrock *et al.*, 2003). The principle of the assay lies at the very core of our understanding of the role of MS channels in cell physiology. Bacterial cells adapt to high osmolarity by accumulating solutes to concentrations that can approach molar ranges within the cytoplasm (Booth *et al.*, 1988). When an organism is transferred into media of higher osmolarity there is an initial transient plasmolysis as water leaves the cell down the osmotic gradient. The cell compensates by accumulating K^+ ions and by either the synthesis or the transport of glutamate as a counter ion (Epstein, 1986). The accumulation of these salts leads to the recovery of water, restoration of turgor, and renewed growth. However, a high salt concentration in the cytoplasm is inimical to growth and a secondary strategy is the gradual replacement of these salts with compatible *organic*[2] solutes (e.g., betaine, proline, ectoine, trehalose). The accumulation of

[2] Organic is emphasized here by italics because of a tendency for authors to refer to K^+ and glutamate as a "compatible solute." Compatible solute has a specific meaning and refers to the ability of some organic solutes to restore the activity to enzymes that have been inhibited, in pure solution, by high salt. All the evidence is that K^+ glutamate accumulation by cells is preferred over inorganic salts such as NaCl, but that it is inhibitory to enzyme function, which can be overcome by the accumulation of organic-compatible solutes. Compatible solute effects on enzymes are discussed more fully elsewhere in this volume.

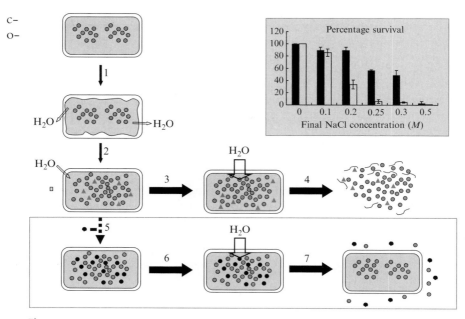

Figure 3.1 Principle of hypoosmotic challenge used to detect MS channel function. The assay depends on the triple MS channel-deficient mutant *E. coli* K-12 MJF465 (MscK⁻, MscS⁻, MscL⁻) (and its derivatives). Cells growing in mid-exponential phase at low osmolarity (≤200 mOsm) have a relatively constant turgor generated primarily through the accumulation of K^+ and glutamate. Transfer to higher osmolarity medium (1) leads to transient loss of turgor, followed by turgor restoration (2) through the increased accumulation of K^+ and glutamate and, in most *E. coli* K-12 strains synthesis of some trehalose, an organic compatible solute (gray triangles). Rapid transfer to low osmolarity medium (3) leads to an inrush of water down the osmotic gradient, causing a rise in turgor. In the absence of significant MS channel function, cell lysis follows (4). The presence of rapidly transported organic compatible solutes in the medium (e.g., betaine) can significantly modify the outcome of the hypoosmotic challenge (5). Some of the intracellular pool of K^+, glutamate, and trehalose is exchanged for betaine (black circles), and subsequent hypoosmotic challenge (6) is survived because of rapid expulsion of the betaine (7). (Inset) Unpublished data for the influence of betaine on the survival of hypoosmotic shock by *E. coli* MJF465. Cells were grown in the presence of 1 mM betaine and 0.5 M NaCl. The cells were diluted rapidly into media containing no betaine and different concentrations of NaCl. Growth with betaine (filled bars); growth in the absence of betaine (open bars) (W. Bartlett, N. R. Stokes, and I. R. Booth, unpublished data).

mpatible organic solutes modifies the water structure and the hydration of proteins, reestablishing the high activity needed for optimum cell function and rapid growth (Elmore *et al.*, 1990; Imhoff, 1986; Winzor *et al.*, 1992). The accumulation of compatible solutes may significantly affect the outcome of hypoosmotic shock assays for MS channel function.

Dilution of cells that have adjusted to high osmolarity into a medium of lower osmolarity generates an immediate increase in transmembrane turgor simply because of the dilution of the external medium. If mixing of bacteria can be affected in a few seconds then the transmembrane pressure can rise by several atmospheres (Fig. 3.1). The physical integrity of the cell can be compromised if the pressure is not dissipated rapidly—this is the specific function of MS channels and this observation provides the basis of the assay. Before providing details of the assay conditions, some further scene setting is required. Growing cells maintain a fine balance of pressure across their cytoplasmic membrane, which keeps the MS channels inactive for the majority of the time. In the patch clamp rig, where the naked cytoplasmic membrane is sealed across the orifice of the patch pipette, a pressure of 70 to 200 mm Hg (Sukharev, 2002; Sukharev et al., 1999) is required to activate MscS and MscL channels.[3] Since 760 mm Hg equals one atmosphere, the minimum net pressure across the bacterial cytoplasmic membrane must be <0.1 atm. The turgor pressure across bacterial cells has been estimated to lie between 2 and 4 atm for E. coli and 20 atm for Staphylococcus aureus (Mitchell and Moyle, 1956). In physiological media growth inhibition is the direct consequence of opening MS channels (Blount et al., 1997). It follows that the turgor pressure across the cytoplasmic membrane must be counterbalanced by the cell wall. In Gram-positive bacteria this resistance is provided by multiple layers of peptidoglycan with intercalated teichoic acids, whereas in Gram-negative bacteria there is a single layer of peptidoglycan anchored to the outer membrane by lipoproteins. The outer layer of lipopolysaccharide with its cross-bridging Ca^{2+} and Mg^{2+} ions provides a further strengthening element. Thus, survival of hypoosmotic shock is likely to depend not just on the activity of the MS channels, but on the intrinsic properties of the cell wall.

The basic assay consists of growing cells into exponential phase at high osmolarity, diluting them into low osmolarity medium, and assaying for either viability or a number of other biochemical parameters (Levina et al., 1999). This method can be decorated in a number of ways to increase the sensitivity or to achieve a degree of automation.

- Cells are grown into exponential phase in a low osmolarity medium to establish a culture of rapidly growing cells. This culture is then used to inoculate fresh growth medium that has been supplemented with NaCl. The dual growth protocol seeks to establish cultures with the maximum numbers of physiologically similar cells (see later). While the cells are still in mid-exponential phase they are diluted rapidly into low osmolarity medium and are sampled over a time course (1 to 30 min) for viable cells

---

[3] Each channel has a defined pressure sensitivity; in E. coli MscM requires the least pressure and then the sequence of opening as pressure is increased is MscK, MscS, and then finally MscL.

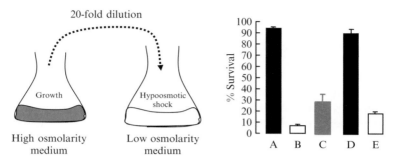

Figure 3.2 (Left) Hypoosmotic shock: Cells are initially grown in high osmolarity medium, which may be based on a minimal salt medium or a complex medium, such as Luria broth. When cells have achieved mid-exponential growth ($OD_{650} \approx 0.2$), they are diluted 20-fold into low osmolarity medium, which may be as simple as distilled water or growth medium with a lower concentration of NaCl. A control dilution into identical high osmolarity medium is undertaken in parallel (not shown). (Right) Survival expected in different situations is depicted: A and B, MJF465; C, D, and E, MJF465 transformed with a plasmid expressing an MS gene, under LacI control. A, control, no hypoosmotic shock; B, 0.3 M shock; C, native *E. coli* MscS uninduced (basal expression frequently occurs from LacI-based plasmids and so some MS protein is made); and D, fully induced native MscS [0.3 mM isopropyl-β-D-thiogalactoside, 30 min prior to harvest]; note that in this expression system the MS channel may be expressed to 10 to 30 times the level of the chromosomal copy in a wild-type *E. coli* strain; E, a fully induced mutant channel that has either a GOF or a LOF mutation (see text).

Table 3.1 Modifications to downshock assay for channel function

Modification	Assay parameter	References
Cell lysis	Light scattering	Levina *et al.* (1999)
Cell lysis	A_{260}/A_{280} cell-free supernatant	Levina *et al.* (1999)
Ethidium bromide	Fluorescence of cell-free supernatant	Powl *et al.* (2003)
Live/dead dyes	Red/green fluorescence	Maurer and Dougherty (2001)
MTSET$^+$ in Cys mutants	Survival	Batiza *et al.* (2002)
Enzyme release	Fluorescence or enzyme assay	None

or other measurable parameters (Fig. 3.2; Table 3.1). In parallel, dilution of the culture into identical high osmolarity medium is undertaken to provide control values for viability and other measurables.

- When measuring viability it is critical that serial dilution and recovery of cells submitted to hypoosmotic shock is undertaken on low osmolarity medium. However, the colony count for the control is undertaken on high osmolarity medium to ensure that these cells have not been subjected to hypoosmotic shock during the assessment of viability.
- Frequently used growth media are Luria broth (yeast extract, tryptone, and NaCl) (LK medium, which replaces the NaCl [5 g/liter] with equimolar KCl [6.4 g/liter], can also be used) and minimal glucose-based salts medium. Although minimal medium provides slower growth, we have usually found greater inter-experiment reproducibility with this medium. Extending the analysis of hypoosmotic shock viability to include analysis of actual cell lysis is dependent on using minimal medium to avoid the optical interference caused by components of complex medium (Levina *et al.*, 1999; Maurer and Dougherty, 2001). We have routinely used McIvaine's buffer (Jordan *et al.*, 1999) as the basis for the minimal growth medium because it allows the assay to be conducted over a wide range of pH without loss of buffering capacity and avoiding changing the composition too drastically.
- Where possible, organic-compatible solutes should not be present in the growth medium, which may be one cause of variation in the outcome of Luria broth experiments where there are low concentrations of betaine and free proline that can be accumulated under high salt conditions. Studies in the authors' laboratory have shown that hypoosmotic shock-induced cell lysis in the triple channel mutant *E. coli* MJF465 (MscS$^-$, MscL$^-$, MscK$^-$) can be prevented by growth with compatible solutes prior to shock (W. Bartlett, N. R. Stokes, S. Miller, and I. R. Booth, unpublished data).
- Hypoosmotic shock of *E. coli* MJF465 (MscS$^-$, MscL$^-$, MscK$^-$) leads to cell death arising from lysis of cells (Levina *et al.*, 1999). However, a small percentage of cells survive, and the size of the survivor population may be affected by the nature of the growth medium, the magnitude of the osmotic shock, and the precise degree of control over preparation of the inoculum and the test cultures. Reproducible growth of the culture to be shocked is critical. Cells that have been grown at high osmolarity in broth to OD$_{650}$ ≈0.2 to 0.3 are suitable for experiments; cultures that have proceeded beyond this stage of growth will show a low tendency to exhibit cell death and lysis when subjected to hypoosmotic shock. It is often not appreciated that growth in LB medium comprises several growth phases and that the later growth phases are not suitable for experimentation.
- The size of the dilution step is important! Analysis of *E. coli* MJF465 has shown that the step change in osmolarity must be ≥ 0.2 M NaCl to induce cell lysis (Levina *et al.*, 1999). We have found no systematic evidence that the actual starting and finishing osmolarities make a significant difference

to the observed percentage of surviving cells, although colony growth at higher osmolarities can be slow, which may increase errors. However, increasing the size of the osmotic shock can lower the number of surviving channel-free cells.[4]

- To assay for function in a putative MS channel protein, one must first clone the gene using standard protocols and then induce its expression during the growth phase at high osmolarity. It is preferable to avoid T7-based expression systems; mutant strains that allow the use of these plasmids for hypoosmotic shock assays have been created, but they are less satisfactory for the survival assay (M. D. Edwards, S. Black, W. Bartlett, and I. R. Booth, unpublished data). The expression of the protein and its integration into the membrane must be checked by Western blotting and, where necessary, flotation methods for membrane preparation to ensure that the protein is actually integrated into the membrane. It is important to recognize that only a small number of channels are required for complete restoration of survival of the mutant MJF465. Single mutants that lack either the chromosomally encoded MscS or MscL channels survive hypoosmotic shock perfectly well because of the functional redundancy of the two channels (Levina *et al.*, 1999). The abundance of these proteins is low and is lower than can be achieved by controlled expression of the majority of cloning vectors. Consequently, it is valuable to assess the MS activity of a putative channel both allowing basal expression from the plasmid and with induction to attain higher expression levels (Miller *et al.*, 2003; Schumann *et al.*, 2004).
- The survival assay alone is subject to significant variability and to experimental error. It is not unusual for the error in measuring the surviving population to be >15%. The assay is excellent for detecting MS channel function, but it is not a subtle experimental measurement. Interpreting small differences between mutant channels by this method is at best hazardous.

Using the protocol described earlier and taking the necessary precautions, MS channel homologues from a range of organisms have been tested using the hypoosmotic shock assay.

3. Protocol

1. Cells are grown aerobically at 37° in McIlvaine's citrate–phosphate buffer, pH 7 (containing per liter: 8.58 g Na_2HPO_4, 1.34 g citric acid,

---

[4] Note that we have rigorously established that the surviving cells are not mutants that have been adapted to the loss of the channels, but represent physiological variants. There is no doubt that survival of hypoosmotic shock could be used to select for resistant mutants and that such an approach would be very informative. However, it would require a secondary screen to detect those few cells that are actually mutants among the many that survive.

0.87 g K_2HPO_4), supplemented with (per liter) 1 g NH_4SO_4, 0.1 g $MgSO_4$, and 0.002 g $(NH_4)_2SO_4 \cdot FeSO_4 \cdot 6H_2O$ (added from 100× concentrated stock dissolved in 0.1 M HCl), 0.001 g thiamine, and 2 g glucose (Levina *et al.*, 1999). This medium has a final osmolarity of 220 mOsm.

2. A single colony is used to inoculate a 5-ml overnight culture in a 25-ml MacCartney in which the glucose concentration is lowered to 0.4 g/liter to limit growth to $OD_{650} \approx 0.4$ after overnight growth. The next morning the glucose is increased to 0.2 g/liter, and the culture is returned to the shaking incubator. After a 1-h incubation to restore exponential growth the culture is diluted 10-fold into fresh growth medium in a 100-ml Erlenmeyer flask and grown to mid-exponential phase ($OD_{650} \approx 0.35$ to 0.4). The culture is then diluted into fresh prewarmed growth medium supplemented with 0.5 M NaCl (to obtain faster growth one can lower the NaCl to 0.3 M) and incubated in the shaking incubator until $OD_{650} \approx 0.2$ to 0.3.

3. Hypoosmotic shock is achieved by diluting the culture 20-fold into fresh prewarmed McIlvaines buffer containing supplements in the presence (control) or absence of 0.5 M NaCl (shock). Note that one can vary the NaCl concentration at this point to achieve a range of osmotic shocks.

4. To determine viability, 50-μl samples are taken at timed intervals and diluted serially in the same growth medium, and four 5-μl samples are spotted onto Luria broth plates in either the presence (control) or the absence of 0.5 M NaCl (shock). To allow for the greatest accuracy in the final count of colonies, the 5-μl spot is gently spread using the Gilson tip from which the sample is dispensed.

5. Plates are incubated at 37° overnight (shock) or for up to 36 h (control) and the colonies counted. On high salt medium growth is somewhat slower and to achieve accurate colony counts one must wait longer than for the test samples.

6. A useful modification of this assay protocol is the combined acid and osmotic shock (Levina *et al.*, 1999; McLaggan *et al.*, 2002). *E. coli* K-12 cells survive quite well when incubated at pH 3.6, but not at lower pH values. If a MS channel opens, however, the cytoplasm is flooded with low pH buffer, causing cell death through protein unfolding and so on. By combining the hypoosmotic shock with a shift to pH 3.6, one can obtain an indication of channel gating. Note that if the channel has shown no protective effect in the aforementioned protocol, one must limit the change in osmolarity to one that does not compromise cell integrity. McIlvaine's buffer can be adjusted to acid pH by varying the ratio of citric acid and phosphate. The technique is otherwise the same but the range of serial dilutions that must be assayed is greater because acid hypoosmotic shock kills $\geq$ 99.99% of cells if the channel has fired but less than 1% if the channel is closed.

4. VARIANTS ON THE BASIC VIABILITY ASSAY

One of the most useful augments to the basic technique is to measure cell lysis (Levina *et al.*, 1999). This can be particularly helpful to distinguish between mechanisms of cell killing when one is comparing mutant channels with the native channel. Cells of *E. coli* K-12 lacking functional channels will lyse when subjected to a large hypoosmotic shock. Lysis is indicated by two parameters: decrease in the light scattering of the cell suspension (OD_{650}) and increase in absorption (A_{260}/A_{280}) of the supernatant after centrifugation (full-speed Jouan microcentrifuge 30 s) to remove cells and cell debris (Levina *et al.*, 1999). Others have modified this technique by adding ethidium bromide (0.5 μg/ml) to the shock medium (Powl *et al.*, 2003). The effect is to intercalate the ethidium ions into the released DNA. Samples prepared in this way are stored in the dark for 45 min prior to centrifugation to remove cells and debris, and the fluorescence of the supernatant is measured by excitation at 254 nm and recording emission at 632 nm. One can systematically extend this method to any readily measurable assay for an enzyme that is released by cell lysis.

High-throughput fluorescence methods have also been described for identifying channel mutants (Maurer and Dougherty, 2001). In this development the cloned channel is expressed in *E. coli* MJF465 and the cells are subjected to various degrees of hypoosmotic shock into minimal medium containing propidium iodide (PI) and SYTO 9 dyes (Live/Dead BacLight^{TM} assay). The samples are then incubated at 37° in an incubator. Both 96- and 384-well microtiter plate designs are used to read the samples in a fluorescence plate reader with excitation at 480 and 490 nm and emission at 500 and 635 nm, respectively, for PI and SYTO 9. The ratio of green/red (500/635 nm) plotted as a function of hypoosmotic shock correlates with the character of channel mutations (Maurer and Dougherty, 2001). The channel-free strain exhibits high green/red fluorescence until sufficient dilution is achieved to cause cell lysis. When the wild-type channel is introduced, a high green/red ratio is maintained until the highest dilution when activation of the MscL channels may lead to depolarization of the membrane, causing increased accumulation of PI. Mutant channels exhibiting significant loss of function (LOF) show properties intermediate between the wild-type and the channel-free mutant. In contrast, cells expressing channels that possess gain-of-function (GOF) mutations exhibit low green/red ratios at all dilutions, suggesting that the inappropriate gating of the channels depolarizes the membrane, causing high PI uptake. This approach offers the potential for automated screening for mutant channels. How well the method deals with subtle phenotypes is not clear and there may be a need for a refinement of the assay to identify a range of mutants.

5. ANALYZING MUTANT CHANNELS

The analysis of native channels has limited benefits, that is, the identification of homologues that possess significant channel activity. There is such a multiplicity of homologues in bacteria, archaea, and plant organelles that it remains a valid exercise to define important structure–function correlates. The diversity of sizes among MscL homologues is limited, and it has been reproducibly observed that MscL proteins from a range of organisms can be functional when expressed in *E. coli* (Moe *et al.*, 1998). However, the diversity among MscS proteins is immense, and few homologues have been shown to function in *E. coli* irrespective of whether the assay is cell-based or electrophysiological. This changes the focus of assay systems toward their deployment in the analysis of mutant channels.

The availability of crystal structures and simple methods for site-directed mutagenesis (such as QuikChange™) has facilitated the development of structure–function analysis greatly. Particular interest has surrounded the prediction of the closed–open structural transition and therefore prediction of the open channel conformation. MscS and MscL generate channels predicted to be between 14 and 30 Å in diameter, which requires significant structural change in both proteins relative to the crystal form (Sukharev, 2002; Sukharev *et al.*, 1999). The analysis of mutants is often aided by simple cell-based screens of function. This can then be augmented by mutation-dependent assays.

Straightforward analysis of channel function by survival of hypoosmotic shock is not feasible when mutations are introduced into the channel protein. Loss of protective ability is seen with both GOF and LOF mutant channels, but for different reasons (Miller *et al.*, 2003). LOF mutants simply fail to release solutes, causing cell death via lysis. However, GOF mutants will gate at the lowest applied hypoosmotic pressure and their "inappropriate" activity is sufficient to inhibit colony formation even when plated onto low osmolarity growth medium. Blount and colleagues (1997) were the first to illustrate this with MscL channel mutants. Expression of some MscL mutants inhibited growth of the cells on both liquid and solid medium simply because the cells cannot sustain cellular homeostasis in the presence of the channels. The loss of homeostasis is a function of both the level of expression and the propensity of the mutant channel to gate at the normal levels of turgor pressure found in growing *E. coli* incubated in low osmolarity (≈200 mOsm) medium. Lowering channel expression can be sufficient to block the toxicity associated with channel activity, which can form the basis of other assays (see later). An important element in this analysis is that there appears to be a degree of stochasticity in the survival patterns. Thus, the strength of a GOF phenotype can be revealed by plating serial dilutions of cells expressing the mutant channels onto low osmolarity medium. Severe GOF mutations are associated with failure of even low dilutions of the cell suspension to form colonies. Other mutations may allow colony formation, but the range of dilutions at which they are observed is much narrower than for the native channel expressed from a similar plasmid. In other words, some cells retain sufficient activity to start the formation of the colony whereas the majority does not. The implication is that there is interplay among expression level, channel activity, and other factors to determine the culturability. Finally, it is worth noting, at least for MscS mutant channels, that their expression can modify the physical appearance of the colonies—sometimes colonies can appear translucent rather than the normal opaque form (S. Miller, S. Black, W. Bartlett, and I. R. Booth, unpublished data).

Cell lysis assays and their derivatives can be used to differentiate between GOF and LOF mutants. LOF mutants offer little or no protection from hypoosmotic shock and consequently cell death is associated with lysis. In contrast, GOF mutant channels cause cell death without compromising structural integrity and thus cause no lysis. For this reason a simple assay for cell integrity in parallel with measurement of viability will illuminate the cause of cell death and show which mutants are LOF and which are GOF. A further development of this protocol has been described by our group. A new *E. coli* strain was created to study GOF alleles in the cloned MscK channel (Li *et al.*, 2007). The strain lacks the major potassium transport systems and has a deletion of MscK; as a consequence the strain cannot grow rapidly on media containing ≤ 20 mM K$^+$. GOF alleles in the MscK, but not the native MscK, channel suppressed the K$^+$ growth phenotype. A parallel for this observation suggests that this might be generalized to other MS channels. A chromosomal MscL mutation, N14D, which is a known GOF allele from studies of the cloned gene, also showed suppression of the K$^+$ transport defect (Buurman *et al.*, 2004). One can extend this approach to any system where the cell seeks to exclude or take up solutes, as the overactivity of an MS channel will modify the magnitude of the proton motive force, thus affecting the ability to exclude specific ions or solutes, for example, antibiotics, nutrients, and toxic ions. In this respect it is interesting to note that the multiplicity of efflux systems and channels has made direct efflux assays of limited value in analyzing MS channels. However, assays based on release or retention of solutes may provide valuable selection mechanisms for the identification of novel mutants.

A development on the use of mutant channels deserves mention. Batiza and colleagues (2002) combined the introduction of single Cys residues into Cys-free channels with sulfhydryl reagents to demonstrate the gating of the MscL channel in cells that also possess MscS. Theoretically, the high abundance of MscS coupled with the potential solute flow could be sufficient to ease the turgor pressure during hypoosmotic shock. This can certainly be inferred from the observation that mutants lacking MscL show no significant sensitivity to hypoosmotic shock. However, using a Leu19Cys mutant, Batiza *et al.* (2002) showed that MscL opened even in the presence of MscS, but that the osmotic shift required in the assay was diminished if MscS was absent. The basis of the assay is that the Leu19Cys mutant protein can react with [2(triethylammonium)ethyl]methylsulfonate ($MTSET^+$), which results in the introduction of a positive charge into each subunit near the seal of the channel (Batiza *et al.*, 2002). This reaction predisposes the channel to gate (GOF) and prevents the channel from closing once open. In the experimental system cells were subjected to hypoosmotic shock in the presence of $MTSET^+$ and survival was recorded. Strains expressing the native MscL channel or lacking MscL survived hypoosmotic shock in the presence of $MTSET^+$. However, the Leu19Cys mutant was relatively insensitive to $MTSET^+$ during isoosmotic dilution ($\approx 80\%$ cell survival) but suffered complete loss of viability when subjected to hypoosmotic shock. This example demonstrated how mutagenesis can be coupled with cell survival assays to generate new insights.

REFERENCES

Bakker, E. P., Booth, I. R., Dinnbier, U., Epstein, W., and Gajewska, A. (1987). Evidence for multiple K^+ export systems in *Escherichia coli*. *J. Bacteriol.* **169**, 3743–3749.

Bass, R. B., Strop, P., Barclay, M., and Rees, D. C. (2002). Crystal structure of *Escherichia coli* MscS, a voltage-modulated and mechanosensitive channel. *Science* **298**, 1582–1587.

Batiza, A. F., Kuo, M. M. C., Yoshimura, K., and Kung, C. (2002). Gating the bacterial mechanosensitive channel MscL *in vivo*. *Proc. Natl. Acad. Sci. USA* **99**, 5643–5648.

Blount, P., Schroeder, M. J., and Kung, C. (1997). Mutations in a bacterial mechanosensitive channel change the cellular response to osmotic stress. *J. Biol. Chem.* **272**, 32150–32157.

Blount, P., Sukharev, S. I., Moe, P. C., Martinac, B., and Kung, C. (1999). Mechanosensitive channels of bacteria. *Methods Enzymol.* **294**, 458–482.

Booth, I. R., Cairney, J., Sutherland, L., and Higgins, C. F. (1988). Enteric bacteria and osmotic-stress: An integrated homeostatic system. *J. Appl. Bacteriol.* **65**, S35–S49.

Buurman, E. T., McLaggan, D., Naprstek, J., and Epstein, W. (2004). Multiple paths for aberrant transport of K^+ in *Escherichia coli*. *J. Bacteriol.* **186**, 4238–4245.

Elmore, M. J., Lamb, A. J., Ritchie, G. Y., Douglas, R. M., Munro, A., Gajewska, A., and Booth, I. R. (1990). Activation of potassium efflux from *Escherichia coli* by glutathione metabolites. *Mol. Microbiol.* **4**, 405–412.

Epstein, W. (1986). Osmoregulation by potassium transport in *Escherichia coli*. *FEMS Microbiol. Rev.* **39**, 73–78.

Folgering, J. H., Kuiper, J. M., de Vries, A. H., Engberts, J. B., and Poolman, B. (2004). Lipid-mediated light activation of a mechanosensitive channel of large conductance. *Langmuir* **20**, 6985–6987.

Folgering, J. H., Moe, P. C., Schuurman-Wolters, G. K., Blount, P., and Poolman, B. (2005). *Lactococcus lactis* uses MscL as its principal mechanosensitive channel. *J. Biol. Chem.* **280**, 8784–8792.

Haswell, E. S., and Meyerowitz, E. M. (2006). MscS-like proteins control plastid size and shape in *Arabidopsis thaliana*. *Curr. Biol.* **16**, 1–11.

Imhoff, J. F. (1986). Osmoregulation and compatible solutes in eubacteria. *FEMS Microbiol. Rev.* **39**, 57–66.

Jordan, S. L., Glover, J., Malcolm, L., Thomson-Carter, F. M., Booth, I. R., and Park, S. F. (1999). Augmentation of killing of *Escherichia coli* O157 by combinations of lactate, ethanol, and low-pH conditions. *Appl. Environ. Microbiol.* **65**, 1308–1311.

Koo, S. P., Higgins, C. F., and Booth, I. R. (1991). Regulation of compatible solute accumulation in *Salmonella typhimurium*: Evidence for a glycine betaine efflux system. *J. Gen. Microbiol.* **137**, 2617–2625.

Levina, N., Totemeyer, S., Stokes, N. R., Louis, P., Jones, M. A., and Booth, I. R. (1999). Protection of *Escherichia coli* cells against extreme turgor by activation of MscS and MscL mechanosensitive channels: Identification of genes required for MscS activity. *EMBO J.* **18**, 1730–1737.

Li, C., Edwards, M. D., Hocherl, J., Roth, J., and Booth, I. R. (2007). Identification of mutations that alter the gating of the *E. coli* mechanosensitive channel protein, MscK. *Mol. Microbiol.* **64**, 560–574.

Martinac, B., Buehner, M., Delcour, A. H., Adler, J., and Kung, C. (1987). Pressure-sensitive ion channel in *Escherichia coli*. *Proc. Natl. Acad. Sci. USA* **84**, 2297–2301.

Maurer, J. A., and Dougherty, D. A. (2001). A high-throughput screen for MscL channel activity and mutational phenotyping. *Biochim. Biophys. Acta* **1514**, 165–169.

McLaggan, D., Jones, M. A., Gouesbet, G., Levina, N., Lindey, S., Epstein, W., and Booth, I. R. (2002). Analysis of the kefA2 mutation suggests that KefA is a cation-specific channel involved in osmotic adaptation in *Escherichia coli*. *Mol. Microbiol.* **43**, 521–536.

Miller, S., Bartlett, W., Chandrasekaran, S., Simpson, S., Edwards, M., and Booth, I. R. (2003). Domain organization of the MscS mechanosensitive channel of *Escherichia coli*. *EMBO J.* **22**, 36–46.

Mitchell, P., and Moyle, J. (1956). Osmotic function and structure in bacteria. *In* "Bacterial Anatomy" (E. Spooner and B. Stocker, eds.), Vol. 6, pp. 150–180. Cambridge University Press, Cambridge.

Moe, P. C., Blount, P., and Kung, C. (1998). Functional and structural conservation in the mechanosensitive channel MscL implicates elements crucial for mechanosensation. *Mol. Microbiol.* **28**, 583–592.

Nottebrock, D., Meyer, U., Kramer, R., and Morbach, S. (2003). Molecular and biochemical characterization of mechanosensitive channels in *Corynebacterium glutamicum*. *FEMS Microbiol. Lett.* **218**, 305–309.

Pivetti, C. D., Yen, M. R., Miller, S., Busch, W., Tseng, Y. H., Booth, I. R., and Saier, M. H. (2003). Two families of mechanosensitive channel proteins. *Microbiol. Mol. Biol. Rev.* **67**, 66–85.

Powl, A. M., East, J. M., and Lee, A. G. (2003). Lipid-protein interactions studied by introduction of a tryptophan residue: The mechanosensitive channel MscL. *Biochemistry* **42**, 14306–14317.

Schumann, U., Edwards, M. D., Li, C., and Booth, I. R. (2004). The conserved carboxy-terminus of the MscS mechanosensitive channel is not essential but increases stability and activity. *FEBS Lett.* **572**, 233–237.

Steinbacher, S., Bass, R., Strop, P., and Rees, D. C. (2007). Structures of the prokaryotic mechanosensitive channels MscL and MscS. *Curr. Topics Membr.* **58**, 1–24.

Sukharev, S. (2002). Purification of the small mechanosensitive channel of *Escherichia coli* (MscS): The subunit structure, conduction, and gating characteristics in liposomes. *Biophys. J.* **83**, 290–298.

Sukharev, S. I., Blount, P., Martinac, B., Blattner, F. R., and Kung, C. (1994). A large-conductance mechanosensitive channel in *E. coli* encoded by *mscL* alone. *Nature* **368,** 265–268.

Sukharev, S. I., Sigurdson, W. J., Kung, C., and Sachs, F. (1999). Energetic and spatial parameters for gating of the bacterial large conductance mechanosensitive channel, MscL. *J. Gen. Physiol.* **113,** 525–540.

Touze, T., Gouesbet, G., Boiangiu, C., Jebbar, M., Bonnassie, S., and Blanco, C. (2001). Glycine betaine loses its osmoprotective activity in a *bspA* strain of *Erwinia chrysanthemi*. *Mol. Microbiol.* **42,** 87–99.

van den Bogaart, G., Krasnikov, V., and Poolman, B. (2006). Dual-color fluorescence-burst analysis to probe protein efflux through the mechanosensitive channel MscL. *Biophys. J.* **92,** 1233–1240.

Winzor, C. L., Winzor, D. J., Paleg, L. G., Jones, G. P., and Naidu, B. P. (1992). Rationalization of the effects of compatible solutes on protein stability in terms of thermodynamic nonideality. *Arch. Biochem. Biophys.* **296,** 102–107.

CHAPTER FOUR

CONTROL OF CELL CYCLE IN RESPONSE TO OSMOSTRESS: LESSONS FROM YEAST

J. Clotet* *and* F. Posas[†]

Contents

Abstract

To maximize the probability of survival and proliferation, cells coordinate various intracellular activities in response to changes in the extracellular environment. Eukaryotic cells transduce diverse cellular stimuli by multiple mitogen-activated protein kinase (MAPK) cascades. Exposure of cells to stress results in rapid activation of a highly conserved family of MAPKs, known as stress-activated protein kinases (SAPKs). Activation of SAPKs results in the generation of a set of adaptive responses that leads to the modulation of several aspects of cell physiology essential for cell survival, such as gene expression, translation, and morphogenesis. This chapter proposes that regulation of cell cycle progression is another general stress response critical for cell survival. Studies from yeast, both *Schizosaccharomyces pombe* and *Saccharomyces cerevisiae*, have served to start understanding how SAPKs control cell cycle progression in response to stress.

* Department of Molecular and Cellular Biology, Universitat Internacional de Catalunya (UIC), Sant Cugat del Vallès, Spain
† Cell Signaling Unit, Departament de Ciències Experimentals i de la Salut, Universitat Pompeu Fabra (UPF), Barcelona, Spain

Methods in Enzymology, Volume 428
ISSN 0076-6879, DOI: 10.1016/S0076-6879(07)28004-8

1. INTRODUCTION

Cells are exposed constantly to stress situations such as changes in temperature, pH, radiation, availability of nutrients, and access to oxygen. Cells are able to coordinate intracellular activities in order to respond to these stresses, which allow them to adapt to the new environmental conditions. One of the most common stresses is the change of osmotic pressure. When cells are submitted to hyperosmotic shock, these cells lose cellular water by passive diffusion. Yeast cells have developed a battery of mechanisms to fight against water loss. Accumulation of osmolytes has a central role in osmoadaption; in *Saccharomyces cerevisiae*, this osmolyte is glycerol (de Nadal *et al.*, 2002; Hohmann, 2002; Yancey *et al.*, 1982).

The Hog1 stress-activated protein kinase cascade, a prototypical eukaryotic mitogen-activated protein kinase (MAPK) pathway in *S. cerevisiae*, was discovered by its involvement in osmoadaptation (Brewster *et al.*, 1993; Gustin *et al.*, 1998); similarly, the Sty1 MAPK was discovered in *Schizosaccharomyces pombe*. A consensus idea of osmoadaptation is conserved across eukaryotes, and thus yeasts are an ideal model system to study these processes.

The cell cycle process, by which one cell grows and divides into two daughter cells, consists of four phases: G_1 (in which cells grow and make the decision to commit cell division), S (when chromosomes replicate), G_2 (when cells prepare the next phase), and M (a complex phase in which chromosomes separate and cells divide). After the M phase, cells enter again into G_1, hence completing a "cycle."

Cell cycle progression is often modulated by signal transduction pathways as a result of either intracellular stimuli (e.g., by the activation of mitotic spindle assembly checkpoint) or external factors such as mating pheromone. This chapter proposes the existence of a new checkpoint, the osmocheckpoint, which consists of a complex response controlled by stress-activated protein kinases (SAPKs), that affects several phases of the cell cycle to allow adaptation to external osmolarity.

2. REGULATION OF THE YEAST CELL CYCLE

In *S. cerevisiae*, a highly regulated and complex network of proteins governs the cell cycle process, although major events are controlled by a single CDK (Cdc28) whose regulation is achieved mainly through the synthesis and degradation of cyclins and inhibitors. At the beginning of G_1, Cln3 cyclin is sequestered in the cytoplasm by the Whi3 retention factor (Belli *et al.*, 2001; Gari *et al.*, 2001); later, the nuclear concentration of Cln3 increases in relationship to the total cell mass (Chen *et al.*, 2000). When the level of the Cln3/Cdc28 complex is higher than a certain threshold, it triggers the phosphorylation of Whi5 (an orthologue of the mammalian Rb protein). This event allows the transcription of a second wave of cyclins (Cln1, Cln2, Clb5, and Clb6) by the activation of the SBF and the MBF transcription complex. The activation and accumulation of Cln1,2/Cdc28 engage two processes: the initiation of bud formation and phosphorylation of Sic1, which leads to its ubiquitination by the SCF/Cdc4 complex (Verma *et al.*, 1997) and its destruction by the proteasome. When Sic1 is degraded, Clb5,6/Cdc28 activity rises abruptly and drives cells into S phase.

The progression into G_2 phase depends mainly on another wave of cyclin production: Clb1 and Clb2. The Mcm1/SFF (Mcm1/Fkh2/Ndd1) complex is the transcription factor of *CLB1* and *CLB2* (Althoefer *et al.*, 1995; Jorgensen and Tyers, 2000; Maher *et al.*, 1995). The transcription of *CLB1* and *CLB2* is autocatalytic because it promotes its own transcription.

The activity of the Clb2/Cdc28 complex is controlled by inhibitory phosphorylation at a conserved tyrosine in the N terminus of Cdc28 by the kinase Swe1 and the reverse phosphatase Mih1. Swe1 does not play an important role during vegetative growth (Amon *et al.*, 1992), but it has been demonstrated that it responds to cytoskeletal perturbations (McNulty and Lew, 2005).

At the end of the cell cycle, when DNA is fully replicated and all chromosomes are aligned on the metaphase plate, Cdc20 is activated (Hwang *et al.*, 1998) and activation of Cdc20/APC promotes degradation of part of the Clb2 protein (Baumer *et al.*, 2000). Later, when the nucleus migrates into the daughter cell, the Cdh1/APC destroys the remaining Clb2 protein. This event represents the exit from mitosis and the start of a new cycle.

In the fission yeast *S. pombe*, cell cycle progression is also controlled by a single CDK (Cdc2) associated with several cyclins. During G_1, Cdc2 is associated with the cyclin Puc1(+) (a close relative of Cln proteins in budding yeast), which regulates the length of G_1 (Martin-Castellanos *et al.*, 2000). It is worth noting that puc1 is not the unique cyclin that controls G_1 because mitotic cyclins are also playing some role in G_1 (see later). Also, Cdc2/Puc1 may be important in downregulating the rum1 Cdk inhibitor (the homologue of *S. cerevisiae* Sic1) at the end of G_1 (Martin-Castellanos *et al.*, 2000).

The onset of the S phase in fission yeast is regulated positively at Start by another cyclin: cig2, a B-type cyclin that performs similar functions to Clb5, Clb6 in *S. cerevisiae* (Mondesert *et al.*, 1996). Similarly to *S. cerevisiae*, rum1 may transiently inhibit cig2-associated cdc2 activity until the critical cell size required for Start is reached (Martin-Castellanos *et al.*, 1996).

During G_2, two B-type cyclins promote progression of the cell cycle: cig1 and cdc13 (cig1 has no obvious homologues in *S. cerevisiae*, but cdc13 corresponds to the Clb1 and the Clb2 mitotic B-type cyclins). Correspondingly, the Cdc2 mitotic complexes are regulated negatively by the Wee1 protein kinase. Phosphorylation of Cdc2 on tyrosine-15 by Wee1 is critical for the inhibition of the complex, whereas dephosphorylation of these residues by the Cdc25 phosphatase is the key event governing the initiation of mitosis (Gould and Nurse, 1989; Millar et al., 1991). Therefore, the balance between the levels of Wee1 and Cdc25 sets a threshold for the activity of the cdc13–cdc12 and cig1–cdc13 complexes that determines the correct timing of M-phase initiation.

At the end of mitosis, Cdc13 and Cig1 are partially degraded, possibly by the Slp1–APC/C complex (the equivalent of Cdc20/APC complex in S. cerevisiae). The remaining activity is necessary for the progression in G_1 phase and is fully degraded by the Ste9/Srw1 associated with the APC/C (the homologous complex of Cdh1/APC) (Blanco et al., 2000). Then, the complete destruction of mitotic cyclins occurs during the G_1 phase, when cells start a new cycle.

3. OSMOSTRESS RESPONSE IN BUDDING AND FISSION YEAST

Exposure of cells to osmostress results in rapid activation of a highly conserved family of MAPKs, known as stress-activated protein kinases. Activation of SAPKs elicits the program for cell adaptation, which includes transcriptional and translational regulation of the genome allowing long-term adaptation. Moreover, the same program also produces a short-term adaptation effect, such as glycerol accumulation (Albertyn et al., 1994), or the reestablishment of ionic balance (Proft and Struhl, 2004). The accumulation of compatible solutes, such as glycerol, is a ubiquitous mechanism in cellular osmoregulation. Yeast cells also control glycerol accumulation via the Fps1p-mediated export of glycerol (Tamas et al., 1999).

The SAPK pathways are extremely conserved among eukaryotes and are composed of a tier of three consecutively activated kinases: a MAP kinase kinase kinase (MAPKKK), a MAP kinase kinase (MAPKK), and a MAP kinase (MAPK). Once phosphorylated, the MAPK concentrates in the nucleus, where it can phosphorylate its protein targets on serine/threonine followed by a proline (S/T-P). However, a portion of activated MAP kinase is retained in the cytoplasm to regulate cytosolic events (Reiser et al., 1999). Pathways of MAP kinase are controlled negatively by protein phosphatases acting on two levels: the MAPKK (serine-threonine phosphatases) or only on the MAP kinase (serine-threonine phosphatases and tyrosine phosphatases) (Keyse, 2000). There are different sensing mechanisms upstream of

MAP kinase and generally include receptor tyrosine-kinases (in animal systems), G protein-coupled receptors, phosphorelay systems, and others.

In *S. cerevisiae*, the high osmolarity glycerol (HOG) pathway is the best-characterized system implicated in osmostress. As it happens with other MAPK pathways, the central core of the yeast HOG pathway comprises a layer of three MAPKKKs (Ssk2, Ssk22, and Ste11), which are responsible for activation of the unique MAPKK of the pathway Pbs2. Afterward, Pbs2 activates the Hog1 MAPK by phosphorylation (Brewster *et al.*, 1993) (Fig. 4.1). Hog1 is a homologue of the p38 and c-Jun N-terminal kinase (JNK) families of SAPKs. Hog1 phosphorylation promotes a fast and pronounced concentration of Hog1 in the nucleus, whereas in basal conditions, distribution of Hog1 is between the nucleus and the cytosol (Ferrigno *et al.*, 1998; Reiser *et al.*, 1999).

Two branches activate the HOG pathway through the activation of either the Ssk2 and Ssk22 or the Ste11 MAPKKKs: the Sln1 branch and the Sho1 branch (Maeda *et al.*, 1995).

The Sln1 branch involves a "two-component" osmosensor, composed of the Sln1-Ypd1-Ssk1 proteins that funnel the signal to the Ssk2/22 MAPKKKs. Unexpectedly, because Sln1 acts as a negative regulator of the HOG pathway, deletion of *SLN1* leads to a lethal phenotype because the pathway is activated sustainedly (Maeda *et al.*, 1994). Once dephosphorylated,

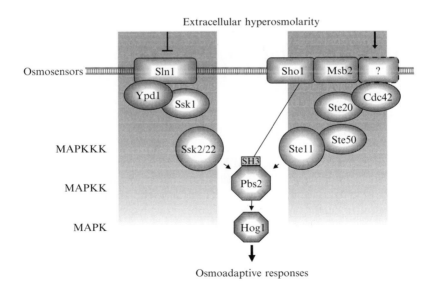

Figure 4.1 The Hog1 pathway is composed of a central core of MAP kinases and two osmosensing branches: Sln1 and Sho1-Msb2. The signal converges at the MAPKK Pbs2. Once Hog1 is active, different osmoadaptive responses are unleashed.

Ssk1 binds to the N-terminal regulatory domain of Ssk2, leading to Ssk2 activation (Posas and Saito, 1998).

A second mechanism for Pbs2 activation involves several proteins: the transmembrane protein Sho1, the mucin-like protein Msb2 (Albertyn *et al.*, 1994), the MAPKKK Ste11, the Ste11-binding protein Ste50, the Ste20 p21-activated kinase, and the small GTPase Cdc42 (Posas *et al.*, 1998), but the complete mechanism used by this osmosensor is still unknown.

In fission yeast, the Sty1 SAPK pathway is linked to osmostress adaptation. The central elements of this pathway are the MAPK Sty1 (also known as Spc1), the MAPKK Wis1, and the two MAPKKK Wak1 and Win1 (Wilkinson and Millar, 1998). Cells deleted for *sty1* or *wis1* are highly elongated as a consequence of a delay in the timing of mitotic initiation, which is exacerbated in response to stress. The pathway is activated by a sensor composed of three histidine kinases (Phk1/Mak2, Phk2/Mak3, and Phk3/Mak1)(Aoyama *et al.*, 2001) and two response regulators (Mcs4 and Prr1) (Ohmiya *et al.*, 1999; Shieh *et al.*, 1997). This pathway is also inactivated by protein phosphatases. A number of effectors of the Sty1 SAP kinase have been identified, including the Atf1, Pcr1, and Pap1 transcription factors, which are homologues of mammalian ATF-2 and c-Jun, respectively (Shiozaki and Russell, 1996; Wilkinson *et al.*, 1996). In addition, Sty1 binds and phosphorylates two downstream kinases, Cmk2 and Srk1 (Sty1-regulated kinase), which are related to the mammalian calmodulin-dependent and MAPKAP kinases.

4. CONTROL OF THE G_1 PHASE BY SAPKs

Activation of Hog1 by exposure to the cells to high osmolarity (0.4 M NaCl) or because of activation of upstream components of the MAPK pathway (i.e., Sln1 inactivation) results in cell cycle arrest in G_1 (Escote *et al.*, 2004). Hog1-mediated G_1 arrest is partially due to the downregulation of Cln1, Cln2, and Clb5 expression. The exact nature of the mechanism that represses the expression of SBF and MBF promoters under osmotic stress is still not known. Cln1,2/Cdc28 activities are necessary to reach a threshold of Sic1 phosphorylation, the subsequent ubiquitination by the Cdc4 ubiquitin ligase, and the posterior degradation by the proteasome. Therefore, the downregulation of cln production might explain, at least in part, the delay in S-phase entry as a consequence of an increased accumulation of Sic1.

Hog1 arrests cells in G_1 by an alternative mechanism: the direct phosphorylation of Sic1 (Escote *et al.*, 2004). Hog1 interacts physically with Sic1 *in vivo* and *in vitro* and, upon osmostress, phosphorylates a single residue (Thr 173) at the carboxyl terminus of Sic1, which results in Sic1 stabilization and

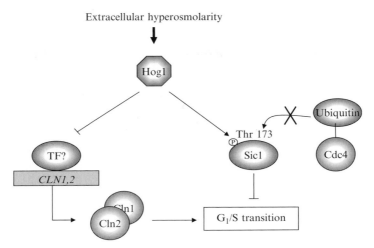

Extracellular hyperosmolarity

Figure 4.2 Hog1 activation prevents entry into S phase by a dual mechanism. The Sic1 CDK inhibitor is responsible for inhibition of the Cdc28/Clb5,6 complexes to block cell cycle entry into S phase. At G_1, persistent accumulation of cyclins induces phosphorylation of Sic1. Multiple phosphorylated Sic1 is then recognized by the Cdc4 ubiquitin ligase of the SCF complex and is targeted for degradation. Activation of Hog1 inhibits CLN1 and CLN2 expression and, in addition, phosphorylates T173 on Sic1 directly, which interferes with binding to Cdc4 and its degradation.

inhibition of cell cycle progression. Phosphorylation at threonine 173 interferes somehow with the binding of Sic1 with the Cdc4 (Fig. 4.2). Therefore, sic1 is targeted by Hog1 through two independent mechanisms: downregulation of Cln levels and direct phosphorylation of T173. This dual mechanism of regulation raises the following question. What is the advantage of a redundant mechanism to regulate entry into S phase upon osmostress? This system could be envisioned as a two-step mechanism: an initial step to cause immediate arrest by acting directly over the CDK inhibitor (Sic1 phosphorylation) and then a decrease on Cln transcription that would be essential in maintaining the arrest for a prolonged time. This scenario could also satisfy the critical step of preventing entry into S phase for different type of cells. Cells at early G_1 could be arrested easily by the simple decrease on Cln levels, whereas cells in advanced G_1, with high levels of Clns, would require a more direct effect over the Sic1 inhibitor to restrict their entry into S phase. Quantitative experiments performed in our laboratory support this hypothesis: cells stressed later in G_1 with high levels of Cln still stabilize Sic1 and are arrested in G_1 (unpublished data). Alternatively, and based on the fact that Clns and Sic1 might control different phenomena during G_1 progression, Clns regulate SPB duplication and bud emergence and Sic1 controls the exact timing of DNA replication. Activation of Hog1 could control all the processes that occur in G_1. Hog1 could act on Cln proteins to avoid early budding under stress conditions and, specifically, on Sic1 to allow replication only when cells were adapted. Indeed, cells that carry a copy of Sic1 nonphosphorylatable by Hog1 (substitution of Thr173 by Ala) have a little arrest in G_1 upon osmotic stress, but result in genomic instability (Escote et al., 2004). These results suggest that the increase of Sic1 stability by Cln1,2/Cdc28 downregulation under osmotic stress is not sufficient to ensure the correct entry into S phase. Direct Hog1-mediated phosphorylation of Sic1 could offer this extra mechanism to ensure the exact timing of G_1/S transition.

5. CONTROL OF THE G_2 PHASE BY SAPKS

G_2 transition depends on the activity of CDK associated with mitotic cyclins. Wee kinases are negative regulators of such activity and control the length of G_2 phase under specific conditions. In S. pombe, the Wee1 kinase allows entry into mitosis when cells reach a minimum threshold size and, in S. cerevisiae, Swe1 activates the morphogenetic checkpoint. The physiological role of this later checkpoint is under discussion (Keaton and Lew, 2006; Kellogg, 2003; McNulty and Lew, 2005) and is not clear whether it responds to cytoskeletal perturbations or monitors the size of the bud like in S. pombe cells. The molecular basis of the morphogenetic checkpoint seems firmly established: the septin ring formed on the neck between mother and daughter cell recruits Hsl1 (a septin-dependent protein kinase), which interacts with Hsl7 (a protein that serves as a holder to recruit Swe1); this complex is necessary to localize Swe1 to the neck and target it for destruction, as well as to activate Cdc28/Clb2. When bud formation (or growth) is impaired, Swe1 remains active and cells do not progress into mitosis (McMillan et al., 1999).

Activation of Hog1 upon osmostress induces a delay in G_2 by the decrease of Clb2/Cdc28 activity and by the downregulation of CLB2 transcription (Alexander et al., 2001; Clotet et al., 2006). The mechanisms that Hog1 uses to downregulate Clb2 transcription are not known, but this could be a secondary effect caused by the decrease on Clb2/Cdc28 activity. Instead, the mechanism required to decrease Clb2 activity seems clear: Hog1 is acting over the machinery of the morphogenetic checkpoint that controls Swe1 levels. Activated Hog1 interacts and directly phosphorylates Hsl1 in a residue within the Hsl7-docking site, which promotes the delocalization of Hsl7 from the neck that results in Swe1 accumulation (Clotet et al., 2006). Upon Hog1 activation, cells containing a nonphosphorylatable Hsl1 are unable to promote Hsl7 delocalization, fail to accumulate Swe1, and arrest at G_2. Thus they are sensitive to osmotic stress.

It is worth noting that the same machinery used by the cells to monitor morphogenetic defects can integrate other signals. Interestingly, whereas the morphogenetic checkpoint monitors the absence of a septin ring, which produces the absence of Hsl1, Hsl7, and Swe1 at the bud neck, the osmocheckpoint specifically delocalizes Hsl7 without affecting Hsl1 or septins (Clotet *et al.*, 2006).

In *S. pombe*, the Skb1 protein has been described as the homologue of Hsl7 in *S. cerevisiae*. In response to osmostress, Skb1 delocalizes from the cell ends and nuclei; in addition, *skb1* deletion strains are osmosensitive (Bao *et al.*, 2001). However, there is not a perfect homologue of Hsl1, and the exact mechanism by which Skb1 is delocalized and confers osmoresistance has not been described. Another important difference is the fact that Skb1 has been described as a factor that regulates Wee1 positively. This is the opposite of what has been described in *S. cerevisiae* and mammalian cells in which it has been demonstrated that Skb1Hs plays a role in destabilizing Wee1 (Lew, 2003).

Another interesting mechanism to slow down G_2 progression in response to osmotic stress consists of regulation of the Cdc25 phosphatase by Srk1 (López-Avilés *et al.*, 2005). Cdc25 is the key factor governing the initiation of mitosis that reverses inhibition of the Cdc2–mitotic complex because of phosphorylation by the Wee1 kinase (Gould and Nurse, 1989; Millar *et al.*, 1991). Srk1 is related to the mammalian MAPKAP kinases and its overexpression causes cell cycle arrest in late G_2 phase. Srk1 interacts with and phosphorylates Cdc25. After exposure of cells to KCl, Sty1 phosphorylates Srk1 and the catalytic activity of Srk1 increases more than fivefold. Phosphorylation of Cdc25 by Srk1 causes Cdc25 to bind to Rad24, a 14-3-3 protein family member, which helps retain Cdc25 in the cytoplasm and protect it from degradation. These results suggest that Sty1 SAP kinase is the effector of a checkpoint pathway that controls Cdc25 activity in response to osmotic stress.

6. CONTROL OF THE ANAPHASE–TELOPHASE TRANSITION

Fission yeast has several mechanisms of surveillance that control the transition between different phases of the cycle. For example, during anaphase, the spindle assembly checkpoint blocks the separation of chromosomes until the chromosomes are aligned, attached, and tensed in the middle of the spindle. This checkpoint has been described in many organisms with the same elements: (1) a cohesin complex, an element that makes cohesion between sister chromatids, which includes the Scc1/Rad21 protein; (2) the separase Cut1, a large molecular weight protease, which specifically cleaves Rad21; and (3) the securin Cut2, a chaperon and inhibitor of the Cut1 separase, which is a substrate of the E3-ligase APC/C. When the chromosomes

are perfectly aligned, APC is activated and polyubiquitinates Cut2 for destruction by the proteasome. Then, the separase (Cut1) is active and cleaves the
cohesin subunit Scc1/Rad21. This later event marks the transition between
anaphase and telophase.

In *S. pombe*, it has been described that SAPK Sty1 affects the levels and
phosphorylation of securin under osmotic stress (Kawasaki *et al.*, 2006).
Stress causes a transient cell cycle arrest in anaphase, and this arrest is
dependent neither on the DNA damage checkpoint nor on the mitotic
spindle checkpoint but requires the Spc1/Sty1 SAP kinase. The increase on
Cut2 protein after the arrest caused by osmostress requires the presence of
Sty1 SAP kinase, but the exact mechanism of this control is still unknown
(Kawasaki *et al.*, 2006). All these data suggest that SAP kinases could also
control the anaphase–telophase transition under osmotic stress.

7. REGULATION OF EXIT FROM MITOSIS BY SAPKs

The exit from mitosis after chromosome segregation is triggered by
the activation of Tem1, a G protein located in the spindle pole body
(Morgan, 1999). When cells undergo anaphase, the spindle pole body enters
into the daughter cell where the GEF for Tem1, Lte1, is localized (Pereira
et al., 2000). Tem1 then binds to and activates the cdc15 kinase and the
mitotic exit network (MEN), which leads to activation of the phosphatase
cdc14. This protein phosphatase is tightly regulated by the competitive
inhibitor Cfi1/Net1, which holds Cdc14 inactive in the nucleolus during
G_1 until nearly mitosis. Cdc14 is released by the anaphase release (FEAR)
network and by the MEN and spreads throughout the nucleus and cytoplasm to induce exit from mitosis, promoting the destruction of the remaining B-type cyclins via activation of the Cdh1–APC complex. Cdc14 also
promotes the accumulation of Sic1 and directly dephosphorylates substrates
for the CDK1 (Stegmeier and Amon, 2004).

The MEN pathway turns itself off to facilitate future G_1 events by activation of the antagonistic MEN pathway (AMEN) by, at least in part, induction
of the Amn1 protein that binds directly to Tem1 and prevents its association
with its target kinase Cdc15 (Wang *et al.*, 2003). Thus, the AMEN pathway is
the last step in the process that allows reentry into the next cell cycle.

Exit from mitosis is a key step in regulation of the cell cycle and could be
targeted by the action of SAPK under osmotic stress. This idea is sustained
by results shown by Amon and colleagues (1992), who found that under
hypertonic stress MEN mutants exit from mitosis in a manner dependent on
the Hog1 stress-activated kinase. In such MEN mutants, the HOG pathway
drives exit from mitosis by promoting activation of the MEN effector, the
protein phosphatase Cdc14, although the exact mechanism of such an effect
remains unclear (Reiser *et al.*, 2006).

8. CONCLUSIONS AND PERSPECTIVES

Exposure of the cells to stress induces a rapid activation of the highly conserved family of MAPKs, known as SAPKs. Activation of SAPKs results in the generation of a set of adaptive responses that lead to the modulation of several aspects of the cell physiology essential for cell survival, such as gene expression, translation, and morphogenesis. This chapter proposed that regulation of the cell cycle is a key general stress response essential for cell survival.

We have shown that the variety of molecular strategies that SAPKs use for modulating the timing of cell cycle progression is of great complexity: downregulation of different cyclins, stabilization of inhibitors, and delocalization of key molecules. We also showed that SAPKs are vigilant guards capable of acting immediately in different phases of the cell cycle. In contrast to internal signals, which produce limited checkpoint activation of specific phases of the cycle, stress stimuli activate a general checkpoint in multiple parts of the cell cycle. As expected for a SAPK that protects cells from external insults, Hog1 is able to induce a controlled arrest in several parts of the cell cycle: in G_1, S, G_2, and exit from mitosis (Fig. 4.3). Hence, we propose the existence of an osmocheckpoint: a complex response that could offer protection to cells in multiple stages of the cell cycle.

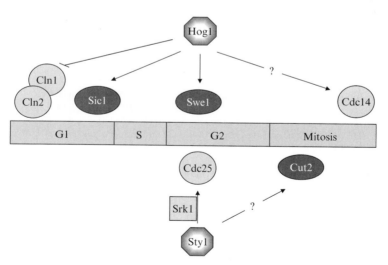

Figure 4.3 The osmocheckpoint modulates cell cycle transitions through SAPK activation. SAPKs from different yeasts act on several factors that control cell cycle progression positively (clear circles) or negatively (dark ellipses). Question marks indicate that the exact mechanism of action is not known. This figure suggests that SAPKs are able to protect cells from osmoinsults in any phase of the cell cycle. (See color insert.)

Again, the use of unicellular models has been shown to be very useful in uncovering complex responses. From yeast studies, we propose that a similar response to stress might be coordinated by SAPKs in mammals. Actually, different reports and unpublished observations indicate that different types of mammalian cells arrest at several stages of the cell cycle upon osmostress (Gustin *et al.*, 1998). Results from yeast suggest a conserved mechanism within more complex eukaryotic organisms.

REFERENCES

Albertyn, J., Hohmann, S., Thevelein, J. M., and Prior, B. A. (1994). GPD1, which encodes glycerol-3-phosphate dehydrogenase, is essential for growth under osmotic stress in *Saccharomyces cerevisiae*, and its expression is regulated by the high-osmolarity glycerol response pathway. *Mol. Cell. Biol.* **14,** 4135–4144.

Alexander, M. R., Tyers, M., Perret, M., Craig, B. M., Fang, K. S., and Gustin, M. C. (2001). Regulation of cell cycle progression by Swe1p and Hog1p following hypertonic stress. *Mol. Biol. Cell.* **12,** 53–62.

Althoefer, H., Schleiffer, A., Wassmann, K., Nordheim, A., and Ammerer, G. (1995). Mcm1 is required to coordinate G2-specific transcription in *Saccharomyces cerevisiae*. *Mol. Cell. Biol.* **15,** 5917–5928.

Amon, A., Surana, U., Muroff, I., and Nasmyth, K. (1992). Regulation of p34CDC28 tyrosine phosphorylation is not required for entry into mitosis in *S. cerevisiae*. *Nature* **355,** 368–371.

Aoyama, K., Aiba, H., and Mizuno, T. (2001). Genetic analysis of the his-to-asp phosphorelay implicated in mitotic cell cycle control: Involvement of histidine-kinase genes of *Schizosaccharomyces pombe*. *Biosci. Biotechnol. Biochem.* **65,** 2347–2352.

Bao, S., Qyang, Y., Yang, P., Kim, H., Du, H., Bartholomeusz, G., Henkel, J., Pimental, R., Verde, F., and Marcus, S. (2001). The highly conserved protein methyltransferase, Skb1, is a mediator of hyperosmotic stress response in the fission yeast *Schizosaccharomyces pombe*. *J. Biol. Chem.* **276,** 14549–14552.

Baumer, M., Braus, G. H., and Irniger, S. (2000). Two different modes of cyclin clb2 proteolysis during mitosis in *Saccharomyces cerevisiae*. *FEBS Lett.* **468,** 142–148.

Belli, G., Gari, E., Aldea, M., and Herrero, E. (2001). Osmotic stress causes a G1 cell cycle delay and downregulation of Cln3/Cdc28 activity in *Saccharomyces cerevisiae*. *Mol. Microbiol.* **39,** 1022–1035.

Blanco, M. A., Sanchez-Diaz, A., de Prada, J. M., and Moreno, S. (2000). APC(ste9/srw1) promotes degradation of mitotic cyclins in G(1) and is inhibited by cdc2 phosphorylation. *EMBO J.* **19,** 3945–3955.

Brewster, J. L., de Valoir, T., Dwyer, N. D., Winter, E., and Gustin, M. C. (1993). An osmosensing signal transduction pathway in yeast. *Science* **259,** 1760–1763.

Chen, K. C., Csikasz-Nagy, A., Gyorffy, B., Val, J., Novak, B., and Tyson, J. J. (2000). Kinetic analysis of a molecular model of the budding yeast cell cycle. *Mol. Biol. Cell.* **11,** 369–391.

Clotet, J., Escote, X., Adrover, M. A., Yaakov, G., Gari, E., Aldea, M., de Nadal, E., and Posas, F. (2006). Phosphorylation of Hsl1 by Hog1 leads to a G2 arrest essential for cell survival at high osmolarity. *EMBO J.* **25,** 2338–2346.

de Nadal, E., Alepuz, P. M., and Posas, F. (2002). Dealing with osmostress through MAP kinase activation. *EMBO Rep.* **3,** 735–740.

Escote, X., Zapater, M., Clotet, J., and Posas, F. (2004). Hog1 mediates cell-cycle arrest in G1 phase by the dual targeting of Sic1. *Nat. Cell Biol.* **6,** 997–1002.

Ferrigno, P., Posas, F., Koepp, D., Saito, H., and Silver, P. A. (1998). Regulated nucleo/cytoplasmic exchange of HOG1 MAPK requires the importin beta homologs NMD5 and XPO1. *EMBO J.* **17,** 5606–5614.

Gari, E., Volpe, T., Wang, H., Gallego, C., Futcher, B., and Aldea, M. (2001). Whi3 binds the mRNA of the G1 cyclin CLN3 to modulate cell fate in budding yeast. *Genes Dev.* **15,** 2803–2808.

Gould, K. L., and Nurse, P. (1989). Tyrosine phosphorylation of the fission yeast cdc2+ protein kinase regulates entry into mitosis. *Nature* **342,** 39–45.

Gustin, M. C., Albertyn, J., Alexander, M., and Davenport, K. (1998). MAP kinase pathways in the yeast *Saccharomyces cerevisiae. Microbiol. Mol. Biol. Rev.* **62,** 1264–1300.

Hohmann, S. (2002). Osmotic adaptation in yeast control of the yeast osmolyte system. *Int. Rev. Cytol.* **215,** 149–187.

Hwang, L. H., Lau, L. F., Smith, D. L., Mistrot, C. A., Hardwick, K. G., Hwang, E. S., Amon, A., and Murray, A. W. (1998). Budding yeast Cdc20: A target of the spindle checkpoint. *Science* **279,** 1041–1044.

Jorgensen, P., and Tyers, M. (2000). The fork'ed path to mitosis. *Genome Biol.* **1,** Reviews 1022.

Kawasaki, Y., Nagao, K., Nakamura, T., and Yanagida, M. (2006). Fission yeast MAP kinase is required for the increased securin–separase interaction that rescues separase mutants under stresses. *Cell. Cycle* **5,** 1831–1839.

Keaton, M. A., and Lew, D. J. (2006). Eavesdropping on the cytoskeleton: Progress and controversy in the yeast morphogenesis checkpoint. *Curr. Opin. Microbiol.* **9,** 540–546.

Kellogg, D. R. (2003). Wee1-dependent mechanisms required for coordination of cell growth and cell division. *J. Cell. Sci.* **116,** 4883–4890.

Keyse, S. M. (2000). Protein phosphatases and the regulation of mitogen-activated protein kinase signalling. *Curr. Opin. Cell Biol.* **12,** 186–192.

Lew, D. J. (2003). The morphogenesis checkpoint: How yeast cells watch their figures. *Curr. Opin. Cell Biol.* **15,** 648–653.

López-Avilés, S., Grande, M., Gonzalez, M., Helgesen, A. L., Alemany, V., Sanchez-Piris, M., Bachs, O., Millar, J. B., and Aligue, R. (2005). Inactivation of the Cdc25 phosphatase by the stress-activated Srk1 kinase in fission yeast. *Mol. Cell.* **17,** 163–164.

Maeda, T., Takekawa, M., and Saito, H. (1995). Activation of yeast PBS2 MAPKK by MAPKKKs or by binding of an SH3-containing osmosensor. *Science* **269,** 554–558.

Maeda, T., Wurgler-Murphy, S. M., and Saito, H. (1994). A two-component system that regulates an osmosensing MAP kinase cascade in yeast. *Nature* **369,** 242–245.

Maher, M., Cong, F., Kindelberger, D., Nasmyth, K., and Dalton, S. (1995). Cell cycle-regulated transcription of the CLB2 gene is dependent on Mcm1 and a ternary complex factor. *Mol. Cell. Biol.* **15,** 3129–3137.

Martin-Castellanos, C., Blanco, M. A., de Prada, J. M., and Moreno, S. (2000). The puc1 cyclin regulates the G1 phase of the fission yeast cell cycle in response to cell size. *Mol. Biol. Cell* **11,** 543–554.

Martin-Castellanos, C., Labib, K., and Moreno, S. (1996). B-type cyclins regulate G1 progression in fission yeast in opposition to the p25rum1 cdk inhibitor. *EMBO J.* **15,** 839–849.

McMillan, J. N., Longtine, M. S., Sia, R. A., Theesfeld, C. L., Bardes, E. S., Pringle, J. R., and Lew, D. J. (1999). The morphogenesis checkpoint in *Saccharomyces cerevisiae*: Cell cycle control of Swe1p degradation by Hsl1p and Hsl7p. *Mol. Cell. Biol.* **19,** 6929–6939.

McNulty, J. J., and Lew, D. J. (2005). Swe1p responds to cytoskeletal perturbation, not bud size, in *S. cerevisiae. Curr. Biol.* **15,** 2190–2198.

Millar, J. B., McGowan, C. H., Lenaers, G., Jones, R., and Russell, P. (1991). p80cdc25 mitotic inducer is the tyrosine phosphatase that activates p34cdc2 kinase in fission yeast. *EMBO J.* **10**, 4301–4309.

Mondesert, O., McGowan, C. H., and Russell, P. (1996). Cig2, a B-type cyclin, promotes the onset of S in *Schizosaccharomyces pombe.. Mol. Cell. Biol.* **16**, 1527–1533.

Morgan, D. O. (1999). Regulation of the APC and the exit from mitosis. *Nat. Cell. Biol.* **1**, E47–E53.

Ohmiya, R., Kato, C., Yamada, H., Aiba, H., and Mizuno, T. (1999). A fission yeast gene (prr1(+)) that encodes a response regulator implicated in oxidative stress response. *J. Biochem.* **125**, 1061–1066.

Pereira, G., Hofken, T., Grindlay, J., Manson, C., and Schiebel, E. (2000). The Bub2p spindle checkpoint links nuclear migration with mitotic exit. *Mol. Cell* **6**, 1–10.

Posas, F., and Saito, H. (1998). Activation of the yeast SSK2 MAP kinase kinase kinase by the SSK1 two-component response regulator. *EMBO J.* **17**, 1385–1394.

Posas, F., Witten, E. A., and Saito, H. (1998). Requirement of STE50 for osmostress-induced activation of the STE11 mitogen-activated protein kinase kinase kinase in the high-osmolarity glycerol response pathway. *Mol. Cell. Biol.* **18**, 5788–5796.

Proft, M., and Struhl, K. (2004). MAP kinase-mediated stress relief that precedes and regulates the timing of transcriptional induction. *Cell* **118**, 351–361.

Reiser, V., D'Aquino, K. E., Ee, L. S., and Amon, A. (2006). The stress-activated mitogen-activated protein kinase signaling cascade promotes exit from mitosis. *Mol. Biol. Cell* **17**, 3136–3146.

Reiser, V., Ruis, H., and Ammerer, G. (1999). Kinase activity-dependent nuclear export opposes stress-induced nuclear accumulation and retention of Hog1 mitogen-activated protein kinase in the budding yeast *Saccharomyces cerevisiae. Mol. Biol. Cell* **10**, 1147–1161.

Shieh, J. C., Wilkinson, M. G., Buck, V., Morgan, B. A., Makino, K., and Millar, J. B. (1997). The Mcs4 response regulator coordinately controls the stress-activated Wak1-Wis1-Sty1 MAP kinase pathway and fission yeast cell cycle. *Genes Dev.* **11**, 1008–1022.

Shiozaki, K., and Russell, P. (1996). Conjugation, meiosis, and the osmotic stress response are regulated by Spc1 kinase through Atf1 transcription factor in fission yeast. *Genes Dev.* **18**, 2276–2288.

Stegmeier, F., and Amon, A. (2004). Closing mitosis: The functions of the Cdc14 phosphatase and its regulation. *Annu. Rev. Genet.* **38**, 203–232.

Tamas, M. J., Luyten, K., Sutherland, F. C., Hernandez, A., Albertyn, J., Valadi, H., Li, H., Prior, B. A., Kilian, S. G., Ramos, J., Gustafsson, L., Thevelein, J. M., and Hohmann, S. (1999). Fps1p controls the accumulation and release of the compatible solute glycerol in yeast osmoregulation. *Mol. Microbiol.* **31**, 1087–1104.

Verma, R., Annan, R. S., Huddleston, M. J., Carr, S. A., Reynard, G., and Deshaies, R. J. (1997). Phosphorylation of Sic1p by G1 cdk required for its degradation and entry into S phase. *Science* **278**, 455–460.

Wang, Y., Shirogane, T., Liu, D., Harper, J. W., and Elledge, S. J. (2003). Exit from exit: Resetting the cell cycle through Amn1 inhibition of G protein signaling. *Cell* **112**, 697–709.

Wilkinson, M. G., and Millar, J. B. (1998). SAPKs and transcription factors do the nucleo-cytoplasmic tango. *Genes Dev.* **12**, 1391–1397.

Wilkinson, M. G., Samuels, M., Takeda, T., Toone, W. M., Shieh, J. C., Toda, T., Millar, J. B., and Jones, N. (1996). The Atf1 transcription factor is a target for the Sty1 stress-activated MAP kinase pathway in fission yeast. *Genes Dev.* **10**, 2289–2301.

Yancey, P., Clark, M., Hand, S., Bowlus, R. D., and Somero, G. (1982). Living with water stress: Evolution of osmolyte systems. *Science* **217**(4566), 1214–1222.

CHAPTER FIVE

BACTERIAL OSMOSENSING TRANSPORTERS

Janet M. Wood

Contents

Abstract

Cells faced with dehydration because of increasing extracellular osmotic pressure accumulate solutes through synthesis or transport. Water follows, restoring cellular hydration and volume. Prokaryotes and eukaryotes possess arrays of osmoregulatory genes and enzymes that are responsible for solute accumulation under osmotic stress. In bacteria, osmosensing transporters can detect

Department of Molecular and Cellular Biology, University of Guelph, Guelph, Ontario, Canada

Methods in Enzymology, Volume 428
ISSN 0076-6879, DOI: 10.1016/S0076-6879(07)28005-X

increasing extracellular osmotic pressure and respond by mediating the uptake of organic osmolytes compatible with cellular functions ("compatible solutes"). This chapter reviews concepts and methods critical to the identification and study of osmosensing transporters. Like some experimental media, cytoplasm is a "nonideal" solution so the estimation of key solution properties (osmotic pressure, osmolality, water activity, osmolarity, and macromolecular crowding) is essential for studies of osmosensing and osmoregulation. Because bacteria vary widely in osmotolerance, techniques for its characterization provide an essential context for the elucidation of osmosensory and osmoregulatory mechanisms. Powerful genetic, molecular biological, and biochemical tools are now available to aid in the identification and characterization of osmosensory transporters, the genes that encode them, and the osmoprotectants that are their substrates. Our current understanding of osmosensory mechanisms is based on measurements of osmosensory transporter activity performed with intact cells, bacterial membrane vesicles, and proteoliposomes reconstituted with purified transporters. In the quest to elucidate the structural mechanisms of osmosensing and osmoregulation, researchers are now applying the full range of available biophysical, biochemical, and molecular biological tools to osmosensory transporter prototypes.

1. Introduction

1.1. Bacterial osmotolerance, halotolerance, and osmoregulation

Bacteria vary widely in osmotolerance. Their survival and growth may depend on the rate at which osmotic pressure varies and on the amplitude and duration of osmotic pressure changes. For example, obligate halophiles survive and grow only if the salinity (and hence the osmotic pressure) of their environment is consistently very high. Other bacteria tolerate environments with routinely variable osmotic pressures. For example, some *Vibrio* species inhabit marine estuaries. Uropathogenic *Proteus* species and *Escherichia coli* strains colonize mammalian urinary tracts. Osmotic pressure variations are routine in both habitats.

Analysis of the osmoregulatory mechanisms of a bacterium begins with delineation of its osmotolerance and halotolerance (see Glossary in Table 5.1). Halotolerance is more widely documented than osmotolerance because salts are the predominant solutes in many natural environments. However, both properties should be defined, as the mechanisms conferring osmotolerance and halotolerance are not identical (e.g., Kanesaki *et al.*, 2002; Han *et al.*, 2005).

As extracellular osmotic pressure rises or falls, transmembrane water fluxes tend to concentrate or dilute the cytoplasm, threatening cell structure

Table 5.1 Vocabulary of bacterial osmotolerance, osmosensing, and osmoregulation

Compatible solute	Cytoplasmic *cosolvent* whose level can be modulated over a broad range without disrupting cellular functions. See also *osmolyte*.
Cosolvent	A solute that affects the properties of water as a solvent.
Excluded volume	Φ, the fraction of solution volume that is inaccessible to macromolecules because it is occupied by other macromolecules. See also *macromolecular crowding*.
Halotolerance	The *salinity* range of the media that support growth of a particular organism. For example, marine bacteria are halotolerant because they grow in sea water, which contains 3–5% salt.
Halophile	Halophiles are organisms that have a specific requirement for sodium *and* grow optimally at high salinity. Mild, moderate, and extreme halophiles grow optimally in media with low (1–6%), moderate (6–15%), and high (15–30%) salinities, respectively.
Ionic strength	A measure of the collective impact of ions on the activities of individual ions in a solution. Defined as $1/2 \sum (m_i z_i^2)$, where a solution contains i ions, m_i are their molalities (moles/kg solvent), and z_i are their charges.
Macromolecular crowding	The tendency of macromolecules to influence biochemical equilibria (e.g., folding, interactions of macromolecules) or reaction rates (e.g., enzyme activities) by occupying space in a solution. See also *excluded volume*.
Osmolality	The *osmotic pressure* at a particular temperature (Π/RT). Osmolality can be measured but not calculated.
Osmolarity	The sum of the concentrations of osmotically active solutes in solution. The osmolarity is an estimate of the *osmolality*.

(*continued*)

Table 5.1 (*continued*)

Osmolyte	A small organic molecule that accumulates in cells and protects cellular components against denaturing environmental stresses (Bolen, 2001). See also *compatible solute*.
Osmoprotectant	A compound that stimulates bacterial growth at a high *osmolality* (but not a low osmolality) when provided in the growth medium.
Osmoregulation	Physiological processes that mitigate changes in cell structure and function caused by changes in extracellular *osmotic pressure*.
Osmosensor	A protein that detects changes in *water activity* (direct osmosensing) or resulting changes in cell structure or composition (indirect osmosensing) and directs osmoregulatory responses.
Osmotic pressure	The hydrostatic pressure that arises in an aqueous solution because it is bounded by rigid walls and separated from pure water by a semipermeable membrane. Defined as $-(RT/V_w) \ln a_w$, where R is the gas constant, T is the temperature (Kelvin), V_w is the partial molar volume of water, and a_w is the water activity.
Osmotolerance	The *osmolality* range of the media that support growth of a particular organism. For example, *Escherichia coli* is moderately *osmotolerant*, growing in media with osmolalities in the range 0.1 to 2 mol/kg.
Salinity	The salt concentration of a solution (usually given as weight percent).
Turgor pressure	The hydrostatic pressure difference that balances the osmotic pressure difference between the cell interior and exterior in walled cells, rendering the activities of intracellular and extracellular water equal at equilibrium.

and function. Halophilic organisms survive only in high salt media, including those up to 5 M in NaCl. In organisms with "salt in cytoplasm" adaptation, dehydration is forestalled as K^+ salts accumulate in the cytoplasm and macromolecules are specialized to fold and function only in this saline environment (Pflüger and Müller, 2004; Soppa, 2006). Many organisms forestall or reverse osmotically induced water flux by actively adjusting the distributions of organic "compatible solutes" across the cytoplasmic membrane (Wood, 1999). Osmosensors and osmoregulators concerned with compatible solute accumulation are the primary focus of this chapter.

As osmotic pressure increases, solutes accumulate in the cytoplasm through synthesis or transport and water follows, restoring cellular hydration and volume (Fig. 5.1). K^+ salts (e.g., K glutamate) can serve this purpose but

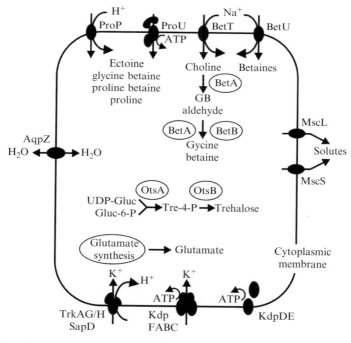

Figure 5.1 Osmoregulatory systems of *Escherichia coli*. Aquaporin AqpZ mediates transmembrane water flux. K^+ transporters TrkA(G/H)/SapD and KdpFABC mediate K^+ accumulation in response to high osmotic pressure. KdpDE is a two-component regulatory system that controls *kdpFABC* transcription in response to K^+ supply and osmotic stress. Suppression of glutamate catabolism leads to its accumulation as K^+ counterion. Transporters ProP, ProU, BetT, and BetU mediate organic osmolyte accumulation at high osmotic pressure. ProU is an orthologue of OpuA from *L. lactis*. BetT and BetU are orthologues of BetP from *C. glutamicum*. Enzymes BetA and BetB mediate glycine betaine synthesis from choline, and enzymes OtsA and OtsB mediate trehalose synthesis from glucose at high osmotic pressure. Mechanosensitive channels MscL and MscS mediate solute efflux in response to decreasing osmotic pressure.

most bacteria extend their osmotolerance by accumulating organic osmolytes that restore cellular hydration more effectively (Cayley and Record,
2003) and act as protein stabilizers (Bolen, 2001). As in eukaryotes, these
compounds include polyols such as trehalose, amino acids such as proline,
and amino acid derivatives such as glycine betaine and ectoine (Pflüger and
Müller, 2004). As osmotic pressure decreases, mechanosensitive channels
mediate solute release and water follows. Each organism studied in depth
possesses multiple osmoregulatory systems (see Fig. 5.1 for the osmoregulatory systems of *E. coli*). They appear to be functionally redundant. The
enumeration of osmoregulatory systems is now facilitated by available
DNA and protein sequence data. However more traditional tools remain
important, as osmoregulatory transporters with novel sequences continue to
emerge (e.g., that of TRAP transporter TeaABC [Gramman *et al.*, 2002]).

1.2. Osmosensing and osmosignaling

An osmosensor is a protein that detects changes in water activity, or resulting changes in cell structure or composition, and directs osmoregulatory
responses. Operationally, osmosensors are identified as proteins that respond
with the same activity or output signal to isotonic aqueous solutions that
differ in solute composition (e.g., those containing an electrolyte such as
NaCl or a nonelectrolyte such as sucrose). Most detailed analyses of osmosensing and osmosignaling have focused on membrane-based osmosensors
from moderately osmotolerant Eubacteria. The identified osmoregulatory
systems include osmosensory transporters, the histidine kinase components
of two-component transcriptional regulatory systems, and mechanosensitive channels (Table 5.2). In addition, osmotic pressure may control bacterial transcription without benefit of these sensory or signal transduction
mechanisms.

Prototypical osmosensory transporters ProP, BetP, and OpuA can detect
osmotic pressure changes and respond by mediating osmoprotectant uptake
without the assistance of other proteins. Membrane-embedded histidine
kinases KdpD and EnvZ detect changes in osmotic pressure and other
signals and then respond by directing cognate response regulators to modulate transcription. KdpD directs KdpE to stimulate transcription of the genes
encoding K^+-motive ATPase KdpFABC. EnvZ directs OmpR to modulate transcription of many genes, among them those encoding the outer
membrane porins OmpC and OmpF. Mechanosensitive channels MscL and
MscS open in response to in-plane membrane strain, releasing solutes and
thereby triggering water efflux. Study of these bacterial systems, each
successfully purified and reconstituted in artificial phospholipid vesicles
(proteoliposomes), is revealing fundamental principles of osmosensing and
osmosignaling.

Table 5.2 Prototypical bacterial osmosensors[a]

Name	Function	Protein family	Source organism	E. coli orthologue(s)	Osmotic phenotype[b]
Trk	Sensing and K$^+$ uptake	NA[c]	Escherichia coli	NA	Yes
ProP	Sensing and H$^+$-osmoprotectant symport	Major facilitator superfamily	E. coli	NA	Yes
BetP	Sensing and Na$^+$-osmoprotectant symport	Betaine/carnitine/choline transporter family	Corynebacterium glutamicum	BetT, BetU	Yes
OpuA	Sensing and ATP-dependent osmoprotectant transport	ABC transporter family	Lactococcus lactis	ProU	Yes
KdpD	Sensing, autophosphorylation, and phosphotransfer to KdpE	Histidine kinase	E. coli	NA	Yes
EnvZ	Sensing, autophosphorylation, and phosphotransfer to OmpR	Histidine kinase	E. coli	NA	No
MscL	Sensing and solute release	Mechanosensitive channel	E. coli	NA	Yes
MscS	Sensing and solute release	Mechanosensitive channel	E. coli	NA	Yes

[a] Progress toward understanding these systems is described elsewhere. For Trk, see Stumpe et al. (1996). For ProP, BetP, and OpuA, see Morbach and Krämer (2002), Poolman et al. (2004), Krämer and Morbach (2004), and Wood (2006). For KdpD and EnvZ, see Pratt et al. (1996), Jung and Altendorf (2002), Epstein (2003), and Heermann and Jung (2004). For MscL and MscS, see Martinac (2001).

[b] As discussed in the listed reviews, bacterial growth at high osmolality can be impaired by deficiencies in ProP, BetP, OpuA, or KdpD, but not EnvZ. E. coli lacking both MscS and MscL are sensitive to osmotic downshocks (Levina et al., 1999).

[c] Not applicable.

Osmotic pressure is not solute specific. All solutes are also cosolvents that contribute to the osmotic pressure as they modulate water activity (defined further later). A direct osmosensor would detect water activity, just as ligand-specific receptors (chemosensors) detect ligand activity (usually approximated by ligand concentration). However, osmotic shifts alter many cellular properties. Thus, an indirect osmosensor could detect an osmotic shift-induced change in cell volume, turgor pressure, or membrane strain, or in the concentration of an individual solute, the ionic strength, or the crowding of macromolecules in the cytoplasm (Wood, 1999). Osmosensors are interesting because they may be designed to detect solvent properties such as hydrostatic pressure, water activity (to which all solutes contribute), ionic strength (to which all electrolytes contribute), or macromolecular crowding (to which all macromolecules contribute). Mechanosensitive channels are indirect osmosensors that open in response to a hydrostatic pressure-induced membrane strain.

This chapter includes methods for the analysis of osmosensing derived from studies of the prototypical osmosensing transporters ProP, BetP, and OpuA. The activity of each transporter varies as the osmotic pressure of the external medium is adjusted with chemically diverse, membrane-impermeant solutes. Each can act as both an osmosensor and an osmoregulator, as this behavior is observed when the proteins are present in intact cells and retained if they are purified and reconstituted in proteoliposomes. Their activation coincides with the osmotic inactivation of other membrane enzymes such as LacY, the lactose permease of *E. coli*, that are not osmoregulators (Wood, 2006). This makes sense. Osmoregulation is necessary because all cell constituents are affected by increasing osmotic pressure. Systems such as ProP, BetP, and OpuA can both sense this problem and respond, correcting cellular hydration. But are these systems direct or indirect osmosensors? If indirect, what cellular property do they detect?

Solvents internal and external to proteoliposomes can be manipulated at will, with or without imposing osmotic gradients that would alter the topology of the bounding membrane (Rübenhagen *et al.*, 2001). Proteoliposomes were used to show that transporters BetP, OpuA, and ProP can be activated in the absence of osmotic shifts so they do not respond to changes in membrane strain and they are probably not designed to regulate turgor pressure (Culham *et al.*, 2003; van der Heide *et al.*, 2001; Rübenhagen *et al.*, 2001). As in cells, all three respond to the osmotic pressure (not the chemical composition) of the external solvent as it is adjusted with membrane-impermeant solutes. However, all three respond with specificity to the composition of the internal solvent. This observation has led to divergent models for osmosensing. ProP is proposed to be a direct osmosensor whose activity is determined by its own hydration (Wood, 2006). In contrast, BetP and OpuA are proposed to be indirect osmosensors, detecting cytoplasmic K^+ concentration (Rübenhagen *et al.*, 2000, 2001;

Schiller *et al.*, 2004a, 2006) and ionic strength (Biemans-Oldehinkel and Poolman, 2003; Biemans-Oldehinkel *et al.*, 2006), respectively. All available biophysical, biochemical, and molecular biological tools are now being applied to these osmosensory transporters to further delineate the structural mechanisms by which they sense and respond to osmotic pressure changes.

 ## 2. ESTIMATION OF KEY SOLUTION PROPERTIES AND OF TURGOR PRESSURE

2.1. Osmotic pressure, water activity, osmolality, and osmolarity

Measurements of the osmotic pressures of microbiological and biochemical media and estimates of those quantities for the water within living cells are fundamental to the study of osmosensing and osmoregulation. Osmotic pressure and water activity are related by

$$\Pi = -(RT/\bar{V}_w)\ln a_w \qquad (5.1)$$

where Π is the osmotic pressure, R is the gas constant, T is the temperature, $\bar{V}_w$ is the partial molar volume of water, and a_w is the water activity. The osmotic pressure of an aqueous solution can be determined by measuring the vapor pressure of water in equilibrium with that solution because

$$a_w = P/P^* \qquad (5.2)$$

where P and P^* are the vapor pressures of the solution and pure water, respectively. The osmotic pressure is often expressed as the osmolality (Π/RT), the osmotic pressure at a particular temperature. The osmolarity, an approximation for the osmolality, is the sum of the concentrations of osmotically active solutes in a solution. Relevant units and unit conversion factors are summarized in Table 5.3.

An ideal solution is one for which, over the full range of solute concentrations, the vapor pressure of the solution (P) is the product of the mole fraction of the solvent in the solution and the vapor pressure of the pure solvent (P^*) (Sweeney and Beuchat, 1993). Real solutions approach ideality only as the solute concentration approaches zero. Nonideality can arise because solutes interact with the solvent or occupy appreciable space within the solution. Water interacts powerfully with biomolecules and the cytoplasm is a crowded milieu in which macromolecules (collectively) occupy close to half the available volume. Thus the aqueous solutions in biological systems and many experimental media are nonideal, and the osmotic

Table 5.3 Quantities relevant to osmotic calculations

Quantity	Symbol	Units, values, and conversions
Osmotic pressure	Π	atmosphere (atm) pascal (Pa) (1 atm = 1.01325×10^5 Pa) dyne · cm^{-2} (1 atm = 1.01325×10^6 $\quad$ dyne · cm^{-2})
Osmolality	Π/RT	mol kg^{-1}
Osmolarity		molar
		0.082054 liter · atm · mol^{-1} · °K^{-1}
Gas constant	R	8.3144×10^7 ergs · mol^{-1} · °K^{-1}
		8.3144×10^7 dyne · cm · mol^{-1} · °K^{-1}
Temperature	T	Degrees Kelvin (°K) (°K = °C + $\quad$ 273.15)
		Degrees centigrade (°C)
Partial molar volume of water	$\bar{V}_w$	0.01801 liter mol^{-1}

pressures of solutions relevant to the study of osmosensing and osmoregulation must be measured. Sweeney and Beuchat (1993) have provided a useful review of principles and techniques for the measurement of osmotic pressure, concluding that vapor pressure osmometry is the method of choice. Many laboratories use osmometers marketed by Wescor, Inc. (Logan, UT, www.wescor.com).

2.2. Ionic strength

Whereas all solutes contribute to the osmotic pressure of a solution, only ionic solutes contribute to the ionic strength (I), which can be calculated as

$$I = 1/2\Sigma(m_i z_i^2) \tag{5.3}$$

where a solution contains i ions, m_i are their molalities (moles/kg solvent), z_i are their charges, and summation is over the "i" ions. Ions in solution modulate macromolecular interactions by reducing the strength of electrostatic interactions between macromolecular surfaces. Because ionic strength varies as the square of ion charge, polyvalent ions contribute much more strongly to the ionic strength than monovalent ions, and the effects of monovalent and polyvalent ions are usually compared to determine whether the ionic strength is implicated in biochemical processes (e.g., Mahmood et al., 2006).

2.3. Macromolecular crowding, volume exclusion, and protein hydration

Macromolecular crowding is relevant to osmosensing and osmosignaling because crowding can change dramatically as cellular hydration changes in response to osmotic stress, significantly altering the structures, aggregation, and functions of individual macromolecules in the cytoplasm and cytoplasmic membrane (Minton, 2006). The fraction of solution volume occupied by a crowding agent such as a protein or nucleic acid (and hence the volume fraction from which other macromolecules are excluded) can be calculated as

$$\Phi = cv \tag{5.4}$$

where c is the mass concentration of the crowding agent (grams of solute per milliliter of solution) and v is the partial specific volume of the crowding agent (milliliter of solute per gram of solute). The excluded volume fraction of the non-nucleoid cytoplasm in E. coli has been estimated as 0.3 to 0.4, and such crowding could be simulated by a globular protein with a molecular mass close to 75 kDa at a concentration of 0.34 g/ml (Zhang et al., 1996; Zimmerman and Trach, 1991). Experiments designed to analyze the impact of crowding on macromolecular structure and function require crowding agents that are highly soluble and do not interact (specifically or nonspecifically) with the molecule under study. Bovine serum albumin, cyanomethemoglobin, and dextran have been used for this purpose (Minton, 1998, 2006; Zimmerman and Minton, 1993).

Large solutes also affect protein structure and aggregation by being sterically excluded from water-filled clefts within or between proteins, hence causing osmotically induced protein dehydration (Parsegian et al., 1995, 2000). Poly(ethylene)glycols (PEGs) are sometimes used as crowding agents and may also act via steric exclusion. Both mechanisms can be probed effectively with PEGs available from Polypure AS (Oslo, Norway; http://www.polypure.no/) as monodisperse preparations with systematically varying molecular sizes (e.g., Culham et al., 2003). However, PEGs must be used with care because they can interact with proteins (Bhat and Timasheff, 1992); those with molecular masses greater than approximately 3 kDa can catalyze membrane fusion (Lentz and Lee, 1999).

2.4. Turgor pressure

Solute accumulation within walled cells generates turgor pressure (ΔP), the hydrostatic pressure difference that balances the osmotic pressure difference between the cell interior and the cell exterior:

$$\Delta P = P_i - P_o = \Pi_i - \Pi_o \qquad (5.5)$$

where P is hydrostatic pressure. The turgor pressures of large walled cells (e.g., those of plants and eukaryotic microorganisms) can be measured directly because the cells can be impaled by pressure probes. Because bacteria are too small to be studied in this way, indirect methods must be used to estimate bacterial turgor pressure (for a review of these methods, see Wood, 1999). It has been shown that the turgor pressure maintained by *E. coli* cells decreased when they were cultivated in media of increasing osmotic pressure (Cayley *et al.*, 2000), and the osmosensory transporters that are the subject of this chapter can activate in the absence of turgor pressure changes (Krämer and Morbach, 2004; Poolman *et al.*, 2004). However, the maintenance of turgor pressure is still believed to be an objective of osmoregulation by bacteria (Epstein, 2003).

3. Assessment of Osmotolerance

Bacterial osmotolerance is usually defined as the osmolality range of the media that support bacterial survival and growth. Osmotolerance must be cited with respect to specific growth media because it varies with growth medium phase (solid versus liquid) and composition. For example, most bacteria grow to a higher osmolality in the common, rich microbiological medium LB (which contains osmoprotectants such as proline and glycine betaine) than in minimal salts media devoid of osmoprotectants. The relationships between osmolality and NaCl supplementation for LB medium (Miller, 1972) and MOPS medium (Neidhardt *et al.*, 1974) are cited in Table 5.4.

Osmotolerance is usually estimated by growing bacteria in liquid batch culture. The osmolalities of liquid media are readily measurable and multiple growth parameters characterize batch cultures of suspended (planktonic) bacteria. For example, medium osmolality can affect duration of the lag phase, exponential growth rate, growth yield in stationary phase, and long-term survival. Assessment of population density after a single, arbitrary growth period may fail to detect differential effects of osmoregulatory and osmoadaptive defects on these parameters. Growth parameters obtained after subculturing from a medium of one osmolality directly into those of other osmolalities are sensitive to the rates of adaptive processes. Experiments can also be designed to allow adaptation prior to the measurement of growth parameters. The ability of bacteria to tolerate osmotic pressure fluctuations imposed at different rates or frequencies could be tested via chemostat culture. In their natural habitats, many bacteria grow on surfaces

Table 5.4 Osmolalities of NaCl-supplemented microbiological media

NaCl $(M)^a$	0	0.2	0.3	0.4	0.6	0.8
Osmolality of LB (mol/kg)^b	0.4	—	1.0	—	1.6	—
Osmolality of MOPS (mol/kg)^c	0.20	0.56	—	0.94	1.41	1.77

^{a} The listed NaCl was added to that present in the base media [10 g/liter for LB (Miller, 1972) and 50 mM for MOPS (Neidhardt et al., 1974)].

^{b} LB medium (Miller, 1972) contains tryptone (a tryptic hydrolysate of casein, 10 g/liter), yeast extract (5 g/liter), and NaCl (10 g/liter). Note that LB medium formulations vary in NaCl content, and hence osmolality.

^{c} MOPS is a minimal salts medium designed for the growth of enterobacteria (Neidhardt et al., 1974). For these osmolality measurements, MOPS medium was supplemented with D-glucose (11 mM) as carbon source and NH$_4$Cl (9.5 mM) as nitrogen source.

as biofilms. New tools can now be applied to study osmoregulation by biofilm bacteria (see *Methods in Enzymology*, Volumes 336 and 337).

Light scattering is usually used as a measure of bacterial population density in liquid media. The relationship between light scattering and bacterial population density is affected by medium refractive index (in turn a function of solute content) and by bacterial aggregation, size, shape, and refractive index (all of which can be affected by osmotic pressure and osmotic adaptation). For a discussion of light scattering as a tool for the assessment of bacterial numbers and bacterial cell structure, see Wood (1999). Light scattering should therefore be supplemented with viable counts (enumeration of colonies arising when aliquots of culture dilutions are spread on petri plates) or measurements of the protein content of cell suspensions for studies of osmotolerance and osmoregulation (e.g., Culham et al., 2001; Romantsov et al., 2007).

4. IDENTIFICATION OF OSMOPROTECTANTS, COMPATIBLE SOLUTES, AND ORGANIC OSMOLYTES

Osmoprotectant activity can be measured by testing the abilities of compounds to stimulate bacterial growth in high (but not low) osmolality media (e.g., MacMillan et al., 1999; Peddie et al., 1994, 1998). Osmoprotection has been used as a bioassay to identify new osmoprotectants (e.g., Chambers and Kunin, 1987).

Osmoprotectants may accumulate in cells as, or be converted to, compatible solutes (sometimes denoted organic osmolytes). Nuclear magnetic resonance (NMR) spectroscopy of intact cells or cell extracts is the least biased method for compatible solute identification (Galinski and Oren, 1991; Galinski *et al.*, 1985; Motta *et al.*, 2004; Smith and Smith, 1989). The signatures of compatible solutes are prominent in NMR spectra of cells and crude cell extracts because compatible solutes accumulate to high cytoplasmic levels. Once compatible solute accumulation is detected, extract fractionation and additional analytical techniques may be required to identify novel compatible solutes. Because NMR spectroscopy is not quantitative, solute-specific techniques are used to determine the cytoplasmic concentrations of particular compatible solutes. This may require extract fractionation and the use of solute-specific detection methods (e.g., Kunte *et al.*, 1993; MacMillan *et al.*, 1999), as well as the concomitant measurement of cytoplasmic volume (Stock *et al.*, 1977) (for reviews of the latter techniques, see Wood [1999] and Cayley *et al.* [2000]).

5. IDENTIFICATION AND DISTRIBUTION OF OSMOSENSORY AND OSMOREGULATORY TRANSPORTERS

5.1. Transporter identification

In principle, many approaches can be used to detect genes and enzymes responsible for osmosensing and osmoregulatory transport. They include the following.

Approach 1: Selection of mutants resistant to toxic osmoprotectant analogues.

Approach 2: Screening of genomic DNA libraries for functional complementation of transporter defects in osmoregulation-deficient bacterial hosts.

Approach 3: Targeted disruption and complementation of putative osmosensory or osmoregulatory transporter genes, identified via sequence similarity.

Approach 4: Identification of genes whose expression is modulated by osmotic pressure.

5.1.1. Approach 1

Because organisms with transporter defects are resistant to toxic analogues of their substrates, such analogues can be used to select bacteria with transporter gene defects. For example, derivatives of *E. coli* K-12 harboring mutant alleles of the genes encoding proline/glycine betaine transporters PutP, ProP, and ProU were isolated by sequential selection with toxic proline/betaine analogues 3,4-dehydroproline and L-azetidine-2-carboxylic acid (Wood, 1988). This approach remains useful if the molecular biological tools or sequence data required for Approaches 2 to 4 are not available. It is not contingent on the phenotypes of existing, osmoregulation-deficient hosts or subject to problems of gene expression *in trans*. Substrate structure–activity relationships for the osmoregulatory transporters of *E. coli* and other bacteria have been characterized extensively (Gouesbet *et al.*, 1994; Jebbar *et al.*, 1992; MacMillan *et al.*, 1999; Peddie *et al.*, 1994, 1998, 1999; Randall *et al.*, 1995, 1996).

5.1.2. Approach 2

Available bacterial hosts can be used as vehicles to screen DNA libraries for genes that functionally complement osmoregulatory transporter defects. For example, a bank of genes from a target organism can be prepared for expression in *E. coli* and then introduced to one of the *E. coli* K-12 strains listed in Table 5.5, which also lists cognate media that are then used to select bacteria expressing heterologous osmoregulatory transporter genes. These selection systems are based on the following principles.

Proline is an osmoprotectant for many organisms and a glycine betaine analogue. Some osmoregulatory transporters accept proline, among other substrates such as glycine betaine and ectoine (MacMillan *et al.*, 1999). Thus, selection for complementation of proline uptake defects can be used to identify transporters that also mediate osmoprotectant uptake. Because proline can serve as an osmoprotectant, a protein constituent, and a nitrogen or carbon source for *E. coli*, proline transporters can be identified by taking advantage of diverse mutants and cognate selective media. *E. coli* WG350 lacks the three known proline transporters (PutP, ProP, and ProU), as well as proline dehydrogenase (PutA). The ability of *E. coli* to use proline as an osmoprotectant can be exploited by selecting derivatives of *E. coli* WG350 that grow on salt-supplemented petri plates only in the presence of proline. *E. coli* WG389 is a derivative of *E. coli* WG350 that is unable to synthesize proline [it grows only in media supplemented with proline at a very high level (25 mM)]. The requirement for proline as a protein constituent can be exploited by selecting derivatives of *E. coli* WG389 that grow on minimal medium supplemented with proline at a low level (25 μM) but not on proline-free medium. *E. coli* WG379 is a derivative of *E. coli* WG350 that retains proline dehydrogenase (PutA). It can use proline as a sole nitrogen source only if a complementing proline transporter is provided. This property can be exploited by selecting derivatives of *E. coli* WG379 that grow on minimal medium with proline as a sole nitrogen source. The selection based on the proline auxotrophy of *E. coli* WG389 has proven cleanest in our hands, however.

Betaine transporter genes can be selected by seeking osmoprotection of *E. coli* MKH13 (which lacks betaine transporters ProP, ProU, and BetT) (Table 5.5). In principle, either WG350 or MKH13 could be used with the appropriately supplemented media to identify genes encoding transporters

Table 5.5 *Escherichia coli* strains used to identify osmoprotectant transporter genes by functional complementation

E. coli strain	Genotype	Selection principle and medium	Ref.[d]	Applications[e]
WG350	F⁻ *trp rpsL thi* Δ(*putPA*)101 Δ(*proP-mel*)212 Δ(*proU*)600	Osmoprotection[a]: 0.6 M NaCl 1 mM L-proline	1	1
WG389	F⁻ *trp rpsL thi* Δ(*brnQ phoA proC*) Δ(*putPA*)101 Δ(*proP-mel*)212 Δ(*proU*)600	Auxotrophy[a]: 25 µM L-proline	1	2,3
WG379	F⁻ *trp rpsL thi putP3*::Tn5 Δ(*proP-mel*)212 Δ(*proU*)600	Proline catabolism[a]: 17 mM L-proline 54 mM glycerol[a,b]	1	—
MKH13	F⁻ *araD139* Δ(*argF-lac*)U169 *rpsL150 relA1 deoC1 ptsF25 rbsR flbB5301* Δ(*putPA*)101 Δ(*proP*)2 Δ(*proU*)608	Osmoprotection[c]: 0.8 M NaCl 8 mM glycine betaine	2,3	3–7

[a] These media are based on MOPS medium (Neidhardt *et al.*, 1974) supplemented with D-glucose (2 mg/ml) as carbon source and NH$_4$Cl (9.5 mM) as nitrogen source, as well as L-tryptophan (245 µM) and L-thiamine (1 µg/ml), to meet auxotrophic requirements.

[b] In this medium, L-proline replaces NH$_4$Cl and glycerol replaces D-glucose.

[c] This medium is based on M9 medium (Smith and Levine, 1964) supplemented with D-glucose (5 mg/ml) as carbon source and NH$_4$CL (18 mM) as nitrogen source, as well as arginine, isoleucine, and valine (0.4 mg/ml), to meet osmotic stress-induced auxotrophic requirements.

[d] The selections are described in references 1 (Culham *et al.*, 1993), 2 (Peter *et al.*, 1996), and 3 (Kempf and Bremer, 1995).

[e] Applications of these selection systems are described in references 1 (Culham *et al.*, 1993), 2 (Peter *et al.*, 1997), 3 (Peter *et al.*, 1998), 4 (Peter *et al.*, 1996), 5 (Sleator *et al.*, 1999), 6 (Kempf and Bremer, 1995), and 7 (Kappes *et al.*, 1996).

with other osmoprotectant specificities. This approach could be broadened by exploiting existing, osmoregulation-deficient derivatives of the gram-positive bacteria *Bacillus subtilis* (Kempf and Bremer, 1998), *Corynebacterium glutamicum* (Morbach and Krämer, 2002), and *Listeria monocytogenes* (Sleator *et al.*, 2003). It has the potential to identify structurally novel transporters for known osmoprotectants and transporters with new osmoprotectant specificities. In principle, it can also be used to identify eukaryotic systems if they are expressed and functional in a bacterial host.

5.1.3. Approach 3

Existing osmoprotectant transporter protein sequences can now be used as probes to identify putative homologues encoded by published DNA sequences. Table 5.6 lists currently known transporter prototypes: well-characterized, structurally distinct bacterial systems for which functional homologues are known (with the exception of TeaABC). These protein sequences can be used as queries in homology (BLAST) searches (McGinnis and Madden, 2004) (http://www.ncbi.nlm.nih.gov/BLAST/). Identified genetic loci are tested for osmoregulatory activity by targeted gene disruption and/or complementation using the selective media described earlier and by biochemical characterization (see later).

The characterization of transporters identified via sequence homology is critical, as functionally related transporters are not identified easily via overall sequence similarity. Within each transporter class, membrane-embedded amino acid sequences may share significant homology regardless of the substrate specificity or physiological role of the transporter. For example, transporters ProP of *E. coli* and *C. glutamicum* are functionally homologous osmosensory betaine transporters that share only 39% sequence identity (Peter *et al.*, 1998). Transporters KgtP and ShiA of *E. coli* are also quite similar in sequence to ProP (30 and 27% identity, respectively), but KgtP is an α-ketoglutarate transporter (Seol and Shatkin, 1993) and ShiA is a shikimate transporter (Whipp *et al.*, 1998). The tertiary structure of *E. coli* ProP has been modeled successfully on the crystal structure of anion anti-porter GlpT from *E. coli*, although they share only 17% sequence identity (Wood *et al.*, 2005). KgtP, ShiA, and GlpT are neither osmosensors nor osmoregulators. The genome of *E. coli* K-12 encodes 80 known or putative ABC transporters (Blattner *et al.*, 1997), all of which share significant sequence similarity, but only one of them is known to be osmoregulatory (ProU [proteins ProXYZ]).

Searches for full-length sequence similarity are usually refined by seeking evidence for particular conservation of structurally or functionally important residues and sequence motifs. Among the transporters listed in Table 5.6, residues critical for osmoprotectant (substrate) binding are known only for the soluble, periplasmic-binding protein component OpuAC of ABC transporter OpuA in *B. subtilis* (Horn *et al.*, 2006).

Table 5.6 Osmoprotectant transporter prototypes

Name	Class	Accession number(s) (GI)	Organism	Ref.[a]	E. coli homologue
ProP	Major facilitator superfamily (MFS)	ProP: 147357	*Escherichia coli*	1	N/A
BetP	Betaine–choline–carnitine family (BCCT)	BetP: 1705453	*Corynebacterium glutanicum*	2	BetT, BetU
OpuE[b]	Sodium solute symporter superfamily (SSSS)	OpuE: 2114423	*Bacillus subtilis*	3	PutP[c]
OpuA	ATP-binding cassette (ABC) transporter family	OpuAA: 7212813 OpuABC: 7212814	*Lactococcus lactis*	4	ProU
TeaABC[b]	Tripartite ATP-independent periplasmic (TRAP) transporter family	TeaA: 18031990 TeaB: 18031991 TeaC: 18031992	*Halomonas elongata*	5	None known

[a] The references are (1) Culham et al. (1993), (2) Peter et al. (1996), (3) von Blohn et al. (1997), (4) Bouvier et al. (2006), and (5) Gramman et al. (2002).

[b] OpuE and TeaABC mediate osmoprotective solute uptake. They are not known to act as osmosensors because they have not been purified and examined *in vitro*.

[c] Despite being an OpuE homologue, PutP of *E. coli* is not an osmoregulatory transporter (Milner et al., 1987).

Among residues implicated in energy transduction, the highly conserved Walker A/B and other motifs in the ABC subunits of ABC transporters can be identified readily (Davidson and Chen, 2004), but no unusual features of those motifs have been reported to occur only in osmosensory transporters. The putative H^+ or Na^+ translocating residues of secondary transporters cannot yet be predicted, nor have signatures distinctive among osmosensing secondary transporters been identified.

Current studies on osmosensing by transporters ProP, BetP, and OpuA focus on their extended C-terminal domains (Wood, 2006). The α-helical character seems to be the key determinant for osmosensing by the extended C-terminal region of BetP (Schiller et al., 2006). The C-terminal α-helical coiled-coil domain, present in one of two subgroups of ProP orthologues, adjusts the osmolality range over which the transporter activates (Poolman et al., 2004; Tsatskis et al., 2005). Cystathionine-β-synthase domains have been implicated in osmosensing by OpuA from Lactococcus lactis. In addition, a downstream anionic segment is suggested to play a role analogous to that of the coiled-coil domain in ProP (Biemans-Oldehinkel et al., 2006; Mahmood et al., 2006). Thus current research is revealing sequence features related to osmosensing and the osmoregulation of transporter activity.

5.1.4. Approach 4

Bacterial osmoregulation was one of the first biological phenomena to be approached using "omic" tools. Banks of E. coli strains in which genes were fused at random to the "indicator" lacZ or phoA were screened to identify genes whose expression responded to osmotic stress (Gowrishankar, 1985; Gutierrez et al., 1987). Today DNA microarrays are used to screen transcriptomes and two-dimensional gel electrophoresis is used to screen proteomes for osmotic stress-induced or -repressed genes and proteins. Organisms already examined in these ways include Bacillus subtilis (Hoper et al., 2006; Steil et al., 2002), E. coli (Weber and Jung, 2002; Weber et al., 2006), Listeria monocytogenes (Duche et al., 2006), Pseudomonas aeruginosa (Aspedon et al., 2006), and Synechocystis (Fulda et al., 2006; Marin et al., 2004). These approaches are also being used to explore the phenotypes associated with defects in known or putative osmoregulatory genes (Lee et al., 2005). Volumes 410 and 411 of the Methods in Enzymology series are devoted to the technology and applications of DNA microarrays. Techniques for proteomic analysis have also been discussed (Stults and Arnott, 2005).

5.2. Transporter gene distribution

Once genes encoding osmosensors and osmoregulators are identified, their distribution and genomic localization among isolates of a particular bacterium can be determined. This can be accomplished by analyzing polymerase chain reaction (PCR)-generated amplicons obtained with genomic DNAs from diverse bacterial isolates as templates (Culham *et al.*, 1994; Ly *et al.*, 2004). For example, genomic sequencing has revealed large differences among naturally occurring *E. coli* isolates, many apparently arising via lateral gene transfer and some correlated with bacterial virulence (Dobrindt *et al.*, 2002). PCR-based analysis of the occurrence of the osmoregulatory transporters illustrated in Fig. 5.1 showed that *proP*, *proU*, *betTIBA*, and *trkAH* are present in all tested *E. coli* genomes. In contrast, *betU* and *trkAG* were present only sporadically and the surrounding DNA contained signatures characteristic of lateral gene transfer (Ly *et al.*, 2004). *E. coli* shows clonal evolution, isolates representative of particular evolutionary clones are available, and particular clones are associated with particular pathotypes. Thus such studies can reveal the evolutionary origins of these stress responses and their relationships to *E. coli* virulence.

6. *In Vitro* Systems

Osmosensory transporters are integral membrane proteins. As a result, their activities can only be studied using membrane-bounded experimental systems, including intact cells, bacterial membrane vesicles, and proteoliposomes. Evidence indicates key roles for membrane lipids in osmosensing and osmosignaling. Thus the following sections summarize methods for the preparation of membrane vesicles and proteoliposomes, the analysis of membrane composition, and the measurement of osmosensory transporter activity.

6.1. Membrane vesicle preparation

Membrane vesicles (bacterial cell "ghosts") have been key tools for the analysis of bacterial membrane proteins since Kaback and colleagues developed techniques for their preparation in the 1960s. To prepare membrane vesicles, cells are plasmolyzed, treated with EDTA and lysozyme to compromise the integrity of the murein layer (releasing periplasmic contents), lysed osmotically to release cytoplasmic contents, harvested, and washed by repeated centrifugation (Kaback, 1971, 1974; Konings, 1977; Konings *et al.*, 1973). Extensive characterization has shown that, correctly prepared, each ghost is derived from a single cell and retains the topology of its cytoplasmic membrane (Altendorf and Staehelin, 1974). However such preparations must be used with care, as topology can be disrupted during vesicle preparation and handling.

6.2. Proteoliposome preparation

Proteoliposomes can be used to attribute particular functions to particular membrane proteins. Thus, proteoliposomes were used to show that transporters ProP (Racher et al., 1999), BetP (Rübenhagen et al., 2000), and OpuA (van der Heide and Poolman, 2000) can sense osmotic stress and respond by mediating osmoprotectant uptake without the help of other proteins. Proteoliposomes are prepared by purifying an integral membrane protein and reconstituting it with membrane lipids to form topologically closed vesicles. In brief, proteoliposomes containing bacterial osmosensory transporters are prepared as follows (Poolman et al., 2005; Rigaud and Levy, 2003; Rigaud et al., 1995). First, liposomes (large unilamellar phospholipid vesicles) are prepared by delivering phospholipid from organic solvent solution onto a glass surface by rotary evaporation, hydrating the lipid with a desired buffer, dialyzing extensively, and extruding the resulting preparation through a microporous membrane to obtain monodisperse vesicles (Mui et al., 2003; White et al., 2000). In parallel, inverted membrane vesicles are prepared from bacteria expressing the protein of interest (Altendorf and Staehelin, 1974; Rosen and Tsuchiya, 1979), usually with an engineered tag for purification (Bornhorst and Falke, 2007; Skerra and Schmidt, 2000), the vesicles are solubilized with detergent, and the protein is purified in the presence of detergent. Because high-level membrane protein overexpression is usually deleterious or even lethal to bacterial growth, tightly controlled, moderate overexpression (e.g., using the pBAD series of plasmid vectors [Guzman et al., 1995; Racher et al., 1999]) is usually more successful (Saidijam et al., 2005). Because membrane proteins are notoriously unstable in the detergent-solubilized state, they must be purified quickly and reconstituted immediately with lipid. To create proteoliposomes, the lipid vesicles are destabilized by adding detergent and mixed with the purified membrane protein. Detergent is gradually removed with an affinity resin, causing the protein to insert in the vesicle membrane (Rigaud and Levy, 2003). The proteoliposomes are harvested by centrifugation.

6.3. Membrane composition and topology in cells, membrane vesicles, and proteoliposomes

The levels of transporter proteins in cells and membrane vesicles are usually determined by Western blotting, often using antibodies that recognize introduced tags. For example, commercially available enzyme-conjugated antibodies recognize the His_6 tag that is also used for affinity purification. If the purified transporter is available as a standard, densitometry can be used to render this approach quantitative (e.g., Culham et al., 2000). The protein content of proteoliposomes is determined with the Schaffner–Weissmann assay, which incorporates a sample delipidation step (Schaffner and Weissmann, 1973). Various approaches have been used to determine the topology of membrane proteins in proteoliposomes (Rigaud and Levy, 2003). These include covalent labeling of introduced Cys residues by membrane-impermeant reagents (Jung *et al.*, 1998), as well as limited proteolysis (Bogdanov *et al.*, 2005) and measurement of antibody reactivity in intact versus disrupted proteoliposomes (Schiller *et al.*, 2004b).

To assess the membrane environment of an osmosensory transporter, the phospholipid head group distribution of cells can be determined by thin-layer chromatography of polar lipid extracts prepared after bacterial growth for several generations in the presence of ^{33}P- or ^{32}P-orthophosphate (Tsatskis *et al.*, 2005). Such analyses only approximate membrane protein environments, as phospholipids are not distributed randomly over the surfaces of bacterial cells (Matsumoto *et al.*, 2006). To adjust their phospholipid composition, proteoliposomes can be fused with liposomes of defined phospholipid composition by coextrusion through microporous membranes (van der Heide *et al.*, 2001). The sizes and size distributions of liposomes and proteoliposomes can be determined by dynamic light-scattering spectrometry (Ertel *et al.*, 1993; Rigaud and Levy, 2003; White *et al.*, 1996, 2000).

7. MEASUREMENT OF OSMOSENSORY TRANSPORTER ACTIVITY

Well-established procedures are used to measure active solute uptake by bacteria, membrane vesicles, and proteoliposomes (Kaback, 1974; Poolman *et al.*, 2005; Viitanen *et al.*, 1986; Voegele *et al.*, 1995). In most cases, the uptake of radiolabeled substrates is estimated by scintillation counting after cells, vesicles, or proteoliposomes are separated from their suspending assay medium by filtration.

Membrane vesicles can be used to compare the activities of transporters, coexpressed by the source bacteria, that have not been affected adversely by purification and reconstitution. Further, secondary transport can be energized by providing electron donor/acceptor systems to the membrane-integral respiratory chain. However, rates of energized solute uptake via transporters in membrane vesicles reflect the impacts of osmotic stress on both respiration and transport.

In proteoliposomes, secondary transport must be energized by artificially imposing a membrane potential ($\Delta\psi$), a pH gradient (ΔpH), and/or a Na^+ gradient (ΔpNa) (Viitanen *et al.*, 1986; Rottenberg, 1986). The imposed gradients can be rendered independent of any applied osmotic stress if adjustments are made to correct for the impact of vesicle shrinkage on lumenal ion concentrations (and hence transmembrane ion gradients) (Racher et al., 2001). For ABC transporters, ATP provided internally via an ATP-regenerating system can be used to power substrate uptake via transporters with the ABC domain on the internal membrane surface; ATP provided in the external medium can be used to power substrate efflux via transporters with the ABC domain on the external membrane surface (Poolman et al., 2005). Proteoliposomes have been critical to our under-standing of osmosensory transporters, in part because their lumenal contents can be adjusted at will. This adjustment is achieved by modifying the external medium and then extruding the suspended proteoliposomes through a microporous filter. This procedure renders the membrane leaky, achieving equilibration of low and high molecular weight solutes between the proteoliposome lumen and the external medium.

Osmosensory transporters are distinguished by activity that increases, in cells, membrane vesicles and proteoliposomes, as the osmolality of the external medium is raised by adding chemically diverse, membrane-impermeant solutes. The osmotic activation of osmosensory transporter ProP from *E. coli* is illustrated in Fig. 5.2. Sigmoid osmotic activation profiles are also observed for other osmosensory transporters. Because the biophysical basis for ProP activation is not fully understood, such data are fit to the empirical relationship:

$$a_0 = A_{max}[1 + \exp(-(\Pi - \Pi_{1/2})/RTB))]^{-1} \qquad (5.6)$$

where Π is the osmotic pressure of the transport assay medium, a_0 is the initial rate of proline uptake measured with medium of osmolality Π/RT, A_{max} is the uptake rate that would be observed at infinite medium osmolal-ity, $\Pi_{1/2}/RT$ is the medium osmolality yielding half-maximal transporter activity, and B is a constant inversely related to the slope of the response curve. Thus each data set can be characterized by parameters A_{max}, $\Pi_{1/2}/RT$, and B. These parameters vary with the experimental system (cells, membrane vesicles, or proteoliposomes), the phospholipid compositions of bacterial and proteoliposome membranes, the composition of the lumenal solvent in proteoliposomes, and modifications to transporter structure (e.g., individual amino acid substitutions, peptide deletions, and structural varia-tions of the C-terminal domains among ProP orthologues) (Wood, 2006).

Study of such variations has yielded different views of the structural mechanisms of osmosensing and osmoregulation of transporter activity for ProP, BetP, and OpuA (Wood, 2006). OpuA is currently viewed as a sensor of lumenal (and hence cytoplasmic) ionic strength (Biemans-Oldehinkel et al., 2006; Mahmood et al., 2006), BetP is viewed as a K^+ sensor (Schiller et al., 2004a,b), and ProP may detect its own hydration (Culham et al., 2003; Wood, 2006).

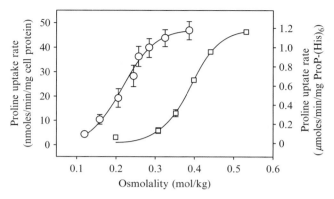

Figure 5.2 Osmotic activation profiles for ProP-$(His)_6$ in cells (○) and proteoliposomes
(□) are similar. *E. coli* was cultivated, washed, and resuspended in NaCl-free MOPS minimal medium, which has an osmolality of 0.12 mol/kg. ProP-$(His)_6$ was purified and reconstituted in proteoliposomes. The initial rate of proline uptake via ProP-$(His)_6$ (a_0) was
measured using assay media adjusted with NaCl to the indicated osmolalities and radiolabeled proline at a concentration of 200 μM. Data for cells and for proteoliposomes are
derived from Culham *et al.* (2003) and Racher *et al.* (2001), respectively. Regression lines
were obtained using Eq. (6) as model. Reproduced with permission from Culham
et al. (2003) ©2003 American Chemical Society and Racher *et al.* (2001) ©2001 American
Chemical Society.

 8. Conclusion

Multiple transporters contribute to cellular osmoregulation in all bacteria
studied thoroughly to date. The osmoregulatory transporter family will
likely continue to grow, as very large numbers of putative bacterial transporters have been identified via genomic sequencing. Does this apparent
functional redundancy merely reflect the importance of osmoregulation for
bacteria or are these transporters specialized in ways that are not yet known?
Could they promote bacterial growth in specific locations by mediating the
accumulation of osmolytes that are present only in certain environmental
niches? Osmolyte transporters can confer tolerance to stresses other than
osmotic pressure changes, including cold tolerance (Brigulla *et al.*, 2003;
Ko *et al.*, 1994; Ozcan *et al.*, 2005). Perhaps some transporters identified
experimentally as osmoregulators are designed to fulfill other physiological
roles.

Currently the full range of genetic, molecular biological, biochemical,
and biophysical techniques are being applied to transporters BetP, OpuA,
and ProP to determine which cellular properties they sense and to delineate
the structural mechanism(s) of osmosensing. This work also suggests that
osmosensing can be distinguished from osmotic adaptation, or tuning of the

absolute osmolality at which each transporter responds (Mahmood *et al.*, 2006; Tsatskis *et al.*, 2005). Insights derived from these studies may help elucidate other osmosensory mechanisms in prokaryotes and eukaryotes.

REFERENCES

Altendorf, K., and Staehelin, L. A. (1974). Orientation of membrane vesicles from *Escherichia coli* as detected by freeze-cleave electron microscopy. *J. Bacteriol.* **117,** 888–899.

Aspedon, A., Palmer, K., and Whiteley, M. (2006). Microarray analysis of the osmotic stress response in *Pseudomonas aeruginosa. J. Bacteriol.* **188,** 2721–2725.

Bhat, R., and Timasheff, S. N. (1992). Steric exclusion is the principle source of the preferential hydration of proteins in the presence of polyethylene glycols. *Protein Sci.* **1,** 1133–1143.

Biemans-Oldehinkel, E., Mahmood, N. A., and Poolman, B. (2006). A sensor for intracellular ionic strength. *Proc. Natl. Acad. Sci. USA* **103,** 10624–10629.

Biemans-Oldehinkel, E., and Poolman, B. (2003). On the role of the two extracytoplasmic substrate-binding domains in the ABC transporter OpuA. *EMBO J.* **22,** 5983–5993.

Blattner, F. R., Plunkett, G., III, Bloch, C. A., Perna, N. T., Burland, V., Riley, M., Collado-Vides, J., Glasner, J. D., Rode, C. K., Mayhew, G., Gregor, J., Davis, N. W., Kirkpatrick, H. A., Goeden, M. A., Rose, D., Mau, B., and Shao, Y. (1997). The complete genome sequence of *Escherichia coli* K-12. *Science* **277,** 1453–1462.

Bogdanov, M., Zhang, W., Xie, J., and Dowhan, W. (2005). Transmembrane protein topology mapping by the substituted cysteine accessibility method (SCAM(TM)): Application to lipid-specific membrane protein topogenesis. *Methods* **36,** 148–171.

Bolen, D. W. (2001). Protein stabilization by naturally occurring osmolytes. *Methods Mol. Biol.* **168,** 17–36.

Bornhorst, J. A., and Falke, J. J. (2007). Purification of proteins using polyhistidine affinity tags. *Methods Enzymol.* **326,** 245–254.

Bouvier, J., Bordes, P., Romeo, Y., Fourcans, A., Bouvier, I., and Gutierrez, C. (2006). Characterization of OpuA, a glycine-betaine uptake system of *Lactococcus lactis. J. Mol. Microbiol. Biotechnol.* **2,** 199–205.

Brigulla, M., Hoffmann, T., Krisp, A., Völker, A., Bremer, E., and Völker, U. (2003). Chill induction of the SigB-dependent general stress response in *Bacillus subtilis* and its contribution to low-temperature adaptation. *J. Bacteriol.* **185,** 4305–4314.

Cayley, D. S., Guttman, H. J., and Record, M. T., Jr. (2000). Biophysical characterization of changes in amounts and activity of *Escherichia coli* cell and compartment water and turgor pressure in response to osmotic stress. *Biophys. J.* **78,** 1748–1764.

Cayley, S., and Record, M. T., Jr. (2003). Roles of cytoplasmic osmolytes, water and crowding in the response of *Escherichia coli* to osmotic stress: Biophysical basis of osmoprotection by glycine betaine. *Biochemistry* **42,** 12596–12609.

Chambers, S. T., and Kunin, C. M. (1987). Isolation of glycine betaine and proline betaine from human urine: Assessment of their role as osmoprotective agents for bacteria and the kidney. *J. Clin. Invest.* **79,** 731–737.

Culham, D. E., Emmerson, K. S., Lasby, B., Mamelak, D., Steer, B. A., Gyles, C. L., Villarejo, M., and Wood, J. M. (1994). Genes encoding osmoregulatory proline/glycine betaine transporters and the proline catabolic system are present and expressed in diverse clinical *Escherichia coli* isolates. *Can. J. Microbiol.* **40,** 397–402.

Culham, D. E., Henderson, J., Crane, R. A., and Wood, J. M. (2003). Osmosensor ProP of *Escherichia coli* responds to the concentration, chemistry and molecular size of osmolytes in the proteoliposome lumen. *Biochemistry* **42,** 410–420.

Culham, D. E., Lasby, B., Marangoni, A. G., Milner, J. L., Steer, B. A., van Nues, R. W., and Wood, J. M. (1993). Isolation and sequencing of *Escherichia coli* gene *proP* reveals unusual structural features of the osmoregulatory proline/betaine transporter, ProP. *J. Mol. Biol.* **229**, 268–276.

Culham, D. E., Lu, A., Jishage, M., Krogfelt, K. A., Ishihama, A., and Wood, J. M. (2001). The osmotic stress response and virulence in pyelonephritis isolates of *Escherichia coli*: Contributions of *rpoS*, *proP*, *proU* and other systems. *Microbiology* **147**, 1657–1670.

Culham, D. E., Tripet, B., Racher, K. I., Voegele, R. T., Hodges, R. S., and Wood, J. M. (2000). The role of the carboxyl terminal a-helical coiled-coil domain in osmosensing by transporter ProP of *Escherichia coli*. *J. Mol. Recog.* **13**, 1–14.

Davidson, A. L., and Chen, J. (2004). ATP-binding cassette transporters in bacteria. *Annu. Rev. Biochem.* **73**, 241–268.

Dobrindt, U., Hentschel, U., Kaper, J. B., and Hacker, J. (2002). Genome plasticity in pathogenic and nonpathogenic enterobacteria. *Curr. Top. Microbiol. Immunol.* **264**, 157–175.

Duche, O., Tremoulet, F., Namane, A., and Labadie, J. (2006). A proteomic analysis of the salt stress response of *Listeria monocytogenes*. *FEMS Microbiol. Lett.* **215**, 183–188.

Epstein, W. (2003). The roles and regulation of potassium in bacteria. *Prog. Nucleic Acid Res. Mol. Biol.* **75**, 293–320.

Ertel, A., Marangoni, A. G., Marsh, J., Hallett, F. R., and Wood, J. M. (1993). Mechanical properties of vesicles. I. Coordinated analysis of swelling and lysis. *Biophys. J.* **64**, 426–434.

Fulda, S., Mikkat, S., Huang, F., Huckauf, J., Marin, K., Norling, B., and Hagemann, M. (2006). Proteome analysis of salt stress response in the cyanobacterium *Synechocystis* sp. strain PCC 6803. *Proteomics* **6**, 2733–2745.

Galinski, E. A., and Oren, A. (1991). Isolation and structure determination of a novel compatible solute from the moderately halophilic purple sulfur bacterium *Ectothiorhodospira marismortui*. *Eur. J. Biochem.* **198**, 593–598.

Galinski, E. A., Pfeiffer, H. P., and Trüper, H. G. (1985). 1,4,5,6-Tetrahydro-2-methyl-4-pyrimidinecarboxylic acid: A novel cyclic amino acid from halophilic phototrophic bacteria of the genus *Ectothiorhodospira*. *Eur. J. Biochem.* **149**, 135–139.

Gouesbet, G., Jebbar, M., Talibart, R., Bernard, T., and Blanco, C. (1994). Pipecolic acid is an osmoprotectant for *Escherichia coli* taken up by the general osmoporters ProU and ProP. *Microbiology* **140**, 2415–2422.

Gowrishankar, J. (1985). Identification of osmoresponsive genes in *Escherichia coli*: Evidence for participation of potassium and proline transport systems in osmoregulation. *J. Bacteriol.* **164**, 434–445.

Gramman, K., Volke, A., and Kunte, H. J. (2002). New type of osmoregulated solute transporter identified in halophilic members of the bacteria domain: TRAP transporter TeaABC mediates uptake of ectoine and hydroxyectoine in *Halomonas elongata* DSM 2581(T). *J. Bacteriol.* **184**, 3078–3085.

Gutierrez, C., Barondess, J., Manoil, C., and Beckwith, J. (1987). The use of transposon Tn*phoA* to detect genes for cell envelope proteins subjected to a common regulatory stimulus. *J. Mol. Biol.* **195**, 289–297.

Guzman, L.-M., Belin, D., Carson, M. J., and Beckwith, J. (1995). Tight regulation, modulation, and high-level expression by vectors containing the arabinose P_{BAD} promoter. *J. Bacteriol.* **177**, 4121–4130.

Han, Y., Zhou, D., Pang, X., Zhang, L., Song, Y., Tong, Z., Bao, J., Dai, E., Wang, J., Guo, Z., Zhai, J., Du, Z., Wang, X., Wang, J., Huang, P., and Yang, R. (2005). Comparative transcriptome analysis of *Yersinia pestis* in response to hyperosmotic and high-salinity stress. *Res. Microbiol.* **156**, 403–415.

Heermann, R., and Jung, K. (2004). Structural features and mechanisms for sensing high osmolarity in microorganisms. *Curr. Opin. Microbiol.* **7**, 168–174.

Hoper, D., Bernhardt, J., and Hecker, M. (2006). Salt stress adaptation of *Bacillus subtilis*: A physiological proteomics approach. *Proteomics* **6**, 1550–1562.

Horn, C., Sohn-Basser, L., Breed, J., Welte, W., Schmitt, L., and Bremer, E. (2006). Molecular determinants for substrate specificity of the ligand-binding protein OpuAC from *Bacillus subtilis* for the compatible solutes glycine betaine and proline betaine. *J. Mol. Biol.* **357**, 592–606.

Jebbar, M., Talibart, R., Gloux, K., Bernard, T., and Blanco, C. (1992). Osmoprotection of *Escherichia coli* by ectoine: Uptake and accumulation characteristics. *J. Bacteriol.* **174**, 5027–5035.

Jung, H., Tebbe, S., Schmid, R., and Jung, K. (1998). Unidirectional reconstitution and characterization of purified Na$^+$/proline transporter of *Escherichia coli*. *Biochemistry* **37**, 11083–11088.

Jung, K., and Altendorf, K. (2002). Towards an understanding of the molecular mechanisms of stimulus perception and signal transduction by the KdpD/KdpE system of *Escherichia coli*. *J. Mol. Microbiol. Biotechnol.* **4**, 223–228.

Kaback, H. R. (1971). Bacterial membranes. *Methods Enzymol.* **22**, 99–120.

Kaback, H. R. (1974). Transport in isolated bacterial membrane vesicles. *Methods Enzymol.* **31**, 698–709.

Kaneschi, Y., Suzuki, I., Allakhverdiev, S. I., Mikami, K., and Murata, N. (2002). Salt stress and hyperosmotic stress regulate the expression of different sets of genes in *Synechocystis* sp. PCC 6803. *Biochem. Biophys. Res. Commun.* **290**, 339–348.

Kappes, R. M., Kempf, B., and Bremer, E. (1996). Three transport systems for the osmoprotectant glycine betaine operate in *Bacillus subtilis*: Characterization of OpuD. *J. Bacteriol.* **178**, 5071–5079.

Kempf, B., and Bremer, E. (1995). OpuA, an osmotically regulated binding protein-dependent transport system for the osmoprotectant glycine betaine in *Bacillus subtilis*. *J. Biol. Chem.* **270**, 16701–16713.

Kempf, B., and Bremer, E. (1998). Uptake and synthesis of compatible solutes as microbial stress responses to high osmolality environments. *Arch. Microbiol.* **170**, 319–330.

Ko, R., Smith, L. T., and Smith, G. M. (1994). Glycine betaine confers enhanced osmotolerance and cryotolerance on *Listeria monocytogenes*. *J. Bacteriol.* **176**, 426–431.

Konings, W. N. (1977). Active transport of solutes in bacterial membrane vesicles. *Adv. Microb. Physiol.* **15**, 175–251.

Konings, W. N., Bisschop, A., Veenhuis, M., and Vermeulen, C. A. (1973). New procedure for the isolation of membrane vesicles of *Bacillus subtilis* and an electron microscopy study of their ultrastructure. *J. Bacteriol.* **116**, 1456–1465.

Krämer, R., and Morbach, S. (2004). BetP of *Corynebacterium glutamicum*, a transporter with three different functions: Betaine transport, osmosensing, and osmoregulation. *Biochim. Biophys. Acta* **1658**, 31–36.

Kunte, H. J., Galinski, E. A., and Trüper, H. G. (1993). A modified FMOC method for the detection of amino acid type osmolytes and tetrahydropyrimidines (ectoines). *J. Microbiol. Methods* **17**, 129–136.

Lee, E. J., Karoonuthaisiri, N., Kim, H. S., Park, J. H., Cha, C. J., Kao, C. M., and Roe, J. H. (2005). A master regulator sigmaB governs osmotic and oxidative response as well as differentiation via a network of sigma factors in *Streptomyces coelicolor*. *Mol. Microbiol.* **57**, 1252–1264.

Lentz, B. R., and Lee, J. K. (1999). Poly(ethylene glycol) (PEG)-mediated fusion between pure lipid bilayers: A mechanism in common with viral fusion and secretory vesicle release? *Mol. Membr. Biol.* **16**, 279–296.

Levina, N., Tötemeyer, S., Stokes, N. R., Louis, P., Jones, M. A., and Booth, I. R. (1999). Protection of *Escherichia coli* cells against extreme turgor by activation of MscS and MscL mechanosensitive channels: Identification of genes required for MscS activity. *EMBO J.* **18,** 1730–1737.

Ly, A., Henderson, J., Lu, A., Culham, D. E., and Wood, J. M. (2004). The osmoregulatory systems of *Escherichia coli*: Identification of BCCT family member BetU and distributions of *betU* and *trkG* among pathogenic and non-pathogenic isolates. *J. Bacteriol.* **186,** 296–306.

MacMillan, S. V., Alexander, D. A., Culham, D. E., Kunte, H. J., Marshall, E. V., Rochon, D., and Wood, J. M. (1999). The ion coupling and organic substrate specificities of osmoregulatory transporter ProP in *Escherichia coli*. *Biochim. Biophys. Acta* **1420,** 30–44.

Mahmood, N. A., Biemans-Oldehinkel, E., Patzlaff, J. S., Schuurman-Wolters, G. K., and Poolman, B. (2006). Ion specificity and ionic strength dependence of the osmoregulatory ABC transporter OpuA. *J. Biol. Chem.* **281,** 29830–29839.

Marin, K., Kanesaki, Y., Los, D. A., Murata, N., Suzuki, I., and Hagemann, M. (2004). Gene expression profiling reflects physiological processes in salt acclimation of *Synechocystis* sp. strain PCC 6803. *Plant Physiol.* **136,** 3290–3300.

Martinac, B. (2001). Mechanosensitive channels in prokaryotes. *Cell Physiol. Biochem.* **11,** 61–76.

Matsumoto, K., Kusaka, J., Nishibori, A., and Hara, H. (2006). Lipid domains in bacterial membranes. *Mol. Microbiol.* **61,** 1110–1117.

McGinnis, S., and Madden, T. L. (2004). BLAST: At the core of a powerful and diverse set of sequence analysis tools. *Nucleic Acids Res.* **32,** W20–W25.

Miller, J. H. (1972). "Experiments in Molecular Genetics." Cold Spring Harbor Laboratory, Cold Spring Harbor, NY.

Milner, J. L., McClellan, D. J., and Wood, J. M. (1987). Factors reducing and promoting the effectiveness of proline as an osmoprotectant in *Escherichia coli* K12. *J. Gen. Microbiol.* **133,** 1851–1860.

Minton, A. P. (1998). Molecular crowding: Analysis of effects of high concentrations of inert cosolutes on biochemical equilibria and rates in terms of volume exclusion. *Methods Enzymol.* **295,** 127–149.

Minton, A. P. (2006). How can biochemical reactions within cells differ from those in test tubes? *J. Cell Sci.* **119,** 2863–2869.

Morbach, S., and Krämer, R. (2002). Body shaping under water stress: Osmosensing and osmoregulation of solute transport in bacteria. *Chembiochem.* **3,** 384–397.

Motta, A., Romano, I., and Gambacorta, A. (2004). Rapid and sensitive NMR method for osmolyte determination. *J. Microbiol. Methods* **58,** 289–294.

Mui, B., Chow, L., and Hope, M. J. (2003). Extrusion technique to generate liposomes of defined size. *Methods Enzymol.* **367,** 3–14.

Neidhardt, F. C., Bloch, P. L., and Smith, D. F. (1974). Culture medium for enterobacteria. *J. Bacteriol.* **119,** 736–747.

Ozcan, N., Kramer, R., and Morbach, S. (2005). Chill activation of compatible solute transporters in *Corynebacterium glutamicum* at the level of transport activity. *J. Bacteriol.* **187,** 4752–4759.

Parsegian, V. A., Rand, R. P., and Rau, D. C. (1995). Macromolecules and water: Probing with osmotic stress. *Methods Enzymol.* **259,** 43–95.

Parsegian, V. A., Rand, R. P., and Rau, D. C. (2000). Osmotic stress, crowding, preferential hydration, and binding: A comparison of perspectives. *Proc. Natl. Acad. Sci. USA* **97,** 3987–3992.

Peddie, B. A., Lever, M., Hayman, C. M., Randall, K., and Chambers, S. T. (1994). Relationship between osmoprotection and the structure and intracellular accumulation of betaines by *Escherichia coli*. *FEMS Microbiol. Lett.* **120,** 125–132.

Peddie, B. A., Lever, M., Randall, K., and Chambers, S. T. (1999). Osmoprotective activity, urea protection, and accumulation of hydrophilic betaines in *Escherichia coli* and *Staphylococcus aureus*. *Antonie van Leeuwenhoek* **75**, 183–189.

Peddie, B. A., Wongshe, J., Randall, K., Lever, M., and Chambers, S. T. (1998). Osmoprotective properties and accumulation of betaine analogues by *Staphylococcus aureus*. *FEMS Microbiol. Lett.* **160**, 25–30.

Peter, H., Bader, A., Burkovski, A., Lambert, C., and Krämer, R. (1997). Isolation of the putP gene of *Corynebacterium glutamicum* and characterization of a low-affinity uptake system for compatible solutes. *Arch. Microbiol.* **168**, 143–151.

Peter, H., Burkovski, A., and Krämer, R. (1996). Isolation, characterization, and expression of the *Corynebacterium glutamicum* betP gene, encoding the transport system for the compatible solute glycine betaine. *J. Bacteriol.* **178**, 5229–5234.

Peter, H., Weil, B., Burkovski, A., Krämer, R., and Morbach, S. (1998). *Corynebacterium glutamicum* is equipped with four secondary carriers for compatible solutes: Identification, sequencing, and characterisation of the proline/ectoine uptake system ProP and the ectoine/proline/glycine betaine carrier EctP. *J. Bacteriol.* **180**, 6005–6012.

Pflüger, K., and Müller, V. (2004). Transport of compatible solutes in extremophiles. *J. Bioenerg. Biomembr.* **36**, 17–24.

Poolman, B., Doeven, M. K., Geertsma, E. R., Biemans-Oldehinkel, E., Konings, W. N., and Rees, D. C. (2005). Functional analysis of detergent-solubilized and membrane-reconstituted ATP-binding cassette transporters. *Methods Enzymol.* **400**, 429–459.

Poolman, B., Spitzer, J. J., and Wood, J. M. (2004). Bacterial osmosensing: Roles of membrane structure and electrostatics in lipid-protein and protein-protein interactions. *Biochim. Biophys Acta* **1666**, 88–104.

Pratt, L. A., Hsing, W. H., Gibson, K. E., and Silhavy, T. J. (1996). From acids to *osmZ*: Multiple factors influence synthesis of the OmpF and OmpC porins in *Escherichia coli*. *Mol. Microbiol.* **20**, 911–917.

Racher, K. I., Culham, D. E., and Wood, J. M. (2001). Requirements for osmosensing and osmotic activation of transporter ProP from *Escherichia coli*. *Biochemistry* **40**, 7324–7333.

Racher, K. I., Voegele, R. T., Marshall, E. V., Culham, D. E., Wood, J. M., Jung, H., Bacon, M., Cairns, M. T., Ferguson, S. M., Liang, W.-J., Henderson, P. J. F., White, G., and Hallett, F. R. (1999). Purification and reconstitution of an osmosensor: Transporter ProP of *Escherichia coli* senses and responds to osmotic shifts. *Biochemistry* **38**, 1676–1684.

Randall, K., Lever, M., Peddie, B. A., and Chambers, S. T. (1995). Competitive accumulation of betaines by *Escherichia coli* K-12 and derivative strains lacking betaine porters. *Biochim. Biophys. Acta* **1245**, 116–120.

Randall, K., Lever, M., Peddie, B. A., and Chambers, S. T. (1996). Natural and synthetic betaines counter the effects of high NaCl and urea concentrations. *Biochim. Biophys. Acta* **1291**, 189–194.

Rigaud, J. L., and Levy, D. (2003). Reconstitution of membrane proteins into liposomes. *Methods Enzymol.* **372**, 65–86.

Rigaud, J. L., Pitard, B., and Levy, D. (1995). Reconstitution of membrane proteins into proteoliposomes: Application to energy transducing membrane proteins. *Biochim. Biophys. Acta* **1231**, 223–246.

Romantsov, T., Helbig, S., Culham, D. E., Gill, C., Stalker, L., and Wood, J. M. (2007). Cardiolipin synthesis is osmoregulated and cardiolipin promotes polar localization of osmosensory transporter ProP in *Escherichia coli*. *Mol. Microbiol.* In press.

Rosen, B. P., and Tsuchiya, T. (1979). Preparation of everted membrane vesicles from *Escherichia coli* for the measurement of calcium transport. *Methods Enzymol.* **56**, 233–241.

Rottenberg, H. (1986). Energetics of proton transport and secondary transport. *Methods Enzymol.* **125**, 3–15.

Rübenhagen, R., Morbach, S., and Krämer, R. (2001). The osmoreactive betaine carrier BetP from *Corynebacterium glutamicum* is a sensor for cytoplasmic K^+. *EMBO J.* **20,** 5412–5420.

Rübenhagen, R., Roensch, H., Jung, H., Krämer, R., and Morbach, S. (2000). Osmosensor and osmoregulator properties of the betaine carrier BetP from *Corynebacterium glutamicum* in proteoliposomes. *J. Biol. Chem.* **275,** 735–741.

Saidijam, M., Bettaney, K. E., Szakonyi, G., Psakis, G., Shibayama, K., Suzuki, S., Clough, J. L., Blessie, V., Abu-bakr, A., Baumberg, S. G., Meuller, J., Hoyle, C. K., *et al.* (2005). Active membrane transport and receptor proteins from bacteria. *Biochem. Soc. Trans.* **33,** 867–872.

Schaffner, W., and Weissmann, C. (1973). Rapid, sensitive, and specific method for the determination of protein in dilute solution. *Anal. Biochem.* **56,** 502–514.

Schiller, D., Krämer, R., and Morbach, S. (2004a). Cation specificity of osmosensing by the betaine carrier BetP of *Corynebacterium glutamicum*. *FEBS Lett.* **563,** 108–112.

Schiller, D., Ott, V., Krämer, R., and Morbach, S. (2006). Influence of membrane composition on osmosensing by the betaine carrier BetP from *Corynebacterium glutamicum*. *J. Biol. Chem.* **281,** 7737–7746.

Schiller, D., Rübenhagen, R., Krämer, R., and Morbach, S. (2004b). The C-terminal domain of the betaine carrier BetP of *Corynebacterium glutamicum* is directly involved in sensing K^+ as an osmotic stimulus. *Biochemistry* **43,** 5583–5591.

Seol, W., and Shatkin, A. J. (1993). Membrane topology model of *Escherichia coli* α-ketoglutarate permease by PhoA fusion analysis. *J. Bacteriol.* **175,** 565–567.

Skerra, A., and Schmidt, T. G. (2000). Use of the Strep-Tag and streptavidin for detection and purification of recombinant proteins. *Methods Enzymol.* **326,** 271–304.

Sleator, R. D., Francis, G. A., O'Beirne, D., Gahan, C. G. M., and Hill, C. (2003). Betaine and carnitine uptake systems in *Listeria monocytogenes* affect growth and survival in foods and during infection. *J. Appl. Microbiol.* **95,** 839–846.

Sleator, R. D., Gahan, C. G. M., Abee, T., and Hill, C. (1999). Identification and disruption of BetL, a secondary glycine betaine transport system linked to the salt tolearnce of *Listeria monocytogenes* LO28. *Appl. Environ. Microbiol.* **65,** 2078–2083.

Smith, H. O., and Levine, M. (1964). Two sequential repressions of DNA synthesis in the establishment of lysogeny by phage P22 and its mutants. *Proc. Natl. Acad. Sci. USA* **52,** 356–363.

Smith, L. T., and Smith, G. M. (1989). An osmoregulated dipeptide in stressed *Rhizobium meliloti*. *J. Bacteriol.* **171,** 4714–4717.

Soppa, G. (2006). From genomes to function: Haloarchaea as model organisms. *Microbiology* **152,** 585–590.

Steil, L., Hoffmann, T., Budde, I., Volker, U., and Bremer, E. (2002). Genome-wide transcriptional profiling analysis of adaptation of *Bacillus subtilis* to high salinity. *J. Bacteriol.* **185,** 6358–6370.

Stock, J. B., Rauch, B., and Roseman, S. (1977). Periplasmic space in *Salmonella typhimurium* and *Escherichia coli*. *J. Biol. Chem.* **252,** 7850–7861.

Stults, J. T., and Arnott, D. (2005). Proteomics. *Methods Enzymol.* **402,** 245–289.

Stumpe, S., Schlösser, A., Schleyer, M., and Bakker, E. P. (1996). K^+ circulation across the prokaryotic cell membrane: K^+-uptake systems. *In* "Handbook of Biological Physics" (W. N. Konings, H. R. Kaback, and J. S. Lolkema, eds.), pp. 473–499. Elsevier Science BV, Amsterdam, The Netherlands.

Sweeney, T. E., and Beuchat, C. A. (1993). Limitations of methods of osmometry: Measuring the osmolality of biological fluids. *Am. J. Physiol.* **264,** R469–R480.

Tsatskis, Y., Khambati, J., Dobson, M., Bogdanov, M., Dowhan, W., and Wood, J. M. (2005). The osmotic activation of transporter ProP is tuned by both its C-terminal coiled-coil and osmotically induced changes in phospholipid composition. *J. Biol. Chem.* **280,** 41387–41394.

van der Heide, T., and Poolman, B. (2000). Osmoregulated ABC-transport system of *Lactococcus lactis* senses water stress via changes in the physical state of the membrane. *Proc. Natl. Acad. Sci. USA* **97,** 7102–7106.

van der Heide, T., Stuart, M. C. A., and Poolman, B. (2001). On the osmotic signal and osmosensing mechanism of an ABC transport system for glycine betaine. *EMBO J.* **20,** 7022–7032.

Viitanen, P., Newman, M. J., Foster, D. L., Wilson, T. H., and Kaback, H. R. (1986). Purification, reconstitution and characterization of the lac permease of *Escherichia coli*. *Methods Enzymol.* **125,** 429–452.

Voegele, R. T., Marshall, E. V., and Wood, J. M. (1995). Membrane permeability and transport. In "Bioenergetics: A Practical Approach" (G. C. Brown, and C. E. Cooper, eds.), pp. 17–37. Oxford University Press, Oxford.

von Blohn, C., Kempf, B., Kappes, R. M., and Bremer, E. (1997). Osmostress response in *Bacillus subtilis*: Characterization of a proline uptake system (OpuE) regulated by high osmolarity and the alternative transcription factor sigma B. *Mol. Microbiol.* **25,** 175–187.

Weber, A., and Jung, K. (2002). Profiling early osmostress-dependent gene expression in *Escherichia coli* using DNA microarrays. *J. Bacteriol.* **184,** 5502–5507.

Weber, A., Kogl, S. A., and Jung, K. (2006). Time-dependent proteome alterations under osmotic stress during aerobic and anaerobic growth in *Escherichia coli*. *J. Bacteriol.* **188,** 7165–7175.

Whipp, M. J., Camakaris, H., and Pittard, A. J. (1998). Cloning and analysis of the *shiA* gene, which encodes the shikimate transport system of *Escherichia coli* K-12. *Gene* **209,** 185–192.

White, G., Pencer, J., Nickel, B. G., Wood, J. M., and Hallett, F. R. (1996). Optical changes in unilamellar vesicles experiencing osmotic stress. *Biophys. J.* **71,** 2701–2715.

White, G. F., Racher, K. I., Lipski, A., Hallett, F. R., and Wood, J. M. (2000). Physical properties of liposomes and proteoliposomes prepared from *Escherichia coli* polar lipids. *Biochim. Biophys. Acta* **77915,** 1–12.

Wood, J. M. (1988). Proline porters effect the utilization of proline as nutrient or osmoprotectant for bacteria. *J. Membr. Biol.* **106,** 183–202.

Wood, J. M. (1999). Osmosensing by bacteria: Signals and membrane-based sensors. *Microbiol. Mol. Biol. Rev.* **63,** 230–262.

Wood, J. M. (2006). Osmosensing by bacteria. *Sci. STKE* **357,** 43.

Wood, J. M., Culham, D. E., Hillar, A., Liu, F., Vernikovska, Ya.I., Boggs, J. M., and Keates, R. A. B. (2005). Structural model for the osmosensor, transporter, and osmoregulator ProP of *Escherichia coli*. *Biochemistry* **44,** 5634–5646.

Zhang, W. T., Capp, M. W., Bond, J. P., Anderson, C. F., and Record, M. T., Jr. (1996). Thermodynamic characterization of interactions of native bovine serum albumin with highly excluded (glycine betaine) and moderately accumulated (urea) solutes by a novel application of vapor pressure osmometry. *Biochemistry* **35,** 10506–10516.

Zimmerman, S. B., and Minton, A. P. (1993). Macromolecular crowding: Biochemical, biophysical and physiological consequences. *Annu. Rev. Biophys. Biomol. Struct.* **22,** 27–65.

Zimmerman, S. B., and Trach, S. O. (1991). Estimation of macromolecule concentrations and excluded volume effects for the cytoplasm of *Escherichia coli*. *J. Mol. Biol.* **222,** 599–620.

CHAPTER SIX

Plant Gene Networks in Osmotic Stress Response: From Genes to Regulatory Networks

Lam-Son Phan Tran,* Kazuo Nakashima,* Kazuo Shinozaki,[†,‡] *and* Kazuko Yamaguchi-Shinozaki*,[‡,§]

Contents

* Biological Resources Division, Japan International Research Center for Agricultural Sciences, Tsukuba, Ibaraki, Japan
† RIKEN Plant Science Center, Tsurumi-ku, Yokohama, Japan
‡ Core Research for Evolution Science and Technology, Japan Science and Technology, Kawaguchi, Saitama, Japan
§ Laboratory of Plant Molecular Physiology, Graduate School of Agricultural and Life Sciences, The University of Tokyo, Bunkyo-ku, Tokyo, Japan

Methods in Enzymology, Volume 428
ISSN 0076-6879, DOI: 10.1016/S0076-6879(07)28006-1

Abstract

Because of their sessile nature, plants grown in a dynamic climate have evolved a range of adaptations that enable them to survive in various environmental stress conditions during growth and development. Plants respond to environmental stresses at both cellular and molecular levels by altering the expression of many genes via a complexity of signaling pathways. These pathways begin with signal perception and end with the expression of stress-responsive target genes. Ultimately, the selective upregulation of target genes leads to the alteration of physiological response so as to confer tolerance of the stress. In the signal transduction network, various regulatory and functional proteins function collectively to ensure survival of the plants. This chapter summarizes the methodology used to dissect gene regulatory networks involved in the response to osmotic stresses, such as drought and high salinity.

1. INTRODUCTION

Plants are constantly exposed to environmental stresses that frequently impose constraints on growth and productivity. Among the adverse environmental factors commonly encountered by land plants are extremes in temperature and osmotic stress, which includes drought and high salinity. The early events of plant adaptation to environmental stresses are the perception and the subsequent stress–signal transduction. This in turn leads to the activation of various physiological and metabolic responses, including stress–responsive gene expression. Research focused on this area using *Arabidopsis* as a model plant resulted in spectacular progress and a remarkable insight of defense mechanisms that the plants activate to augment tolerance to the severe stress conditions (Bray, 2004; Nakashima and Yamaguchi–Shinozaki, 2006; Shinozaki and Yamaguchi–Shinozaki, 2000;

Thomashow, 1999; Yamaguchi-Shinozaki and Shinozaki, 2005, 2006; Zhang *et al.*, 2004; Zhu, 2002).

In *Arabidopsis*, molecular and biochemical studies have suggested a model that consists of both abscisic acid (ABA)-dependent and ABA-independent regulatory pathways for gene expression in response to drought and high salinity stress. This conclusion was reached due to the observations that some stress-inducible genes respond to ABA, whereas others do not (Shinozaki and Yamaguchi-Shinozaki, 2000) (Fig. 6.1). It is well known that many of the stress-related genes can be induced by exogenous ABA; however, this does not necessarily imply that all these genes are also regulated by ABA *in vivo*. This conclusion is supported by the fact that a number of these genes were also induced by cold or drought in ABA-deficient (*aba*) or ABA-insensitive (*abi*) mutants (Ingram and Bartels, 1996; Shinozaki and Yamaguchi-Shinozaki, 2000; Shinozaki *et al.*, 2003). Several reports have demonstrated that cross talk exists between the ABA-dependent and ABA-independent regulatory pathways (Fig. 1; Haake *et al.*, 2002; Narusaka *et al.*, 2003).

The fundamental goal for plant biology is to gain a better understanding of the signaling pathways controlling the molecular and cellular mechanisms underlying stress acclimation. Once this knowledge has been obtained, it

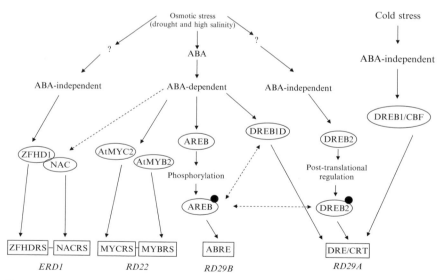

Figure 6.1 Specificity and cross talk of regulatory networks of gene expression in response to osmotic and cold stresses. *cis* elements and transcription factors (TFs) are shown in boxes and ovals, respectively. Black circles indicate the modification of TFs. Dotted arrow indicates possible regulation. Dotted double arrows indicate possible cross talk.

Identification of stress-inducible genes
Differential screening of cDNA
Homology-based PCR cloning
Microarray

Characterization of stress-inducible genes
Structural analyses and classification of genes
Studies of gene expression under various stress conditions using Northern blot
analysis or real-time PCR analysis
Studies of developmental expression Northern blot analysis or real-time PCR analysis

Identification of *cis*-acting elements in promoter region of gene of interest
Isolation of promoter region (ca. 2-2.5 kb from ATG)
  · PCR-based cloning if the genomic sequence data are available
  · Cloning of genomic DNA from genomic library if the genomic sequence
    data is not available *in silico* analysis of promoter region
  · To search for known *cis*-elements
  · To search for conserved sequence(s) if a number of genes are under investigation
Deletion and base-substitution analyses using *GUS*- or *LUC*- containing binary pBI-based
or pGreen-based vector

Identification of *tran*-acting elements (transcription factors)
Homology-based PCR cloning
Yeast one-hybrid system
DNA-ligand binding
Verification of the binding specificity both *in vivo* and *in vitro* using yeast one-hybrid
system and EMSA

Chracterization of transcription factors
Structural analyses and classification
Localization of DNA-binding domain using yeast one-hybrid system and/or EMSA
Localization of activation domain using protoplast transactivation assay
Possible interaction of transcription factors using yeast two-hybrid system and pull-down assay

Functional analysis of transcription factors *in planta*
Possible modification of transcription factors such as phophorylation
Gain-of-function study of transcription factor in plant
  · Morphological phenotype of transgenic plants overexpressing the transcription factor
  · Stress-tolerant ability of transgenic plants overexpressing the transcription factor
  · Response to ABA of transgenic plants overexpressing the transcription factor
  · Microarray
Loss-of-function study of transcription factor
  · Morphological phenotype of the mutant plant
  · Stress-tolerant ability of the mutant plant
  · Response to ABA of the mutant plant
  · Microarray

Figure 6.2 Scheme for study of regulatory networks of gene involved in stress
signaling.

is hoped that it can be used to achieve improvements in crop production.
This chapter highlights the methods used to study signaling pathways
operating in stress–affected cells and mutual interactions between these
pathways. The steps are summarized in a flow diagram as illustrated in
Fig. 6.2.

2. IDENTIFICATION OF STRESS-RESPONSIVE GENES

As the first step for studying signaling pathways, we initially used differential hybridization to screen stress-responsive genes. More than 50 independent cDNAs have been identified from *Arabidopsis*, which were confirmed to respond to osmotic and cold stresses by RNA gel blot hybridization (Shinozaki and Yamaguchi-Shinozaki, 1997, 2000). With the help of powerful microarray technology, a large array of genes—299 drought-inducible genes, 213 high salinity stress-inducible genes, and 54 cold-inducible genes—induced by the stress conditions were discovered by transcriptome analysis using a cDNA microarray containing approximately 7000 independent full-length *Arabidopsis* cDNA clones (Seki *et al.*, 2004; Shinozaki *et al.*, 2003). Upon release of the Affimetrix 22K Gene Chip, more stress-inducible genes were identified, and data are now available on numerous public Web sites, for example, TAIR (http://www.arabidopsis.org/) and Genevestigator (https://www.genevestigator.ethz.ch/).

The products of stress-inducible genes have been classified into two groups by their functions: (i) the group of functional proteins that directly protect cells from stresses by the production of important metabolic proteins and (ii) the group of regulatory proteins that regulate gene expression and signal transduction in the stress response. The first group includes enzymes required for biosynthesis of various osmoprotectants and enzymes for fatty acid metabolisms, late embryogenesis–abundant proteins, antifreeze proteins, chaperones, water channel proteins, sugar and proline transporters, detoxification enzymes, proteinase inhibitors, ferritin, and lipid-transfer proteins. The second group consists of transcription factors (TFs), protein kinases, enzymes involved in phosphoinositide metabolism, and enzymes required for the synthesis of the plant hormone abscisic acid (Nakashima and Yamaguchi-Shinozaki, 2006; Yamaguchi-Shinozaki and Shinozaki, 2006). As for the timing of the induction of stress-inducible genes, those encoding the regulatory proteins are logically induced rapidly and transiently in response to drought, high salinity, and cold stresses. Meanwhile, those encoding functional proteins accumulate slowly and gradually within 10 h subsequent to stress treatments (Fowler and Thomashow, 2002; Kreps *et al.*, 2002; Seki *et al.*, 2002; Vogel *et al.*, 2005).

2.1. Plant preparation and treatments

All of the methods that we use routinely in our laboratory have been described previously in detail in Nakashima *et al.* (1997).

2.2. RNA purification

Total RNA is extracted from plants using TRIzol reagent according to the supplier's instructions (Takara, Japan, or equivalent). For homogenizing plant material, plants and four steel balls are placed in a 2-ml Eppendorf tube and homogenized using a Shake Master version 1.2 (BMS-12). Alternatively, RNA can be extracted from plants using the ATA method. However, because ATA inhibits the activity of reverse transcriptase, the RNA cannot be used for real-time polymerase chain reaction (RT-PCR) and cDNA synthesis. For a detailed protocol of this method, see Nakashima and Yamaguchi-Shinozaki (2002).

2.3. Preparation of cDNA library and screening using differential hybridization

The standard procedure can be followed as described in Yamaguchi-Shinozaki *et al.* (1992).

2.4. *Arabidopsis* full-length cDNA microarray analysis

Total RNA is isolated from *Arabidopsis* plants using TRIzol reagent (Invitrogen, Carlsbad, CA), and 1 mg of total RNA is used for the isolation of mRNA by PolyATtract mRNA isolation systems (Promega, Madison, WI). Microarray analysis is performed as described previously using 1 μg of the isolated mRNA (Seki *et al.*, 2002). Alternatively, the detailed methods for 7K transcriptome profiling can be obtained from RIKEN URL (http://rarge.gsc.riken.jp/microarray/microarray.pl). mRNA samples from plants of interest are labeled fluorescently with Cy5-dUTP, and samples from control plants are labeled with Cy3-dUTP. λ-DNA (Takara, Tokyo, Japan) is employed as an internal control because its fluorescence level is almost the same in the two conditions. In order to assess the reproducibility of microarray analysis, each experiment should be repeated at least three times.

2.5. Agilent *Arabidopsis* 2 oligo microarray analysis

Total RNA samples are extracted from *Arabidopsis* plants using TRIzol reagent according to the supplier's instructions (Invitrogen). For each biological replicate, material from 10 plants is pooled to make a single sample for RNA purification. These isolated total RNA extracts are then used to prepare Cy5- and Cy3-labeled cDNA probes for microarray experiments using Agilent *Arabidopsis* 2 Oligo Microarray (Agilent Technologies, Inc.). All microarray experiments, including data analysis, are carried out according to the manufacturer (http://www.chem.agilent.com/script/generic.asp?lpage=11617&indcol=Nandprodcol=Y). The reproducibility of the microarray analysis can be assessed by incorporating a dye swap into the experimental plan. Under our experimental conditions, genes showing a signal value <1000 in both Cy3 and Cy5 channels of the control plants are not considered for analysis because expression of these genes is not always detected reproducibly by RNA gel blot analysis. In addition, only genes with P values <0.001 are selected for further analysis. According to published data, most genes with changes in expression of >3 are confirmed clearly and reproducibly by RNA gel blot or real-time quantitative RT-PCR analyses (Fujita *et al.*, 2004; Maruyama *et al.*, 2004; Rabbani *et al.*, 2003; Tran *et al.*, 2004, 2007). Feature extraction and image analysis software (version A.6.1.1; Agilent Technologies) is used to locate and delineate every spot in the array and to integrate the intensity, filtering, and normalization of each spot using the Lowess method. We perform gene clustering analysis with Genespring 6.1 (Silicon Genetics).

2.6. RNA gel blot analyses

We follow a method with minor modifications that was described previously in Nakashima and Yamaguchi-Shinozaki (2002). We perform prehybridization and hybridization at 65° instead of 42°.

2.7. Real-time PCR

For primer design we use ProbeFinder software available at the Universal ProbeLibrary Assay Design Center (https://www.roche-applied-science.com/servlet/RCConfigureUser?URL=StoreFramesetView&storeId=10202&catalogId=10202&langId=1&countryId=us). In our laboratory, we use a 7500 real-time PCR system (Applied Biosystems, Foster City, CA), and we conduct real-time PCR reactions according to the manufacturer's instruction using the SYBR Premix Ex Taq kit (Takara, Japan) or an equivalent.

3. IDENTIFICATION OF *CIS*-ACTING REGULATORY ELEMENTS: USE OF *GUS*, *LUC*, AND *GFP* REPORTER GENES

To understand the molecular switches of stress-inducible genes in the signal pathways, it is important to identify the *cis*-acting element located in the promoter regions of stress-responsive genes. The putative promoter region of a gene of interest, and its deleted and base-substituted fragments, can be fused to either the β-glucuronidase (*GUS*) or the firefly luciferase (*LUC*) reporter gene. For analysis of *cis*-regulatory elements involved in stress-responsive transcription in transgenic plants, the luciferase system is thought to be more sensitive, but less stable than the *GUS* system (Nakashima and Yamaguchi-Shinozaki, 2002). These constructs are then introduced to *Arabidopsis* for expression analyses, through which the areas of the promoter region important in stress-inducible gene expression can be localized. Using this strategy, we have succeeded in identifying several key *cis* elements involved in plants response to stress. For example, *GUS* reporter promoter analysis of the *RD29A/COR78/LTI78* gene, which is induced not only by osmotic stress such as drought and high salinity, but also by low-temperature stress, showed that a 9-bp conserved sequence, "TACCGA-CAT," known as the dehydration-responsive element (DRE), is an essential *cis*-acting element for regulating the *RD29A* gene of *Arabidopsis* in the ABA-independent response to dehydration and cold (Fig. 6.1) (Yamaguchi-Shinozaki and Shinozaki, 1994). The promoter region of the *ERD1* gene, which was used as a representative to study another ABA-independent pathway (Fig. 6.1), was analyzed precisely using the *LUC* reporter system (Simpson *et al.*, 2003). Results indicated that *ERD1* expression during dehydration depends on the integrity of both the 14-bp *rps1* site 1-like sequence, "CACTAAATTGTCAC," and the "CATGTG" MYC-like motif in the promoter region. Using a similar approach, we also identified *cis*-acting elements MYCRS and MYBRS in *RD22* and ABRE in *RD29B* (Abe *et al.*, 1997; Uno *et al.*, 2000). The drought induction of *RD22* and *RD29B* was mediated by ABA, which plays important roles in adapting vegetative tissues to osmotic stresses such as drought and high salinity. Thus, these two genes were used for key studies of the two ABA-dependent pathways, respectively (Fig. 6.1) (Abe *et al.*, 1997, 2003; Uno *et al.*, 2000). The list of *cis*-acting elements identified to date by our group and other groups is available in Yamaguchi-Shinozaki and Shinozaki (2005).

In addition, the promoter–*GUS* and promoter–*LUC* reporter systems can be used for the screening of mutants that showed altered signal transduction pathways in response to various stresses. This type of genetic screen represents an important advance in genetic technologies. Cao *et al.* (1994) made use of transgenic *Arabidopsis* expressing the *BGL2–GUS* construct, which provided a visible phenotype for mutants impaired in the induction of systemic-acquired resistance after pathogen attack. By using this method, they were able to successfully identify the *npr1* mutation that yielded important information regarding the key role of NPR1 in salicylic acid signaling. However, Ishitani *et al.* (1997) used transgenic *Arabidopsis* expressing a *RD29A–LUC* construct to identify a large number of mutants by monitoring alterations in the pattern of *LUC* expression relative to the wild-type plants under various abiotic stress conditions.

The jellyfish green fluorescent protein (GFP) possesses a number of desirable traits as a universal reporter, especially in living tissues. As a result, GFP has a broad range of applications as a protein marker in molecular studies. Common applications for GFP include monitoring gene expression, signal transduction, cotransfection, transformation, protein trafficking and localization, protein–protein interaction, cell separation and purification, and cell lineage in higher plants (Chiu *et al.*, 1996; Fujita *et al.*, 2004; Nakashima *et al.*, 1997; Sheen *et al.*, 1995). Expression of *GFP* is cell autonomous and independent of cell type and location. As a result, fusion proteins can be easily made to provide a fluorescent tag without disturbing the native protein function and compartmentation (Sheen *et al.*, 1995). Sakuma *et al.* (2006) have constructed a synthetic sGFP (S65T) fusion vector and have successfully studied the subcellular accumulation pattern of the DREB2A protein.

3.1. Methods for promoter analysis and screening of mutants using GUS reporter gene

For standard methods of construction of promoter–*GUS* constructs, *Arabidopsis* transformation procedures, and GUS and protein assays of transgenic plants, please refer to Nakashima and Yamaguchi-Shinozaki (2002). A detailed method describing the screening of mutants using the promoter–*GUS* system can be obtained from Cao *et al.* (1994).

3.2. Methods for promoter analysis and screening of mutants using LUC reporter gene

Descriptions for standard methods of creating promoter–*LUC* constructs, the LUC assay of transgenic plants, please refer to the protocol used for the study of *ERD1* promoter as described by Simpson *et al.* (2003). A detailed method describing the screening of mutants using the promoter–*LUC* system can be obtained from Ishitani *et al.* (1997).

3.3. Methods for application of GFP reporter gene

A number of methods of regarding the application of *GFP* gene can be followed in Sheen *et al.* (1995). For sGFP fusion vectors and methods used to study subcellular accumulation of a protein of interest, please refer to papers published previously (Fujita *et al.*, 2004; Nakashima *et al.*, 1997; Sakuma *et al.*, 2006).

3.4. *In silico* analysis

This method is very helpful if we want to analyze the promoter regions of many genes, which are induced by the same stress or hormone, for example, ABA. Comparison of promoter regions of these genes using analysis such as MEME (Multiple Em for Motif Elicitation; http://meme.sdsc.edu/meme/intro.html) may identify conserved regions that are candidates for *cis* elements. Identified *cis* elements can then be confirmed by simple base-substitution analyses using the *GUS* or *LUC* system. In turn, this approach will reduce the time and cost of promoter analysis experiments using reporter genes.

3.5. Histochemical assay

In parallel with promoter analysis experiments, the histochemical assay can be performed with stably transformed plants to study both temporal and spatial expression patterns for the gene of interest (Topping and Lindsey, 1997). A detailed protocol for histochemical assay using the *GUS* reporter gene can be obtained from Satoh *et al.* (2002).

4. IDENTIFICATION OF *TRANS*-ACTING REGULATORY ELEMENTS

Various TFs interact with *cis*-acting elements in promoter regions and form a transcriptional initiation complex on the TATA box. The transcriptional initiation complex then activates the RNA polymerase to start the transcription of genes. Several TFs involved in the regulation of stress signaling pathways have been identified and studied using a yeast one-hybrid system or DNA ligand-binding method (Fig. 6.1). cDNAs encoding DRE-binding proteins DREB1/CBF and DREB2 TFs are isolated using a yeast one-hybrid system (Liu *et al.*, 1998; Stockinger *et al.*, 1997). Expression of *DREB1* genes is induced by cold stress, but not by dehydration and high salinity (Liu *et al.*, 1998; Shinwari *et al.*, 1998). In contrast, expression of *DREB2* genes is induced by dehydration and high salinity, but not by cold stress (Liu *et al.*, 1998; Nakashima *et al.*, 2000). Among the eight DREB2-type proteins, DREB2A and DREB2B are major TFs that function in dehydration and high salinity stress responses (Nakashima *et al.*, 2000; Sakuma *et al.*, 2002). The function of DREB2A was analyzed precisely, and results indicated that the DREB2A protein requires a post-translational modification for its activation (Sakuma *et al.*, 2006). We isolated three cDNAs encoding the ANAC019, ANAC055, and ANAC072 TFs that bind to the "ANNNNNTCNNNNNNNNACACG-CATGT" NAC recognition sequence (NACRS). The NACRS contains the CATGTG MYC-like motif, which is found within the *ERD1* promoter. We also isolated a cDNA encoding the zinc finger homeodomain ZFHD1 that binds to the "CACTAAATTGTCAC" *rps1* site 1-like sequence, which is located on the *ERD1* promoter (the *rps1* site 1-like sequence was then renamed ZFHDRS). Expression of all the *ANAC* and *ZFHD1* genes is induced by dehydration and high salinity. The ANAC and ZFHD1 TFs interact together via a protein–protein interaction and are required for enhanced expression of the *ERD1* gene (Simpson *et al.*, 2003; Tran *et al.*, 2004, 2007).

ABRE-binding basic leucine zipper-type proteins, which are known as AREBs/ABFs, have been identified as transcriptional activators in the major

ABA-dependent regulatory system (Choi *et al.*, 2000; Uno *et al.*, 2000). Among the nine members of the AREB family, the function of the AREB1 was deeply elucidated, as it exhibits the highest expression to ABA and dehydration (Fujita *et al.*, 2005). All of the NACs, ZFHD1, and AREBs were isolated using the yeast one-hybrid system. In order to regulate the major ABA-dependent signaling pathway in vegetative tissues under drought, AREB1 needs to undergo an ABA-dependent phosphorylation event (Furihata *et al.*, 2006). The MYC-binding AtMYC2 transcription factor was isolated by the DNA ligand-binding method (Abe *et al.*, 1997). Both the AtMYC2 and the MYB-binding AtMYB2, which was isolated by homology-based PCR cloning, are required for the ABA-dependent induction of the *RD22* gene (Abe *et al.*, 2003). The following sections discuss the practical procedures used to isolate, characterize, and functionally study the TFs in detail.

4.1. Strategy for isolation of cDNAs encoding TFs using yeast one-hybrid system

Twenty grams of whole unstressed or stressed rosette plants grown on GM agar plates for 3 weeks is used to prepare unstressed (control) and stressed cDNA libraries. For the isolation of total RNA and poly(A)$^+$ RNA, we follow the methods described by Yamaguchi-Shinozaki *et al.* (1992). The cDNA fragments are synthesized and cDNA libraries are constructed in the pADGAL4 vector using the HybriZAP-2.1 XR library construction kit and the HybriZAP-2.1 XR cDNA synthesis kit according to the supplier's instructions (http://www.stratagene.com/manuals/235612.pdf) (Stratagene). Under these conditions, the primary libraries contain approximately 5–7 × 10^6 pfu.

To make the yeast reporter strain, we first use PCR to amplify the 50- to 75-bp fragment, which contains a *cis* element at the center and is flanked by the *Hind*III site at its 5′ and 3′ ends. This fragment is ligated into four to six tandemly repeated copies and inserted into the *Hind*III site of the pBluescript II SK- (Stratagene) vector. The fragment containing the tandem copies is then transferred to the pHISi-1 and pLacZi expression vectors (Clontech). The resulting plasmids are cut and simultaneously cointegrated into the yeast strain YM4271 with selection on SD/-Ura/-His. Yeast reporter strains are then tested for background expression of *HIS* and *LacZ* reporter genes. Reporter strains, which form white colonies on filter paper containing X-Gal and cannot grow on 3-aminotriazole (3-AT) (7 to 10 mM is lethal concentration in our cases), are used to screen the cDNA libraries. Recipes of media and methods related to yeast transformation, the testing of new reporter strains, and screening of pADGAL4 cDNA libraries are obtained from the "Yeast Protocol Handbook" (PT-3024–1), which is available on the Clontech Web site (www.clontech.com). A range of yeast transformants from approximately 1 to 5 × 10⁶ are screened according to the manufacturer's protocol (Clontech Matchmaker One-Hybrid System, Palo Alto, CA) using 40 μg of stressed and unstressed pADGAL4-cDNA libraries. Most positive colonies are obtained from the stressed library, and positive clones should grow well on selective plates containing 3-AT (10 mM or higher). A colony-lift filter assay is then performed as described in the "Yeast Protocols Handbook" (Clontech) to verify the DNA–protein interaction. The isolated positive colonies should confer β-galactosidase activity.

4.2. Verification of specific binding of TFs to *cis* elements

Currently, we verify the specific binding of TFs both *in vivo* and *in vitro*. For our *in vivo* experiment we use the yeast one-hybrid system. We mutate the DNA sequence of the *cis* element to construct the mutated yeast reporter strain using the same procedure as described previously. Subsequently, the TF should not bind to the mutated *cis* element, that is, the yeast transformants neither grow on the 3-AT containing plate nor confer β-galactosidase activity in the mutated yeast reporter strain. To validate *in vivo* data with an *in vitro* assay, we use the electrophoresis mobility shift assay (EMSA). DNA fragments encoding TFs are PCR amplified and fused to one of the pGEX-4T series vector (Amersham Biosciences, Buckinghamshire, UK). The recombinant pGEX-4T-TF plasmid is introduced into *E. coli* strain BL21, and GST fusion proteins are produced and purified according to the manufacturer's instructions. For EMSA, DNA fragments containing wild-type or mutated *cis* elements are PCR amplified and cut with *Bam*HI, which has been designed at the two ends. The fragments are then labeled by filling the 5′ overhang with [³²P]dGTP and the Klenow fragment (Takara, Japan, or equivalent) as the following: mix 1 μl of 0.5 mM dATP, 1 μl 0.5 mM of dCTP, 1 μl of 0.5 mM dTTP, 2 μl of 10× Klenow buffer, 1 μl of Klenow fragment, 2 μl of [³²P]dGTP (0.37 MBq/μl), 5 μl of DNA probe (DNA probe, 20 ng/μl, was preheated at 65° for 5 min and cooled to room temperature) and 7 μl of water. The mixture is incubated for 30 min at 37°, the unlabeled ³²P is removed using a Sepharose G-25 spin column (Sigma), and the radioactivity is estimated using a quick counter. The DNA-binding reaction is allowed to proceed for 30 min at 25° in 30 μl of binding buffer [25 mM HEPES/KOH, pH 7.9, 50 mM KCl, 0.5 mM dithiothreitol, 0.5 mM EDTA, 5% glycerol, 1 μg/μl bovine serum albumin, 0.2 μg/μl poly(dI-dC)] that contains the TF fusion protein and 20,000 dpm of the ³²P-labeled probe. The DNA–fusion protein mixture is separated on a 5% polyacrylamide, 0.25 × TBE gel and the gel is then dried at 80° for 1 h before exposing it to an imaging plate (Fuji BAS2000 system or equivalent).

5. Characterization of TFs

5.1. Transient transactivation assay

A transient transactivation assay using protoplasts derived from *Arabidopsis* T87-cultured cells is a very powerful method for characterization and functional analysis of TFs. The vector systems that we generally use in our laboratory are well described in Fujita *et al.* (2005). For protoplast transformation, GUS and LUC assays that methods that we effectively use are described in detail by Satoh *et al.* (2004).

5.2. Nature of the TF

A TF that binds to a *cis* element and activates the stress–inducible induction of a downstream gene is expected to function as a transcriptional activator. We can quickly test whether a TF is a transcriptional activator or not in yeast by using the yeast one–hybrid system or in plants by using a protoplast transactivation assay. As for the yeast system, the cDNA encoding the candidate TF was subcloned into the yeast expression vector YepGAP and the resulting plasmid was introduced into the respective yeast reporter strain that had been used for TF screening. If the TF is able to activate *HIS* and *LacZ* reporter genes, it can then be concluded that the TF functions as an activator (Liu *et al.*, 1998).

For the plant system, we coexpressed the cDNA encoding the TF and the *GUS* reporter gene in *Arabidopsis* protoplasts. The reporter gene had been placed under the control of one or several copies of the *cis* element and the minimal promoter of a gene of interest in order to confirm whether the TF is able to induce the promoter–*GUS* reporter construct in the transactivation assay. Using this experiment we can also confirm quickly whether the TF is truly involved in regulation of the downstream target gene (Liu *et al.*, 1998).

5.3. Localization of DNA-binding domain

For identifying a DNA-binding domain, both the EMSA and the yeast one–hybrid system can be used. In our laboratory, we initially perform a deletion analysis of the TF first in yeast. A series of fusions of the GAL4 activation domain and different portions of TF were designed and assayed for their ability to bind to the *cis* element and activate the *HIS3* and *lacZ* reporter genes in the yeast reporter strain used for screening of the TF. Those fragments able to activate reporter genes in yeast were then fused to the GST protein in a pGEX-4T vector for EMSA as a means to confirm the *in vivo* results. The method for analysis using EMSA has been described previously (Tran *et al.*, 2007).

5.4. Localization of activation domain

A transient transactivation assay can be used for localization of the activation domain. First, we need to construct the effector plasmids that are driven by the 35S promoter, carrying fusion genes that consist of the GAL4 DNA–binding domain and different portions of the TF. The resultant plasmids are then cotransfected into *Arabidopsis* protoplasts with a reporter plasmid, GAL4–GUS, which contains nine copies of a GAL4-binding site fused to the minimal promoter of CaMV 35S and *GUS*. The effector plasmid that is able to transactivate the reporter gene harbors the activation domain of the TF. Alternatively, the native DNA-binding domain of the TF, and the respective promoter–*GUS*, can be used instead of the GAL4 DNA-binding domain and the GAL4-binding site. Examples for the application of this method can be seen in Fujita *et al.* (2005). Although a similar system in a yeast system can also be used, it poses a problem in that the TF that activates in yeast does not always activate in plants.

5.5. Interaction of TFs

To detect the possible interaction of TFs, *in vivo* yeast two–hybrid and *in vitro* pull–down assays can be used. Tran *et al.* (2007) described the vectors in detail that were used in the yeast two–hybrid system to detect the interaction between ANAC and ZFHD1 TFs. For the *in vitro* pull–down assay, we fused the TFs in the pGEX-4T and pET-28 (Novagen, Germany) vectors. Recombinant protein purifications, pull–down reactions, and Western blotting were performed as described previously in Zhang *et al.* (2005).

6. Functional Analysis of TFs *In Planta*

6.1. Post-translational modification of TF by phosphorylation

In order to become activated, some TFs require post-translational modifications. We have observed that overexpression of DREB2A and AREB1 did not activate the expression of downstream genes (Liu *et al.*, 1998; Uno *et al.*, 2000). These peculiar data suggested that these TFs require post-translational modification for their activation. The protoplast transactivation assay can serve as an effective approach to identify the regulatory regions of TFs that require modification. By this method, Sakuma *et al.* (2006) and Fujita *et al.* (2005) have shown that the (136–165 amino acid) region of DREB2A and the QT region of AREB1 are regulatory domains of

DREB2A and AREB1, respectively. Furihata *et al.* (2006) used both a protoplast transactivation and an in–gel kinase assay to further demonstrate that the ABA-dependent phosphorylation of Ser/Thr residues in the conserved regions of AREB1 is required for the activation of AREB1. Methods for in–gel kinase assay can be obtained from Furihata *et al.* (2006).

6.2. Gain-of-function study of TFs in plants

In our next line of investigation following the characterization of isolated TFs, we usually overexpress the cDNA encoding TFs to study their *in vivo* function in plants. This ectopic overexpression method is generally used, especially when knockout mutants for loss-of-function studies are not available. Both the pBI-based (Abe *et al.*, 1997) and the pGreen-based vectors (http://www.pgreen.ac.uk/a_hom.htm) are suitable for the overexpression of genes in *Arabidopsis*. We normally confirm the expression levels of the transgene driven by a CaMV 35S promoter by RNA gel blot hybridization or real-time PCR. In general, independent transgenic lines that exhibit the highest levels of transgene expression are generally chosen for further studies. In some difficult and complicated cases, use of the constitutive 35S promoter leads to a dwarf phenotype or the generation of only a few transgenic plants (Liu *et al.*, 1998; Sakamoto *et al.*, 2004). As a means to overcome this problem, stress–inducible promoters, such as the *RD29A* promoter or a DEX-inducible promoter, can be used (Kasuga *et al.*, 1999; Sakuma *et al.*, 2006; Xie *et al.*, 2000). The application of stress–inducible promoters, therefore, also has high impact on the genetic engineering of stress-tolerant plants (Kasuga *et al.*, 1999; Sakuma *et al.*, 2006). Methods for morphological phenotyping and analyses of tolerance to stresses such as drought, high salinity, and cold have been described previously. These physiological methods include assessments of ion leakage, water loss, standardized water content, and stomatal aperture. We have also described methods to measure the contents of compatible osmolytes, such as proline and various sugars, and to also direct tests on soil-grown plants under various stress conditions (Fujita *et al.*, 2005; Ito *et al.*, 2006; Kasuga *et al.*, 1997; Kuromori *et al.*, 2006; Liu *et al.*, 1998; Nakashima *et al.*, 1997). Analysis of response to ABA using seed germination assay, root growth assay, and stomatal aperture was followed as described in Osakabe *et al.* (2005). If conditions allow, microarray analysis of the transgenic plants can be used to detect the putative target genes of the TFs.

6.3. Loss-of-function study of TFs in plants

A direct method for investigating the biological function of every gene is to assess the phenotype of loss-of-function mutants. For research in the *Arabidopsis* model system, various sources of insertional knockout mutants are readily available. For example, these materials can be accessed by the public at the following Web addresses: http://www.arabidopsis.org/abrc/ or http://www.brc.riken.jp/inf/en/. Loss-of-function mutants, from insertional mutagenesis or transposable elements, have provided a tremendous amount of information regarding the role of some genes. Despite these advance, however, the assignment of function to a large portion of genes remains unknown. If knockout mutant lines are not available, the accumulation of mRNA corresponding to genes of interest may also be selectively suppressed by antisense or RNA interference (RNAi) techniques. As an example for the suppression of a target gene using the antisense technique, please refer to Osakabe *et al.* (2002). For RNAi, we are currently using the vector system developed by CSIRO (Wesley *et al.*, 2001). Information regarding the vector system can be accessed at http://www.pi.csiro.au/ tech_licensing_biol/genesilencingvectors.htm. For researchers who are using rice as their model system, another RNAi vector system has been developed by Miki and Shimamoto (2004) and can effectively provide transient suppression of gene function in rice. Detailed information about this vector system and its application can be obtained from the same reference.

A new method has been developed for loss-of-function studies (Hiratsu *et al.*, 2003). This method, which is termed the dominant repression technique, was developed to overcome the functional compensation that usually occurs when using insertion mutant lines due to redundancy of homologous genes (Zhang, 2003). Hiratsu *et al.* (2003) showed that the expression of specific target genes was suppressed dominantly by translational fusions between several TFs and a repression domain (RD) derived from the EAR motif of SUPERMAN, a TFIIIA-type zinc finger repressor. This dominant suppression occurred even in the presence of redundant TFs. Our group successfully applied this method to suppress the function of RD26 and AREB1 TFs (Fujita *et al.*, 2004, 2005). Detailed information, including plasmid construction methods, can be obtained from Fujita *et al.* (2005). The same approaches described earlier were used to assess the phenotype of knockout or suppressed plants.

ACKNOWLEDGMENTS

This work was supported in part by the Program for Promotion of Basic Research Activities for Innovative Biosciences (BRAIN). It was also supported in part by a project grant from the Ministry of Agriculture, Forestry and Fisheries, Japan. L.-S. P. Tran was supported by a JIRCAS fellowship under the JIRCAS Visiting Research Fellowship Program at Tsukuba.

REFERENCES

Abe, H., Yamaguchi-Shinozaki, K., Urao, T., Iwasaki, T., Hosokawa, D., and Shinozaki, K. (1997). Role of *Arabidopsis* MYC and MYB homologs in drought- and abscisic acid-regulated gene expression. *Plant Cell* **9**, 1859–1868.

Abe, H., Urao, T., Ito, T., Seki, M., Shinozaki, K., and Yamaguchi-Shinozaki, K. (2003). *Arabidopsis* AtMYC2 (bHLH) and AtMYB2 (MYB) function as transcriptional activators in abscisic acid signaling. *Plant Cell* **15**, 63–78.

Bray, E. A. (2004). Genes commonly regulated by water-deficit stress in *Arabidopsis* thaliana. *J. Exp. Bot.* **55**, 2331–2341.

Cao, H., Bowling, S. A., Gordon, S., and Dong, X. (1994). Characterization of an *Arabidopsis* mutant that is nonresponsive in to inducers of systemic acquired resistance. *Plant Cell* **6**, 1583–1592.

Chiu, W.-L, Niwa, Y., Zeng, W., Hirano, T., Kobayashi, H., and Sheen, J. (1996). Engineered GFP as a vital reporter in plants. *Curr. Biol.* **6**, 325–330.

Choi, H., Hong, J., Ha, J., Kang, J., and Kim, S. Y. (2000). ABFs, a family of ABA-responsive element binding factors. *J. Biol. Chem.* **275**, 1723–1730.

Fowler, S., and Thomashow, M. F. (2002). *Arabidopsis* transcriptome profiling indicates that multiple regulatory pathways are activated during cold acclimation in addition to the CBF cold response pathway. *Plant Cell* **14**, 1675–1690.

Fujita, M., Fujita, Y., Maruyama, K., Seki, M., Hiratsu, K., Ohme-Takagi, M., Tran, L.-S. P., Yamaguchi-Shinozaki, K., and Shinozaki, K. (2004). A dehydration-induced NAC protein, RD26, is involved in a novel ABA-dependent stress-signaling pathway. *Plant J.* **39**, 863–876.

Fujita, Y., Fujita, M., Satoh, R., Maruyama, K., Parvez, M. M., Seki, M., Hiratsu, K., Ohme-Takagi, M., Shinozaki, K., and Yamaguchi-Shinozaki, K. (2005). AREB1 is a transcription activator of novel ABRE-dependent ABA signaling that enhances drought stress tolerance in *Arabidopsis*. *Plant Cell* **17**, 3470–3488.

Furihata, T., Maruyama, K., Fujita, Y., Umezawa, T., Yoshida, R., Shinozaki, K., and Yamaguchi-Shinozaki, K. (2006). ABA-dependent multisite phosphorylation regulates the activity of a transcription activator AREB1. *Proc. Natl. Acad. Sci. USA* **103**, 1988–1993.

Haake, V., Cook, D., Riechmann, J. L., Pineda, O., Thomashow, M. F., and Zhang, J. Z. (2002). Transcription factor CBF4 is a regulator of drought adaptation in *Arabidopsis*. *Plant Physiol.* **130**, 639–648.

Hiratsu, K., Matsui, K., Koyama, T., and Ohme-Takagi, M. (2003). Dominant repression of target genes by chimeric repressors that include the EAR motif, a repression domain, in *Arabidopsis*. *Plant J.* **34**, 733–739.

Ingram, J., and Bartels, D. (1996). The molecular basis of dehydration tolerance in plants. *Annu. Rev. Plant Physiol. Plant Mol. Biol.* **47**, 377–403.

Ishitani, M., Xiong, L., Stevenson, B., and Zhu, J. K. (1997). Genetic analysis of osmotic and cold stress signal transduction in *Arabidopsis*: Interactions and convergence of abscisic acid-dependent and abscisic acid-independent pathways. *Plant Cell* **9**, 1935–1949.

Ito, Y., Katsura, K., Maruyama, K., Taji, T., Kobayashi, M., Seki, M., Shinozaki, K., and Yamaguchi-Shinozaki, K. (2006). Functional analysis of rice DREB1/CBF-type transcription factors involved in cold-responsive gene expression in transgenic rice. *Plant Cell Physiol.* **47**, 141–153.

Kasuga, M., Liu, Q., Miura, S., Yamaguchi-Shinozaki, K., and Shinozaki, K. (1999). Improving plant drought, salt, and freezing tolerance by gene transfer of a single stress-inducible transcription factor. *Nat. Biotechnol.* **17**, 287–291.

Kreps, J. A., Wu, Y., Chang, H. S., Zhu, T., Wang, X., and Harper, J. F. (2002). Transcriptome changes for *Arabidopsis* in response to salt, osmotic, and cold stress. *Plant Physiol.* **130,** 2129–2141.

Kuromori, T., Wada, T., Kamiya, A., Yuguchi, M., Yokouchi, T., Imura, Y., Takabe, H., Sakurai, T., Akiyama, K., Hirayama, T., Okada, K., and Shinozaki, K. (2006). A trial of phenome analysis using 4000 Ds-insertional mutants in gene-coding regions of *Arabidopsis. Plant J.* **47,** 640–651.

Liu, Q., Kasuga, M., Sakuma, Y., Abe, H., Miura, S., Yamaguchi-Shinozaki, K., and Shinozaki, K. (1998). Two transcription factors, DREB1 and DREB2, with an EREBP/AP2 DNA binding domain separate two cellular signal transduction pathways in drought- and low-temperature-responsive gene expression, respectively, in *Arabidopsis. Plant Cell* **10,** 1391–1406.

Maruyama, K., Sakuma, Y., Kasuga, M., Ito, Y., Seki, M., Goda, H., Shimada, Y., Yoshida, S., Shinozaki, K., and Yamaguchi-Shinozaki, K. (2004). Identification of cold-inducible downstream genes of the *Arabidopsis* DREB1A/CBF3 transcriptional factor using two microarray systems. *Plant J.* **38,** 982–993.

Miki, D., and Shimamoto, K. (2004). Simple RNAi vectors for stable and transient suppression of gene function in rice. *Plant Cell Physiol.* **45,** 490–495.

Nakashima, K., Kiyosue, T., Yamaguchi-Shinozaki, K., and Shinozaki, K. (1997). A nuclear gene encoding a chloroplast-targeted Clp protease regulatory subunit homolog is not only induced by water stress but also developmentally up-regulated during senescence in *Arabidopsis* thaliana. *Plant J.* **12,** 851–861.

Nakashima, K., Shinwari, Z. K., Sakuma, Y., Seki, M., Miura, S., Shinozaki, K., and Yamaguchi-Shinozaki, K. (2000). Organization and expression of two *Arabidopsis* DREB2 genes encoding DRE-binding proteins involved in dehydration- and high-salinity-responsive gene expression. *Plant Mol. Biol.* **42,** 657–665.

Nakashima, K., and Yamaguchi-Shinozaki, K. (2002). Use of β-glucuronidase to show dehydration and high-salt gene expression. *In* "Molecular Methods of Plant Analysis: Testing for Genetic Manipulation in Plants" (J. F. Jackson, H. F. Linsken, and R. B. Inman, eds.), Vol. 22, pp. 37–61. Springer-Verlag, Berlin.

Nakashima, K., and Yamaguchi-Shinozaki, K. (2006). Regulons involved in osmotic stress-responsive and cold stress-responsive gene expression in plants. *Physiol. Plant.* **126,** 62–71.

Narusaka, Y., Nakashima, K., Shinwari, Z. K., Sakuma, Y., Furihata, T., Abe, H., Narusaka, M., Shinozaki, K., and Yamaguchi-Shinozaki, K. (2003). Interaction between two *cis*-acting elements, ABRE and DRE, in ABA-dependent expression of *Arabidopsis* rd29A gene in response to dehydration and high-salinity stresses. *Plant J.* **34,** 137–148.

Osakabe, Y., Maruyama, K., Seki, M., Satou, M., Shinozaki, K., and Yamaguchi-Shinozaki, K. (2005). Leucine-rich repeat receptor-like kinase1 is a key membrane-bound regulator of abscisic acid early signaling in *Arabidopsis. Plant Cell* **17,** 1105–1119.

Osakabe, Y., Miyata, S., Urao, T., Seki, M., Shinozaki, K., and Yamaguchi-Shinozaki, K. (2002). Overexpression of *Arabidopsis* response regulators, ARR4/ATRR1/IBC7 and ARR8/ATRR3, alters cytokinin responses differentially in the shoot and in callus formation. *Biochem. Biophys. Res. Commun.* **293,** 806–815.

Rabbani, M. A., Maruyama, K., Abe, H., Khan, M. A., Katsura, K., Ito, Y., Yoshiwara, K., Seki, M., Shinozaki, K., and Yamaguchi-Shinozaki, K. (2003). Monitoring expression profiles of rice genes under cold, drought, and high-salinity stresses and abscisic acid application using cDNA microarray and RNA gel-blot analyses. *Plant Physiol.* **133,** 1755–1767.

Sakamoto, H., Maruyama, K., Sakuma, Y., Meshi, T., Iwabuchi, M., Shinozaki, K., and Yamaguchi-Shinozaki, K. (2004). *Arabidopsis Cys2/His2-type* zinc-finger proteins function as transcription repressors under drought-, cold-, and high-salinity-stress conditions. *Plant Physiol.* **136,** 2734–2746.

Sakuma, Y., Liu, Q., Dubouzet, J. G., Abe, H., Shinozaki, K., and Yamaguchi-Shinozaki, K. (2002). DNA-binding specificity of the ERF/AP2 domain of *Arabidopsis* DREBs, transcription factors involved in dehydration- and cold-inducible gene expression. *Biochem. Biophys. Res. Commun.* **290,** 998–1009.

Sakuma, Y., Maruyama, K., Osakabe, Y., Qin, F., Seki, M., Shinozaki, K., and Yamaguchi-Shinozaki, K. (2006). Functional analysis of an *Arabidopsis* transcription factor, DREB2A, involved in drought-responsive gene expression. *Plant Cell* **18,** 1292–1309.

Satoh, R., Fujita, Y., Nakashima, K., Shinozaki, K., and Yamaguchi-Shinozaki, K. (2004). A novel subgroup of bZIP proteins function as transcriptional activators in hypoosmolarity-responsive expression of the *ProDH* gene in *Arabidopsis*. *Plant Cell Physiol.* **45,** 309–317.

Satoh, R., Nakashima, K., Seki, M., Shinozaki, K., and Yamaguchi-Shinozaki, K. (2002). ACTCAT, a novel cis-acting element for proline- and hypoosmolarity-responsive expression of the *ProDH* gene encoding proline dehydrogenase in *Arabidopsis*. *Plant Physiol.* **130,** 709–719.

Seki, M., Narusaka, M., Ishida, J., Nanjo, T., Fujita, M., Oono, Y., Kamiya, A., Nakajima, M., Enju, A., Sakurai, T., Satou, M., Akiyama, K., *et al.* (2002). Monitoring the expression profiles of 7000 *Arabidopsis* genes under drought, cold and high-salinity stresses using a full-length cDNA microarray. *Plant J.* **31,** 279–292.

Seki, M., Satou, M., Sakurai, T., Akiyama, K., Iida, K., Ishida, J., Nakajima, M., Enju, A., Narusaka, M., Fujita, M., Oono, Y., Kamei, A., Yamaguchi-Shinozaki, K., and Shinozaki, K. (2004). RIKEN *Arabidopsis* full-length (RAFL) cDNA and its applications for expression profiling under abiotic stress conditions. *J. Exp. Bot.* **55,** 213–223.

Sheen, J., Hwang, S., Niwa, Y., Kobayashi, H., and Galbraith, D. W. (1995). Green-fluorescent protein as a new vital marker in plant cells. *Plant J.* **8,** 777–784.

Shinozaki, K., and Yamaguchi-Shinozaki, K. (1997). Gene expression and signal transduction in water-stress response. *Plant Physiol.* **115,** 327–334.

Shinozaki, K., and Yamaguchi-Shinozaki, K. (2000). Molecular responses to dehydration and low temperature: Differences and cross-talk between two stress signaling pathways. *Curr. Opin. Plant Biol.* **3,** 217–223.

Shinozaki, K., Yamaguchi-Shinozaki, K., and Seki, M. (2003). Regulatory network of gene expression in the drought and cold stress responses. *Curr. Opin. Plant Biol.* **6,** 410–417.

Shinwari, Z. K., Nakashima, K., Miura, S., Seki, M., Yamaguchi-Shinozaki, K., and Shinozaki, K. (1998). An *Arabidopsis* gene family encoding DRE/CRT binding proteins involved in low-temperature-responsive gene expression. *Biochem. Biophys. Res. Commun.* **250,** 161–170.

Simpson, S. D., Nakashima, K., Narusaka, Y., Seki, M., Shinozaki, K., and Yamaguchi-Shinozaki, K. (2003). Two different novel *cis*-acting elements of *erd1*, a *clpA* homologous *Arabidopsis* gene function in induction by dehydration stress and dark-induced senescence. *Plant J.* **33,** 259–270.

Stockinger, E. J., Gilmour, S. J., and Thomashow, M. F. (1997). *Arabidopsis* thaliana CBF1 encodes an AP2 domain-containing transcription activator that binds to the C-repeat/DRE, a cis-acting DNA regulatory element that stimulates transcription in response to low temperature and water deficit. *Proc. Natl. Acad. Sci. USA* **94,** 1035–1040.

Thomashow, M. F. (1999). Plant cold acclimation: Freezing tolerance genes and regulatory mechanisms. *Annu. Rev. Plant Physiol. Plant Mol. Biol.* **50,** 571–599.

Topping, J. F., and Lindsey, K. (1997). Molecular characterization of transformed plants. *In* "Plant Molecular Biology: A Laboratory Manual" (M. S. Clark, ed.), pp. 427–442. Springer, Berlin.

Tran, L.-S. P., Nakashima, K., Sakuma, Y., Osakabe, Y., Qin, F., Simpson, S. D., Maruyama, K., Fujita, Y., Shinozaki, K., and Yamaguchi-Shinozaki, K. (2007). Co-expression of the stress-inducible zinc finger homeodomain ZFHD1 and NAC transcription factors enhances expression of the *ERD1* gene in *Arabidopsis. Plant J.* **49,** 46–63.

Tran, L.-S. P., Nakashima, K., Sakuma, Y., Simpson, S. D., Fujita, Y., Maruyama, K., Fujita, M., Seki, M., Shinozaki, K., and Yamaguchi-Shinozaki, K. (2004). Isolation and functional analysis of *Arabidopsis* stress-inducible NAC transcription factors that bind to a drought-responsive *cis*-element in the early responsive to dehydration stress 1 promoter. *Plant Cell* **16,** 2481–2498.

Uno, Y., Furihata, T., Abe, H., Yoshida, R., Shinozaki, K., and Yamaguchi-Shinozaki, K. (2000). *Arabidopsis* basic leucine zipper transcription factors involved in an abscisic acid-dependent signal transduction pathway under drought and high-salinity conditions. *Proc. Natl. Acad. Sci. USA* **97,** 11632–11637.

Vogel, J. T., Zarka, D. G., Van Buskirk, H. A., Fowler, S. G., and Thomashow, M. F. (2005). Roles of the CBF2 and ZAT12 transcription factors in configuring the low temperature transcriptome of *Arabidopsis. Plant J.* **41,** 195–211.

Wesley, S. V., Helliwell, C. A., Smith, N. A., Wang, M. B., Rouse, D. T., Liu, Q., Gooding, P. S., Singh, S. P., Abbott, D., Stoutjesdijk, P. A., Robinson, S. P., Gleave, A. P., *et al.* (2001). Construct design for efficient, effective and high-throughput gene silencing in plants. *Plant J.* **27,** 581–590.

Xie, Q., Frugis, G., Colgan, D., and Chua, N. H. (2000). *Arabidopsis* NAC1 transduces auxin signal downstream of TIR1 to promote lateral root development. *Genes Dev.* **14,** 3024–3036.

Yamaguchi-Shinozaki, K., Koizumi, M., Urao, S., and Shinozaki, K. (1992). Molecular cloning and characterization of 9 cDNAs for genes that are responsive to desiccation in *Arabidopsis* thaliana: Sequence analysis of one cDNA clone that encodes a putative transmembrane channel protein. *Plant Cell Physiol.* **33,** 217–224.

Yamaguchi-Shinozaki, K., and Shinozaki, K. (1994). A novel *cis*-acting element in an *Arabidopsis* gene is involved in responsiveness to drought, low-temperature, or high-salt stress. *Plant Cell* **6,** 251–264.

Yamaguchi-Shinozaki, K., and Shinozaki, K. (2005). Organization of *cis*-acting regulatory elements in osmotic- and cold-stress-responsive promoters. *Trends Plant Sci.* **10,** 88–94.

Yamaguchi-Shinozaki, K., and Shinozaki, K. (2006). Transcriptional regulatory networks in cellular responses and tolerance to dehydration and cold stresses. *Annu. Rev. Plant Biol.* **57,** 781–803.

Zhang, J. Z. (2003). Overexpression analysis of plant transcription factors. *Curr. Opin. Plant Biol.* **6,** 430–440.

Zhang, J. Z., Creelman, R. A., and Zhu, J. K. (2004). From laboratory to field: Using information from *Arabidopsis* to engineer salt, cold, and drought tolerance in crops. *Plant Physiol.* **135,** 615–621.

Zhang, X., Garreton, V., and Chua, N. H. (2005). The AIP2 E3 ligase acts as a novel negative regulator of ABA signaling by promoting ABI3 degradation. *Genes Dev.* **19,** 1532–1543.

Zhu, J. K. (2002). Salt and drought stress signal transduction in plants. *Annu. Rev. Plant. Physiol. Plant Mol. Biol.* **53,** 247–273.

CHAPTER SEVEN

OSMOSENSING BY INTEGRINS IN RAT LIVER

Freimut Schliess *and* Dieter Häussinger

Contents

Abstract

Changes in hepatocyte hydration are induced not only by ambient hypo- or hyperosmolarity, but also under isosmotic condition by hormones, substrates, and oxidative stress. The perfused rat liver is a well-established intact organ model with preservation of the three-dimensional hepatocyte anchoring to the extracellular matrix and/or adjacent cells, parenchymal cell polarity, liver cell heterogeneity, acinar construction, and gene expression gradients. Originally, data from the perfused rat liver indicated that changes of cell hydration independent of their origin critically contribute to the control of autophagic proteolysis and canalicular bile acid excretion. Meanwhile, the concept that cell hydration changes trigger signal transduction processes that control metabolism, gene expression, transport, and the susceptibility to stress is well accepted. This chapter summarizes evidence obtained from experiments with the perfused rat liver that integrins are osmosensors in the liver and thereby

Clinic for Gastroenterology, Hepatology, and Infectiology, Heinrich-Heine-University, Düsseldorf, Germany

Methods in Enzymology, Volume 428
ISSN 0076-6879, DOI: 10.1016/S0076-6879(07)28007-3

critically contribute to the Src- and MAP-kinase-dependent inhibition of auto-phagic proteolysis, stimulation of canalicular taurocholate excretion, and regulatory volume decrease as induced by hypoosmotic swelling. Moreover, integrin-dependent sensing of hepatocyte swelling is essential for signaling and proteolysis inhibition by insulin and glutamine. These findings define a novel role of integrins in insulin and glutamine signaling and set an example for mechanotransduction as an integral part of overall growth factor and nutrient signaling.

1. INTEGRINS IN MECHANOTRANSDUCTION AND GROWTH FACTOR SIGNALING

Anchoring of cells to the extracellular matrix (ECM) is a critical condition not only for three-dimensional (3D) tissue organization, but also to ensure the entire response to growth factors, nutrients, and mechanical forces. Integrins comprise a family of transmembrane glycoproteins that bind to ECM proteins or to counterreceptors presented by adjacent cells. Two integrin subunits (α and β type, respectively) form heterodimers with each subunit having a large extracellular domain, a single transmembrane passage, and usually a short cytoplasmic domain (for detailed reviews, see Aplin *et al.*, 1998; Cukierman *et al.*, 2002; Giancotti and Ruoslahti, 1999; Hynes, 2002; Ingber, 2003; Stupack, 2005; Vogel, 2006). Integrins expressed by liver cells include $\alpha_1\beta_1$, $\alpha_5\beta_1$, and $\alpha_9\beta_1$ (Carloni *et al.*, 1996, 2001; Hsu *et al.*, 2001; Torimura *et al.*, 2001).

With respect to ligand binding, low- and high-affinity states of integrins exist, and the switch to high affinity can be induced by either extracellular triggers (binding of divalent cations, ligands, or antibodies) or in response to signals produced within the cell. Ligand binding to $\alpha_5\beta_1$ and many other integrins depends on presentation of the RGD motif by the ligand, although other motifs and surrounding sequences may also be crucial (Ruoslahti, 1996). The cytoplasmic domains bind proteins associated to the cytoskeleton and are connected to signaling proteins frequently organized within focal adhesions. Integrin-dependent cell adhesion is followed by activation of the focal adhesion-dependent kinase and Src-type kinases, which in turn are involved in the activation of signaling pathways, for example, toward MAP-kinase activation (outside-in signaling).

There is growing evidence that mechanical forces, as produced by shear stress, gravity, and osmotic stress, critically affect cell biology. According to the tensegrity concept (Ingber, 1997, 2006), because of an isometric tension, changes in cellular geometry or size alter the balance of forces within the cellular cytoskeletal network and by this way modify the activity and toponymy of signaling proteins. Apart from actin microfilaments, microtubules contribute significantly to the cytoskeletal prestress (Wang et al., 2001).

Integrin-mediated adhesion to the ECM is crucial in mechanotransduction. Thus, binding of specific ECM ligands to integrins essentially contributes to signal transduction as induced by shear stress in vascular endothelial cells (Jalali et al., 2001; Li et al., 2005). Further, integrins are involved in the stimulation of glutamine uptake and glycogen synthesis in cultured rat skeletal muscle cells as induced by hypoosmotic swelling (Low and Taylor, 1998; Low et al., 1997). In kidney cells, hyperosmotic shrinkage increases expression of the β_1 integrin subunit and the tetraspan interaction partner CD9 (Sheikh-Hamad et al., 1997). The absence of $\alpha_1\beta_1$ integrin in kidney cells impairs hyperosmotic activation of the tonicity enhancer-binding protein (TonEBP) and consequently prevents osmolyte transporter expression and uptake of protective organic osmolytes (Moeckel et al., 2006).

A complex interplay exists between integrin- and growth factor receptor-mediated signal transduction (Cabodi et al., 2004; Chan et al., 2006; Schwartz and Ginsberg, 2002). The engagement of integrins by cell adhesion may transactivate growth factor receptors. For example, integrin-mediated epidermal growth factor (EGF) receptor activation increases MAP-kinase activity and cell survival in human skin fibroblasts and ECV304 endothelial cells in response to adhesion (Moro et al., 1998). Further, β_1 integrin activation by stretch via EGF receptor activation triggers an outwardly rectifying chloride current (Browe and Baumgarten, 2006), which can be mimicked by EGF and hypoosmolarity, respectively (Varela et al., 2004). Similar to adhesion and shear stress, anisosmotic swelling or shrinkage can activate growth factor receptors in a ligand-independent manner (Franco et al., 2004; Rosette and Karin, 1996). In cultured rat hepatocytes, moderate hyperosmolarity increased EGF receptor tyrosine phosphorylation in a NADPH oxidase-dependent manner (Reinehr et al., 2006). In this case the activated EGF receptor tyrosine phosphorylates and recruits CD95, which is then targeted to the plasma membrane. There, the death-inducing signaling complex is formed, which leaves the hepatocytes in a proapoptotic state (Reinehr et al., 2002, 2004a,b, 2007). Whether integrins are involved in the anisosmotic activation of growth factor and cytokine receptors is currently unknown.

Integrins may also modify growth factor signaling. Thus, shear stress synergizes with insulin-like growth factor (IGF)-1-induced signaling by an integrin-dependent recruitment of the tyrosine phosphatases SHP-1 and -2 away from the IGF-1 receptor, thereby increasing MAP-kinase activation and proliferation of osteosarcoma cells (Kapur et al., 2005). However, integrin-mediated activation of a T-cell protein tyrosine phosphatase was reported to counteract EGF-induced EGF receptor activation (Mattila *et al.*, 2005).

Integrin signaling may also localize downstream of ligand-dependent growth factor receptor activation. One example is the decrease of $\alpha_5\beta_1$-mediated cell adhesion by IGF-1, which increases breast cancer cell motility (Lynch *et al.*, 2005). Similarly, glutamine decreases integrin-dependent adhesion of human colon carcinoma cell lines to the ECM, which increases proliferation and dedifferentiation of these cells (Turowski *et al.*, 1994).

Short synthetic peptides containing the RGD motif are well-established tools used to address the integrin dependence of cellular responses. When immobilized to a surface they promote cell attachment and when administrated in solution they prevent attachment that would otherwise occur (Ruoslahti, 1996). Studies on shear-stressed cells indicated that the dynamic connection and disconnection of integrin–ligand associations depend on the presence of unoccupied ligand-binding sites and are essential for integrin-dependent mechanotransduction (Jalali *et al.*, 2001). By occupying free ligand-binding sites, RGD peptides may interfere with integrin activation.

2. CELL HYDRATION AND CELL FUNCTION

Already moderate and well-tolerated cell volume changes are induced within minutes by changes of ambient osmolarity, but also by hormones, amino acids, second messengers, and oxidative stress. Cell swelling due to a net uptake of inorganic and organic osmolytes is obligatory during the cell cycle, whereas cell shrinkage as a result of osmolyte release may play a role in apoptosis (apoptotic volume decrease) (Lang *et al.*, 1998; Rozengurt, 1986; Schliess and Häussinger, 2005). Cell volume changes are registered by osmo-sensing structures, thereby activating signals that contribute to the control and regulation of metabolism and gene expression (Burg *et al.*, 1997; Häussinger, 1996; Häussinger and Schliess, 1999; McManus *et al.*, 1995). Anisosmotically exposed cells and tissues are a widely used experimental paradigm for the study of osmosensing and the osmosensitivity of signal transduction, metabolic pathways, and gene expression. Regarding osmosensing processes and structures, macromolecular crowding, stretch-activated ion channels, the cytoskeleton, intracellular organelles, and autocrine stimulation of signal transduction by release of mediators such as ATP were considered (Hamill and Martinac, 2001; Hoffmann, 2000; Loomis *et al.*, 1997; Wang *et al.*, 1996). For example, hepatocyte dehydration is sensed by endosomes, which acidify rapidly and produce an acid sphingomyelinase-, ceramide-, and PKCζ-dependent

NAPDH oxidase-catalyzed generation of reactive oxygen species as signaling metabolites (Reinehr et al., 2006).

Liver cell hydration-dependent pathways and functions have been reviewed extensively elsewhere (Häussinger, 1996; Häussinger and Lang, 1991; Lang et al., 1998; Schliess and Häussinger, 2000, 2005). Changes in cell hydration within a narrow, physiological range markedly affect hepatic carbohydrate and protein metabolism, as well as bile flow and pH regulation. A multitude of cellular signal transduction components, which are affected by anisosmotic cell swelling or shrinkage, has been characterized and it is meanwhile well established that "osmosignaling" is integrated into the overall context of hormone- and nutrient-induced signal transduction (Häussinger and Schliess, 1999). Hypoosmotic perfusion of rat liver rapidly activates mitogen-activated protein kinases of the p38- and extracellular signal-activated protein kinase (Erk) type, but not c-Jun-N-terminal kinases (Häussinger et al., 1999; Kurz et al., 2001). As has been shown, inhibition of autophagic proteolysis and regulatory volume decrease in response to hepatocyte swelling depends on the p38 signal, whereas stimulation of bile acid secretion requires both Erk- and p38-dependent signaling (Häussinger et al., 1999; Kurz et al., 2001)(Fig. 7.1).

3. THE PERFUSED RAT LIVER: AN INTACT ORGAN MODEL OF THE LIVER

The perfused rat liver is an intact organ model that, compared to cultured hepatocytes, preserves tissue structure, 3D hepatocyte anchoring to the ECM, polarity of the parenchymal cells, liver cell heterogeneity, acinar construction, and gene expression gradients along the acinus (Sies, 1978). The perfused rat liver is a well-established experimental system used to study hepatic signal transduction, gene expression, metabolism, hepatobiliary transport, and detoxification. A detailed description of the experimental setup is presented elsewhere (vom Dahl and Häussinger, 1997). This section briefly outlines methods used for the investigation of integrin-dependent osmosensing in rat liver.

3.1. Measurement of cell hydration

The intracellular water space in perfused rat liver is calculated from the difference of washout profiles of simultaneously infused [^{14}C]urea (has access to extra- and intracellular water spaces) and [^{3}H]inulin (has access only to the extracellular water space) as described elsewhere (vom Dahl et al., 1991a). Following equilibration of both tracers within the intracellular and extracellular compartments, tracer infusion is stopped and the effluent

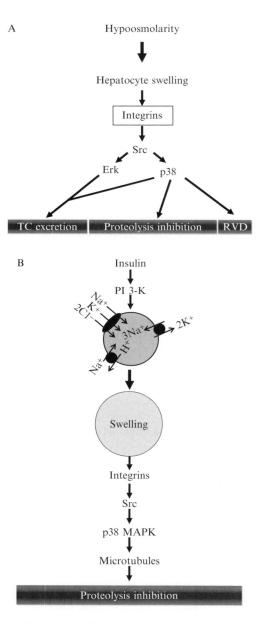

Figure 7.1 Integrin-dependent cell volume sensing in hypoosmotic and insulin signaling in perfused rat liver. (A) Hypoosmotic cell swelling is registered by integrins, leading to a Src-dependent activation of Erk- and p38-type MAP kinases. Signals triggered by Erk and p38 may converge and are required to produce the hypoosmotic increase of

perfusate is collected for another 5 min, a time period allowing complete washout of both isotopes. The effluent is monitored for the total sum of released ^3H and ^{14}C label by dual spectrum scintillation spectrometry. From the $^3H/^{14}C$ ratio under steady-state conditions, a putative ^{14}C-accessible but not ^3H-accessible space is calculated. This space reflects the cellular hydration state. In livers from fed animals, the cell water content under control conditions is 548 ± 10 $\mu l/g$ wet weight (vom Dahl et al., 1991a).

3.2. Measurement of proteolysis

The rate of proteolysis is assessed by measuring the release of ^3H from perfused rat livers after prelabeling of liver proteins in vivo by intraperitoneal injection of L-[4,5-^3H]leucine 16 to 20 h before starting the perfusion experiment (Häussinger et al., 1990a). Following a preperfusion period, the release of trichloroacetic acid-soluble ^3H label into the effluent perfusate reaches a steady state and is by more than 95% associated with effluent leucine. Because leucine is neither synthesized nor catabolized in liver, the effluent radioactivity derives from intracellular protein breakdown. The influent perfusate is supplemented with 100 $\mu mol/liter$ unlabeled leucine in order to prevent reutilization of [^3H]leucine for protein synthesis. To correct for different labeling of hepatic proteins after intraperitoneal injection in the individual perfusion experiment, the rate of proteolysis is set to 100% under normosmotic control conditions, and changes of proteolysis are determined 30 min after institution of the respective condition when a new steady state has been reached.

3.3. Measurement of potassium fluxes

The effluent K^+ concentration is monitored continuously with a potassium-sensitive electrode (vom Dahl et al., 1991c), and stimulus-induced K^+ fluxes are determined as described (Häussinger et al., 1990a).

---

canalicular taurocholate (TC) excretion. The p38 mediates hypoosmotic inhibition of autophagic proteolysis. Also, the regulatory volume decrease (RVD) induced by hypoosmotic swelling depends on p38 activation. Downstream or at the side of MAP-kinase activation microtubules are involved in hypoosmotic proteolysis inhibition and stimulation of canalicular bile acid excretion (not shown). (B) Integrin-dependent cell volume sensing integrates into overall insulin signaling. Insulin via Ptdins 3-kinase and a concerted activation of $Na^+/K^+/2Cl^-$ cotransport, Na^+/H^+ exchange, and Na^+/K^+-ATPase induce a net uptake of K^+, contributing to hepatocyte swelling. Hepatocyte swelling by insulin is registered by the integrins, and integrin signaling via Src and p38 mediates the microtubule-dependent proteolysis inhibition by the hormone.

3.4. Measurement of canalicular bile acid excretion

For determination of taurocholate excretion into bile, the bile duct is cannulated and bile is collected in 1-min samples. Bile flow is determined gravimetrically. [^{3}H]Taurocholate, together with unlabeled taurocholate, is infused, and biliary taurocholate excretion is calculated by measuring radioactivity in the bile samples on the basis of the specific radioactivity in the influent perfusate as described (Häussinger *et al.*, 1992; Noe *et al.*, 1996).

3.5. Tissue processing for Western blot analysis

Small liver pieces are excised in a way that keeps the portal pressure largely constant and then homogenized in lysis buffer as described (Häussinger *et al.*, 1999). The setting of perfused rat liver allows the intraexperimental follow-up of protein phosphorylation.

4. ACTIVATION OF β_1 INTEGRIN SUBUNIT BY HYPOOSMOLARITY AND INSULIN IN PERFUSED RAT LIVER

Sections from perfused livers were analyzed for β_1 immunoreactivity using antibodies recognizing the active β_1 integrin subunit and the $\alpha_5\beta_1$ integrin irrespective of its activation state (Schliess *et al.*, 2004).

The $\alpha_5\beta_1$ integrin in perfused rat liver is localized predominantly in the plasma membrane. Hypoosmolarity or insulin has no effect on the expression and localization of $\alpha_5\beta_1$, but increases immunoreactivity of the activated β_1 integrin subunit. The latter is largely confined to the plasma membrane, as indicated by superimposition of activated β_1 and cortical actin staining. Furosemide, a loop diuretic inhibiting $Na^+/K^+/2Cl^-$ cotransport, prevents the insulin-induced appearance of activated β_1 immunoreactivity, but is without effect on $\alpha_1\beta_5$ immunoreactivity and localization. This suggests that K^+ uptake via the $Na^+/K^+/2Cl^-$ cotransporter and hepatocyte swelling in response to insulin (Häussinger and Lang, 1992) trigger β_1 integrin activation in perfused rat liver.

5. INTEGRINS IN THE REGULATION OF AUTOPHAGIC PROTEOLYSIS IN THE LIVER

A close correlation exists between autophagic proteolysis and the hepatocellular hydration state (Häussinger *et al.*, 1991; vom Dahl and Häussinger, 1996). Like hypoosmotic swelling, glutamine, ethanol, and insulin in perfused rat liver potently inhibit autophagic proteolysis (Hallbrucker et al., 1991a,b,c; Häussinger et al., 1990a; vom Dahl and Häussinger, 1998; vom Dahl et al., 1991b), and it was shown that proteolysis inhibition by glutamine, ethanol, and insulin largely depends on the degree of cell swelling induced by these compounds (vom Dahl and Häussinger, 1996). Proteolysis inhibition by insulin in perfused rat liver is abolished if insulin–induced swelling is blocked by loop diuretics, glucagon, starvation, hyperosmolarity, or PI 3–kinase inhibitors (Hallbrucker et al., 1991a; Schliess et al., 2001; vom Dahl et al., 1991b). Colchicine blocks swelling-dependent proteolysis inhibition by hypoosmolarity, insulin, and glutamine without affecting the cell volume increase, indicating that microtubules are involved in coupling swelling to proteolysis inhibition (vom Dahl et al., 1995).

The p38 MAP kinase plays a key role in swelling-dependent inhibition of hepatic proteolysis. Already 250 nM of the p38 inhibitor SB203580 potently abolishes proteolysis inhibition, but not the cell swelling induced by hypoosmolarity, insulin, and glutamine in perfused rat liver, indicating that p38 inhibition uncouples cell swelling from proteolysis inhibition (Häussinger et al., 1999). In the case of insulin it was shown that antagonizing cell swelling by bumetanide, PI 3–kinase inhibitors, and hyperosmolarity largely inhibited p38 activation and proteolysis inhibition (Häussinger and Lang, 1992; Häussinger et al., 1999; Schliess et al., 2001).

The integrin antagonistic hexapeptide GRGDSP but not the inactive GRGESP potently inhibits the antiproteolytic response to hypoosmolarity, insulin, and glutamine, respectively (Schliess et al., 2004; vom Dahl et al., 2003). Likewise, inhibition of the tyrosine kinase Src with PP2 prevents proteolysis inhibition by hypoosmolarity, insulin, and glutamine (Schliess et al., 2004; vom Dahl et al., 2003). Both GRGDSP and PP2 also inhibit the hypoosmotic decrease of the fractional volume of autophagic vacuoles (vom Dahl et al., 2003). In the case of hypoosmolarity and insulin it was shown that both GRGDSP and PP2 prevent p38 activation and that GRGDSP abolishes the activating Src phosphorylation, suggesting Src to act here as an effector of integrins localized upstream of p38 (Schliess et al., 2004; vom Dahl et al., 2003).

GRGDSP and PP2 do not decrease cell swelling by hypoosmolarity, insulin, and glutamine and thus, like SB203580 (Häussinger et al., 1999), uncouple swelling from proteolysis inhibition (Schliess et al., 2004; vom Dahl et al., 2003). It should be noted that the initial steps of insulin signaling, that is, insulin–induced tyrosine phosphorylation of IRβ, PKB activation, K^+ uptake, and hepatocyte swelling, remain unaffected by GRGDSP and PP2 (Schliess et al., 2004).

Of interest, infusion of latrunculin B at a dose that induces sustained cholestasis (vom Dahl et al., 2003) and provokes disappearance of polymerized actin in immunofluorescence-labeled thin sections of rat livers was without effect on proteolysis inhibition as induced by either hypoosmolarity or insulin (Schliess *et al.*, 2004; vom Dahl *et al.*, 2003), indicating that stable actin microfilaments are not required for mediating integrin-dependent proteolysis inhibition. To what extent the swelling-dependent integrin activation in hepatocytes is coupled to microtubular changes is currently unknown.

Swelling-independent proteolysis inhibition by phenylalanine is insensitive to GRGDSP, PP2, colchicine, and SB203580, respectively (Häussinger *et al.*, 1999; vom Dahl *et al.*, 1995, 2003), indicating that the inhibitors do not interfere nonspecifically with autophagy and autophagic proteolysis. However, GRGDSP, PP2, and SB203580 potently inhibited swelling-independent proteolysis inhibition by tauroursodesoxycholate (TUDC) (Häussinger *et al.*, 2003; Kurz *et al.*, 2001). The mechanisms underlying integrin engagement by TUDC are currently unknown. This suggests that nonosmotic stimuli can also activate the integrin system, thereby producing a "swelling signal" toward proteolysis inhibition.

6. INTEGRINS IN THE REGULATION OF CANALICULAR BILE ACID EXCRETION IN THE LIVER

Bile formation is an osmotic process brought about by the vectorial transport of solutes from the sinusoidal space to the canalicular lumen (Müller and Jansen, 1998). A major site of control is the excretion of cholephilic compounds across the canalicular membrane of the hepatocyte by transport ATPases, including the conjugate export pump multidrug resistance-associated protein (MRP)2 for the transport of glutathione and glucuronide conjugates and the bile salt export pump (BSEP) for the excretion of conjugated bile acids (Häussinger *et al.*, 2000). As shown in the perfused rat liver, canalicular secretion by these transporters is strongly regulated by the hepatocellular hydration state (Häussinger *et al.*, 1992). Roughly, an increase of cell water by 10% doubles the transport capacity for taurocholate into bile within minutes. This occurs in a microtubule-dependent way (Häussinger *et al.*, 1993) and is explained by the rapid insertion of MRP2 and BSEP transporter molecules, which are stored underneath the canalicular membrane inside the cell, into the canalicular membrane (Kubitz *et al.*, 1997, 1999; Kurz *et al.*, 2001; Schmitt *et al.*, 2000, 2001).

Inhibition of integrins, Src, p38, and Erk signaling by GRGDSP, PP2, SB203590, and PD098059, respectively, in the perfused rat liver revealed that an integrin-dependent Src activation essentially mediates the activation of both p38 and Erk-type MAP kinases by hypoosmolarity (Häussinger *et al.*, 2003; Kurz *et al.*, 2001; Noé *et al.*, 1996). Moreover, these inhibitors potently blocked the hypoosmotic increase in taurocholate excretory capacity (Häussinger *et al.*, 2003; Kurz *et al.*, 2001). Interestingly, the TUDCA-induced increase of canalicular taurocholate excretion accompanied by hepatocyte swelling (Häussinger *et al.*, 1992) was also sensitive to these inhibitors.

7. INTEGRINS IN HEPATOCYTE VOLUME REGULATION

Hypoosmolarity in hepatocytes stimulates the release of cellular K^+, Cl^-, and HCO_3^- (Graf and Häussinger, 1996; Haddad and Graf, 1989; Häussinger *et al.*, 1990b) and of organic osmolytes such as taurine and betaine, leading to a regulatory volume decrease (RVD) (Warskulat *et al.*, 1997; Wettstein *et al.*, 1998). Inhibition of p38 by SB203580 significantly increases hypoosmotic cell swelling in perfused rat liver by attenuating the volume-regulatory K^+ release, indicating an involvement of the p38 in RVD (Häussinger *et al.*, 1999; vom Dahl *et al.*, 2001). A role of p38 in RVD was also reported for cultured rat hepatocytes (Feranchak *et al.*, 2001).

Also, the integrin antagonistic GRGDSP and the inhibitor of Src-type kinases PP2 in perfused rat liver significantly increases hypoosmotic swelling and decreases the volume-regulatory K^+ release (vom Dahl *et al.*, 2003), suggesting that integrin-dependent volume sensing is essential to trigger RVD.

8. CONCLUSION

The findings summarized in this chapter indicate that hypoosmotic swelling and insulin in perfused rat liver lead to the activation of β_1 integrin and that cell volume sensing by integrins triggers a Src-dependent activation of MAP kinases that mediates proteolysis inhibition, an increase of canalicular taurocholate excretion, and RVD in response to hypoosmotic swelling. Moreover, it was shown that integrin-dependent registration of hepatocyte swelling ("osmosensing") integrates into the overall context of signal transduction by insulin and glutamine, respectively, and is therefore essential for generating the entire metabolic response to these compounds. These findings explain why both hypoosmotic swelling and insulin are ineffective in triggering proteolysis inhibition in suspended hepatocytes (Blommaart *et al.*, 1995; Meijer *et al.*, 1993; vom Dahl *et al.*, 1995). In suspended cells, swelling-dependent signaling may be incomplete because of the disruption of hepatocyte/matrix interactions, which implies impaired volume sensing by the integrins. It also may explain why hepatocytes regain their functional osmosensitivity when taken into culture for 24 h (vom Dahl *et al.*, 1995, 2003). The findings also provide novel aspects on the pathogenesis of insulin resistance, which should occur when insulin-induced hepatocyte swelling or its sensing of it is impaired. Such conditions comprise dehydration and hyperosmolarity, hyperglucagonemia, uremia, and conditions with increased oxidative stress, all of which lead to a decrease of cell hydration (Schliess and Häussinger, 2003). Also, the long-known prodiabetic action of loop diuretics may find its partial explanation in an inhibition of hepatic $Na^+/K^+/2Cl^-$ cotransporter NKCC1, that is, the electrolyte transport system that mediates insulin-induced hepatocyte swelling (Hallbrucker et al., 1991a; Häussinger and Lang, 1992; Schliess et al., 2001, 2002). Insulin resistance is found frequently in liver cirrhosis, and the possibility has not yet been considered that alterations of the ECM in cirrhosis may impair volume sensing and signaling via the integrin system, which is normally required for proper insulin action.

As mentioned earlier, different modes of integrin involvement in growth factor receptor signaling are well established. The registration of insulin-induced cell swelling defines a novel role of integrins in growth factor signaling and sets an example for mechanotransduction to be an integrated part of growth factor signaling.

REFERENCES

Aplin, A. E., Howe, A., Alahari, S. K., and Juliani, R. L. (1998). Signal transduction and signal modulation by cell adhesion receptors: The role of integrins, cadherins, immunoglobulin-cell adhesion molecules, and selectins. *Pharmacol. Rev.* **50,** 197–263.

Blommaart, E. F., Luiken, J. J., Blommaart, P. J., van Woerkom, G. M., and Meijer, A. J. (1995). Phosphorylation of ribosomal protein S6 is inhibitory for autophagy in isolated rat hepatocytes. *J. Biol. Chem.* **270,** 2320–2326.

Browe, D. M., and Baumgarten, C. M. (2006). EGFR kinase regulates volume-sensitive chloride current elicited by integrin stretch via PI-3K and NADPH oxidase in ventricular myocytes. *J. Gen. Physiol* **127,** 237–251.

Burg, M. B., Kwon, E. D., and Kultz, D. (1997). Regulation of gene expression by hypertonicity. *Annu. Rev. Physiol.* **59,** 437–455.

Cabodi, S., Moro, L., Bergatto, E., Boeri, E. E., Di, S. P., Turco, E., Tarone, G., and Defilippi, P. (2004). Integrin regulation of epidermal growth factor (EGF) receptor and of EGF-dependent responses. *Biochem. Soc. Trans.* **32,** 438–442.

Carloni, V., Mazzocca, A., Pantaleo, P., Cordella, C., Laffi, G., and Gentilini, P. (2001). The integrin, alpha6beta1, is necessary for the matrix-dependent activation of FAK and MAP kinase and the migration of human hepatocarcinoma cells. *Hepatology* **34,** 42–49.

Carloni, V., Romanelli, R. G., Pinzani, M., Laffi, G., and Gentilini, P. (1996). Expression and function of integrin receptors for collagen and laminin in cultured human hepatic stellate cells. *Gastroenterology* **110,** 1127–1136.

Chan, P. C., Chen, S. Y., Chen, C. H., and Chen, H. C. (2006). Crosstalk between hepatocyte growth factor and integrin signaling pathways. *J. Biomed. Sci.* **13,** 215–223.

Cukierman, E., Pankov, R., and Yamada, K. M. (2002). Cell interactions with three-dimensional matrices. *Curr. Opin. Cell Biol.* **14,** 633–639.

Feranchak, A. P., Berl, T., Capasso, J., Wojtaszek, P. A., Han, J., and Fitz, J. G. (2001). p38 MAP kinase modulates liver cell volume through inhibition of membrane Na^+ permeability. *J. Clin. Invest.* **108,** 1495–1504.

Franco, R., Lezama, R., Ordaz, B., and Pasantes-Morales, H. (2004). Epidermal growth factor receptor is activated by hyposmolarity and is an early signal modulating osmolyte efflux pathways in Swiss 3T3 fibroblasts. *Pflüg. Arch.* **447,** 830–839.

Giancotti, F. G., and Ruoslahti, E. (1999). Integrin signaling. *Science* **285,** 1028–1032.

Graf, J., and Häussinger, D. (1996). Ion transport in hepatocytes: Mechanisms and correlations to cell volume, hormone actions and metabolism. *J. Hepatol.* **24,** 53–77.

Haddad, P., and Graf, J. (1989). Volume-regulatory K^+ fluxes in the isolated perfused rat liver: Characterization by ion transport inhibitors. *Am. J. Physiol.* **257,** G357–G363.

Hallbrucker, C., vom Dahl, S., Lang, F., Gerok, W., and Häussinger, D. (1991a). Inhibition of hepatic proteolysis by insulin: Role of hormone-induced alterations of the cellular K^+ balance. *Eur. J. Biochem.* **199,** 467–474.

Hallbrucker, C., vom Dahl, S., Lang, F., Gerok, W., and Häussinger, D. (1991b). Modification of liver cell volume by insulin and glucagon. *Pflüg. Arch.* **418,** 519–521.

Hallbrucker, C., vom Dahl, S., Lang, F., and Häussinger, D. (1991c). Control of hepatic proteolysis by amino acids: The role of cell volume. *Eur. J. Biochem.* **197,** 717–724.

Hamill, O. P., and Martinac, B. (2001). Molecular basis of mechanotransduction in living cells. *Physiol. Rev.* **81,** 685–740.

Häussinger, D. (1996). The role of cellular hydration in the regulation of cell function. *Biochem. J.* **313,** 697–710.

Häussinger, D., Hallbrucker, C., Saha, N., Lang, F., and Gerok, W. (1992). Cell volume and bile acid excretion. *Biochem. J.* **288,** 681–689.

Häussinger, D., Hallbrucker, C., vom Dahl, S., Decker, S., Schweizer, U., Lang, F., and Gerok, W. (1991). Cell volume is a major determinant of proteolysis control in liver. *FEBS Lett.* **283,** 70–72.

Häussinger, D., Hallbrucker, C., vom Dahl, S., Lang, F., and Gerok, W. (1990a). Cell swelling inhibits proteolysis in perfused rat liver. *Biochem. J.* **272,** 239–242.

Häussinger, D., Kurz, A. K., Wettstein, M., Graf, D., vom Dahl, S., and Schliess, F. (2003). Involvement of integrins and Src in tauroursodeoxycholate-induced and swelling-induced choleresis. *Gastroenterology* **124,** 1476–1487.

Häussinger, D., and Lang, F. (1991). Cell volume in the regulation of hepatic function: A mechanism for metabolic control. *Biochim. Biophys. Acta* **1071,** 331–350.

Häussinger, D., and Lang, F. (1992). Cell volume and hormone action. *Trends Pharmacol. Sci.* **13,** 371–373.

Häussinger, D., Saha, N., Hallbrucker, C., Lang, F., and Gerok, W. (1993). Involvement of microtubules in the swelling-induced stimulation of transcellular taurocholate transport in perfused rat liver. *Biochem. J.* **291,** 355–360.

Häussinger, D., and Schliess, F. (1999). Osmotic induction of signaling cascades: Role in regulation of cell function. *Biochem. Biophys. Res. Commun.* **255,** 551–555.

Häussinger, D., Schliess, F., Dombrowski, F., and vom Dahl, S. (1999). Involvement of p38 MAPK in the regulation of proteolysis by liver cell hydration. *Gastroenterology* **116,** 921–935.

Häussinger, D., Schmitt, M., Weiergräber, O., and Kubitz, R. (2000). Short-term regulation of canalicular transport. *Semin. Liver Dis.* **20,** 307–321.

Häussinger, D., Stehle, T., and Lang, F. (1990b). Volume regulation in liver: Further characterization by inhibitors and ionic substitutions. *Hepatology* **11,** 243–254.

Hoffmann, E. K. (2000). Intracellular signalling involved in volume regulatory decrease. *Cell Physiol. Biochem.* **10,** 273–288.

Hsu, S. L., Cheng, C. C., Shi, Y. R., and Chiang, C. W. (2001). Proteolysis of integrin alpha5 and beta1 subunits involved in retinoic acid-induced apoptosis in human hepatoma Hep3B cells. *Cancer Lett.* **167,** 193–204.

Hynes, R. O. (2002). Integrins: Bidirectional, allosteric signaling machines. *Cell* **110**, 673–687.

Ingber, D. E. (1997). Tensegrity: The architectural basis of cellular mechanotransduction. *Annu. Rev. Physiol.* **59**, 575–599.

Ingber, D. E. (2003). Mechanosensation through integrins: Cells act locally but think globally. *Proc. Natl. Acad. Sci. USA* **100**, 1472–1474.

Ingber, D. E. (2006). Cellular mechanotransduction: Putting all the pieces together again. *FASEB J.* **20**, 811–827.

Jalali, S., del Pozo, M. A., Chen, K., Miao, H., Li, Y., Schwartz, M. A., Shyy, J. Y., and Chien, S. (2001). Integrin-mediated mechanotransduction requires its dynamic interaction with specific extracellular matrix (ECM) ligands. *Proc. Natl. Acad. Sci. USA* **98**, 1042–1046.

Kapur, S., Mohan, S., Baylink, D. J., and Lau, K. H. (2005). Fluid shear stress synergizes with insulin-like growth factor-I (IGF-I) on osteoblast proliferation through integrin-dependent activation of IGF-I mitogenic signaling pathway. *J. Biol. Chem.* **280**, 20163–20170.

Kubitz, R., D'Urso, D., Keppler, D., and Häussinger, D. (1997). Osmodependent dynamic localization of the multidrug resistance protein 2 in the rat hepatocyte canalicular membrane. *Gastroenterology* **113**, 1438–1442.

Kubitz, R., Wettstein, M., Warskulat, U., and Häussinger, D. (1999). Regulation of the multidrug resistance protein 2 in the rat liver by lipopolysaccharide and dexamethasone. *Gastroenterology* **116**, 401–410.

Kurz, A. K., Graf, D., Schmitt, M., vom Dahl, S., and Häussinger, D. (2001). Tauroursodesoxycholate-induced choleresis involves p38 MAPK activation and translocation of the bile salt export pump in rats. *Gastroenterology* **121**, 407–419.

Lang, F., Busch, G. L., Ritter, M., Völkl, H., Waldegger, S., Gulbins, E., and Häussinger, D. (1998). Functional significance of cell volume regulatory mechanisms. *Physiol. Rev.* **78**, 247–306.

Li, Y. S., Haga, J. H., and Chien, S. (2005). Molecular basis of the effects of shear stress on vascular endothelial cells. *J. Biomech.* **38**, 1949–1971.

Loomis, W. F., Shaulsky, G., and Wang, N. (1997). Histidine kinases in signal transduction pathways of eukaryotes. *J. Cell Sci.* **110**, 1141–1145.

Low, S. Y., Rennie, M. J., and Taylor, P. M. (1997). Involvement of integrins and the cytoskeleton in modulation of skeletal muscle glycogen synthesis by changes in cell volume. *FEBS Lett.* **417**, 101–103.

Low, S. Y., and Taylor, P. M. (1998). Integrin and cytoskeletal involvement in signalling cell volume changes to glutamine transport in rat skeletal muscle. *J. Physiol.* **512**, 481–485.

Lynch, L., Vodyanik, P. I., Boettiger, D., and Guvakova, M. A. (2005). Insulin-like growth factor I controls adhesion strength mediated by alpha5beta1 integrins in motile carcinoma cells. *Mol. Biol. Cell* **16**, 51–63.

Mattila, E., Pellinen, T., Nevo, J., Vuoriluoto, K., Arjonen, A., and Ivaska, J. (2005). Negative regulation of EGFR signalling through integrin-alpha1beta1-mediated activation of protein tyrosine phosphatase TCPTP. *Nat. Cell Biol.* **7**, 78–85.

McManus, M., Churchwell, K. B., and Strange, K. (1995). Regulation of cell volume in health and desease. *N. Engl. J. Med.* **333**, 1260–1265.

Meijer, A. J., Gustafson, L. A., Luiken, J. J., Blommaart, P. J., Caro, L. H., van Woerkom, G. M., Spronk, C., and Boon, L. (1993). Cell swelling and the sensitivity of autophagic proteolysis to inhibition by amino acids in isolated rat hepatocytes. *Eur. J. Biochem.* **215**, 449–454.

Moeckel, G. W., Zhang, L., Chen, X., Rossini, M., Zent, R., and Pozzi, A. (2006). Role of integrin alpha1beta1 in the regulation of renal medullary osmolyte concentration. *Am. J. Physiol. Renal Physiol.* **290**, F223–F231.

Moro, L., Venturino, M., Bozzo, C., Silengo, L., Altruda, F., Beguinot, L., Tarone, G., and Defilippi, P. (1998). Integrins induce activation of EGF receptor: Role in MAP kinase induction and adhesion-dependent cell survival. *EMBO J.* **17**, 6622–6632.

Müller, M., and Jansen, P. L. M. (1998). Molecular aspects of hepatobiliary transport. *Am. J. Physiol.* **272**, G1285–G1303.

Noé, B., Schliess, F., Wettstein, M., Heinrich, S., and Häussinger, D. (1996). Regulation of taurocholate excretion by a hypoosmolarity-activated signal transduction pathway in rat liver. *Gastroenterology* **110**, 858–865.

Reinehr, R., Becker, S., Braun, J., Eberle, A., Grether-Beck, S., and Häussinger, D. (2006). Endosomal acidification and activation of NADPH oxidase isoforms are upstream events in hyperosmolarity-induced hepatocyte apoptosis. *J. Biol. Chem.* **281**, 23150–23166.

Reinehr, R., Becker, S., Höngen, A., and Häussinger, D. (2004a). The Src family kinase yes triggers hyperosmotic activation of the epidermal growth factor receptor and CD95. *J. Biol. Chem.* **279**, 23977–23987.

Reinehr, R., Görg, B., Becker, S., Qvartskhava, N., Bidmon, H. J., Selbach, O., Haas, H. L., Schliess, F., and Häussinger, D. (2007). Hypoosmotic swelling increases oxidative stress by NADPH oxidase in cultured astrocytes and vital brain slices. *GLIA* **55**, 758–771.

Reinehr, R., Görg, B., Höngen, A., and Häussinger, D. (2004b). CD95-tyrosine nitration inhibits hyperosmotic and CD95 ligand-induced CD95 activation in rat hepatocytes. *J. Biol. Chem.* **279**, 10364–10373.

Reinehr, R. M., Graf, D., Fischer, R., Schliess, F., and Häussinger, D. (2002). Hyperosmolarity triggers CD95 membrane trafficking and sensitizes rat hepatocytes towards CD95L-induced apoptosis. *Hepatology* **36**, 602–614.

Rosette, C., and Karin, M. (1996). Ultraviolet light and osmotic stress: Activation of the JNK cascade through multiple growth factor and cytokine receptors. *Science* **274**, 1194–1197.

Rozengurt, E. (1986). Early signals in the mitogenic response. *Science* **234**, 161–166.

Ruoslahti, E. (1996). RGD and other recognition sequences for integrins. *Annu. Rev. Cell Dev. Biol.* **12**, 697–715.

Schliess, F., and Häussinger, D. (2000). Cell hydration and insulin signalling. *Cell Physiol. Biochem.* **10**, 403–409.

Schliess, F., and Häussinger, D. (2003). Cell volume and insulin signalling. *Int. Rev. Cytol.* **225**, 187–228.

Schliess, F., and Häussinger, D. (2005). The cellular hydration state: Role in apoptosis and proliferation. *Signal Transduct.* **6**, 297–302.

Schliess, F., Reissmann, R., Reinehr, R., vom Dahl, S., and Häussinger, D. (2004). Involvement of integrins and Src in insulin signaling towards autophagic proteolysis in rat liver. *J. Biol. Chem.* **279**, 21294–21301.

Schliess, F., Schäfer, C., Fischer, R., Lordnejad, M. R., and Häussinger, D. (2002). Expression and regulation of the $Na^+/K^+/2Cl^-$ cotransporter NKCC1 in rat liver and human HuH-7 hepatoma cells. *Arch. Biochem. Biophys.* **401**, 187–197.

Schliess, F., vom Dahl, S., and Häussinger, D. (2001). Insulin resistance by loop diuretics and hyperosmolarity in perfused rat liver. *Biol. Chem.* **382**, 1063–1069.

Schmitt, M., Kubitz, R., Lizun, S., Wettstein, M., and Häussinger, D. (2001). Regulation of the dynamic localization of the rat Bsep gene-encoded bile salt export pump by anisoosmolarity. *Hepatology* **33**, 509–518.

Schmitt, M., Kubitz, R., Wettstein, M., vom Dahl, S., and Häussinger, D. (2000). Retrieval of the mrp2 gene encoded conjugate export pump from the canalicular membrane contributes to cholestasis induced by tert-butyl hydroperoxide and chloro-dinitrobenzene. *Biol. Chem.* **381**, 487–495.

Schwartz, M. A., and Ginsberg, M. H. (2002). Networks and crosstalk: Integrin signalling soreads. *Nat. Cell Biol.* **4**, E65–E68.

Sheikh-Hamad, D., Suki, W. N., and Zhao, W. (1997). Hypertonic induction of the cell adhesion molecule beta 1-integrin in MDCK cells. *Am. J. Physiol.* **273,** C902–C908.

Sies, H. (1978). The use of perfusion of liver and other organs for the study of microsomal electron-transport and cytochrome P-450 systems. *Methods Enzymol.* **52,** 48–59.

Stupack, D. G. (2005). Integrins as a distinct subtype of dependence receptors. *Cell Death Differ.* **12,** 1021–1030.

Torimura, T., Ueno, T., Kin, M., Harada, R., Nakamura, T., Kawaguchi, T., Harada, M., Kumashiro, R., Watanabe, H., Avraham, R., and Sata, M. (2001). Autocrine motility factor enhances hepatoma cell invasion across the basement membrane through activation of beta1 integrins. *Hepatology* **34,** 62–71.

Turowski, G. A., Rashid, Z., Hong, F., Madri, J. A., and Basson, M. D. (1994). Glutamine modulates phenotype and stimulates proliferation in human colon cancer cell lines. *Cancer Res.* **54,** 5974–5980.

Varela, D., Simon, F., Riveros, A., Jorgensen, F., and Stutzin, A. (2004). NAD(P)H oxidase-derived H_2O_2 signals chloride channel activation in cell volume regulation and cell proliferation. *J. Biol. Chem.* **279,** 13301–13304.

Vogel, V. (2006). Mechanotransduction involving multimodular proteins: Converting force into biochemical signals. *Annu. Rev. Biophys. Biomol. Struct.* **35,** 459–488.

vom Dahl, S., Hallbrucker, C., Lang, F., Gerok, W., and Häussinger, D. (1991a). A non-invasive technique for cell volume determination in perfused rat liver. *Biol. Chem. Hoppe Seyler* **372,** 411–418.

vom Dahl, S., Hallbrucker, C., Lang, F., Gerok, W., and Häussinger, D. (1991b). Regulation of liver cell volume and proteolysis by glucagon and insulin. *Biochem. J.* **278,** 771–777.

vom Dahl, S., Hallbrucker, C., Lang, F., and Häussinger, D. (1991c). Regulation of cell volume in the perfused rat liver by hormones. *Biochem. J.* **280,** 105–109.

vom Dahl, S., and Häussinger, D. (1996). Nutritional state and the swelling-induced inhibition of proteolysis in perfused rat liver. *J. Nutr.* **126,** 395–402.

vom Dahl, S., and Häussinger, D. (1997). Experimental methods in hepatology. Guidelines of the German Association for the Study of the Liver (GASL). Liver perfusion: Technique and applications. *Z. Gastroenterol.* **35,** 221–226.

vom Dahl, S., and Häussinger, D. (1998). Bumetanide-sensitive cell swelling mediates the inhibitory effect of ethanol on proteolysis in rat liver. *Gastroenterology* **114,** 1046–1053.

vom Dahl, S., Schliess, F., Graf, D., and Häussinger, D. (2001). Role of p38MAPK in cell volume regulation of perfused rat liver. *Cell Physiol. Biochem.* **11,** 285–294.

vom Dahl, S., Schliess, F., Reissmann, R., Görg, B., Weiergräber, O., Kocalkova, M., Dombrowski, F., and Häussinger, D. (2003). Involvement of integrins in osmosensing and signaling toward autophagic proteolysis in rat liver. *J. Biol. Chem.* **278,** 27088–27095.

vom Dahl, S., Stoll, B., Gerok, W., and Häussinger, D. (1995). Inhibition of proteolysis by cell swelling in the liver requires intact microtubular structures. *Biochem. J.* **308,** 529–536.

Wang, N., Naruse, K., Stamenovic, D., Fredberg, J. J., Mijailovich, S. M., Tolic-Norrelykke, I. M., Polte, T., Mannix, R., and Ingber, D. E. (2001). Mechanical behavior in living cells consistent with the tensegrity model. *Proc. Natl. Acad. Sci. USA* **98,** 7765–7770.

Wang, Y., Roman, R., Lidofsky, S. D., and Fitz, J. G. (1996). Autocrine signaling through ATP release represents a novel mechanism for cell volume regulation. *Proc. Natl. Acad. Sci. USA* **93,** 12020–12025.

Warskulat, U., Wettstein, M., and Häussinger, D. (1997). Osmoregulated taurine transport in H4IIE rat hepatoma cells and perfused rat liver. *Biochem. J.* **321,** 683–690.

Wettstein, M., Weik, C., Holneicher, C., and Häussinger, D. (1998). Betaine as an osmolyte in rat liver: Metabolism and cell-to-cell interactions. *Hepatology* **27,** 787–793.

CHAPTER EIGHT

HYPEROSMOTIC ACTIVATION OF THE CD95 SYSTEM

Roland Reinehr *and* Dieter Häussinger

Contents

Abstract

Cell shrinkage, nuclear condensation, DNA fragmentation, and apoptotic body formation are hallmarks of programmed apoptotic cell death. Herein, apoptotic volume decrease (AVD) is an early and ubiquitous event. Conversely, in hepatocytes, hyperosmotic cell shrinkage leads to an activation of the CD95 death receptor system, which involves CD95 tyrosine phosphorylation, CD95 oligomerization, and subsequent trafficking of the CD95 to the plasma membrane, and sensitizes hepatocytes toward CD95 ligand (CD95L)-induced apoptosis. Early signaling events leading to CD95 activation by hyperosmolarity have been identified. In hepatocytes, hyperosmotic exposure induces an almost instantaneous acidification of an acidic sphingomyelinase (ASM) containing endosomal compartment, which is followed by an increase in the intracellular ceramide concentration. Inhibition of anion channels or the vacuolar-type H^+-ATPase abolishes not only endosomal acidification and subsequent ceramide generation, but also the otherwise observed hyperosmotically induced generation of reactive oxygen species (ROS) by NADPH oxidase isoforms. Hyperosmolarity-induced ROS formation then leads to a Src-family kinase Yes-mediated activation of the epidermal growth factor receptor (EGFR) and to an activation of the c-Jun-N-terminal kinase (JNK). JNK then provides a signal for CD95/EGFR association and subsequent CD95 tyrosine phosphorylation, which is mediated by the EGFR tyrosine kinase activity. CD95 tyrosine phosphorylation then allows for CD95 receptor oligomerization, translocation of the CD95/EGFR protein complex to the plasma membrane, and formation of the death inducing

Clinic for Gastroenterology, Hepatology, and Infectiology, Heinrich-Heine-University Düsseldorf, Germany

Methods in Enzymology, Volume 428
ISSN 0076-6879, DOI: 10.1016/S0076-6879(07)28008-5

signaling complex (DISC). Mild hyperosmotic exposure, that is, 405 mosmol/liter, does not lead to a reduction of cell viability, even if DISC formation and subsequent caspase 8 and 3 activation occur, but sensitizes hepatocytes to CD95L-induced apoptosis. However, activation of the CD95 system by a more severe hyperosmotic challenge (>505 mosmol/liter) is followed by execution of the apoptotic cell death. Other covalent modifications of CD95, such as CD95 tyrosine nitration or CD95 serine/threonine phosphorylation, were shown to inhibit the CD95 activation process.

1. INTRODUCTION

Apoptosis is characterized by cell shrinkage, nuclear condensation, DNA fragmentation, and apoptotic body formation. These features separate apoptosis from other forms of cell death, such as necrosis (Kerr et al., 1974). It is well known that shrunken hepatocytes are more susceptible to stress-induced cell death, which may explain the clinical observation that transplantation of livers from hypernatremic donors shows high rates of primary graft dysfunction. Apoptotic volume decrease (AVD) is an early and ubiquitous event in apoptotic cell death (Bortner and Cidlowski, 2001, 2002; Lang et al., 2000a; Yu et al., 2001). Ion fluxes and cellular mechanisms of volume regulation participate in the regulation of AVD as well as in apoptosis induction. For example, in Jurkat cells, CD95 activation inhibits Na^+/H^+-antiport (Lang et al., 2000a), the Na^+/K^+-ATPase (Nobel et al., 2000), activates an outwardly rectifying chloride channel (Szabo et al., 1998), and stimulates efflux of taurine (Lang et al., 1998), resulting in AVD. Furthermore, inhibition of AVD by inhibition of volume-sensitive K^+ or Cl^- channels (Maeno et al., 2000) or by elevation of extracellular K^+ (Gomez-Angelats et al., 2000; Thompson et al., 2001) counteracts apoptosis, whereas inhibition of Na^+/H^+ exchange accelerates apoptotic cell death (Lang et al., 2000a).

Apparently, in hepatocytes, hyperosmolarity-induced CD95 activation and CD95 activation induced by the CD95 ligand (CD95L) or proapoptotic bile salts share almost the same signaling events, which finally lead to activation of the apoptotic machinery. Thus the question arises at which point in the respective intracellular signaling cascades induced by these distinct stimuli does signal transduction converge? One early signaling event shared by all these stimuli was shown to be an almost instantaneously generated oxidative stress response (Reinehr et al., 2003a,b).

In rat hepatocytes, oxidative and osmotic stresses are apparently closely interlinked. On the one hand, hyperosmotic cell shrinkage generates oxidative stress (Reinehr et al., 2003a,b) and on the other, oxidative stress induces cell shrinkage (Becker et al., 2007a; Hallbrucker et al., 1994;

Saha *et al.*, 1993). Apparently the balance between intracellular metabolic H_2O_2 generation and its removal by detoxication systems such as catalase and glutathione peroxidase may be one determinant for hepatocellular K^+ balance and, accordingly, cell volume. Hyperosmolarity, as well as other proapoptotic stimuli, such as CD95L or proapoptotic bile salts, was reported to trigger an almost instantaneous oxidative stress response in rat hepatocytes (Reinehr *et al.*, 2003a,b). Like tumor necrosis factor-α (Stadler *et al.*, 1992), proapoptotic bile acids were shown to induce oxygen radical formation by mitochondria (Krähenbühl *et al.*, 1994; Sokol *et al.*, 1995, 1998); however, this may represent a downstream consequence, not the cause of CD95 activation. Similar considerations may apply for the oxygen radical formation due to an endoplasmic reticulum stress response, which is induced by hydrophobic bile acids (Crowley *et al.*, 2000; Sokol *et al.*, 1998). In rat hepatocytes, NADPH oxidase isoforms, which are found in many tissues and exhibit sequence homology to the classical phagocyte NADPH oxidase gp91^{phox} (Nox 2) (Lambeth, 2002), have been identified as the source of the instantaneous reactive oxygen species generation upon proapoptotic stimulation, that is, hyperosmolarity, CD95L, or proapoptotic bile salts (Reinehr *et al.*, 2005a,b, 2006). NADPH oxidase isoforms, which are homologues of the gp91^{phox}, are called Nox and Duox (Edens *et al.*, 2001; Lambeth, 2002; Lambeth *et al.*, 2000) and participate in a variety of signal transduction cascades (Lambeth *et al.*, 2000). NADPH oxidases are activated by a self-assembly, regulated by proteins such as p47^{phox}, p67^{phox}, and Rac (Bokoch *et al.*, 2002; Nauseef *et al.*, 2004; Pani *et al.*, 2001; Vignais, 2002). Thereby, p47^{phox} is thought to be critical for normal NADPH oxidase function because p47^{phox} acts as an adapter protein, which facilitates stimulus-induced binding of p67^{phox} to the enzyme complex (Nauseef *et al.*, 2004; Vignais, 2002). Evidence has been given that the small G protein Rac acts as an upstream signaling event of hyperosmotically induced caspase 3 activation in NIH 3T3 fibroblasts (Friis *et al.*, 2005).

Within the last decade, the contribution of AVD and ROS formation to apoptotic signal transduction has been under intense investigation. Hyperosmolarity mimics AVD and induces ROS formation, thereby stimulating CD95-mediated apoptosis in hepatocytes; other death receptors were also shown to be activated upon hyperosmotic exposure. For example, hyperosmotic exposure can induce JNK activation and TNF receptor clustering (Rosette and Karin, 1996). This, however, is cell type specific and may depend on the efficacy of volume-regulatory increase mechanisms (Bortner and Cidlowski, 1996; Reinehr *et al.*, 2003b). As outlined in more detail later, hyperosmotic cell shrinkage in rat hepatocytes triggers ROS formation, which leads to CD95 activation, oligomerization, and subsequent membrane trafficking, thereby sensitizing hepatocytes toward CD95L-induced apoptosis (Reinehr *et al.*, 2002, 2003b).

2. HYPEROSMOTIC ACTIVATION OF THE CD95 SYSTEM IN HEPATOCYTES

Whereas CD95 in normosmotically exposed rat hepatocytes is localized inside the cells and exhibits no detectable membrane localization, this death receptor is rapidly targeted to the plasma membrane in response to proapoptotic stimuli, such as hyperosmotic exposure, CD95L, or bile acids (Eberle *et al.*, 2005, 2007; Reinehr and Häussinger, 2004; Reinehr *et al.*, 2002, 2003a,b, 2004a,b,c). An easy way to visualize CD95 translocation to the plasma membrane is to detect CD95 by immunocytochemistry under permeabilizing conditions, which allow for intracellular and membrane CD95 staining, compared to nonpermeabilizing conditions, which allow for plasma membrane CD95 staining only (Reinehr *et al.*, 2002). Other ways to detect CD95 membrane trafficking are Western blot analysis of membrane and cytosolic fractions using ultracentrifugation (Reinehr *et al.*, 2003b) or transfection of a yellow fluorescent protein (YFP)-coupled CD95 receptor and subsequent analysis of living cells by fluorescence microscopy (Reinehr *et al.*, 2004c) (Figs. 8.1 and 8.2).

Hyperosmotic CD95 activation and subsequent translocation of the CD95 to the plasma membrane are complex processes. Hyperosmotic exposure leads to an almost instantaneous increase of ceramide levels measured by lipid extraction and subsequent high-performance thin-layer chromatography, which was sensitive to inhibition of the acidic sphingomyelinase (ASM) using inhibitors such as AY9944 or desipramine or by ASM protein knockdown using ASM-specific antisense oligonucleotides (Reinehr *et al.*, 2006).

Hyperosmotic hepatocyte shrinkage increases the cytosolic Cl^- concentration (Graf and Häussinger, 1996; Häussinger *et al.*, 1990; Lang *et al.*, 1993) not only because of the almost immediate osmotic water efflux in response to hyperosmolarity, which has a concentrative effect on cytosolic chloride concentration, but also because of the ionic mechanisms of volume regulatory increase. The latter involves a hyperosmotic activation of Na^+/H^+ exchange and Na^+/K^+-ATPase together with HCO_3^-/Cl^- exchange, resulting in net accumulation of Na^+, K^+, and Cl^- in rat hepatocytes (Graf and Häussinger, 1996; Häussinger *et al.*, 1990; Lang *et al.*, 1993). Evidence has shown that hyperosmotic ASM activation is probably regulated by changes in the intracellular chloride concentration ($[Cl^-]_i$). In hepatocytes, ASM localizes at the cellular membrane or in the lysosomes (Goni and Alonso, 2002; Ohanian and Ohanian, 2001), but also in an early endosomal compartment with an apparent pH_{ves} of about 6.0 (Reinehr *et al.*, 2006). In order to detect pH changes in this presumably endosomal compartment, hepatocytes were allowed to endocytose FITC-coupled dextran molecules

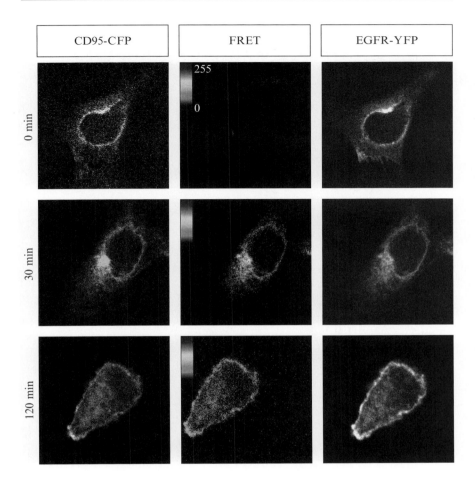

Figure 8.1   Hyperosmolarity-induced CD95-CFP/EGFR-YFP association in Huh7 cells. Huh7 hepatoma cells were cotransfected with CD95-CFP and EGFR-YFP and then exposed to hyperosmolarity (405 mosmol/liter) for 0, 30, and 120 min. FRET pictures were taken according to Eberle *et al.* (2005) and then normalized with respect to FRET efficiencies (scale indicates FRET efficiency from blue/0 to red/255). Hyperosmolarity induced an intracellular FRET signal within 30 min, indicating an intracellular CD95-CFP/EGFR-YFP association, and within 120 min a membrane translocation of the protein complex, as indicated by a FRET signal at the plasma membrane. (See color insert.)

with an average molecular mass of 70 kDa for 60 min. The intracellular vesicular compartment reached by FITC–dextran under these conditions was shown to reflect an early endosomal compartment (Schreiber *et al.*, 1994, 1995, 1996). Hyperosmolarity induces an almost instantaneous acidification of this ASM-containing endosomal compartment in rat hepatocytes

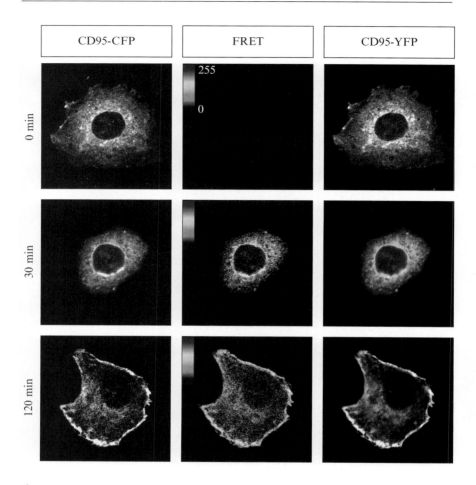

Figure 8.2 Hyperosmolarity-induced CD95–CFP/CD95–YFP oligomerization in Huh7 cells. Huh7 hepatoma cells were cotransfected with CD95–CFP and CD95–YFP and then exposed to hyperosmolarity (405 mosmol/liter) for 0, 30, and 120 min. FRET pictures were taken according to Eberle *et al.* (2007) and then normalized with respect to FRET efficiencies (scale indicates FRET efficiency from blue/0 to red/255). Hyperosmolarity induced an intracellular FRET signal within 30 min, indicating an intracellular CD95–CFP/CD95–YFP association, and within 120 min a FRET signal at the plasma membrane, suggestive for a translocation of the former formed protein complex. (See color insert.)

(Reinehr *et al.*, 2006). ASM exhibits a pH optimum around 5 (Marchesini and Hannun, 2004; Yoshida *et al.*, 1985), and acidity is thought to increase the affinity of ASM to its substrate (Callahan *et al.*, 1983). Thus, a hyperosmotically induced shift of an apparent endosomal pH_{ves} from 6.0 to 5.6 may activate ASM. Hyperosmolarity-induced endosomal acidification was sensitive to bafilomycin, an inhibitor of vacuolar-type H^+-ATPase (vH^+-ATPase), and the anion channel blocker 4,4′-diisothiocyano-stilbene-2,2′-disulfonic acid disodium salt. Because the vesicular proton pump is activated directly by Cl^- (Moriyama and Nelson, 1987; Pazoles et al., 1980; Zeuzem et al., 1992), one may suggest that the hyperosmotically induced increase in the cytosolic Cl^- concentration may increase the activity of the vacuolar H^+-ATPase and that the resulting endosomal acidification may activate ASM, which then finally leads to an increase in the intracellular ceramide concentration (Reinehr et al., 2006).

In hepatocytes, several types of intracellular chloride channels have been identified at functional, pharmacological, and molecular levels (Li and Weinman, 2002). At least five members of the ClC-type chloride channel family (ClC3–7) are expressed in intracellular vesicles, including endosomes (Jentsch et al., 2005), and are also present in hepatocytes (Li and Weinman, 2002). Involvement of ClC3 in endosomal acidification by Cl^- shunting of the interior-positive membrane potential created by the vacuolar proton pump has been demonstrated in a study using control versus ClC3-deficient hepatocytes (Hara-Chikuma et al., 2005).

Apart from hyperosmolarity, CD95L (Reinehr and Häussinger, unpublished data) and proapoptotic bile salts (Becker et al., 2007b) also elicit an increase in $[Cl^-]_i$ as detected by fluorescence measurements at the single cell level using the chloride-sensitive dye N-(ethoxycarbonylmethyl)-6-methoxyquinolinium bromide. This increase in $[Cl^-]_i$ was again paralleled by endosomal acidification and subsequent ASM-mediated ceramide generation (Reinehr and Häussinger, unpublished data; Becker et al., 2007b).

In addition to ceramide generation, hyperosmotic exposure rapidly stimulates the production of reactive oxygen species (Reinehr et al., 2003b) as an initial signaling event, which triggers CD95 activation and its translocation to the plasma membrane. ROS formation upon hyperosmotic exposure was measured by the use of 5-(and 6-)-chloromethyl-2′,7′-dichlorodihydrofluorescein diacetate fluorescence at the single cell level. Activation of the NADPH oxidase underlies the rapid oxidative stress response of hepatocytes after hyperosmotic exposure (Reinehr et al., 2006), CD95L (Reinehr et al., 2005a), or hydrophobic bile acids (Becker et al., 2007b; Reinehr et al., 2005b). Rat hepatocytes express mRNAs for Nox 1, 2, and 4 and Duox 1 and 2 as well as Nox 2 and $p47^{phox}$ protein (Reinehr et al., 2005a,b). Hyperosmolarity rapidly induces an activating $p47^{phox}$-serine phosphorylation (Reinehr et al., 2006), which is known to trigger the self-assembly of the NADPH-oxidase subunits and subsequent ROS generation. Immunoprecipitation of the $p47^{phox}$ protein allows for detection of serine phosphorylation of the latter protein by Western blot analysis as a surrogate marker for NADPH oxidase activation (Bataller et al., 2003; Reinehr et al., 2005a,b, 2006). Hyperosmotic $p47^{phox}$-serine phosphorylation was sensitive to inhibitors of PKCζ (Reinehr et al., 2006), which is known to target the $p47^{phox}$

protein (El Benna *et al.*, 1996). PKCζ is known to be activated by ceramide (Mathias *et al.*, 1998). Therefore, one functional consequence of the ASM-mediated increase in intracellular ceramide levels upon hyperosmotic exposure is an activation of PKCζ, as demonstrated by its translocation to the plasma membrane using immunocytochemical PKCζ staining (Reinehr *et al.*, 2006). Thus it is likely that hyperosmolarity induces a rapid activation of ASM and a subsequent increase in intracellular ceramide levels, which triggers PKCζ activation and subsequent p47^{phox}-serine phosphorylation, which leads to NADPH-oxidase activation and ROS generation (Reinehr *et al.*, 2006).

Hyperosmolarity-induced ROS generation was identified as an important upstream event that triggers activation of both the Src-family kinase Yes and the c-Jun-N-terminal kinases. In order to detect Yes activation, the Yes protein was immunoprecipitated and activating phosphorylation on position Y^{418} was detected by subsequent Western blot analysis using phospho-specific antibodies (Piiper *et al.*, 2003, Reinehr *et al.*, 2004a,b, 2005a). Activated Yes was further demonstrated to associate with the EGFR upon hyperosmotic exposure. This protein/protein interaction was shown by Yes immunoprecipitation and subsequent Western blot analysis for EGFR in the immunoprecipitated sample as a surrogate marker for the association of the latter proteins. In the following, the EGFR becomes tyrosine phosphorylated by Yes, as shown by its SU6656 sensitivity on position Y^{845} (according to the nomenclature of the human EGFR), followed by autophosphorylation at position Y^{1173} (according to the nomenclature of the human EGFR) and EGFR activation (Reinehr *et al.*, 2004a,b, 2005a).

Whereas Yes activates the EGFR, the JNK signal is delayed and mediates an association of the EGFR with CD95, as shown by coimmunoprecipitation and fluorescence resonance energy transfer (FRET) studies, respectively (Eberle *et al.*, 2005, 2007; Reinehr *et al.*, 2002, 2003a,b, 2004a,b). In order to gain further insight into EGFR/CD95 interactions in response to hyperosmolarity, the FRET technique was used for the investigation of protein/protein interactions in the intact cell. A so-called *FRET effect* occurs when two proteins, which are coupled to corresponding fluorescent proteins, for example, yellow (YFP) and cyan fluorescent protein (CFP), interact and thereby get as close as 3 to 10 nm. Under these conditions, the light emitted upon excitation of the CFP moiety will excite the YFP moiety and yield the so-called *FRET signal* (Siegel *et al.*, 2000). FRET experiments in CD95-YFP/EGFR-CFP cotransfected human hepatoma (Huh7) cells, which lack endogenous CD95 expression (Seki *et al.*, 1999), revealed an intracellular CD95/EGFR association within 30 min, which was followed by a translocation to the plasma membrane of the protein complex after 120 min (Eberle *et al.*, 2005, 2007) (Fig. 8.1). After association of the CD95 and EGFR, CD95 becomes tyrosine phosphorylated by the EGFR tyrosine kinase activity, which provides a signal for a microtubule-dependent targeting of the EGFR/CD95 complex to the plasma membrane (Eberle *et al.*, 2005, 2007), where formation of the death-inducing signaling complex occurs. DISC formation was demonstrated by CD95 immunoprecipitation and subsequent Western blot analysis for caspase 8 and FADD in the immunoprecipitated samples in order to demonstrate the recruitment of the latter proteins to the CD95 death receptor. DISC formation is then followed by an activation of caspase 8 and 3, as measured by the respective fluorimetric capase activity assays (Reinehr *et al.*, 2002). Covalent modifications of the CD95, that is, tyrosine phosphorylation (Reinehr *et al.*, 2003a,b), serine/threonine phosphorylation (Reinehr and Häussinger, 2004; see later) or tyrosine nitration (Reinehr *et al.*, 2004c; see later), were detected by CD95 immunoprecipitation and subsequent Western blot analysis using antiphosphotyrosine, -serine, -threonine, or nitrotyrosine-specific antibodies, respectively.

Human CD95 contains three tyrosine residues: one in the extracellular domain (Y^{91}) and two within the death domain (Y^{232} and Y^{291}) (Oehm *et al.*, 1992). It has been demonstrated by single, double, and triple mutations (Y $\rightarrow$ F) of fluorescent CD95 constructs using side-directed mutagenesis (Eberle *et al.*, 2005, 2007) that phosphorylation of both tyrosine residues within the death domain (Y^{232} and Y^{291}) is critical not only for hyperosmolarity-induced CD95 membrane trafficking and DISC formation (Eberle *et al.*, 2005), but also for oligomerization of the CD95 receptor, which precedes CD95 membrane translocation and DISC formation (Eberle *et al.*, 2007). In order to demonstrate whether CD95 tyrosine phosphorylation is not only a prerequisite for CD95 membrane trafficking and DISC formation, but also for CD95 oligomerization, the FRET technique was again used (Fig. 8.2). Cotransfection of CD95-YFP and CD95-CFP constructs into Huh7, which lack endogenous CD95 expression (Seki *et al.*, 1999), allows for detection of CD95/CD95 oligomerization, which was shown to take place within 30 min already in the cytosol. In addition, CD95/CD95 oligomerization was demonstrated to depend on tyrosine phosphorylation of both tyrosine residues within the death domain, that is, Y^{232} and Y^{291}, of at least one CD95, as demonstrated by the use of single, double, and triple mutations (Y $\rightarrow$ F) of YFP- and CFP-coupled CD95 constructs, respectively (Eberle *et al.*, 2007).

A similar sequence of events is induced upon treatment of hepatocytes with CD95L (Reinehr *et al.*, 2005a) or hydrophobic bile acids (Becker *et al.*, 2007b; Reinehr *et al.*, 2005b), and in contrast to mild hyperosmolarity, that is, 405 literl/liter, apoptosis is executed as detected by terminal deoxynucleotidyl transferase-mediated X-dUTP nick-end labeling of FITC-conjugated deoxyuridine triphosphate or annexin V assay under these conditions (Reinehr *et al.*, 2002, 2003a,b). Although the shrinkage of hepatocytes induced by a mild hyperosmotic challenge already targets CD95 to the plasma membrane and leads to DISC formation with activation of caspase 8 and 3, this does not lead to hepatocyte apoptosis, suggesting that death-inducing signals are antagonized by yet unknown survival signals. However, hyperosmolarity sensitizes hepatocytes to apoptosis induced by the CD95 ligand under these conditions (Reinehr et al., 2002, 2003b). These findings suggest that already mild dehydrating conditions may be sufficient to trigger CD95 membrane trafficking and to sensitize hepatocytes toward apoptotic cell death. In this context it is interesting to note that Kupffer cells were shown to express the membrane-bound CD95L in response to endotoxin (Müschen et al., 1998, 1999), which binds to its receptor only when presented at the hepatocyte surface (Reinehr et al., 2002), and might explain why shrunken hepatocytes are more susceptible to stress-induced cell death. However, hyperosmolarity exceeding 505 mosmol/liter not only leads to CD95 activation, but also induces apoptotic cell death (Reinehr et al., 2006).

As outlined earlier, hyperosmotic CD95 tyrosine phosphorylation is an essential step for CD95 oligomerization and membrane targeting, DISC formation, and apoptosis induction. In line with this, pharmacological maneuvers, which prevent hyperosmotic CD95 tyrosine phosphorylation, such as inhibitors of anion channels, vacuolar-type H^+-ATPase, acidic sphingomyelinase, PKCζ, p47^{phox}, Yes, JNK, and EGFR activation, abolish translocation of the CD95 to the plasma membrane. In addition, taurine and glutamine antagonize hyperosmotic CD95 targeting to the plasma membrane, probably because of an attenuation of hyperosmotic shrinkage (Reinehr et al., 2002, 2003b).

Inhibition of hyperosmotic CD95 activation and sensitization toward apoptosis are found following other covalent modifications of the CD95. For example, cyclic AMP, which exerts antiapoptotic effects in a variety of cell types, including hepatocytes (Webster and Anwer, 1998), inhibits hyperosmotic CD95 activation in hepatocytes. This is, on the one hand, explained by a protein kinase A (PKA)-dependent inhibition of the hyperosmotic EGFR activation, which prevents CD95 tyrosine phosphorylation, and, on the other hand, by a PKA-dependent serine/threonine phosphorylation of the CD95 (Reinehr and Häussinger, 2004). Serine/threonine phosphorylation of CD95 was shown to act as an internalization signal (Reinehr and Häussinger, 2004), whereas CD95 tyrosine phosphorylation is a signal for membrane targeting (Reinehr et al., 2003b). Another covalent modification of the CD95 is nitration of CD95 tyrosine residues (Reinehr et al., 2004c), which was demonstrated to occur in response to endotoxin or peroxynitrite ($ONOO^-$). $ONOO^-$-induced CD95 tyrosine nitration prevents CD95 tyrosine phosphorylation in response to hyperosmolarity (Reinehr et al., 2004c). Conversely, when CD95 tyrosine phosphorylation was induced prior to $ONOO^-$ addition, no CD95 tyrosine nitration was inducible, indicating that CD95 tyrosine phosphorylation and nitration are mutually exclusive (Reinehr et al., 2004c). CD95 tyrosine nitration is also induced *in vivo* following endotoxin administration, probably because of the induction of NOS and MPO (Bogdan, 2001; Reinehr *et al.*, 2004c). Under these conditions, CD95 tyrosine nitration may serve to limit the apoptotic cell death in response to inflammatory stimuli.

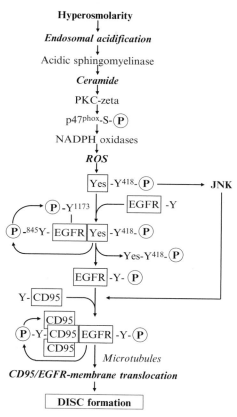

Figure 8.3 Hyperosmolarity-induced CD95 activation in hepatocytes. Our current view is summarized with respect to hyperosmolarity-induced activation of the CD95 system. Hyperosmotic exposure leads to acidification of an endosomal compartment, which activates ASM. The subsequent ceramide formation then activates PKCζ. The latter leads to an activating serine phosphorylation of the NADPH oxidase regulatory subunit p47phox, which is followed by the generation of reactive oxygen species (ROS). ROS formation triggers activation of the Src family kinase Yes, which then associates with and activates the epidermal growth factor receptor (EGFR). EGFR then dissociates from Yes and associates with CD95 in a c-Jun-N-terminal kinase (JNK)-dependent way, which leads to EGFR-tyrosine kinase-mediated CD95 tyrosine phosphorylation, which is a prerequisite for CD95 oligomerization and microtubule-dependent translocation of CD95 to the plasma membrane, which allows for subsequent FADD and caspase 8 recruitment to the CD95, that is, DISC formation.

3. Concluding Remarks

Hyperosmotic hepatocyte shrinkage sensitizes the cells toward CD95 ligand-induced apoptosis by activating the CD95 system. Figure 8.3 summarizes our current view on the signaling events involved in hyperosmotic activation of the CD95 system in hepatocytes. The hyperosmotically induced cascade of signaling events starts with endosomal acidification triggering an ASM-derived increase in the intracellular ceramide concentration, which leads via PKCζ and p47^{phox} to NADPH oxidase activation and subsequent ROS formation as an important upstream event. ROS generation allows via Yes, JNK, and EGFR activation for CD95 tyrosine phosphorylation as a prerequisite for CD95 oligomerization and targeting to the plasma membrane, DISC formation, and apoptosis induction or sensitization. Other covalent modifications, such as CD95 tyrosine nitration or CD95 serine/threonine phosphorylation, can counteract the CD95 activation process. The findings not only provide a mechanistic explanation for the high susceptibility of dehydrated cells for apoptosis, but also give insight into the early signaling events involved in apoptotic cell death in general.

ACKNOWLEDGMENTS

Our studies were supported by Deutsche Forschungsgemeinschaft (DFG) through Sonderforschungsbereich 575 "Experimentelle Hepatologie" (Düsseldorf). Excellent work by Stephan Becker, Andrea Eberle, Boris Görg, Elisabeth Winands, and Claudia Rupprecht is gratefully acknowledged.

REFERENCES

Bataller, R., Schwabe, R. F., Choi, Y. H., Yang, L., Paik, Y. H., Lindquist, J., Qian, T., Schoonhoven, R., Hagedorn, C. H., Lemasters, J. J., and Brenner, D. A. (2003). NADPH oxidase signal transduces angiotensin II in hepatic stellate cells and is critical in hepatic fibrosis. *J. Clin. Invest.* **112,** 1383–1394.

Becker, S., Reinehr, R., Graf, D., vom Dahl, S., and Häussinger, D. (2007a). Hydrophobic bile salts induce hepatocyte shrinkage via NADPH oxidase activation. *Cell Physiol. Biochem.* **19,** 89–98.

Becker, S., Reinehr, R., Grether-Beck, S., Eberle, A., and Häussinger, D. (2007b). Hydrophobic bile salts trigger ceramide formation through endosomal acidification. *Biol. Chem.* **388,** 185–196.

Bogdan, C. (2001). Nitric oxide and the immune response. *Nat. Immunol.* **2,** 907–916.

Bokoch, G. M., and Diebold, B. A. (2002). Current molecular models for NADPH oxidase regulation by Rac GTPase. *Blood* **100,** 2692–2696.

Bortner, C. D., and Cidlowski, J. A. (1996). Absence of volume regulatory mechanisms contributes to the rapid activation of apoptosis in thymocytes. *Am. J. Physiol.* **40,** C950–C961.

Bortner, C. D., and Cidlowski, J. A. (2001). Flow cytometric analysis of cell shrinkage and monovalent ions during apoptosis. *Methods Cell Biol.* **66,** 49–67.

Bortner, C. D., and Cidlowski, J. A. (2002). Apoptotic volume decrease and the incredible shrinking cell. *Cell Death Differ.* **9,** 1307–1310.

Callahan, J. W., Jones, C. S., Davidson, D. J., and Shankaran, P. (1983). The active site of lysosomal sphingomyelinase: Evidence for the involvement of hydrophobic and ionic groups. *J. Neurosci. Res.* **10,** 151–163.

Crowley, C. L., Pyne, C. M., Bernstein, H., Bernstein, C., and Roe, D. (2000). The NAD$^+$ precursors, nicotinic acid and nicotinamide protect cells against apoptosis induced by a multiple stress inducer, deoxycholate. *Cell Death Differ.* **7,** 314–326.

Eberle, A., Reinehr, R., Becker, S., and Häussinger, D. (2005). Fluorescence resonance energy transfer analysis of proapoptotic CD95-EGF receptor interactions in Huh7 cells. *Hepatology* **41,** 315–326.

Eberle, A., Reinehr, R., Becker, S., Keitel, V., and Häussinger, D. (2007). CD95 tyrosine phosphorylation is required for CD95 oligomerization. *Apoptosis* **12,** 719–729.

Edens, W. A., Sharling, L., Cheng, G., Shapira, R., Kinkade, J. M., Lee, T., Edens, H. A., Tang, X., Sullards, C., Flaherty, D. B., Benian, G. M., and Lambeth, J. D. (2001). Tyrosine cross-linking of extracellular matrix is catalyzed by Duox, a multidomain oxidase/peroxidase with homology to the phagocyte oxidase subunit gp91phox. *J. Cell. Biol.* **154,** 879–891.

El Benna, J., Faust, R. P., Johnson, J. L., and Babior, B. M. (1996). Phosphorylation of the respiratory burst oxidase subunit p47phox as determined by two-dimensional phospho-peptide mapping: Phosphorylation by protein kinase C, protein kinase A, and a mitogen-activated protein kinase. *J. Biol. Chem.* **271,** 6374–6378.

Friis, M. B., Friborg, C. R., Schneider, L., Nielsen, M. B., Lambert, I. H., Christensen, S. T., and Hoffmann, E. K. (2005). Cell shrinkage as a signal to apoptosis in NIH 3T3 fibroblasts. *J. Physiol.* **567,** 427–443.

Gomez-Angelats, M., Bortner, C. D., and Cidlowski, J. A. (2000). Protein kinase C (PKC) inhibits fas receptor-induced apoptosis through modulation of the loss of K$^+$ and cell shrinkage: A role for PKC upstream of caspases. *J. Biol. Chem.* **275,** 19609–19619.

Goni, F. M., and Alonso, A. (2002). Sphingomyelinases: Enzymology and membrane activity. *FEBS Lett.* **531,** 38–46.

Graf, J., and Häussinger, D. (1996). Ion transport in hepatocytes: Mechanisms and correlations to cell volume, hormone actions and metabolism. *J. Hepatol.* **24,** 53–77.

Hallbrucker, C., vom Dahl, S., Ritter, M., Lang, F., and Häussinger, D. (1994). Effects of urea on K$^+$ fluxes and cell volume in perfused rat liver. *Pflüg. Arch.* **428,** 552–560.

Hara-Chikuma, M., Yang, B., Sonawane, N. D., Sasaki, S., Uchida, S., and Verkman, A. S. (2005). ClC-3 chloride channels facilitate endosomal acidification and chloride accumulation. *J. Biol. Chem.* **280,** 1241–1247.

Häussinger, D., Stehle, T., and Lang, F. (1990). Volume regulation in liver: Further characterization by inhibitors and ionic substitutions. *Hepatology* **11,** 243–254.

Jentsch, T. J., Poet, M., Fuhrmann, J. C., and Zdebik, A. A. (2005). Physiological functions of CLC Cl- channels gleaned from human genetic disease and mouse models. *Annu. Rev. Physiol.* **67,** 779–807.

Kerr, J. F., Harmon, B., and Searle, J. (1974). An electron-microscope study of cell deletion in the anuran tadpole tail during spontaneous metamorphosis with special reference to apoptosis of striated muscle fibers. *J. Cell Sci.* **14,** 571–585.

Krähenbühl, S., Talos, C., Fischer, S., and Reichen, J. (1994). Toxicity of bile acids on the electron transport chain of isolated liver mitochondria. *Hepatology* **19,** 471–479.

Lambeth, J. D. (2002). Nox/Duox family of nicotinamide adenine dinucleotide (phosphate) oxidases. *Curr. Opin. Hematol.* **9,** 11–17.

Lambeth, J. D., Cheng, G., Arnold, R. S., and Edens, W. A. (2000). Novel homologs of gp91phox. *Trends Biochem. Sci.* **25,** 459–461.

Lang, F., Madlung, J., Bock, J., Lukewille, U., Kaltenbach, S., Lang, K. S., Belka, C., Wagner, C. A., Lang, H. J., Gulbins, E., and Lepple-Wienhues, A. (2000a). Inhibition of Jurkat-T-lymphocyte Na$^+$/H$^+$-exchanger by CD95 (Fas/Apo-1)-receptor stimulation. *Pflüg. Arch.* **440,** 902–907.

Lang, F., Madlung, J., Uhlemann, A. C., Risler, T., and Gulbins, E. (1998). Cellular taurine release triggered by stimulation of the Fas (CD95) receptor in Jurkat lymphocytes. *Pflüg. Arch* **436,** 377–383.

Lang, F., Ritter, M., Volkl, H., and Häussinger, D. (1993). The biological significance of cell volume. *Ren. Physiol. Biochem.* **16,** 48–65.

Li, X., and Weinman, S. A. (2002). Chloride channels and hepatocellular function: Prospects for molecular identification. *Annu. Rev. Physiol.* **64,** 609–633.

Maeno, E., Ishizaki, Y., Kanaseki, T., Hazama, A., and Okada, Y. (2000). Normotonic cell shrinkage because of disordered volume regulation is an early prerequisite to apoptosis. *Proc. Natl. Acad. Sci. USA* **97,** 9487–9492.

Marchesini, N., and Hannun, Y. A. (2004). Acid and neutral sphingomyelinases: Roles and mechanisms of regulation. *Biochem. Cell. Biol.* **82,** 27–44.

Mathias, S., Pena, L. A., and Kolesnick, R. N. (1998). Signal transduction of stress via ceramide. *Biochem. J.* **335,** 465–480.

Moriyama, Y., and Nelson, N. (1987). The purified ATPase from chromaffin granule membranes is an anion-dependent proton pump. *J. Biol. Chem.* **262,** 9175–9180.

Müschen, M., Warskulat, U., Douillard, P., Gilbert, E., and Häussinger, D. (1998). Regulation of CD95 (APO-1/Fas)-receptor and ligand expression by lipopolysaccharide and dexamethasone in parenchymal and nonparenchymal rat liver cells. *Hepatology* **27,** 200–208.

Müschen, M., Warskulat, U., Peters-Regehr, T., Bode, J. G., Kubitz, R., and Häussinger, D. (1999). Involvement of CD95 (Apo-1/Fas)-ligand expressed by rat Kupffer cells in hepatic immunoregulation. *Gastroenterology* **116,** 666–677.

Nauseef, W. M. (2004). Assembly of the phagocyte NADPH oxidase. *Histochem. Cell Biol.* **122,** 277–291.

Nobel, C. S., Aronson, J. K., van den Dobbelsteen, D. J., and Slater, A. F. (2000). Inhibition of Na+/K+-ATPase may be one mechanism contributing to potassium efflux and cell shrinkage in CD95-induced apoptosis. *Apoptosis* **5,** 153–163.

Oehm, A., Behrmann, I., Falk, W., Pawlita, M., Maier, G., Klas, C., Li-Weber, M., Richards, S., Dhein, J., Trauth, J., Pansting, H., and Krammer, P. H. (1992). Purification and molecular cloning of the APO-1 cell surface antigen, a member of the tumor necrosis factor/nerve growth factor receptor superfamily. *J. Biol. Chem.* **267,** 10709–10715.

Ohanian, J., and Ohanian, V. (2001). Sphingolipids in mammalian cell signalling. *Cell. Mol. Life Sci.* **58,** 2053–2068.

Pani, G., Bedogni, B., Colavitti, R., Anzevino, R., Borrello, S., and Galeotti, T. (2001). Cell compartmentalization in redox signaling. *IUBMB Life* **52,** 7–16.

Pazoles, C. J., Creutz, C. E., and Pollard, H. B. (1980). Evidence for direct coupling of proton and anion transport in chromaffin granules. *Ann. N.Y. Acad. Sci.* **358,** 354–355.

Piiper, A., Elez, R., You, S. J., Kronenberger, B., Loitsch, S., Roche, S., and Zeuzem, S. (2003). Cholecystokinin stimulates extracellular signal-regulated kinase through activation of the epidermal growth factor receptor, Yes, and protein kinase C. *J. Biol. Chem.* **278,** 7065–7072.

Reinehr, R., Becker, S., Braun, J., Eberle, A., Grether-Beck, S., and Häussinger, D. (2006). Endosomal acidification and activation of NADPH oxidase isoforms are upstream events in hyperosmolarity-induced hepatocyte apoptosis. *J. Biol. Chem.* **281,** 23150–23166.

Reinehr, R., Becker, S., Eberle, A., Grether-Beck, S., and Häussinger, D. (2005a). Involvement of NADPH oxidase isoforms and Src family kinases in CD95-dependent hepatocyte apoptosis. *J. Biol. Chem.* **280,** 27179–27194.

Reinehr, R., Becker, S., Höngen, A., and Häussinger, D. (2004a). The Src family kinase Yes triggers hyperosmotic activation of the epidermal growth factor receptor and CD95. *J. Biol. Chem.* **279,** 23977–23987.

Reinehr, R., Becker, S., Keitel, V., Eberle, A., Grether-Beck, S., and Häussinger, D. (2005b). Bile salt-induced apoptosis involves NADPH oxidase isoform activation and hepatocyte shrinkage. *Gastroenterology* **129,** 2009–2031.

Reinehr, R., Becker, S., Wettstein, M., and Häussinger, D. (2004b). Involvement of the Src family kinase yes in bile salt-induced apoptosis. *Gastroenterology* **127,** 1540–1557.

Reinehr, R., Fischer, R., Graf, D., Schliess, F., and Häussinger, D. (2002). Hyperosmolarity triggers CD95 membrane trafficking and sensitizes rat hepatocytes towards CD95 ligand-induced apoptosis. *Hepatology* **36,** 602–614.

Reinehr, R., Görg, B., Höngen, A., and Häussinger, D. (2004c). CD95-tyrosine nitration inhibits hyperosmotic and CD95 ligand-induced CD95 activation in rat hepatocytes. *J. Biol. Chem.* **279,** 10364–10373.

Reinehr, R., Graf, D., and Häussinger, D. (2003a). Bile-salt-induced apoptosis involves epidermal growth factor receptor-dependent tyrosine phosphorylation of CD95 in rat hepatocytes. *Gastroenterology* **125,** 839–853.

Reinehr, R., and Häussinger, D. (2004). Inhibition of apoptosis by cyclic AMP involves protein kinase A-dependent serine/threonine phosphorylation of CD95. *Gastroenterology* **126,** 249–262.

Reinehr, R., Schliess, F., and Häussinger, D. (2003b). Hyperosmolarity and CD95L trigger CD95/EGF receptor association and tyrosine phosphorylation of CD95 as prerequisites for CD95 membrane trafficking and DISC formation. *FASEB J.* **17,** 731–733.

Rosette, C., and Karin, M. (1996). Ultraviolet light and osmotic stress: Activation of the JNK cascade through multiple growth factor and cytokine receptors. *Science* **274,** 1194–1197.

Saha, N., Schreiber, R., vom Dahl, S., Lang, F., Gerok, W., and Häussinger, D. (1993). Endogenous hydroperoxide formation, cell volume and cellular K^+ balance in perfused rat liver. *Biochem. J.* **296,** 701–707.

Schreiber, R., and Häussinger, D. (1995). Characterization of the swelling-induced alkalinization of endocytotic vesicles in fluorescein isothiocyanate-dextran-loaded rat hepatocytes. *Biochem. J.* **309,** 19–24.

Schreiber, R., Stoll, B., Lang, F., and Häussinger, D. (1994). Effects of aniso-osmolarity and hydroperoxides on intracellular pH in isolated rat hepatocytes as assessed by (2′,7′)-bis (carboxyethyl)-5(6)-carboxyfluorescein and fluorescein isothiocyanate-dextran fluorescence. *Biochem. J.* **303,** 113–120.

Schreiber, R., Zhang, F., and Häussinger, D. (1996). Regulation of vesicular pH in liver macrophages and parenchymal cells by ammonia and anisotonicity as assessed by fluorescein isothiocyanate-dextran fluorescence. *Biochem. J.* **315,** 385–392.

Seki, S., Kitada, T., Sakaguchi, H., Kawada, N., Iwai, S., Kadoya, H., and Nakatani, K. (1999). Expression of Fas and Bcl-2 proteins and induction of apoptosis in human hepatocellular carcinoma cell lines. *Med. Electron Microsc.* **32,** 199–203.

Siegel, R. M., Chan, F. K., Zacharias, D. A., Swofford, R., Holmes, K. L., Tsien, R. Y., and Lenardo, M. J. (2000). Measurement of molecular interactions in living cells by fluorescence resonance energy transfer between variants of the green fluorescent protein. *Sci. STKE* **2000,** PL1.

Sokol, R. J., McKim, J. M., Goff, M. C., Ruyle, S. Z., Devereaux, M. W., Han, D., Packer, L., and Everson, G. (1998). Vitamin E reduces oxidant injury to mitochondria and the hepatotoxicity of taurochenodeoxycholic acid in the rat. *Gastroenterology* **114,** 164–174.

Sokol, R. J., Winkelhofer-Roob, B. M., Devereaux, M. W., and McKim, J. M. (1995). Generation of hydroperoxides in isolated rat hepatocytes and hepatic mitochondria exposed to hydrophobic bile acids. *Gastroenterology* **109,** 1249–1256.

Stadler, J., Bentz, B. G., Harbrecht, B. G., Di Silvio, M., Curran, R. D., Billiar, T. R., Hoffman, R. A., and Simmons, R. L. (1992). Tumor necrosis factor alpha inhibits hepatocyte mitochondrial respiration. *Ann. Surg.* **216,** 539–546.

Szabo, I., Lepple-Wienhues, A., Kaba, K. N., Zoratti, M., Gulbins, E., and Lang, F. (1998). Tyrosine kinase-dependent activation of a chloride channel in CD95-induced apoptosis in T lymphocytes. *Proc. Natl. Acad. Sci. USA* **95,** 6169–6174.

Thompson, G. J., Langlais, C., Cain, K., Conley, E. C., and Cohen, G. M. (2001). Elevated extracellular K+ inhibits death-receptor-and chemical-mediated apoptosis prior to caspase activation and cytochrome c release. *Biochem. J.* **357,** 137–145.

Vignais, P. V. (2002). The superoxide-generating NADPH oxidase: Structural aspects and activation mechanism. *Cell Mol. Life Sci.* **59,** 1428–1459.

Webster, C. R., and Anwer, M. S. (1998). Cyclic adenosine monophosphate-mediated protection against bile acid-induced apoptosis in cultured rat hepatocytes. *Hepatology* **27,** 1324–1331.

Yoshida, Y., Arimoto, K., Sato, M., Sakuragawa, N., Arima, M., and Satoyoshi, E. (1985). Reduction of acid sphingomyelinase activity in human fibroblasts induced by AY-9944 and other cationic amphiphilic drugs. *J. Biochem.* **98,** 1669–1679.

Yu, S. P., Canzoniero, L. M., and Choi, D. W. (2001). Ion homeostasis and apoptosis. *Curr. Opin. Cell Biol.* **13,** 405–411.

Zeuzem, S., Feick, P., Zimmermann, P., Haase, W., Kahn, R. A., and Schulz, I. (1992). Intravesicular acidification correlates with binding of ADP-ribosylation factor to microsomal membranes. *Proc. Natl. Acad. Sci. USA* **89,** 6619–6623.

CHAPTER NINE

NEW APPROACHES FOR DETERMINING APOPTOTIC VOLUME DECREASE IN CELLS

Carl D. Bortner, Maria I. Sifre, *and* John A. Cidlowski

Contents

Laboratory of Signal Transduction, National Institute of Environmental Health Sciences, Department of Health and Human Services, National Institutes of Health, Research Triangle Park, North Carolina

Methods in Enzymology, Volume 428
ISSN 0076-6879, DOI: 10.1016/S0076-6879(07)28009-7

Abstract

The loss of cell volume or cell shrinkage, termed apoptotic volume decrease (AVD), is a classical characteristic of apoptosis. Microscopy, Coulter electronic sizing, and/or flow cytometry has traditionally been the means to measure this characteristic of apoptosis. Although electronic cell sizing allows for precise determination of changes in cell size, these measurements provide data on the entire population of apoptotic cells. In contrast, flow cytometry examines and separates unique populations of apoptotic cells based on the light-scattering properties of the cells to determine alterations in cellular dimensions. However, this technique does not provide exact measurements of cell size and volume. The Cell Lab Quanta SC flow cytometer combines the power of electronic sizing with the ability to isolate and examine unique populations of apoptotic cells to determine exact changes in cell size as they undergo cell death. This chapter describes several methods for using the Cell Lab Quanta SC to study AVD during apoptosis.

1. INTRODUCTION

A delicate balance between cell growth and cell death occurs continually throughout one's life to maintain cellular homeostasis. For the health of an organism, it is imperative that cells no longer needed in the body be removed efficiently without an inflammatory response in the surrounding cells or tissue. The removal of unwanted or nonessential cells is largely accomplished through a physiological cell death process known as apoptosis (Kerr *et al.*, 1972), which is an inherent program in all cells that upon activation leads to a systematic series of events resulting in the destruction of the cell. Apoptosis, also referred to as programmed cell death, is defined by a unique set of morphological and biochemical characteristics, including cell shrinkage, nuclear condensation, and internucleosomal DNA fragmentation. These features of apoptotic cells, specifically the loss of cell volume or cell shrinkage, can be used to distinguish this mode of cell death from an otherwise accidental form of death known as necrosis. During necrosis the cells swell and eventually lyse, often resulting in an inflammatory response detrimental to the surrounding cells and tissue. Thus, apoptosis is noninflammatory and is an efficient and silent way in which an organism can remove a population of cells at a given time or in response to a given stimulus.

The loss of cell volume or cell shrinkage is an important aspect of apoptosis for the efficient elimination of unwanted cells. Initially described microscopically as "shrinking necrosis" to highlight this unique morphological feature of cell death (Kerr, 1971), more recently the term "apoptotic volume decrease" (AVD) has been used to refer to this normotonic loss of cell volume that distinguishes this mode of cell death (Maeno et al., 2000). Most cells can combat a change in cell volume through the activation of inherent volume regulatory responses, thus regulating the activity of ion channels and other ionic transport pathways to maintain a consistent cell size. However, during apoptosis, this programmed cell death process inhibits the ability of the cells to maintain a specific or normal cell volume. What has become clear since the late 1990s is not just the importance of a loss of cell volume during apoptosis, but the underlying movement of ions that is critical for the programmed cell death process to occur efficiently (Bortner and Cidlowski, 1998, 2004; Lang et al., 2000; Yu and Choi, 2000). Normal homeostatic levels of intracellular potassium and sodium appear to have a repressive effect on the apoptotic process, specifically regarding caspase activation and apoptotic nuclease activity (Bortner et al., 1997; Cain et al, 2001; Hughes et al., 1997; Thompson et al., 2001). Thus for the effective activation of the apoptotic machinery, a decrease in intracellular ions that results in AVD is required.

2. CLASSICAL WAYS TO STUDY CELL SHRINKAGE OR AVD DURING APOPTOSIS

The loss of cell volume or AVD has been a hallmark of apoptosis. Early studies on this mode of cell death examined cell shrinkage mainly through the use of microscopy and electronic sizing. Additionally, isolation of these shrunken cells was accomplished by using density-gradient centrifugation. Thomas and Bell (1981) reported a significant decrease in cell size by electronic sizing in freshly isolated rat thymocytes upon glucocorticoid (dexamethasone) treatment. Radiation-induced cell death of rat thymocytes resulted in a rapid 25% decrease in cell size, followed by a slower reduction in cell volume to 57% their initial cell size compared to untreated controls using Coulter counter sizing (Klassen et al., 1993). Additionally, human eosinophils upon cytokine withdrawal had a 63.2% reduction in cell volume after 48 h as determined by electronic sizing (Beauvais et al., 1995).

Subsequently, flow cytometry has been used to examine not only cell size, but also other apoptotic parameters. As early as 1988, Compton et al. (1988) were the first to use a flow cytometric approach to study glucocorticoid-induced apoptosis in rat thymocytes. Flow cytometry analyzes changes in the light-scattering properties of the cells, specifically in the forward-scattered direction, to determine a change in cell size. Cells that have a greater ability to scatter light in the forward direction are of a larger cell size, whereas cells that have a reduced ability to scatter light in the forward direction are of a smaller or shrunken cell size. Benson et al. (1996) used this method, along with electronic sizing, to show a 58% decrease in

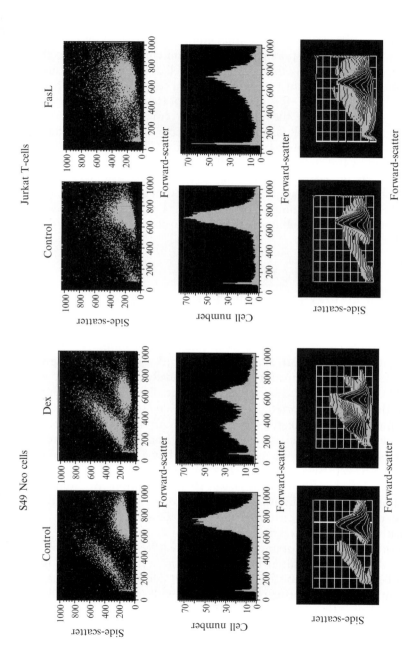

Figure 9.1 Changes in the light-scattering properties of dexamethasone (Dex)–treated S49 Neo– and FasL–treated Jurkat T cells. The observation of normal and shrunken apoptotic cells by flow cytometry is based on changes in the light-scattering properties of the cells. The degree of this change in light scatter is both cell type and stimulus specific. The shrunken population of apoptotic cells can be clearly observed by flow cytometry for S49 Neo cells treated with Dex for 48 h. In both a forward-scatter versus side-scatter dot plot and a forward-scatter histogram, two distinct populations are present comparing Dex–treated cells to control cells. However, similar examination of Jurkat T cells treated with Fas ligand for 4 h shows that the normal and shrunken populations of these cells cannot be clearly distinguished. (See color insert.)

cell size after 48 h following dexamethasone treatment of the T-cell line CEM-C7A.

Flow cytometry has now become a routine method for determining changes in cell size during cell death, along with analyzing various other apoptotic characteristics based on individual fluorescent markers or indicators that are incorporated into the experiment. However, there are limits in the degree of analysis that can be accomplished even with this technology, and a precise measure of cell size cannot be ascertained directly. As shown in Fig. 9.1, depending on the cell type and/or stimulus used to induce apoptosis, unique patterns of forward-scattered light are observed as cells undergo cell death. Treatment of a murine immature T-cell line (S49 Neo cells) with dexamethasone results in two distinct populations of cells: one population of a larger cell size, similar to the control cells, and a second population having a reduction in forward-scattered light, indicating smaller or shrunken cells (Fig. 9.1). However, when Jurkat T cells are induced to undergo apoptosis with Fas ligand, a decrease in cell size is also observed, but a clear distinction between normal and shrunken populations of cells is not evident (Fig. 9.1). Thus only a general conclusion of the presence of a smaller or shrunken population of cells can be accomplished.

The Cell Lab Quanta SC by Beckman Coulter combines the ability to analyze specific fluorescent parameters related to cell death, similar to a conventional flow cytometer, along with simultaneously determining the Coulter volume of these same cells, allowing for precise cell size and volume measurements in apoptotic cells or under other physiological conditions. Thus, this instrument combines the two traditional methods of analyzing AVD into one single unit. The following sections describe the protocols for use of the Cell Lab Quanta SC for assessing changes in cell volume during apoptosis. Here, we analyzed various traditional characteristics of apoptosis, such as externalization of the phosphatidylserine (PS) and caspase activity in relation to exact changes in cell size.

3. Instrument Setup and Calibration for the Cell Lab Quanta SC

The Cell Lab Quanta SC by Beckman Coulter is a unique type of flow cytometer that provides not only the capability of fluorescence detection, but also the flexibility of measuring physical characteristics of a cell, such as electronic or Coulter volume in the same spatial location. This results in the ability to determine a change in cell size or volume with a specific fluorescent characteristic of a single cell. The Cell Lab Quanta SC is equipped with a 488-nm argon laser, along with a mercury arc lamp for ultraviolet (UV) excitable dyes, and three photomultiplier tubes (PMTs) for fluorescent

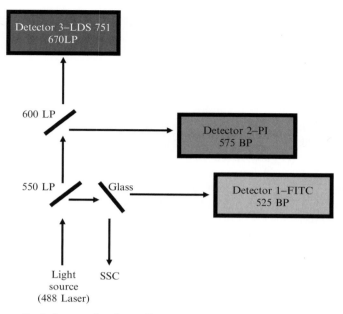

Figure 9.2 Optical setup for the Cell Lab Quanta SC. The Cell Lab Quanta SC is equipped with a 488-nm argon laser, three photomultiplier tubes (PMTs) for fluorescence analysis, and a side-scatter channel. Our optical setup uses a series of long-pass filters to split the emitted fluorescent signal to detector 1 (525 nm BP; FITC), detector 2 (575 nm BP; PI), and detector 3 (670LP; LDS 751). (See color insert.)

analysis. The optical filter setup for the 488-nm laser for examining three independent fluorescent parameters is shown in Fig. 9.2. This section outlines standard methods for setting up and calibrating the Cell Lab Quanta SC for analysis of AVD and fluorescent characteristics using the software specifically designed for this instrument.

3.1. Instrument start-up and calibration

1. Open the Cell Lab Quanta SC Program icon and under the instrument menu choose Start-up.
2. Follow the on-screen instructions for Start-up, which includes a cleaning cycle and turning on the laser.
3. Let the laser warm up for at least 30 min.

3.2. Fluorescence calibration protocol

1. Place 500 µl of filtered Flow Check Beads (FLOW-CHECK Fluorospheres/Flow-Check from Beckman Coulter) using a 25-µm filter tip into a sample cup and place on the instrument.

2. Under protocols, load the Flow Check protocol. This protocol can be previously set to detect FL1, FL2, and FL3 fluorescence and to determine the coefficient of variation (CV) by using regions on the fluorescent histograms. Once created, the protocol can be saved and used for daily quality control of the instrument.

3. Place the sample on the instrument and press Start to run the sample.

4. Adjust the flow rate to around 100 to 150 events per second and verify the CVs for FL1, FL2, and FL3.

5. If the CVs are not below 3, select Auto Optical Alignment under the instrument menu. Perform the optical alignment for the laser using the focus and position parameters.

6. Collect 5000 events.

7. Save and print data.

8. Recover the sample and rinse the instrument by replacing the sample cup with a rinse cup on the instrument.

3.3. Volume calibration protocol

1. Mix 1 ml of sheath fluid with 2 drops of AlignFlow plus flow cytometric alignment beads (6 μm, 488 nm excitation; Molecular Probes) and filter using a 25-μm filter tip into a sample cup.

2. Load the protocol of interest (Annexin, Caspatag, etc.) to calibrate the volume for this specific study. While previous instrument settings are stored with the protocol upon saving the protocol in the instrument, we recommend calibrating the volume with every protocol prior to analysis.

3. Under the Volume menu tab, set the volume to display channels.

4. Set up regions on the electronic volume (EV) histogram and set the statistics to display the linear mean of this parameter. Adjust the EV voltage to the desired value based on the specific type of cell you are analyzing. (Note: for Jurkat T cells we place the EV signal for the 6-μm beads to channels 94 to 96.) Once set, select Calibrate under the volume menu.

5. Under Calibrate, enter the diameter of the beads and the value of EV mean obtained in the statistics under the EV plot, and then press Calibrate.

6. Collect 10,000 events for this calibration sample. Note that if the mean diameter is also selected as one of the statistical EV parameters, a value of 6.00 (the size of the beads) should be displayed.

7. Save and print data.

8. Recover the sample and rinse the instrument by replacing the sample cup with a rinse cup on the instrument.

9. The instrument is now ready for experimental samples.

4. GENERAL CONSIDERATIONS IN STUDYING AVD AND APOPTOSIS ON THE CELL LAB QUANTA SC

Flow cytometers have been traditionally designed to begin analyzing cellular characteristics from the forward-scatter parameter, thus allowing the fluorescent signal to be captured and recorded as the cell passes by the laser once a forward-scatter signal is detected. However, because many current flow cytometers allow the operator to set the initial recording parameter, the fluorescence can be captured and recorded upon the detection of various fluorescent signals. The Cell Lab Quanta SC is also designed to allow the operator to choose which parameter to use as the trigger to capture and record the various fluorescent signals. In studying AVD during apoptosis, we chose to avoid triggering off of the EV (cell size) parameter, as this is a primary characteristic in which we were interested in analyzing. Thus, in all our studies we included a cell-permeant nucleic acid dye known as LDS 751 to trigger the instrument. LDS 751 is a long-wavelength (488 nm excitation; emission maximum of 712 nm) dye that enters and stains all the cells. We detect LDS 751-stained cells in the FL-3 channel (Fig. 9.3), an open channel in the studies outlined later. The incorporation of this dye into our experiments allows as unbiased analysis of the EV of the cell along with the other parameters of interest and increases the signal-to-noise ratio of the system for better–quality fluorescent signals. However, because our primary interest in these studies is changes in cell size, we also used an EV histogram to exclude clumps of cells

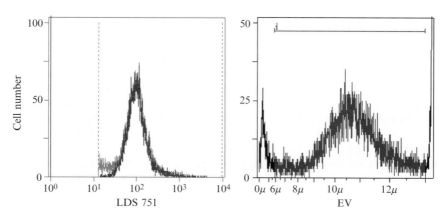

Figure 9.3 Initial gating strategy for examining AVD during apoptosis. LDS 751, a cell-permeant nucleic acid dye, was added to all the samples. For our optical setup, LDS 751 is detected in the FL-3 channel. The addition of LDS 751 allows for a better signal-to-noise ratio, as well as an unbiased analysis of the electronic volume of the cells. Additionally, a gate was used on the electronic volume histogram to eliminate debris and clumps of cells. (See color insert.)

and debris (Fig. 9.3). This allows for the determination of cell size and volume measurements on single cells in regards to other apoptotic parameters of interest.

5. PREPARATION OF CELLS UNDERGOING APOPTOSIS

Apoptosis can occur in most cells given an appropriate cell death stimulus. Treatment and agents such as serum starvation, staurosporine, and etoposide have been common means of inducing apoptosis in a variety of different cell types. For our purpose, we use Jurkat T cells treated with either 25 ng/ml Fas ligand to induce the extrinsic or death receptor pathway or 30 mJ/cm^2 UV to induce the intrinsic or mitochondrial pathway. Jurkat T cells treated with these stimuli are incubated at 37°, 7% CO$_2$ atmosphere for 3 h prior to initiating the specific protocols outlined. In general, any cell type and apoptotic stimulus can be used with these protocols.

6. AVD AND EXTERNALIZATION OF PHOSPHATIDYLSERINE DURING APOPTOSIS

Annexin V conjugates allow the identification of cell surface changes that occur early during the apoptotic process using flow cytometry. PS becomes accessible on the cell surface by remaining exposed on the outer leaflet of the plasma membrane of an apoptotic cell. This event is thought to be important for macrophage recognition of cells undergoing apoptosis (Borisenko et al., 2003; Callahan et al., 2003). The binding of Annexin V to PS is calcium dependent and reversible, although of a very high affinity with a K_d of approximately $5 \times 10^{-10} M$. Different cell types vary in their PS content, along with the amount of PS exposed on the cell surface during apoptosis. Thus, it is necessary to predetermine the amount of Annexin V to use for a given cell type. Typically a 1 to 100 dilution of the Annexin V conjugate is appropriate for most lymphoid–type cells; however, it is our experience that additional Annexin V may be needed for different cell types. Therefore, an initial examination of various concentrations of Annexin V to determine the extent of PS externalization will be beneficial. We routinely use an Annexin V FITC kit from Trevigen (Gaithersburg, MD) in our studies; however, several similar detection kits are available from various manufacturers. The following Annexin V protocol used to examine changes in PS exposure during apoptosis regarding AVD is similar to that suggested by the manufacturer and is designed for using Jurkat T cells.

6.1. Protocol for annexin V FITC/cell size analysis using the Cell Lab Quanta SC

1. After an appropriate time of apoptotic stimulation, place 1 ml of cells in a 6-ml polystyrene Falcon tube and recover by gently centrifuging the cells at 4° for 5 min.
2. Wash the cells once in 500 μl of cold 1× phosphate-buffered saline (PBS) buffer, gently resuspending the cells and then recovering the cells by centrifugation as stated in step 1.
3. For each sample of cells, prepare 100 μl of Annexin V staining solution as described: 10 μl 10× binding buffer, 10 μl 10× propidium iodide (PI), 1 μl Annexin V–FITC, and 79 μl dH$_2$O.
   a. Keep this staining cocktail on ice and in the dark.
   b. Prepare a similar cocktail minus the Annexin V–FITC and PI as a negative control.
4. Gently resuspend the washed cells in 100 μl of the Annexin V staining solution prepared in step 3.
5. Incubate in the dark for 15 min at room temperature.
6. Prepare 400 μl of 1× annexin binding buffer in dH$_2$O for each sample.
7. After 15 min of incubation, add 400 μl of 1× annexin binding buffer to each 100-μl sample.
8. Add 1 μl/ml LDS 751 stock to each tube.
9. Samples are now ready to be examined on the Cell Lab Quanta SC.

7. AVD AND CASPASE ACTIVITY DURING APOPTOSIS

The activation of caspases has been shown to occur under a variety of conditions that induce apoptosis. These proteolytic enzymes cleave numerous protein substrates in response to a proapoptotic signal that results in the disassembly of the cell. Caspases are synthesized as precursors that undergo proteolytic maturation to form an active enzyme complex. We have used the CaspaTag Caspase Activity kit from Intergen (Purchase, NY) to detect these proteases in living cells during apoptosis. The assay is based on carboxyfluorescein-labeled fluoromethyl ketone (FMK) peptide inhibitors of caspases. The inhibitor irreversibly binds to active caspases to distinguish caspase-positive cells from caspase-negative cells by an increase in fluorescence intensity at 530 nm. The following protocol examines caspase activity in regards to AVD during apoptosis.

7.1. Preparing CaspaTag reagents

1. Reconstitute lyophilized FMK peptide in 50 μl of dimethyl sulfoxide (DMSO) resulting in a 150× CaspaTag stock. Aliquots may be stored at −20°.

2. Immediately prior to the experiment, make a 30× working dilution. Dilute the 150× stock 1:5 in PBS, pH 7.4 (one part 150× FMK peptide to four parts PBS). Mix well. Protect from light at all times.
3. Place 10× wash buffer in a 37° water bath for 30 min to dissolve any precipitated salts.
4. Mix the 10× wash buffer thoroughly.
5. Make a 1× solution of the wash buffer by diluting the 10× wash buffer in dH$_2$O and mix thoroughly.

7.2. Protocol for CaspaTag/cell size analysis on the Cell Lab Quanta SC

1. After an appropriate time of apoptotic stimulation, place 900 μl of cells in a 6-ml polystyrene Falcon tube.
2. Add 30 μl of the 30× working dilution FMK peptide directly to the cell suspension and mix gently.
3. Incubate the cells for 1 h at 37°, 7% CO$_2$ atmosphere.
4. Add 2 ml of 1× wash buffer to the labeled cells.
5. Spin down the cells at 400 g for 5 min at room temperature.
6. Remove the supernatant.
7. Resuspend the cells in 1 ml of PBS.
8. Add 2 μl of a 5-mg/ml PI stock solution to each tube.
9. Add 1 μl of a 1-mg/ml LDS 751 stock solution to each tube.
10. Samples are now ready to be examined on the Cell Lab Quanta SC.

8. AVD AND PLASMA MEMBRANE POTENTIAL DURING APOPTOSIS

A depolarization of the plasma membrane potential has been shown to occur under a variety of apoptotic conditions coinciding with AVD (Bortner et al., 2001; Dallaporta et al., 1999; Deckers et al., 1993; Mann and Cidlowski, 2001). We have shown previously that DiBAC$_4$ is most specific to measure changes in the plasma membrane potential (Bortner et al., 2001). DiBAC$_4$ is an anionic oxonal dye that responds with an increase in fluorescent intensity at 530 nm in response to plasma membrane depolarization. This dye is nontoxic to the cells and can be added to the cells prior to apoptotic stimulation. The following protocol examines plasma membrane depolarization in regards to AVD during apoptosis.

8.1. Preparing DiBAC$_4$ reagent

1. Dissolve DiBAC$_4$ in DMSO to result in a 1 mM stock concentration of the dye. This stock can be stored at room temperature. Protect from light at all times.

2. Prepare a 20 μM working stock of $DiBAC_4$ through a dilution of the 1 mM stock using DMSO. This stock can be stored at room temperature.
3. Use $DiBAC_4$ at a final concentration of 150 nM.

8.2. Protocol for $DiBAC_4$/cell size analysis on the Cell Lab Quanta SC

1. Immediately after preparation of the control and apoptotic cells, add $DiBAC_4$ to a final concentration of 150 nM to each sample and incubate at 37°, 7% CO_2 atmosphere.
2. After an appropriate time of apoptotic stimulation, place 1 ml of cells in a 6-ml polystyrene Falcon tube.
3. Add 2 μl of a 5-mg/ml PI stock solution to each tube.
4. Add 1 μl of a 1-mg/mL LDS 751 stock solution to each tube.
5. Samples are now ready to be examined on the Cell Lab Quanta SC.

9. Data Acquisition for the Cell Lab Quanta SC

Once the Cell Lab Quanta SC has been calibrated for fluorescence and EV, and the apoptotic samples to be examined have been prepared, the following steps can be accomplished to record data.

1. Make sure the appropriate instrument protocol and volume calibration have been completed for the current study of interest.
2. Filter the samples using a 25-μm filter tip into a sample cup.
3. Place the sample cup on the instrument and acquire 10,000 events.
4. Save and print data.
5. Recover the sample and rinse the instrument by replacing the sample cup with a rinse cup on the instrument.

10. Data Analysis for AVD and Apoptotic Characteristics on the Cell Lab Quanta SC

The Cell Lab Quanta SC software allows for various types of data analysis, similar to the type of analysis that can be accomplished using most standard flow cytometric software. Off-line compensation of data to correct for emitted fluorescence overlap can be saved to analyze multiple experimental samples. Numerous gates or regions can be drawn on the plots and histograms to isolate a given population of cells. Statistics can be generated to determine the number and percentage of cells in a given population, the mean fluorescent value along with the CV of the fluorescent peak, and detailed information on the EV parameter, including mean diameter of the cells, mean cell volume (MCV), and surface area.

A nonlinear relationship exists between cell diameter and cell volume such that the difference in the diameter between a 4- and a 5-μm cell is 20%. However, the difference in cell volume between these two cells is almost 100%, as a 4-μm cell has a volume of 34 μm^3, whereas a 5-μm cell has a volume of 65 μm^3. Furthermore, a doubling of cell diameter from 4 to 8 μm would result in an increase of cell volume by nearly 800% (34 μm^3 versus 268 μm^3, respectively). The Cell Lab Quanta SC can determine both of these parameters for any given population of cells.

The protocols outlined earlier are designed to use the cell-permeant nucleic acid dye LDS 751 to trigger the instrument for an unbiased determination cellular event. Additionally, a gate or region is set on an EV histogram to exclude clumps of cells and debris from the analysis. These gates are applied to the relevant histograms and dot plots to analyze data. When analyzing the experimental samples, additional gates or regions, including quadrants, can be applied to the histograms or dot plots of interest to determine the specific characteristics of a given population of cells.

When Jurkat T cells are treated with either 25 ng/ml FasL or 30 mJ/cm^2 UV to induce apoptosis and then stained with Annexin V–FITC and PI after 3 h, these cells can be initially examined on an EV versus Annexin V–FITC dot plot (Fig. 9.4). The control sample shows a single population of cells, where as upon apoptotic treatment we observed a unique population of cells that have an increase in Annexin V staining. Additionally, we can examine these cells on an Annexin V–FITC versus PI dot plot, which has been a traditional means for the analysis of these data (Fig. 9.4). Here we observe an increase in Annexin V staining prior to the loss of membrane integrity (increased PI fluorescence) in the treated cell, indicating apoptosis. To obtain an unbiased view of changes in EV, a measure of AVD during apoptosis, a quadrant is drawn on an Annexin V versus PI dot plot (Fig. 9.4) for statistical analysis of EV. Table 9.1 shows the results of this quadrant analysis for AVD in regards to cell diameter, MCV, cell count, and the percentage of cells in each population. In analysis of the Annexin V–negative/PI–negative (double negative) populations for each sample, we observed a significant decrease in both cell diameter and MCV for UV-treated cells (15%) compared to either control or FasL-treated samples, suggesting a loss of cell volume prior to externalization of the phosphatidylserine. Interestingly, only a 3% decrease in MCV is observed when comparing the double negative FasL population to the double negative control cells. Upon treatment with FasL, we observed an approximate 26% decrease in MCV for Annexin V-positive/PI–negative cells compared to double negative control cells. This Annexin V-positive/PI–negative cell population comprised over 24% of the total cell population after 3 h of apoptotic stimulation. Treatment of Jurkat T cells with UV resulted in only

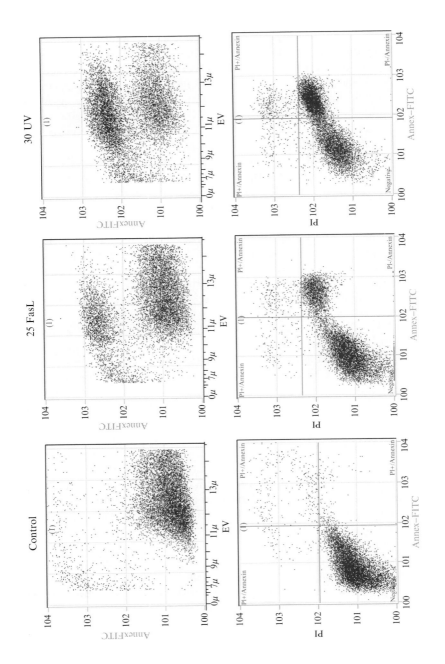

an additional 6% decrease in MCV after 3 h of apoptotic stimulation, as the UV-treated double negative population already had a significant loss of cell volume. Thus, precise changes in AVD can be determined for various individual populations of apoptotic cells.

A similar type of analysis can be applied to apoptotic cells exposed to the CaspaTag reagent. Jurkat T cells induced to undergo apoptosis with either 25 ng/ml FasL or 30 mJ/cm^2 UV in the presence of the CaspaTag reagent for 4 h can be initially examined on an EV versus CaspaTag fluorescence dot plot (Fig. 9.5). Comparable to cells stained with Annexin V–FITC, the CaspaTag control sample shows a single population of cells, whereas apoptotic-stimulated cells show the presence of a unique population with an increase in CaspaTag fluorescence, indicating caspase activity. Furthermore, these samples can be examined on CaspaTag versus PI dot plots (Fig. 9.5). Statistical analysis of EV or AVD for CaspaTag and PI double negative and CaspaTag-positive/PI-negative cells is shown in Table 9.2. A significant decrease in MCV of the double negative UV-treated cells is

Table 9.1 Statistical analysis of changes in AVD for Annexin V-positive and -negative cells during apoptosis

Sample	Region	Diameter (μm)	MCV (μm^3)	Cell count	% gated
Control	Double negative	12.58	1042.5	7489	92.32
Control	Annexin +/PI–	10.96	689.3	261	3.22
25 FasL	Double negative	12.46	1013.5	5747	71.64
25 FasL	Annexin +/PI–	11.29	753.3	1937	24.15
30 UV	Double negative	11.90	883.2	3827	46.64
30 UV	Annexin +/PI–	11.68	833.6	3926	47.84

Figure 9.4 Analysis of Annexin V–FITC and AVD during apoptosis. Jurkat T cells were treated with 25 ng of Fas ligand or 30 mJ/cm^2 of UV for 3 h to induce apoptosis. These samples were then stained with Annexin–FITC and PI and examined on the Cell Lab Quanta SC. A distinct population of Annexin–FITC-positive stained cells can be observed on an electronic volume versus an Annexin–FITC dot plot for the treated samples, indicating externalization of the phosphatidylserine (PS). Examination of these cells on an Annexin–FTIC versus PI dot plot shows that the externalization of PS occurs prior to the loss of membrane integrity, indicative of apoptosis. Quadrants set on the Annexin–FITC versus PI dot plots were used to determine changes in cell size based on the electronic volume parameter. (See color insert.)

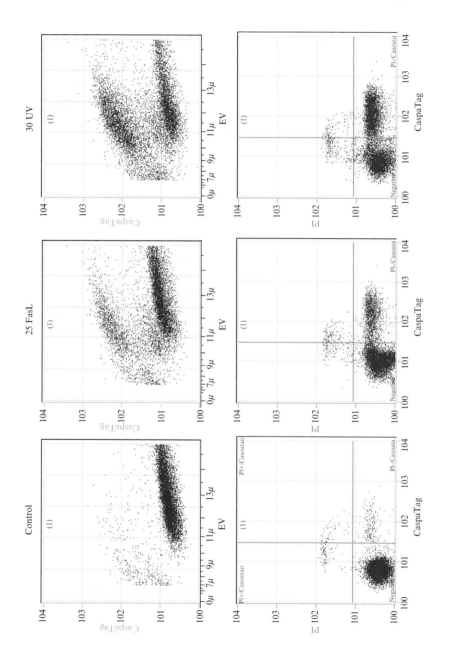

observed (15%), compared to the double negative control or FasL-treated cells. However, an additional 15% decrease in MCV occurs for the Caspa-Tag-positive UV sample. In contrast, apoptotic treatment with FasL resulted in an approximate 25% decrease in MCV for CaspaTag-positive cells compared to either control or FasL-treated double negative cells. Thus, the Cell Lab Quanta SC can determine changes in AVD from a variety of apoptotic characteristics.

Finally, we also examined changes in the plasma membrane potential during apoptosis, specifically cellular depolarization in regards to AVD. Plasma membrane depolarization has been shown to occur early in the apoptotic process (Bortner et al., 2001; Dallaporta et al., 1999; Deckers et al., 1993; Mann and Cidlowski, 2001). We examined control, FasL-, and UV-treated Jurkat T cells for cellular depolarization using the membrane potential dye $DiBAC_4$ over a period of 4 h. Figure 9.6 shows an increase in $DiBAC_4$ fluorescence for cells that have depolarized over time in

Table 9. 2 Statistical analysis of changes in AVD for CaspaTag-positive and -negative cells during apoptosis

Sample	Region	Diameter (μm)	MCV (μm^3)	Cell count	% gated
Control	Double negative	12.81	1101.0	8496	96.08
Control	CaspaTag +/PI-	10.20	554.9	198	2.24
25 FasL	Double negative	12.65	1059.7	6394	77.09
25 FasL	CaspaTag +/PI-	11.53	802.6	1635	19.71
30 UV	Double negative	12.13	934.6	4812	55.89
30 UV	CaspaTag +/PI-	11.51	799.1	3490	40.54

Figure 9.5 Analysis of CaspaTag and AVD during apoptosis. Jurkat T cells were treated with 25 ng of Fas ligand or 30 mJ/cm^2 of UV for 3 h to induce apoptosis. These samples were then stained with CaspaTag reagent for an additional hour prior to the addition of PI and examination on the Cell Lab Quanta SC. A distinct population of CaspaTag positive-stained cells can be observed on an electronic volume versus CaspaTag dot plot for treated samples, indicating caspase activity. Examination of these cells on an Caspa-Tag versus PI dot plot shows that the caspase activity occurs prior to the loss of membrane integrity. Quadrants set on the CaspaTag versus PI dot plot were used to determine changes in cell size based on the electronic volume parameter. (See color insert.)

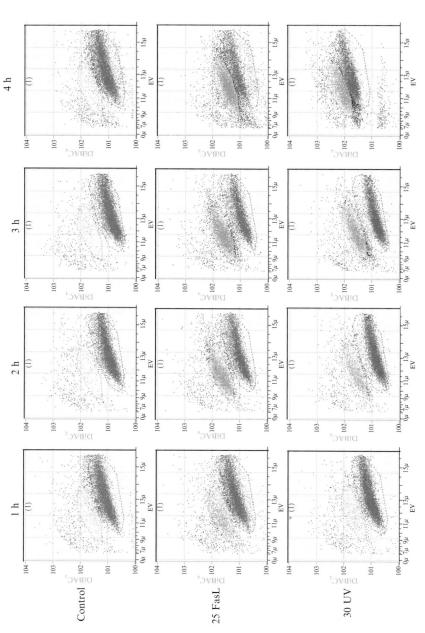

Figure 9.6 Analysis of DiBAC₄ fluorescence and AVD during apoptosis. Jurkat T cells were treated with 25 ng/ml FasL or 30 mJ/cm² UV to induce apoptosis. DiBAC₄ was used to determine changes in the plasma membrane potential over a period of 4 h. Dot plots of electronic volume (EV) versus DiBAC₄ show a time-dependent increase in DiBAC₄ fluorescence for the treated cells (green population). Gates were set on these plots to determine changes in cell size and volume for each population of cells. (See color insert.)

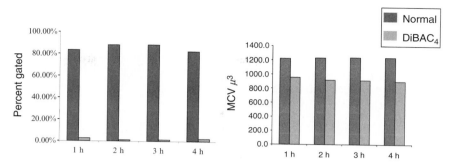

Figure 9.7   Percentage gated and mean cell volume (MCV) for control DiBAC$_4$-stained cells over time. No significant change in the percentage of cells gated or the MCV was observed for the normal (red) population of cells or the small percentage of DiBAC$_4$-posi-tive (green) cells over time, indicating that DiBAC$_4$ is not toxic to the cells. (See color insert.)

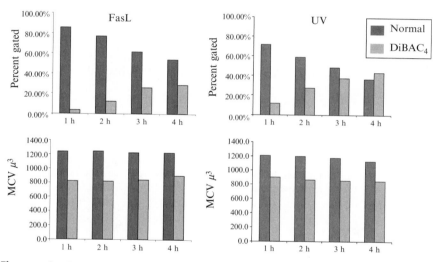

Figure 9.8   Percentage gated and mean cell volume (MCV) for FasL- and UV-treated DiBAC$_4$-stained cells over time. Under each apoptotic condition, an increase in the num-ber of DiBAC$_4$-positive cells, along with a simultaneous decrease in the number of normal (DiBAC$_4$ negative) cells, was observed over a period of 4 h. In contrast, no change in MCV was observed for either FasL- or UV-treated normal or DiBAC$_4$-positive populations of cells, suggesting that the change in MCV remains constant over this period of time. (See color insert.)

response to either FasL or UV treatment. Here, gates were set on an EV versus DiBAC$_4$ fluorescence dot plot to analyze both the percentage of cells and the change in MCV for each population of cells. A constant percentage of normal control cells is observed over the 4-h time span (Fig. 9.7).

Additionally, there is no change in the MCV for the small percentage of DiBAC$_4$-positive cells observed in the control sample. Upon either FasL or UV treatment, an increasing number of cells are observed in a time-dependent manner for the DiBAC$_4$-positive cells (Fig. 9.8). However, the MCV remains constant regardless of the total number of cells examined. Thus, the Cell Lab Quanta SC can also be used to analyze populations of cells over time in regards to AVD during apoptosis.

11. CONCLUSION

The loss of cell volume during apoptosis is a fundamental characteristic of this cell death process. We have shown how AVD can be determined for various subpopulations of apoptotic cells using the Cell Lab Quanta SC flow cytometer. This instrument electronically sizes cells in conjunction with numerous fluorescent parameters to determine precise changes in cell diameter and cell volume. The application of these studies is not limited to only the examination of the externalization of phosphatidylserine, caspase activity, and plasma membrane depolarization, but can be applied to other apoptotic characteristics that can be analyzed through the use of fluorescence.

ACKNOWLEDGMENTS

This research was supported by the Intramural Research Program of the NIH, National Institute of Environmental Health Sciences. We also thank Ernie Thomas at Beckman Coulter for his very helpful discussion and advice.

REFERENCES

Beauvais, F., Michel, L., and Dubertret, L. (1995). Human eosinophils in culture undergo a striking and rapid shrinkage during apoptosis. Role of K$^+$ channels. *J. Leuk. Biol.* **57,** 851–855.

Benson, R. S. P., Heer, S., Dive, C., and Watson, A. J. M. (1996). Characterization of cell volume loss in CEM-C7A cells during dexamethasone-induced apoptosis. *Am. J. Physiol.* **270,** C1190–C1203.

Borisenko, G. G., Matsura, T., Liu, S.-X., Tyurin, V. A., Jianfei, J., Serinkan, F. B., and Kagan, V. E. (2003). Macrophage recognition of externalized phosphatidylserine and phagocytosis of apoptotic Jurkat cells: Existence of a threshold. *Arch. Biochem. Biophys.* **413,** 41–52.

Bortner, C. D., and Cidlowski, J. A. (1998). A necessary role for cell shrinkage in apoptosis. *Biochem. Pharmacol.* **56,** 1549–1559.

Bortner, C. D., and Cidlowski, J. A. (2004). The role of apoptotic volume decrease and ionic homeostasis in the activation and repression of apoptosis. *Pflüg. Arch.* **448,** 313–318.

Bortner, C. D., Gomez-Angelats, M., and Cidlowski, J. A. (2001). Plasma membrane depolarization without repolarization is an early molecular event in anti-Fas-induced apoptosis. *J. Biol. Chem.* **276**, 4304–4314.

Bortner, C. D., Hughes, F. M., Jr., and Cidlowski, J. A. (1997). A primary role for K^+ and Na^+ efflux in the activation of apoptosis. *J. Biol. Chem.* **272**, 32436–32442.

Cain, K., Langlais, C., Sun, X. M., Brown, D. G., and Cohen, G. M. (2001). Physiological concentrations of K^+ inhibit cytochrome c-dependent formation of the apoptosome. *J. Biol. Chem.* **276**, 41985–41990.

Callahan, M. K., Halleck, M. S., Krahling, S., Henderson, A. J., Williamson, P., and Schlegel, R. A. (2003). Phosphatidylserine expression and phagocytosis of apoptotic thymocytes during differentiation of monocytic cells. *J. Leuk. Biol.* **74**, 846–856.

Compton, M. M., Haskill, J. S., and Cidlowski, J. A. (1988). Analysis of glucocorticoid actions on rat thymocyte deoxyribonucleic acid by fluorescence-activated flow cytometry. *Endrocrinology* **122**, 2158–2164.

Dallaporta, B., Marchetti, P., de Pablo, M. A., Maisse, C., Duc, H. T., Metivier, D., Zamzami, N., Geuskens, M., and Kroemer, G. (1999). Plasma membrane potential in thymocyte apoptosis. *J. Immunol.* **162**, 6534–6542.

Deckers, C. L., Lyons, A. B., Samuel, K., Sanderson, A., and Maddy, A. H. (1993). Alternative pathways of apoptosis induced by methyprednisolone and valinomycin analyzed by flow cytometry. *Exp. Cell Res.* **208**, 362–370.

Hughes, F. M., Jr., Bortner, C. D., Purdy, G. D., and Cidlowski, J. A. (1997). Intracellular K^+ suppresses the activation of apoptosis in lymphocytes. *J. Biol. Chem.* **272**, 30567–30576.

Kerr, J. F. (1971). Shrinkage necrosis: A distinct mode of cellular death. *J. Pathol.* **105**, 13–20.

Kerr, J. F., Wyllie, A. H., and Currie, A. R. (1972). Apoptosis: A basic biological phenomenon with wide-ranging implications in tissue kinetics. *Br. J. Cancer* **26**, 239–257.

Klassen, N. V., Walker, P. R., Ross, C. K., Cygler, J., and Lach, B. (1993). Two-stage cell shrinkage and the OER for radiation-induced apoptosis of rat thymocytes. *Int. J. Radiat. Biol.* **64**, 571–581.

Lang, F., Ritter, M., Gamper, N., Huber, S., Fillon, S., Tanneur, V., Lepple-Wienhues, A., Szabo, I., and Gulbins, E. (2000). Cell volume in the regulation of cell proliferation and apoptotic cell death. *Cell. Physiol. Biochem.* **10**, 417–428.

Maeno, E., Ishizaki, Y., Kanaseki, T., Hazama, A., and Okada, Y. (2000). Normotonic cell shrinkage because of disordered volume regulation is an early prerequisite to apoptosis. *Proc. Natl. Acad. Sci. USA* **97**, 9487–9492.

Mann, C. L., and Cidlowski, J. A. (2001). Glucocorticoids regulate plasma membrane potential during rat thymocyte apoptosis *in vivo* and *in vitro*. *Endocrinology* **142**, 421–429.

Thomas, N., and Bell, P. A. (1981). Glucocorticoid-induced cell-size changes and nuclear fragility in rat thymocytes. *Mol. Cell. Endocrinol.* **22**, 71–84.

Thompson, G. J., Langlais, C., Cain, K., Conley, E. C., and Cohen, G. M. (2001). Elevated extracellular [K+] inhibits death-receptor- and chemical-mediated apoptosis prior to caspase activation and cytochrome c release. *Biochem. J.* **357**, 137–145.

Yu, S. P., and Choi, D. W. (2000). Ions, cell volume, and apoptosis. *Proc. Natl. Acad. Sci. USA* **97**, 9360–9362.

CHAPTER TEN

TRANSIENT RECEPTOR POTENTIAL CHANNELS IN MECHANOSENSING AND CELL VOLUME REGULATION

Stine Falsig Pedersen* *and* Bernd Nilius[†,‡]

Contents

Abstract

Transient receptor potential (TRP) channels are unique cellular sensors responding to a wide variety of extra- and intracellular signals, including mechanical and osmotic stress. In recent years, TRP channels from multiple subfamilies have been added to the list of mechano- and/or osmosensitive channels, and it is becoming increasingly apparent that Ca^{2+} influx via TRP channels plays a crucial role in the response to mechanical and osmotic perturbations in a wide range of cell types. Although the events translating mechanical and osmotic stimuli into regulation of TRP channels are still incompletely understood, the specific mechanisms employed vary between different TRP isoforms, and probably include changes in

* Department of Molecular Biology, University of Copenhagen, Copenhagen, Denmark
† Department of Molecular Cell Biology, Division of Physiology KU Leuven, Campus Gasthuisberg, Leuven, Belgium
‡ Corresponding author.

Methods in Enzymology, Volume 428
ISSN 0076-6879, DOI: 10.1016/S0076-6879(07)28010-3

the tension and/or curvature of the lipid bilayer, changes in the cortical cytoskeleton, and signaling events such as lipid metabolism and protein phosphorylation/dephosphorylation. This chapter describes candidate mechanosensitive channels from mammalian TRP subfamilies, discusses inherent and technical issues potentially confounding evaluation of mechano- and/or osmosensitivity, and presents methods relevant to the study of TRP channel regulation by mechanical and osmotic stimuli and involvement in cell volume regulation.

1. Introduction

Transient receptor potential (TRP) channels are unique cellular sensors, the important roles of which include the detection of mechanical forces and of changes in cell volume or intra- or extracellular osmolarity. Based on sequence homology, mammalian TRP channels are divided into six subfamilies: TRPC (canonical), TRPV (vanilloid), TRPM (melastatin), TRPP (polycystin), TRPML (mucolipin), and TRPA (ankyrin). Most TRPs are polymodal channels activated by multiple physical and chemical stimuli through distinct molecular mechanisms (for reviews, see Liedtke, 2005; Liedtke and Kim, 2005; Nilius and Voets, 2005; Nilius *et al.*, 2007; Pedersen *et al.*, 2005; Ramsey *et al.*, 2006; Voets *et al.*, 2005). The first evidence for an involvement of TRP channels in mechanosensation came from mutations in the *Caenorhabditis elegans osm-9* gene, which encodes a TRPV-like channel. Worms with mutations in this channel were shown to exhibit defects in the avoidance reaction to high osmolality and nose touch (Colbert *et al.*, 1997). In recent years, pivotal roles of mammalian TRP channels in mechano- and osmosensing have been demonstrated. This chapter briefly describes these channels and the evidence linking them to mechano- and osmosensing and discusses methods relevant to the study of TRP channel regulation by mechanical and osmotic stimuli and the potential involvement of TRP channels in the regulation of cell volume.

2. General Mechanisms of Mechano- or Osmosensing by Membrane Proteins

Membrane transport proteins appear to sense mechanical forces and/or changes in osmolarity by a number of fundamental mechanisms. These are outlined in Fig. 10.1 and discussed later, focusing on the mechanisms relevant to ion channels in general and TRP channels in particular (for reviews on general mechanisms of mechanotransduction, see Hamill and Martinac, 2001; Kung, 2005; Nicolson, 2005; Perozo, 2006). Any channel embedded in a lipid bilayer is exposed to negative and positive pressures created by the bilayer. Under equilibrium conditions, the conformational energy of the channel matches this energy profile. Any change of this equilibrium can result in modulation of channel activity (Fig. 10.1A). A number of mechanisms have been shown to be involved in mechano- and osmosensing by plasma membrane ion channels. First, a change of the forces acting within the lipid membrane causes conformational changes in the channel protein. In this manner, forces acting at the crucial lipid–protein interface may gate a channel upon membrane stretch or bending (Fig. 10.1B). Second, channel displacement through a tether, for example, a reorganization of the cortical cytoskeleton, can cause a mismatch between the conformation energy of the channel and the intrinsic lipid tension, leading to changes in channel gating (Fig. 10.1C). Third, changes in the membrane content of specific lipids (e.g., phosphatidylcholines, lysophosphatidylcholines, or arachidonic acid and its metabolites) can alter the membrane structure around a channel protein in a concave or a convex manner, depending on the shape of the relevant lipid components, thereby inducing energetic mismatch and changes in channel gating (Fig. 10.1D). The two pore potassium channels, TREK-1 and TRAAK, provide interesting examples of gating by changes in membrane lipid composition. A phospholipid sensor controls mechanoactivation of TREK-1 and phosphatidylinositol (4,5)-bisphosphate [PtdIns(4,5)P_2] sensitizes its activation, and TRAAK is activated by convex bending of the membrane in a manner that can be mimicked by the introduction of membrane components causing convex membrane curvature (Chemin *et al.*, 2005; Maingret *et al.*, 1999; Patel and Honore, 2001; Patel *et al.*, 1998). Fourth, mechanical or osmotic stress may trigger intracellular signaling cascades involving, for example, lipid metabolism, which may affect channel gating and/or plasma membrane insertion/retrieval (Fig. 10.1E). Fifth, mechanical or osmotic stress also elicits numerous changes in protein phosphorylation/ dephosphorylation events, several of which have been shown to play a major role in the regulation of mechano- or osmosensing by membrane transport proteins, including TRPs (for specific examples, see later) (Fig. 10.1F). A discussion of the mechanisms by which such signaling events are initiated by mechanical or osmotic stress is beyond the scope of this chapter, but important mechanisms likely involve integrin clustering, cytoskeletal reorganization, and changes in intracellular ionic strength and/or concentrations of macromolecules or specific ions (for a more general description, see Hoffmann and Pedersen, 2006; Kung, 2005). Although not further discussed here, since there is yet no evidence of such a mechanism for TRP channels, it is also notable that at least some membrane transporters appear to respond directly to volume-dependent changes in intracellular ionic strength by ionic strength-dependent interactions of a specific intracellular domain, the cystathionine β-synthase domain, with the lipid bilayer (Biemans–Oldehinkel *et al.*, 2006).

So far, it is difficult to differentiate among mechanisms 1 through 5 just outlined. First, osmotic cell swelling can be inherently associated with

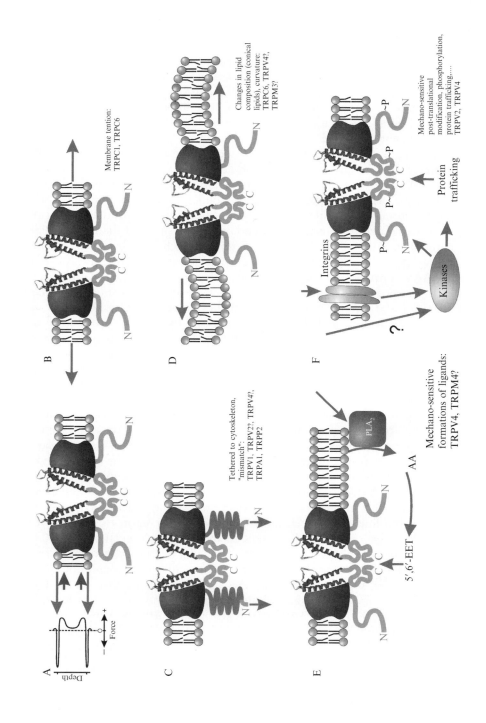

B Membrane tension: TRPC1, TRPC6

D Changes in lipid composition (conical lipids), curvature: TRPC6, TRPV4?, TRPM3?

F Mechano-sensitive post-translational modification, phosphorylation, protein trafficking,.....
TRPV2, TRPV4

Protein trafficking

Integrins

Kinases

A Force
Depth

C Tethered to cytoskeleton, "mismatch": TRPV1, TRPV2?, TRPV4?, TRPA1, TRPP2

E Mechano-sensitive formations of ligands: TRPV4, TRPM4?
PLA₂
AA
5',6'-EET

mechanical stretch and increased membrane tension (mechanism 1). However, in many cell types, cell swelling is associated with unfolding of membrane invaginations, that is, the cortical cytoskeleton is exposed to increased tension and may affect a tethered ion channel (mechanism 2). Also in accordance with this notion, cell volume perturbations elicit substantial rearrangements of the actin-based cytoskeleton (Di Ciano-Oliveira et al., 2006; Pedersen et al., 2001). In contrast, the membrane per se need not be exposed to increased tension until very substantial cell swelling has occurred (Hamill and Martinac, 2001; Maingret et al., 1999; Pedersen et al., 2001). Cell volume changes also elicit a number of signaling events involving the metabolism of membrane components, which may regulate membrane proteins both by altering bilayer curvature per se or by direct interactions of the transporter with the specific lipid messenger in question. Thus, phospholipase A_2 (PLA_2) is activated by cell swelling, leading to a breakdown of phospholipids into arachidonic acid and lysophosphatidic acid, and is conversely inhibited by cell shrinkage, leading to reduced arachidonic acid release (e.g., Basavappa et al., 1998; Pedersen et al., 2006). Moreover, hypotonic cell swelling has been shown to decrease (see Pedersen et al., 2001) and hypertonic cell shrinkage to increase (Nasuhoglu et al., 2002; Yamamoto et al., 2006) cellular $PtdIns(4,5)P_2$ levels. Also the lipid kinase phosphatidylinositol 3-kinase (PI3K) is reported to be volume sensitive (Yamamoto et al., 2006), although the specific effects of volume perturbations on PI3K activity seem to be variable (see Pedersen et al., 2001). As discussed later, arachidonic acid metabolites, $PtdIns(4,5)P_2$, and PI3K are involved in the regulation of several mechano- and osmosensitive TRP channels (Rohacs, 2007), and hence could play a role in messenger-dependent mechano- and osmosensing (mechanism 4). Multiple protein

---

Figure 10.1 Some possible mechanisms of mechano- or osmosensing by TRP channels. (A) Any channel in a plasma membrane is exposed to lateral forces from the bilayer (see energy profile at the left-hand side), which create positive or negative forces (lateral pressure indicated by the arrows) acting on the channel. In equilibrium, the conformational energy of the channel matches this energy profile. Any changes in this equilibrium may cause modulation of channel activity, e.g., gating. (B) Channel gating by changes in bilayer tension, which will alter the tension profile shown in A (referred to in the text as mechanism 1). (C) Channel gating by tethering to cytoskeletal elements, which are reorganized by mechanical or osmotic stress (mechanism 2). (D) Channel gating by changes in membrane curvature (mechanism 3). (E) Channel gating by an intracellular messenger that can be produced by an enzyme acting as a mechano/osmo/volume sensor (e.g., volume-sensitive activation of PLA_2, resulting in the generation of arachidonic acid, which either itself or in the form of its downstream metabolic products gates the channel (mechanism 4). (F) Channel gating as a consequence of regulation by signaling events induced by mechanical or osmotic stress can induce signaling events, e.g., via integrins or kinases, which in turn regulate both the activity of channels resident in the plasma membrane and the membrane insertion/retrieval of channels (mechanism 5). See text for details. (See color insert.)

kinases and phosphatases are regulated in a volume- or mechanosensitive manner (see, e.g., Hoffmann and Pedersen, 2006), and TRP regulation by changes in protein phosphorylation/dephosphorylation may also be relevant to their mechano- and osmosensitivity (mechanism 5). One striking example is the modulation of TRPV4 by kinases of the WNK family. WNK4 downregulates TRPV4 via decreased cell surface expression. Because WNK kinases are activated by anisotonic conditions, they form a link to hypotonic TRPV4 activation (Fu et al., 2006). In yeast cells, Yvc1p, a vacuolar membrane protein with homology to TRPV (Denis and Cyert, 2002), mediates the hyperosmolarity-induced Ca^{2+} release, which might be regulated by intracellular phosphorylation/dephosphorylation signaling cascades (Liedtke and Kim, 2005). This cascade involves MAP kinase activation, which is triggered by Ste20, a shrinkage-activated kinase in yeast (for a review, see Strange et al., 2006). Finally, it is clear that at least some ion channels and volume-sensitive transporters respond to ionic strength rather than cell volume per se; however, the mechanisms remain to be fully elucidated and are not further discussed here (Nilius and Droogmans, 2001; Voets et al., 1999).

As evident from the discussion just given, mechano and osmo/volume/ ionic strength sensitivity are related in an inherently complex manner, yet are obviously not identical entities in terms of the biophysical characterization of membrane transporters. A number of technical aspects further complicate their experimental separation. None of the usually applied methods comply with evaluation of one of the described mechanisms in isolation. For the narrow purpose of patch clamp measurements in artificial lipid membranes—devoid of the cortical cytoskeleton—mechanosensitivity may be defined as activation by negative or positive pressure applied through the patch clamp pipette. However, a problem specifically related to analysis of mechano and osmo/volume sensitivity by the whole cell patch clamp technology is that bath hypotonicity is unavoidably coupled with a constantly increasing cell volume (see later), as well as with changes in intracellular ionic strength and concentrations of macromolecules and probably also with mechanical stimulation by fluid shear stress during cell superfusion (for discussion, see Hamill and Martinac, 2001; Nilius et al., 1998; Voets, 1999).

3. TRP CHANNELS IN MECHANO- AND OSMOSENSING

The ability of cells to sense mechanical stimuli is fundamental to such essential physiological functions as embryonic development, hearing, touch sensitivity, and control of kidney function, vascular tone, and muscle stretch (see, e.g., Hamill and Martinac, 2001; Kung, 2005). Multiple TRP channels from various subfamilies have been shown to be sensitive to various forms of mechanical stress, including fluid shear stress, and increased membrane tension resulting from membrane stretch (Liedtke and Kim, 2005; Nilius *et al.*, 2007; O'Neil and Heller, 2005; Pedersen *et al.*, 2005; Voets *et al.*, 2005). This section briefly reviews the evidence for mechano- or osmosensitivity of individual TRP channels.

TRPC1 has been identified as a mechanosensitive cation channel (MscCa) in *Xenopus laevis* oocytes (Maroto *et al.*, 2005). TRPC1 is also mechanosensitive in liposomes, and hence is probably directly activated by membrane stretch (mechanism 1). Very likely, TRPC1 is involved in stretch-induced muscle damage by elevating $[Ca^{2+}]_i$ upon membrane stretch (Allen *et al.*, 2005).

Another TRPC channel, the receptor-activated, nonselective cation channel TRPC6, is activated by mechanically or osmotically induced membrane stretch/deformation in a manner inhibited by the tarantula peptide toxin GsMTx-4, an inhibitor of mechanosensitive ion channels (Spassova *et al.*, 2006). The activation mechanism is phospholipase C independent. Instead, it seems to depend directly on the lateral-lipid tension and lipid-protein mismatch, such that the stretch-induced reduction in membrane bilayer thickness alters the channel conformation to the open state (mechanism 1) (Spassova *et al.*, 2006). Diacylglycerol, a well-known TRPC6 activator, may act similarly by changing membrane curvature, whereas GsMTx-4 may relieve membrane lipid stress and inhibit channel activation (Spassova *et al.*, 2006). MxA, a member of the dynamin super-family, binds to TRPC6 and connects the channel with the cytoskeleton via the second ankyrin repeat (Lussier *et al.*, 2005). Referring to the description of mechanism 2 given earlier, it has yet to be shown that this coupling to the cytoskeleton may be involved in mechanosensing by TRPC6. Physiologi-cally, the mechanosensitivity of TRPC6 appears to be important in the control of vascular tone in response to increased intravascular pressure (the Baylis effect) (Beech *et al.*, 2004; Inoue *et al.*, 2006; Welsh *et al.*, 2002).

Finally, several TRPCs have been described as stretch-activated chan-nels (SAC) in skeletal muscle (7–8 pS in 100 mM Ca^{2+}), which are involved in the pathophysiology of Duchenne muscular dystrophy. These SACs are inhibited by Gd^{3+}, SKF-96365, and GsMTx4 and are upregulated by IGF-1 (Ducret *et al.*, 2006; Gailly, 2002; Vandebrouck *et al.*, 2002). However, it remains to be analyzed thoroughly whether these channels, as suggested, correspond to TRPC1, TRPC3, and TRPC6 (Gailly, 2002).

A number of TRPV channels are mechano- or volume/osmosensitive. A splice variant of TRPV1, TRPV1b, in which a stretch of 60 amino acids is deleted in the intracellular N-terminal region (Lu *et al.*, 2005), forms stretch-inhibited cation channels, for example, these channels are activated by hypertonic cell shrinkage (Ciura and Bourque, 2006; Naeini *et al.*, 2005). TRPV1b has been shown to mediate the osmosensitivity of arginine/vaso-pressin-releasing neurons in the supraoptic nucleus (Naeini *et al.*, 2005).

TRPV1 is also required for the response of osmosensory neurons in the organum vasculosum lamina terminalis, the primary osmosensor in the brain. In these neurons, hypertonic exposure activates an inward current, which is absent in *trpv1−/−* mice (Ciura and Bourque, 2006).

TRPV2 has been described as a stretch-activated channel functioning as a mechanosensor in vascular smooth muscle cells and also appears to be activated by osmotic cell swelling (Beech *et al.*, 2004; Muraki *et al.*, 2003). Stretch activation of TRPV2 has been assigned a role in skeletal and cardiac muscle degeneration caused by dystrophin–glycoprotein complex disruption (Iwata *et al.*, 2003). The mechanism of activation of TRPV2 in response to cellular stretch/swelling is still a matter of debate. However, it is known that PI3-kinase promotes TRPV2 membrane insertion (Penna *et al.*, 2006); it has also been discussed whether altered cellular PtdIns(4,5)P_2 levels may underlie TRPV2 mechanosensitivity (Yamamoto *et al.*, 2006). Interestingly, TRPV2 interacts with a protein kinase A (PKA)-dependent signaling module, containing PKA and an A-kinase adapter protein (acyl-CoA-binding domain protein ACBD3), and phosphorylation by PKA may modulate mechanoactivation of TRPV2 (Stokes *et al.*, 2004).

TRPV4 was the first TRP channel to be described as a volume-activated, Ca^{2+}-permeable cation channel (Liedtke *et al.*, 2000; Nilius *et al.*, 2001; Strotmann *et al.*, 2000). TRPV4 is activated by osmotic cell swelling in a manner that is not due to direct stretch activation (Liedtke *et al.*, 2000; Strotmann *et al.*, 2000), is independent of intracellular ionic strength and not reproduced by the presence of GTPγS in the pipette solution (Nilius *et al.*, 2001), and is not dependent on N-terminal ankyrin repeats (Liedtke *et al.*, 2000). It has been shown that TRPV4 activation by cell swelling is mediated by the arachidonic acid metabolite 5′,6′-epoxyeicosatrienoic acid (5′,6′-EET) (Vriens *et al.*, 2004, 2005; Watanabe *et al.*, 2003), consistent with the previously demonstrated swelling-induced activation of PLA$_2$ (Kinnunen, 2000; Pedersen *et al.*, 2006). Phosphorylation of TRPV4 at Tyr^{253} in the N-terminal region has also been proposed to play a role in channel activation by hypotonicity, although this finding has been disputed (Cohen, 2005; Xu *et al.*, 2003). Importantly, activation of TRPV4 by cell swelling appears to be modulated by protein–protein interactions. (i) Thus swelling-induced activation of TRPV4 is absent in cystic fibrosis airway epithelial cells, suggesting a functional role of the cystic fibrosis transmembrane regulator (CFTR) in mechanosensing by TRPV4 (Arniges *et al.*, 2004). (ii) TRPV4 activation by hypotonicity in salivary gland epithelial cells appears to depend on the interaction with its binding partner aquaporin 5 (AQP5) rather than on cell swelling directly (Liu *et al.*, 2006). (iii) Pacsin 3, a protein thought to block dynamin-mediated endocytosis, interacts functionally with TRPV4 (Cuajungco *et al.*, 2006), and coexpression with Pacsin 3 increases TRPV4 insertion in the plasma membrane, yet attenuates its activation by cell swelling (D'Hondt and Nilius, unpublished result). This indicates that insertion/retrieval mechanisms may play a role in the regulation of TRPV4 by cell swelling. Interestingly, a similar mechanism may operate in the shear stress–dependent activation of TRPM7 (see later) (Oancea et al., 2006). (iv) Finally, TRPV4 and the polycystic kidney disease protein 2 (TRPP2) colocalize in the kidney and interact in the primary cilium (Giamarchi et al., 2006), apparently forming a novel mechanosensitive complex.

TRPM3 exists as at least 12 splice variants, for which a unified nomenclature is lacking. A long splice variant (1555 amino acids) of TRPM3 forms a constitutively active channel that is further stimulated by muscarinic acetylcholine receptor activation (Grimm et al., 2003; Lee et al., 2003; Oberwinkler et al., 2005), while a shorter (1325 amino acids) human TRPM3 channel is stimulated by hypotonic cell swelling (Grimm et al., 2003) by mechanisms that remain to be elucidated.

TRPM4 has been described as a mechano-/stretch-sensitive channel involved in the control of pressure-induced smooth muscle cell depolarization and myogenic vasoconstriction in cerebral arteries and isolated vascular smooth muscle cells (Dietrich et al., 2006; Earley et al., 2004; Inoue et al., 2004, 2006; Kraft and Harteneck, 2005). However, the mechanism of activation of TRPM4 under these conditions remains to be elucidated in detail.

TRPM7 has been considered as a candidate for mechanosensation in a variety of cell types. TRPM7 has been proposed to be directly activated by cell stretch and potentiated by hypotonic cell swelling (Numata et al., 2006). Moreover, shear stress induces a translocation of TRPM7 to the plasma membrane and a rapid increase in TRPM7 currents (Oancea et al., 2006).

TRPA1 has long been considered a paradigm for a mechanosensitive channel. TRPA1 exhibits 14 N-terminal ankyrin repeats (Lee et al., 2006; Nagata et al., 2005; Story et al., 2003), which have been speculated to act as a gating spring in mechanosensing by this channel (Howard and Bechstedt, 2004; Lee et al., 2006; Nagata et al., 2005; Sotomayor et al., 2005). TRPA1 was proposed to be the mechanically gated transduction channel necessary for the auditory response in mammals (Corey et al., 2004; Gillespie et al., 2005); however, studies in $trpa1^{(-/-)}$ mice revealed no obvious deficits in auditory function (Bautista et al., 2006; Kwan et al., 2006). Although not the auditory transduction channel, TRPA1 may well be a mechanosensory TRP channel involved in other physiological functions (Corey, 2006).

In the context of mechano- and osmosensing, it is noteworthy that in addition to ankyrin repeats, several TRP subfamilies exhibit other conserved domains potentially linking them to the actin-based cytoskeleton. Thus, direct interactions with PDZ scaffold proteins such as ezrin-binding phosphoprotein 50 (EBP50)/Na^+/H^+ exchanger regulatory factor (see Pedersen et al., 2005) link several TRP subfamilies to the ezrin/radixin/moesin proteins (Lockwich et al., 2001), which act as integrators between the F–actin cytoskeleton and integral membrane proteins (e.g., Bretscher *et al.*, 2002) and which have been shown to be volume sensitive (Darborg *et al.*, 2005; Wu *et al.*, 2004).

Finally, the polycystic kidney disease protein 2 (PKD2 or TRPP2) is involved in mechanosensation in the primary cilia of kidney cells. In yeast, a TRPP2-like channel is involved in mechanosensing (sensing of cell shape) (Palmer *et al.*, 2001). TRPP2 forms a complex with polycystin 1–like (PKD1), which is important for TRPP2 regulation by many stimuli; however, data from PKD1 knockout mice also point to a possible mechanosensory role for TRPP2 in the absence of PKD1 (for an excellent review, see Giamarchi *et al.*, 2006). Trafficking of TRPP2 from the endoplasmic reticulum and the Golgi complex to the plasma membrane, where at least some mechanical forces are detected, is directed by the phosphoproteins phosphofurin acidic cluster-1 and -2 and involves binding of these proteins to TRPP2 promoted by its phosphorylation at Ser^{812} by casein-kinase 2 (Köttgen and Walz, 2005; Kottgen *et al.*, 2005). Ca^{2+} influx via TRPP2 translates mechanical deflections of primary cilia into a signaling cascade associated with growth control and differentiation in renal epithelial cells (Nauli and Zhou, 2004; Nauli *et al.*, 2003). A role for TRPP2 in osmosensing has also been proposed (see, e.g., Montalbetti *et al.*, 2005).

4. TRP Channels in Cell Volume Regulation

Cell volume perturbations occur under physiological and pathophysiological conditions in a wide range of cell types, and the ability to regulate cell volume is fundamental to cell function and survival. Following osmotic cell shrinkage or swelling, most cell types are able to regulate their volume in processes termed regulatory volume increase or regulatory volume decrease (RVD), respectively (Hoffmann and Pedersen, 2006; Lang *et al.*, 1998). Obviously, to establish that a given channel actually contributes to cell volume regulation, it is not sufficient to show that it is activated by cell volume perturbations, as the effect of its activity on cell volume regulation must be assessed. Although, as outlined earlier, a number of TRP channels have been shown to be activated by cell swelling (or swelling-induced stretch), direct evidence that TRP channels contribute to RVD is still sparse. In many cell types, increases in $[Ca^{2+}]_i$ contribute to the RVD response after cell swelling, most commonly by the activation of Ca^{2+}-sensitive K^+ channels (see Hoffmann and Pedersen, 2006). This swelling-activated increase in $[Ca^{2+}]_i$ reflects opening of stretch-activated channels, at least some of which have now been shown or proposed to be TRP channels (e.g., Arniges *et al.*, 2004; Maroto *et al.*, 2005; Numata *et al.*, 2006), and hence, the role of TRP channels in volume regulation is generally assumed to be to elicit an increase in $[Ca^{2+}]_i$.

Among the mammalian TRP channels, a role in volume regulation has been documented most thoroughly for TRPV4. As noted earlier, TRPV4 is activated, at least in part, as a consequence of the cell swelling–induced release of 5′,6′-EET (Vriens *et al.*, 2004, 2005; Watanabe *et al.*, 2003). Direct evidence for a role in volume regulation is available for TRPV4, as trpv4-/- mice have a reduced capacity for RVD (Liedtke and Kim, 2005), and exogenous expression of TRPV4 confers osmoregulatory capacity to CHO cells (Becker *et al.*, 2005). Moreover, RVD was reduced markedly in cells with reduced expression of CFTR or AQP5, both of which interact with TRPV4 (Arniges *et al.*, 2004; Liu *et al.*, 2006). In both cases, this defect in RVD was shown to be because of a reduction of the TRPV4-mediated Ca^{2+} entry activated by hypotonic stimulation (Arniges *et al.*, 2004; Liu *et al.*, 2006).

Other TRP channels activated by cell swelling are, as noted previously, TRPV2, TRPM3, probably TRPM7, and TRPP2 (Chen and Barritt, 2003; Grimm *et al.*, 2003; Kraft and Harteneck, 2005; Montalbetti *et al.*, 2005; Muraki *et al.*, 2003; Numata *et al.*, 2006). However, although some mechanisms have been proposed, events leading to the swelling–induced activation of these channels are considerably less well understood than for TRPV4. Nonetheless, for a few of these channels, evidence has been presented for a role in the RVD process based on assessments of cell volume regulation after cell swelling. Hence, knockdown of TRPM7 (Numata *et al.*, 2006), as well as of TRPC1 (Chen and Barritt, 2003), was found to reduce the rate of RVD (although for TRPC1 knockdown, the effect on RVD was marginal).

5. Experimental Procedures

This section describes the foundations and practical procedures for selected methods useful in the evaluation of mechano- and osmosensitivity, as well as of transporter effects on $[Ca^{2+}]_i$ cell volume. Standard patch clamp procedures have been described extensively elsewhere (see, e.g., Hille, 2001; Sakmann and Neher, 1995) and are not detailed here.

5.1. Electrophysiological recordings used to assess mechanosensitivity of TRP channels

5.1.1. Current measurements on TRP channels expressed in mammalian cells

Stretch sensitivity of TRP channels can be measured directly in cells endogenously expressing the channel in question or in an adequate over-expression system. A low DNA concentration should be used for transfection (<2 μg of DNA per well, see also http://www.mirusbio.com/) to reduce the risk of overexpression artifacts. The impact of a particular TRP channel may be studied by using specific modulators, such as 4α PDD for TRPV4 (Liu *et al.*, 2006), specific antisense methods (e.g., Earley *et al.*, 2004), gene silencing methods (see, e.g., Arniges *et al.*, 2004; Numata *et al.*, 2006), and constitutive or conditional knockout models (for a review, see Freichel *et al.*, 2005). Gene silencing has been used as a powerful tool to knock down various TRP channels; however, this approach has to be used cautiously, as off-target gene silencing can present a notable challenge in the interpretation of data from siRNA experiments (Birmingham *et al.*, 2006; Fedorov *et al.*, 2006).

Single-channel recordings to assess channel mechanosensitivity can be performed in the cell-attached and excised inside-out configurations. In these configurations, the amplitude of the single-channel current is measured from the peak-to-peak distance on the amplitude histogram. The open probability (P_o) of the single-channel current can be calculated by dividing the total time spent in the open state by the total time of continuous recording (30–200 s) in patches containing one active channel. To test effects of mechanical stretch, patched membranes are subjected to a pulse of negative pressure applied to the back of the patch pipette by mouth or syringe. The pressure level can be monitored with a manometer or a piezoelectric pressure transducer.

5.1.2. Current measurements on TRP channels reconstituted in liposomes

To study mechanoactivation by membrane stretch directly, avoiding contributions from the cortical cytoskeleton, a liposome preparation may be used. Membrane proteins from an overexpression system (we routinely use HEK or CHO cells) can be reconstituted in liposomes for patch clamp recording (for details, see Hamill *et al.*, 1981; Maroto *et al.*, 2005; Sukharev *et al.*, 1993). To prepare phosphatidylcholine liposomes, phosphatidylcholine is dissolved in chloroform and small aliquots of the lipid are dried under nitrogen, resuspended in 5 mM EDTA, 5 mM EGTA, 1 mM dithiothreitol, 50 mM Tris-HCl, pH 7.4, containing 2% octyl-f3-D-glucopyranoside (OG), and bath sonicated for 5 min. A 200-μl volume (equivalent to 2 mg lipid) is added to the membrane protein to achieve protein:lipid ratios ranging from 1:50 to 1:5000. The mixture is incubated on a platform rocker for 1 h at room temperature. BioBeads are then added to remove the OG, and the suspension is rocked for a further 3 h at room temperature. The liposomes are collected by ultracentrifugation at 90,000g for 30 min and resuspended. Aliquots of the liposomes are spotted onto glass slides and allowed to dehydrate under vacuum at 4° for 6 h, followed by overnight rehydration (for details, see Sukharev *et al.*, 1993).

For liposome recording, standard patch clamp techniques can be used. The liposomes are placed in a patch clamp chamber containing high MgCl$_2$

solution. Seal resistance in the high $MgCl_2$ solution should exceed 20 GΩ. Inside-out patches can be formed by passing the pipette tip briefly through the solution–air interface. As described earlier for cellular systems, channel activation by stretch can be studied by applying negative pressure pulses (for TRPC1, -10 to -200 mm Hg is appropriate) (Maroto et al., 2005; Perozo, 2006).

5.2. Electrophysiological recordings to assess osmosensitivity of TRP channels

5.2.1. Whole cell patch clamp and cell swelling: general considerations

Hypotonic cell swelling in the whole cell patch clamp mode poses the inherent problem that the cell interior will constantly equilibrate with the pipette solution. As a consequence of this, the initial rapid increase in cell volume is followed by a nearly linear continuous increase until the cell ruptures (or extracellular osmolarity is restored), that is, a condition fundamentally different from that resulting from the same transmembrane osmolarity difference in an unpatched cell. In contrast to the cell volume, the current may reach a stationary level; hence, under these conditions, a close correlation between current and cell volume per se does not exist. This behavior can be explained as follows: changes in cell volume over time (dV/dt) are a consequence of the net flux of water across the plasma membrane, which is linearly dependent on the transmembrane osmotic gradient, thus

$$\frac{dV}{dt} = P_W \cdot (c_i - c_o) \tag{10.1}$$

where c_i is the total concentration of solutes of the intracellular medium, c_o is the extracellular medium, and P_w is a constant, here referred to as "water permeation." P_W is the product of the osmotic water permeability of the membrane (P_f), the cell surface (S), and the partial molar volume of water (V_W).

It is assumed that the osmotic gradient driving the water flux is approximated by the difference in total concentration of solutes and that the diffusion rate from pipette to cell is the same for all solute species in the pipette. Because of the combination of the solute exchange with the pipette and the variation in cell volume, c_i changes according to

$$\frac{dc_i}{dt} = \frac{1}{V} \left(k_d \cdot (c_p - c_i) - c_i \cdot \frac{dV}{dt} \right) \tag{10.2}$$

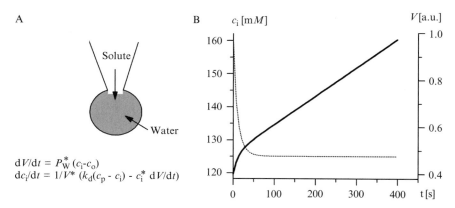

A

B

c_i [mM] V [a.u.]

$$dV/dt = P_W^* (c_i\text{-}c_o)$$
$$dc_i/dt = 1/V^* (k_d(c_p - c_i) - c_i^* \, dV/dt)$$

Figure 10.2   Cell volume and intracellular osmolyte concentration during exposure to hypotonic bath solutions in whole cell patch clamp experiments. If a cell is exposed to a hypotonic solution, water enters the cell and induces a change in cell volume, dV/dt (A). This increase is reflected by the initial change in volume (B, solid line). During the experiment, solutes will enter the cell from the patch pipette, which in turn induces additional water entry. Therefore, the cell volume will increase continuously (B). The intracellular osmolyte concentration, c_i (dotted line), decreases rapidly and reaches a stationary constant value while the volume is constantly increasing. See text for details.

where the constant k_d describes the rate of diffusion of solutes from pipette to cell and c_i and c_p reflect the global concentrations of solutes in the cell and in the pipette, respectively. Thus, hypotonic cell swelling in the whole cell patch clamp mode can be described by a system of two differential equations with the variables c_i and V. Figure 10.2 shows a numerical solution of this system of differential equations. The initial slope of the volume change mainly reflects the influx of water across the plasma membrane, which is faster than the solute exchange via the patch pipette. The subsequent nearly linear increase in volume is a consequence of the exchange of solutes via the patch pipette. Importantly, this model predicts that c_i decreases significantly during an osmotic challenge and reaches a plateau value, despite the continuous increase in cell volume. For further details, see Nilius (2004) and Voets *et al.* (1999).

5.2.2. Practical procedures

Similar to stretch sensitivity, volume/osmosensitivity of TRP channels may be assessed either in cells endogenously expressing the channel in question or after channel expression in a cell type with low endogenous TRP levels such as HEK 293 cells. For measurements of the volume sensitivity of TRP channels in the whole cell patch clamp mode in mammalian cells, we generally employ an isotonic solution containing (in mM) 105 NaCl, 6 CsCl, 5 CaCl$_2$, 1 MgCl$_2$, 10 HEPES, 90 D-mannitol, 10 glucose, buffered pH 7.4 with NaOH (320 $\pm$ 5 mOsm). Cell swelling can thus be induced at constant extracellular ionic strength by omitting the desired amount of mannitol from this solution. The pipette solution for such experiments is generally composed of (in mM) 20 CsCl, 100 Asp, 1 MgCl$_2$, 10 HEPES, 4 Na$_2$ATP, 10 BAPTA, and an equivalent amount of CaCl$_2$ to buffer $[Ca^{2+}]_i$ at an required level (for calculation of free intracellular divalent concentrations, we refer to the CaBuf program ftp://ftp.cc.kuleuven.ac.be/pub/droogmans/cabuf.zip).

5.3. Assessing the contribution of TRP channels to cell volume regulation: simultaneous measurements of $[Ca^{2+}]_i$ and cell volume changes

5.3.1. General considerations

Changes in $[Ca^{2+}]_i$ and cell volume can conveniently be assessed simultaneously in a cell population using the fluorescent Ca^{2+}-sensitive probe Fura-2 in conjunction with large-angle light scattering. Ratiometric measurements of $[Ca^{2+}]_i$ using Fura-2 are widely used and described extensively elsewhere (Grynkiewicz et al., 1985; Tsien, 1989). Large-angle light scattering is a noninvasive and sensitive technique for evaluating volume changes in both adherent and suspended cells, exploiting the fact that the light-scattering properties of cells are altered upon cell volume changes. While smaller particles scatter light in all directions, large particles scatter predominantly in the near-forward direction. A cell scatters light both as a large particle and as a collection of small particles reflecting its internal composition, hence both forward and large-angle light scattering are volume sensitive, although the latter appears to be the more sensitive method (Latimer, 1982; McManus et al., 1993; Meyer and Brunsting, 1975). Generally, large-angle light scattering is inversely related to cell volume, consistent with a dependence on the cellular concentration of scattering particles (Fischbarg et al., 1989; Latimer, 1982; McManus et al., 1993). The only real disadvantage to the light-scattering technique is that absolute cell volumes cannot be obtained in this manner; however, for evaluation of RVD rates, this is rarely a problem.

5.3.2. Practical procedures

Cells are seeded 24 h prior to experiments on 10 $\times$ 50-mm, HCl- and ethanol-washed coverslips to a confluency of 70 to 90% at the time of the experiments. Cells are loaded with Fura-2 by preincubation with 2 μM Fura-2 acetoxymethyl ester (Fura-2-AM) in standard isotonic medium for 20 min at 37° followed by a wash and a 15-min postloading incubation to ensure proper intracellular cleavage of the acetoxymethylester groups to obtain Fura-2.

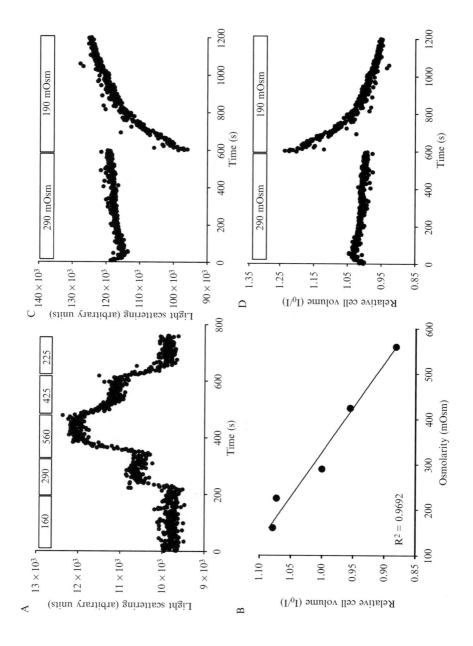

We employ a PTI RatioMaster spectrophotometer equipped with a standard photomultiplier (PMT) system, excitation and emission monochromers for wavelength selection, a 75-W xenon lamp, and a temperature-controlled cuvette house. During experiments, the cuvette is perfused continuously with preheated (37°) standard Ringer's solutions at a speed of about 0.7 ml/min, increasing to 3.5 ml/min during solution changes for rapid complete exchange in the cuvette. The coverslip with the cells is placed at a 50° angle relative to the excitation light, and scattered light is collected at an angle of 90° relative to the excitation light. Excitation and emission wavelengths for light scattering need to be optimized for each cell type used. Generally, the best light-scattering signal is obtained after excitation in the wavelength range of 570 to 600 nm. The emission wavelength is set about 5 nm red shifted to the excitation wavelength to protect the PMT from excitation light. To the extent that the cells swell as near-perfect osmometers, and in the absence of volume regulation, the light-scattering signal will be a linear function of the osmolarity. Practically, this is achieved by very rapid solution changes, such that volume regulation is negligible in the time window studied.

Figure 10.3A and B show an experiment evaluating the linearity of the light-scattering response to osmolarity changes in Ehrlich-Lettre ascites (ELA) murine tumor cells. With these cells, the optimal light-scattering signal in our setup is obtained at 589 nm excitation, and emission is measured at 595 nm. Figure 10.3A is a representative trace, and it is seen that the light-scattering signal is directly related to extracellular osmolarity, that is, inversely related to cell volume. Therefore, data are calculated as the inverse of the light-scattering signal relative to that obtained in the initial, isotonic condition (I_0), that is, $1/(I/I_0)$, or I_0/I. Figure 10.3B shows I_0/I from the experiment in Fig. 10.3A as a function of extracellular osmolarity. Figures 10.3C and 10.3D show the light-scattering signal, and I_0/I, for an

---

Figure 10.3 The use of large-angle light scattering to monitor cell volume changes. (A) Light scattering as a function of extracellular osmolarity in Ehrlich Lettre ascites cells. Cells were seeded 24 h prior to experiments on 10 × 50-mm coverslips to a confluency of about 70% at the time of the experiments. Cells were mounted in a temperature-controlled cuvette in a PTI RatioMaster spectrophotometer and were perfused continuously with preheated (37°) Ringer's solutions, which were changed rapidly to the osmolarity indicated by increasing the perfusion rate from 0.7 to 3.5 ml/min. Excitation was measured at 589 nm excitation, and emission is measured at 595 nm. (B) Data from A were converted to relative cell volumes by calculating the inverse of the light-scattering signal relative to that obtained in the initial, isotonic condition (I_0), i.e., $1/(I/I_0)$, or I_0/I. (C and D) Raw data and I_0/I for an experiment assessing RVD in about 90% confluent ELA cells after a 35% reduction in extracellular osmolarity. The experiment was carried out as described in A. It may be noted that the magnitude of the light-scattering signal, as well as of the relative changes in light scattering, is strongly cell density dependent, hence only populations of equal confluency should be compared.

experiment assessing RVD in ELA cells after a 35% reduction in extracellular osmolarity. RVD may be calculated as the slope of the initial, linear part of the relative cell volume traces following maximal cell swelling (for further examples, see Pedersen et al., 2002).

The simultaneous assessments of Fura-2 fluorescence are carried out by measuring at 510 nm after excitation at 340 and 380 nm. Practically, this is achieved by running continuous cycles of excitation and measurement, such that cells are excited at 589 nm and emission is measured at 595 nm, followed by excitation at 340 and 380 nm, respectively, and emission measurement at 510 nm. Fura-2 data are evaluated as the 340/380-nm ratio after background subtraction and may be converted to $[Ca^{2+}]_i$ values by in vitro calibration as described previously (Grynkiewicz et al., 1985). In our hands, Fura-2 loading has no effect on the light-scattering measurements.

5.4. Combined patch clamp and cell volume measurements

5.4.1. General considerations

A combination of volume measurements and patch clamp can be used on larger cells (e.g., epithelial or endothelial cells) adhering to a coated coverslip. The general problem of using whole cell patch clamp in conjunction with hypotonic cell swelling has already been discussed in detail. It should be carefully taken into account that the cell volume of the patched cells increased constantly and never reached a steady state and evaluated how much this unavoidable pitfall influences the conclusions drawn by the experimentalist.

5.4.2. Practical procedures

We have used a method for simultaneous monitoring of the cell height and whole cell currents. This method is useful for flat and adhering cells (e.g., endothelial or epithelial cells), which do not change the surface area attached to the coverslip during swelling. It is not usable for small, round, nonadhering cells, such as HEK cells. Currents are measured in the normal whole cell patch clamp configuration. We measure simultaneously the cell surface area (CSA) of the patched single cells, which is quantified from digital images recorded with a video camera (Model CF 6, Kappa, Gleichen, Germany) (Voets et al., 1999). Cell borders are traced manually, and the surface of the traced region is determined using IMAGETOOL 1.25 software (University of Texas Health Science Center, San Antonio, TX) (for more details concerning the measurement of cell height without patch clamping, see Van Driessche et al., 1993). CSA is calculated as the average of three independent tracing procedures. Simultaneously, cell thickness, T_c, is monitored constantly as described previously (Van Driessche et al., 1993). Briefly, gelatin-coated coverslips containing nonconfluent endothelial cells are incubated for about 30 min with 4 μl/ml Red Neutravidin-labeled microbeads (F-8775, Molecular Probes), followed by a 15-min washing with microbead-free solution. Visualization of the microbeads is performed using a xenon lamp and the XF40/E filter set (Omega Optical, Brattleboro, VT). Fluorescent images at different vertical positions are recorded with the CF 6 video camera, digitized, and displayed on a video monitor. Vertical displacement is achieved using a low-voltage piezoelectric translator (PIFOC P-721; Physik Instrumente, Waldbronn, Germany). T_c is calculated as the vertical distance between beads on the gelatin surface and on the cell surface, and image analysis is performed online (for details, see Voets et al., 1999).

ACKNOWLEDGMENTS

We thank Dr. Greg Owsianik (Leuven) for his help with Fig. 10.1. Work in the authors' laboratories is supported by the Danish National Research Council (SFP, Grants 21-04-0507 and 272-05-0305) and Human Frontiers Science Programme (HFSP Research Grant Ref. RGP 32/2004), the Belgian Federal Government, the Flemish Government, and the Onderzoeksraad KU Leuven (GOA 2004/07, F.W.O. G. 0136.00; F.W.O. G.0172.03, Interuniversity Poles of Attraction Program, Prime Ministers Office IUAP Nr.3P4/23, Excellentiefinanciering EF/95/010) (BN).

REFERENCES

Allen, D. G., Whitehead, N. P., and Yeung, E. W. (2005). Mechanisms of stretch-induced muscle damage in normal and dystrophic muscle: Role of ionic changes. J. Physiol. **567,** 723–735.

Arniges, M., Vazquez, E., Fernandez-Fernandez, J. M., and Valverde, M. A. (2004). Swelling-activated Ca^{2+} entry via TRPV4 channel is defective in cystic fibrosis airway epithelia. J. Biol. Chem. **279,** 54062–54068.

Basavappa, S., Pedersen, S. F., Jorgensen, N. K., Ellory, J. C., and Hoffmann, E. K. (1998). Swelling-induced arachidonic acid release via the 85-kDa cPLA2 in human neuroblastoma cells. J. Neurophysiol. **79,** 1441–1449.

Bautista, D. M., Jordt, S. E., Nikai, T., Tsuruda, P. R., Read, A. J., Poblete, J., Yamoah, E. N., Basbaum, A. I., and Julius, D. (2006). TRPA1 mediates the inflammatory actions of environmental irritants and proalgesic agents. Cell **124,** 1269–1282.

Becker, D., Blase, C., Bereiter-Hahn, J., and Jendrach, M. (2005). TRPV4 exhibits a functional role in cell-volume regulation. J. Cell Sci. **118,** 2435–2440.

Beech, D. J., Muraki, K., and Flemming, R. (2004). Non-selective cationic channels of smooth muscle and the mammalian homologues of Drosophila TRP. J. Physiol. **559,** 685–706.

Biemans-Oldehinkel, E., Mahmood, N. A., and Poolman, B. (2006). A sensor for intracellular ionic strength. Proc. Natl. Acad. Sci. USA **103,** 10624–10629.

Birmingham, A., Anderson, E. M., Reynolds, A., Ilsley-Tyree, D., Leake, D., Fedorov, Y., Baskerville, S., Maksimova, E., Robinson, K., Karpilow, J., Marshall, W. S., and Khvorova, A. (2006). $3'$ UTR seed matches, but not overall identity, are associated with RNAi off-targets. Nat. Methods **3,** 199–204.

Bretscher, A., Edwards, K., and Fehon, R. G. (2002). ERM proteins and merlin: Integrators at the cell cortex. Nat. Rev. Mol. Cell. Biol. **3,** 586–599.

Chemin, J., Patel, A. J., Duprat, F., Lauritzen, I., Lazdunski, M., and Honore, E. (2005). A phospholipid sensor controls mechanogating of the K^+ channel TREK-1. *EMBO J.* **24**, 44–53.

Chen, J., and Barritt, G. J. (2003). Evidence that TRPC1 (transient receptor potential canonical 1) forms a Ca^{2+}-permeable channel linked to the regulation of cell volume in liver cells obtained using small interfering RNA targeted against TRPC1. *Biochem. J.* **373**, 327–336.

Ciura, S., and Bourque, C. W. (2006). Transient receptor potential vanilloid 1 is required for intrinsic osmoreception in organum vasculosum lamina terminalis neurons and for normal thirst responses to systemic hyperosmolality. *J. Neurosci.* **26**, 9069–9075.

Cohen, D. M. (2005). TRPV4 and the mammalian kidney. *Pflüg. Arch. Eur. J. Physiol.* **451**, 168–175.

Colbert, H. A., Smith, T. L., and Bargmann, C. I. (1997). OSM-9, a novel protein with structural similarity to channels, is required for olfaction, mechanosensation, and olfactory adaptation in *Caenorhabditis elegans. J. Neurosci.* **17**, 8259–8269.

Corey, D. P. (2006). What is the hair cell transduction channel? *J. Physiol.* **576**, 23–28.

Corey, D. P., Garcia-Anoveros, J., Holt, J. R., Kwan, K. Y., Lin, S. Y., Vollrath, M. A., Amalfitano, A., Cheung, E. L., Derfler, B. H., Duggan, A., Geleoc, G. S., Gray, P. A., *et al.* (2004). TRPA1 is a candidate for the mechanosensitive transduction channel of vertebrate hair cells. *Nature* **432**, 723–730.

Cuajungco, M. P., Grimm, C., Oshima, K., D'Hoedt, D., Nilius, B., Mensenkamp, A. R., Bindels, R. J., Plomann, M., and Heller, S. (2006). PACSINs bind to the TRPV4 cation channel: PACSIN 3 modulates the subcellular localization of TRPV4. *J. Biol. Chem.* **281**, 18753–18762.

Darborg, B., Hoffmann, E. K., and Pedersen, S. F. (2005). Causal relationship between NHE1, NKCC1, MAPKs and ERM proteins after osmotic shrinkage in Ehrlich Lettre mouse ascites tumor cells. *FASEB J.* **19**, A115.

Denis, V., and Cyert, M. S. (2002). Internal Ca^{2+} release in yeast is triggered by hypertonic shock and mediated by a TRP channel homologue. *J. Cell Biol.* **156**, 29–34.

Di Ciano-Oliveira, C., Thirone, A. C., Szaszi, K., and Kapus, A. (2006). Osmotic stress and the cytoskeleton: the R(h)ole of Rho GTPases. *Acta Physiol. (Oxf.)* **187**, 257–272.

Dietrich, A., Chubanov, V., Kalwa, H., Rost, B. R., and Gudermann, T. (2006). Cation channels of the transient receptor potential superfamily: Their role in physiological and pathophysiological processes of smooth muscle cells. *Pharmacol. Ther.* **112**, 744–760.

Ducret, T., Vandebrouck, C., Cao, M. L., Lebacq, J., and Gailly, P. (2006). Functional role of store-operated and stretch-activated channels in murine adult skeletal muscle fibres. *J. Physiol.* **575**, 913–924.

Earley, S., Waldron, B. J., and Brayden, J. E. (2004). Critical role for transient receptor potential channel TRPM4 in myogenic constriction of cerebral arteries. *Circ. Res.* **95**, 922–929.

Fedorov, Y., Anderson, E. M., Birmingham, A., Reynolds, A., Karpilow, J., Robinson, K., Leake, D., Marshall, W. S., and Khvorova, A. (2006). Off-target effects by siRNA can induce toxic phenotype. *RNA* **12**, 1188–1196.

Fischbarg, J., Kuang, K. Y., Hirsch, J., Lecuona, S., Rogozinski, L., Silverstein, S. C., and Loike, J. (1989). Evidence that the glucose transporter serves as a water channel in J774 macrophages. *Proc. Natl. Acad. Sci. USA* **86**, 8397–8401.

Freichel, M., Vennekens, R., Olausson, J., Stolz, S., Philipp, S. E., Weissgerber, P., and Flockerzi, V. (2005). Functional role of TRPC proteins in native systems: Implications from knockout and knock-down studies. *J. Physiol.* **567**, 59–66.

Fu, Y., Subramanya, A., Rozansky, D., and Cohen, D. M. (2006). WNK kinases influence TRPV4 channel function and localization. *Am. J. Physiol. Renal Physiol.* **290**, F1305–F1314.

Gailly, P. (2002). New aspects of calcium signaling in skeletal muscle cells: Implications in Duchenne muscular dystrophy. *Biochim. Biophys. Acta* **1600,** 38–44.

Giamarchi, A., Padilla, F., Coste, B., Raoux, M., Crest, M., Honore, E., and Delmas, P. (2006). The versatile nature of the calcium-permeable cation channel TRPP2. *EMBO Rep.* **7,** 787–793.

Gillespie, P. G., Dumont, R. A., and Kachar, B. (2005). Have we found the tip link, transduction channel, and gating spring of the hair cell? *Curr. Opin. Neurobiol.* **15,** 389–396.

Grimm, C., Kraft, R., Sauerbruch, S., Schultz, G., and Harteneck, C. (2003). Molecular and functional characterization of the melastatin-related cation channel TRPM3. *J. Biol. Chem.* **278,** 21493–21501.

Grynkiewicz, G., Poenie, M., and Tsien, R. Y. (1985). A new generation of Ca^{2+} indicators with greatly improved fluorescence properties. *J. Biol. Chem.* **260,** 3440–3450.

Hamill, O. P., and Martinac, B. (2001). Molecular basis of mechanotransduction in living cells. *Physiol. Rev.* **81,** 685–740.

Hamill, O. P., Marty, A., Neher, E., Sakmann, B., and Sigworth, F. J. (1981). Improved patch-clamp techniques for high-resolution current recording from cells and cell-free membrane patches. *Pflüg. Arch. Eur. J. Physiol.* **391,** 85–100.

Hille, B. (2001). "Ionic Channels of Excitable Membranes." Sinauer, Sunderland, MA.

Hoffmann, E. K., and Pedersen, S. F. (2006). Sensors and signal transduction pathways in vertebrate cell volume regulation. *Contrib. Nephrol.* **152,** 54–104.

Howard, J., and Bechstedt, S. (2004). Hypothesis: A helix of ankyrin repeats of the NOMPC-TRP ion channel is the gating spring of mechanoreceptors. *Curr. Biol.* **14,** R224–R226.

Inoue, R., Jensen, L. J., Shi, J., Morita, H., Nishida, M., Honda, A., and Ito, Y. (2006). Transient receptor potential channels in cardiovascular function and disease. *Circ. Res.* **99,** 119–131.

Inoue, R., Morita, H., and Ito, Y. (2004). Newly emerging Ca2+ entry channel molecules that regulate the vascular tone. *Expert Opin. Ther. Targets* **8,** 321–334.

Iwata, Y., Katanosaka, Y., Arai, Y., Komamura, K., Miyatake, K., and Shigekawa, M. (2003). A novel mechanism of myocyte degeneration involving the Ca^{2+}-permeable growth factor-regulated channel. *J. Cell Biol.* **161,** 957–967.

Kinnunen, P. K. (2000). Lipid bilayers as osmotic response elements. *Cell Physiol. Biochem.* **10,** 243–250.

Kottgen, M., Benzing, T., Simmen, T., Tauber, R., Buchholz, B., Feliciangeli, S., Huber, T. B., Schermer, B., Kramer-Zucker, A., Hopker, K., Simmen, K. C., Tschucke, C. C., *et al.* (2005). Trafficking of TRPP2 by PACS proteins represents a novel mechanism of ion channel regulation. *EMBO J.* **24,** 705–716.

Köttgen, M., and Walz, G. (2005). Subcellular localization and trafficking of polycystins. *Pflüg. Arch. Eur. J. Physiol.* **451,** 286–293.

Kraft, R., and Harteneck, C. (2005). TRPM cation channels: An overview. *Pflüg. Arch. Eur. J. Physiol.* **451,** 204–211.

Kung, C. (2005). A possible unifying principle for mechanosensation. *Nature* **436,** 647–654.

Kwan, K. Y., Allchorne, A. J., Vollrath, M. A., Christensen, A. P., Zhang, D. S., Woolf, C. J., and Corey, D. P. (2006). TRPA1 contributes to cold, mechanical, and chemical nociception but is not essential for hair-cell transduction. *Neuron* **50,** 277–289.

Lang, F., Busch, G. L., Ritter, M., Volkl, H., Waldegger, S., Gulbins, E., and Haussinger, D. (1998). Functional significance of cell volume regulatory mechanisms. *Physiol. Rev.* **78,** 247–306.

Latimer, P. (1982). Light scattering and absorption as methods of studying cell population parameters. *Annu. Rev. Biophys. Bioeng.* **11,** 129–150.

Lee, G., Abdi, K., Jiang, Y., Michaely, P., Bennett, V., and Marszalek, P. E. (2006). Nanospring behaviour of ankyrin repeats. *Nature* **440,** 246–249.

Lee, N., Chen, J., Sun, L., Wu, S., Gray, K. R., Rich, A., Huang, M., Lin, J. H., Feder, J. N., Janovitz, E. B., Levesque, P. C., and Blanar, M. A. (2003). Expression and characterization of human transient receptor potential melastatin 3 (hTRPM3). *J. Biol. Chem.* **278,** 20890–20897.

Liedtke, W. (2005). TRPV4 as osmosensor: A transgenic approach. *Pflüg. Arch.* **451,** 176–280.

Liedtke, W., Choe, Y., Marti-Renom, M. A., Bell, A. M., Denis, C. S., Sali, A., Hudspeth, A. J., Friedman, J. M., and Heller, S. (2000). Vanilloid receptor-related osmotically activated channel (VR-OAC), a candidate vertebrate osmoreceptor. *Cell* **103,** 525–535.

Liedtke, W., and Kim, C. (2005). Functionality of the TRPV subfamily of TRP ion channels: Add mechano-TRP and osmo-TRP to the lexicon! *Cell. Mol. Life Sci.* **62,** 2985–3001.

Liu, X., Bandyopadhyay, B., Nakamoto, T., Singh, B., Liedtke, W., Melvin, J. E., and Ambudkar, I. (2006). A role for AQP5 in activation of TRPV4 by hypotonicity: Concerted involvement of AQP5 and TRPV4 in regulation of cell volume recovery. *J. Biol. Chem.* **281,** 15485–15495.

Lockwich, T., Singh, B. B., Liu, X., and Ambudkar, I. S. (2001). Stabilization of cortical actin induces internalization of transient receptor potential 3 (Trp3)-associated caveolar Ca^{2+} signaling complex and loss of Ca^{2+} influx without disruption of Trp3-inositol trisphosphate receptor association. *J. Biol. Chem.* **276,** 42401–42408.

Lu, G., Henderson, D., Liu, L., Reinhart, P. H., and Simon, S. A. (2005). TRPV1b: A functional human vanilloid receptor splice variant. *Mol. Pharmacol.* **67,** 1119–1127.

Lussier, M. P., Cayouette, S., Lepage, P. K., Bernier, C. L., Francoeur, N., St-Hilaire, M., Pinard, M., and Boulay, G. (2005). MxA, a member of the dynamin superfamily, interacts with the ankyrin-like repeat domain of TRPC. *J. Biol. Chem.* **280,** 19393–19400.

Maingret, F., Fosset, M., Lesage, F., Lazdunski, M., and Honore, E. (1999). TRAAK is a mammalian neuronal mechano-gated K^{+} channel. *J. Biol. Chem.* **274,** 1381–1387.

Maroto, R., Raso, A., Wood, T. G., Kurosky, A., Martinac, B., and Hamill, O. P. (2005). TRPC1 forms the stretch-activated cation channel in vertebrate cells. *Nat. Cell Biol.* **7,** 179–185.

McManus, M., Fischbarg, J., Sun, A., Hebert, S., and Strange, K. (1993). Laser light-scattering system for studying cell volume regulation and membrane transport processes. *Am. J. Physiol.* **265,** C562–C570.

Meyer, R. A., and Brunsting, A. (1975). Light scattering from nucleated biological cells. *Biophys. J.* **15,** 191–203.

Montalbetti, N., Li, Q., Gonzalez-Perrett, S., Semprine, J., Chen, X. Z., and Cantiello, H. F. (2005). Effect of hydro-osmotic pressure on polycystin-2 channel function in the human syncytiotrophoblast. *Pflüg. Arch. Eur. J. Physiol.* **451,** 294–303.

Muraki, K., Iwata, Y., Katanosaka, Y., Ito, T., Ohya, S., Shigekawa, M., and Imaizumi, Y. (2003). TRPV2 is a component of osmotically sensitive cation channels in murine aortic myocytes. *Circ. Res.* **93,** 829–838.

Naeini, R. S., Witty, M. F., Seguela, P., and Bourque, C. W. (2005). An N-terminal variant of Trpv1 channel is required for osmosensory transduction. *Nat. Neurosci.* **9,** 93–98.

Nagata, K., Duggan, A., Kumar, G., and Garcia-Anoveros, J. (2005). Nociceptor and hair cell transducer properties of TRPA1, a channel for pain and hearing. *J. Neurosci.* **25,** 4052–4061.

Nasuhoglu, C., Feng, S., Mao, Y., Shammat, I., Yamamato, M., Earnest, S., Lemmon, M., and Hilgemann, D. W. (2002). Modulation of cardiac PIP2 by cardioactive hormones and other physiologically relevant interventions. *Am. J. Physiol. Cell Physiol.* **283**, C223–C234.

Nauli, S. M., Alenghat, F. J., Luo, Y., Williams, E., Vassilev, P., Li, X., Elia, A. E., Lu, W., Brown, E. M., Quinn, S. J., Ingber, D. E., and Zhou, J. (2003). Polycystins 1 and 2 mediate mechanosensation in the primary cilium of kidney cells. *Nat. Genet.* **33**, 129–137.

Nauli, S. M., and Zhou, J. (2004). Polycystins and mechanosensation in renal and nodal cilia. *Bioessays* **26**, 844–856.

Nicolson, T. (2005). Fishing for key players in mechanotransduction. *Trends Neurosci.* **28**, 140–144.

Nilius, B. (2004). Is the volume-regulated anion channel VRAC a "water-permeable" channel? *Neurochem. Res.* **29**, 3–8.

Nilius, B., and Droogmans, G. (2001). Ion channels and their functional role in vascular endothelium. *Physiol. Rev.* **81**, 1415–1459.

Nilius, B., Owsianik, G., Voets, T., and Peters, J. A. (2007). Transient receptor potential channels in disease. *Physiol. Rev.* **87**, 165–217.

Nilius, B., Prenen, J., Voets, T., Eggermont, J., and Droogmans, G. (1998). Activation of volume-regulated chloride currents by reduction of intracellular ionic strength in bovine endothelial cells. *J. Physiol. (Lond.)* **506**, 353–361.

Nilius, B., Prenen, J., Wissenbach, U., Bodding, M., and Droogmans, G. (2001). Differential activation of the volume-sensitive cation channel TRP12 (OTRPC4) and volume-regulated anion currents in HEK-293 cells. *Pflüg. Arch. Eur. J. Physiol.* **443**, 227–233.

Nilius, B., and Voets, T. (2005). Trp channels: A TR(I)P through a world of multifunctional cation channels. *Pflüg. Arch. Eur. J. Physiol.* **451**, 1–10.

Numata, T., Shimizu, T., and Okada, Y. (2006). TRPM7 is a stretch- and swelling-activated cation channel involved in volume regulation in human epithelial cells. *Am. J. Physiol. Cell Physiol.* **292**, C460–C467.

O'Neil, R. G., and Heller, S. (2005). Mechanosensitive nature of TRPV channels. *Pflüg. Arch. Eur. J. Physiol.* **451**, 193–203.

Oancea, E., Wolfe, J. T., and Clapham, D. E. (2006). Functional TRPM7 channels accumulate at the plasma membrane in response to fluid flow. *Circ. Res.* **98**, 245–253.

Oberwinkler, J., Lis, A., Giehl, K. M., Flockerzi, V., and Philipp, S. E. (2005). Alternative splicing switches the divalent cation selectivity of TRPM3 channels. *J. Biol. Chem.* **280**, 22540–22548.

Palmer, C. P., Zhou, X. L., Lin, J., Loukin, S. H., Kung, C., and Saimi, Y. (2001). A TRP homolog in *Saccharomyces cerevisiae* forms an intracellular Ca^{2+}- permeable channel in the yeast vacuolar membrane. *Proc. Natl. Acad. Sci. USA* **98**, 7801–7805.

Patel, A. J., and Honore, E. (2001). Properties and modulation of mammalian 2P domain K^+ channels. *Trends Neurosci.* **24**, 339–346.

Patel, A. J., Honore, E., Maingret, F., Lesage, F., Fink, M., Duprat, F., and Lazdunski, M. (1998). A mammalian two pore domain mechano-gated S-like K^+ channel. *EMBO J.* **17**, 4283–4290.

Pedersen, S. F., Beisner, K. H., Hougaard, C., Willumsen, B. M., Lambert, I. H., and Hoffmann, E. K. (2002). Rho family GTP binding proteins are involved in the regulatory volume decrease process in NIH3T3 mouse fibroblasts. *J. Physiol.* **541**, 779–796.

Pedersen, S. F., Hoffmann, E. K., and Mills, J. W. (2001). The cytoskeleton and cell volume regulation. *Comp. Biochem. Physiol. A Mol. Integr. Physiol.* **130**, 385–399.

Pedersen, S. F., Owsianik, G., and Nilius, B. (2005). TRP channels: An overview. *Cell Calcium* **38**, 233–252.

Pedersen, S. F., Poulsen, K. A., and Lambert, I. H. (2006). Roles of phospholipase A2 isoforms in the swelling- and melittin-induced arachidonic acid release and taurine efflux in NIH3T3 fibroblasts. *Am. J. Physiol. Cell Physiol.* **291**, C1286–C1296.

Penna, A., Juvin, V., Chemin, J., Compan, V., Monet, M., and Rassendren, F. A. (2006). PI3-kinase promotes TRPV2 activity independently of channel translocation to the plasma membrane. *Cell Calcium* **39,** 495–507.

Perozo, E. (2006). Gating prokaryotic mechanosensitive channels. *Nat. Rev. Mol. Cell Biol.* **7,** 109–119.

Ramsey, I. S., Delling, M., and Clapham, D. E. (2006). An introduction to TRP channels. *Annu. Rev. Physiol.* **68,** 619–647.

Rohacs, T. (2007). Regulation of TRP channels by PIP(2). *Pflüg. Arch.* **453,** 753–762.

Sakmann, B., and Neher, E. (1995). "Single-Channel Recording." Plenum Press, New York.

Sotomayor, M., Corey, D. P., and Schulten, K. (2005). In search of the hair-cell gating spring elastic properties of ankyrin and cadherin repeats. *Structure* **13,** 669–682.

Spassova, M. A., Hewavitharana, T., Xu, W., Soboloff, J., and Gill, D. L. (2006). A common mechanism underlies stretch activation and receptor activation of TRPC6 channels. *Proc. Natl. Acad. Sci. USA* **103,** 16586–16591.

Stokes, A. J., Shimoda, L. M., Koblan-Huberson, M., Adra, C. N., and Turner, H. (2004). A TRPV2-PKA signaling module for transduction of physical stimuli in mast cells. *J. Exp. Med.* **200,** 137–147.

Story, G. M., Peier, A. M., Reeve, A. J., Eid, S. R., Mosbacher, J., Hricik, T. R., Earley, T. J., Hergarden, A. C., Andersson, D. A., Hwang, S. W., McIntyre, P., Jegla, T., et al. (2003). ANKTM1, a TRP-like channel expressed in nociceptive neurons, is activated by cold temperatures. *Cell* **112,** 819–829.

Strange, K., Denton, J., and Nehrke, K. (2006). Ste20-type kinases: Evolutionarily conserved regulators of ion transport and cell volume. *Physiology (Bethesda)* **21,** 61–68.

Strotmann, R., Harteneck, C., Nunnenmacher, K., Schultz, G., and Plant, T. D. (2000). OTRPC4, a nonselective cation channel that confers sensitivity to extracellular osmolarity. *Nat. Cell Biol.* **2,** 695–702.

Sukharev, S. I., Martinac, B., Arshavsky, V. Y., and Kung, C. (1993). Two types of mechanosensitive channels in the *Escherichia coli* cell envelope: Solubilization and functional reconstitution. *Biophys. J.* **65,** 177–183.

Tsien, R. Y. (1989). *In* "Fluorescence Microscopy of Living Cells in Culture" (Y. Wang and D. Taylor, eds.). Academic Press, San Diego.

Vandebrouck, C., Martin, D., Colson-Van Schoor, M., Debaix, H., and Gailly, P. (2002). Involvement of TRPC in the abnormal calcium influx observed in dystrophic (mdx) mouse skeletal muscle fibers. *J. Cell Biol.* **158,** 1089–1096.

Van Driessche, W., De Smet, P., and Raskin, G. (1993). An automatic monitoring system for epithelial cell height. *Pflüg. Arch. Eur. J. Physiol.* **425,** 164–171.

Voets, T., Droogmans, G., Raskin, G., Eggermont, J., and Nilius, B. (1999). Reduced intracellular ionic strength as the initial trigger for activation of endothelial volume-regulated anion channels. *Proc. Natl. Acad. Sci. USA* **96,** 5298–5303.

Voets, T., Talavera, K., Owsianik, G., and Nilius, B. (2005). Sensing with TRP channels. *Nat. Biol. Chem.* **1,** 85–92.

Vriens, J., Owsianik, G., Fisslthaler, B., Suzuki, M., Janssens, A., Voets, T., Morisseau, C., Hammock, B. D., Fleming, I., Busse, R., and Nilius, B. (2005). Modulation of the Ca^{2+} permeable cation channel TRPV4 by cytochrome P450 epoxygenases in vascular endothelium. *Circ. Res.* **97,** 908–915.

Vriens, J., Watanabe, H., Janssens, A., Droogmans, G., Voets, T., and Nilius, B. (2004). Cell swelling, heat, and chemical agonists use distinct pathways for the activation of the cation channel TRPV4. *Proc. Natl. Acad. Sci. USA* **101,** 396–401.

Watanabe, H., Vriens, J., Prenen, J., Droogmans, G., Voets, T., and Nilius, B. (2003). Anandamide and arachidonic acid use epoxyeicosatrienoic acids to activate TRPV4 channels. *Nature* **424,** 434–438.

Welsh, D., Morielli, A., Nelson, M., and Brayden, J. (2002). Transient receptor potential channels regulate myogenic tone of resistance arteries. *Circ. Res.* **90,** 248–250.

Wu, K. L., Khan, S., Lakhe-Reddy, S., Jarad, G., Mukherjee, A., Obejero-Paz, C. A., Konieczkowski, M., Sedor, J. R., and Schelling, J. R. (2004). The NHE1 Na^+/H^+ exchanger recruits ezrin/radixin/moesin proteins to regulate Akt-dependent cell survival. *J. Biol. Chem.* **279,** 26280–26286.

Xu, H., Zhao, H., Tian, W., Yoshida, K., Roullet, J.-B., and Cohen, D. M. (2003). Regulation of a TRP channel by tyrosine phosphorylation: Src family kinase-dependent phosphorylation of TRPV4 on Y253 mediates its response to hypotonic stress. *J. Biol. Chem.* **278,** 11520–11527.

Yamamoto, M., Chen, M. Z., Wang, Y. J., Sun, H. Q., Wei, Y., Martinez, M., and Yin, H. L. (2006). Hypertonic stress increases phosphatidylinositol 4,5-bisphosphate levels by activating PIP5KIbeta. *J. Biol. Chem.* **281,** 32630–32638.

CHAPTER ELEVEN

CELL VOLUME REGULATORY ION CHANNELS IN CELL PROLIFERATION AND CELL DEATH

Florian Lang,* Michael Föller,* Karl Lang,* Philipp Lang,*
Markus Ritter,† Alexey Vereninov,‡ Ildiko Szabo,§
Stephan M. Huber,* *and* Erich Gulbins¶

Contents

Abstract

Alterations of cell volume are key events during both cell proliferation and apoptotic cell death. Cell proliferation eventually requires an increase of cell volume, and apoptosis is typically paralleled by cell shrinkage. Alterations of cell volume require the participation of ion transport across the cell membrane, including appropriate activity of Cl^- and K^+ channels. Cl^- channels modify cytosolic Cl^- activity and mediate osmolyte flux, and thus influence cell volume. Most Cl^- channels allow exit of HCO_3^-, leading to cytosolic acidification, which in turn inhibits cell proliferation and favors apoptosis. K^+ exit through K^+ channels decreases cytosolic K^+ concentration, which may sensitize the cell for apoptotic cell death. K^+ channel activity further maintains the cell membrane potential, a critical determinant of Ca^{2+} entry through Ca^{2+} channels. Ca^{2+} may, in addition, enter through Ca^{2+}-permeable cation channels, which, in some cells, are activated by hyperosmotic shock. Increases of cytosolic Ca^{2+} activity may trigger both mechanisms required for cell proliferation and mechanisms, leading to apoptosis. Thereby cell proliferation and apoptosis

* Department of Physiology, University of Tübingen, Germany
† Institute of Physiology and Pathophysiology, Paracelsus Medical University, Salzburg, Austria
‡ Institute of Cytology, Russian Academy of Sciences, St. Petersburg, Russia
§ Department of Biology, University of Padova, Padova, Italy
¶ Department of Molecular Biology, University Duisburg-Essen, Duisburg, Germany

Methods in Enzymology, Volume 428
ISSN 0076-6879, DOI: 10.1016/S0076-6879(07)28011-5

depend on magnitude and temporal organization of Ca^{2+} entry, as well as activity of other signaling pathways. Accordingly, the same ion channels may participate in the stimulation of both cell proliferation and apoptosis. Specific ion channel blockers may thus abrogate both cellular mechanisms, depending on cell type and condition.

1. INTRODUCTION

The adjustment of cell number to functional needs requires an adequate balance between formation of new cells by cell proliferation and their elimination by cell death. Suicidal cell death eliminates abundant and potentially harmful cells (Green and Reed, 1998; Gulbins *et al.*, 2000), which need to be replaced by cell proliferation.

Cell proliferation is stimulated by growth factors (Adams *et al.*, 2004; Bikfalvi *et al.*, 1998; Tallquist and Kazlauskas, 2004), apoptosis by a wide variety of mechanisms, including activation of CD95 (Fillon *et al.*, 2002; Gulbins *et al.*, 2000; Lang *et al.*, 1998b, 1999), somatostatin receptor (Teijeiro *et al.*, 2002), or TNFα receptor (Lang *et al.*, 2002a), by thyroid hormones (Alisi *et al.*, 2005), by lack of growth factors (Sturm *et al.*, 2004), by cell density (Long *et al.*, 2003), by cell adhesion (Davies, 2003; Walsh *et al.*, 2003), or by stressors such as oxidants (Rosette and Karin, 1996), radiation (Rosette and Karin, 1996), inhibition of glutaminase (Rotoli *et al.*, 2005), chemotherapeutics (Cariers *et al.*, 2002; Wieder *et al.*, 2001), energy depletion (Pozzi *et al.*, 2002), choline deficiency (Albright *et al.*, 2005), or osmotic shock (Bortner and Cidlowski, 1998, 1999; Lang *et al.*, 1998a, 2000b; Maeno *et al.*, 2000; Michea *et al.*, 2000; Rosette and Karin, 1996).

To eventually generate daughter cells of similar size as parent cells, cell proliferation needs at some point an increase of cell volume (Lang *et al.*, 1998a). Hallmarks of apoptosis include cell shrinkage (Lang *et al.*, 1998a) and breakdown of phosphatidylserine asymmetry of the plasma membrane (Green and Reed, 1998).

Suicidal cell death is not limited to nucleated cells but may similarly affect erythrocytes (Barvitenko *et al.*, 2005; Rice and Alfrey, 2005). The apoptosis-like death of mature erythrocytes (eryptosis) is characterized by cell shrinkage and breakdown of phosphatidylserine asymmetry, both typical features of apoptosis in nucleated cells (Lang *et al.*, 2003b,c,e).

Cell proliferation and apoptosis both involve and require activation of Cl^- channels, K^+ channels, and Ca^{2+} channels. Ample evidence points to an active role of those channels in the triggering of those two fundamental cellular functions. The following short synopsis compiles evidence for the participation of ion channels in cell proliferation and suicidal cell death. Special emphasis will be placed on cell proliferation of ras oncogene-expressing cells, apoptosis of CD95-triggered Jurkat lymphocytes, and eryptosis of osmotically shrunken erythrocytes.

2. Anion Channels, Osmolyte Transport, and pH Regulation

Anion channels may be activated during cell proliferation (Nilius and Droogmans, 2001; Shen *et al.*, 2000; Varela *et al.*, 2004) and anion channel blockers may interfere with cell proliferation (Jiang *et al.*, 2004; Pappas and Ritchie, 1998; Phipps *et al.*, 1996; Rouzaire-Dubois *et al.*, 2000; Shen *et al.*, 2000; Wondergem *et al.*, 2001). Moreover, cell proliferation may be impaired in cells lacking functional ClC-3 Cl^- channels (Wang *et al.*, 2002). The signaling of cell proliferation may require transient cell shrinkage at some stage, which may be accomplished by activation of Cl^- channels. As intracellular Cl^- activity is usually above electrochemical equilibrium, activation of Cl^- channels leads to Cl^- exit and thus depolarization. If K^+ channels are simultaneously active, the Cl^- exit is paralleled by the exit of K^+. The loss of KCl and osmotically obliged water then leads to cell shrinkage (Lang *et al.*, 1998a). In ras oncogene-expressing cells (Ritter *et al.*, 1993), cell shrinkage is required for the initiation of cytosolic Ca^{2+} oscillations, which are in turn needed for the stimulation of cell proliferation. The initial cell shrinkage is reversed into a later cell swelling, a result of a shifting cell volume regulatory set point toward greater volumes and a subsequent stimulation of Na^+/H^+ exchange and/or $Na^+,K^+,2Cl^-$ cotransport. Activation of Cl^- channels at this later stage may impede cell proliferation.

Activation of Cl^- channels parallels the CD95-induced apoptosis of Jurkat cells (Szabo *et al.*, 1998) and the TNFα- or staurosporine-induced apoptosis of various cell types (Maeno *et al.*, 2000; Okada *et al.*, 2004). Cl^- channels activated during CD95-induced apoptosis are the same as those activated by osmotic cell swelling and participating in regulatory cell volume decrease (Lepple-Wienhues *et al.*, 1998). During both cell swelling (Lepple-Wienhues *et al.*, 1998) and CD95-induced apoptosis (Szabo *et al.*, 1998), the activation of Cl^- channels requires the Src-like kinase Lck[56]. The kinase is in turn activated by ceramide (Gulbins *et al.*, 1997). In lymphocytes from patients with cystic fibrosis the Cl^- channels cannot be opened by protein kinase A but are activated by cell swelling and Lck[56] (Lepple-Wienhues *et al.*, 2001).

Cl^- channel inhibitors may blunt or even disrupt CD95-induced Jurkat cell apoptosis (Szabo *et al.*, 1998), TNFα- or staurosporine-induced apoptosis of various cell types (Maeno *et al.*, 2000; Okada *et al.*, 2004), apoptotic death of cortical neurons (Wei *et al.*, 2004), antimycin A-induced death of proximal renal tubules (Miller and Schnellmann, 1993), GABA-induced enhancement of excitotoxic cell death of rat cerebral neurons (Erdo *et al.*, 1991), cardiomyocyte apoptosis (Takahashi *et al.*, 2005), and eryptosis (Takahashi *et al.*, 2005).

Activation of Cl$^-$ channels leads to cellular loss of KCl and osmotically obliged water and thus to cell shrinkage. Some anion channels further allow exit of organic osmolytes such as taurine (Lang *et al.*, 1998b,e; Moran *et al.*, 2000), an effect contributing to cell shrinkage (Lang *et al.*, 1998a). As organic osmolytes stabilize cellular proteins (Lang *et al.*, 1998a), their loss could destabilize proteins. Inhibition of inositol uptake has indeed been shown to induce renal failure, presumably because of apoptotic death of renal tubular cells (Kitamura *et al.*, 1998).

Many Cl$^-$ channels further allow HCO$_3^-$ exit, leading to cytosolic acidification, a typical feature of cells entering into apoptosis (Lang *et al.*, 2002a; Wenzel and Daniel, 2004). As the DNA-degrading enzyme DNase type II has its pH optimum in the acidic range (for review, see Shrode *et al.*, 1997), acidification is expected to enhance DNA fragmentation. CD95-induced apoptosis is indeed accelerated by the inhibition of Na$^+$/H$^+$ exchange (Lang *et al.*, 2000a).

3. Ca^{2+} AND UNSELECTIVE CATION CHANNELS

Cytosolic Ca^{2+} activity plays a decisive role in the regulation of cell proliferation (Berridge *et al.*, 1998, 2000, 2003; Parekh and Penner, 1997; Santella, 1998; Santella *et al.*, 1998; Whitfield *et al.*, 1995). Growth factors stimulate Ca^{2+} release through activated Ca^{2+} channel I$_{CRAC}$ (Qian and Weiss, 1997), which mediates Ca^{2+} entry, thus triggering and maintaining pulsatile Ca^{2+} release from intracellular stores yielding oscillations of cytosolic Ca^{2+} activity. Those oscillations govern a wide variety of cellular functions (Berridge *et al.*, 1998, 2000, 2003; Parekh and Penner, 1997), including depolymerization of actin filaments (Dartsch *et al.*, 1995; Lang *et al.*, 1992, 2000c; Ritter *et al.*, 1997). The depolymerization of actin filaments results in disinhibition of Na$^+$/H$^+$ exchanger and/or Na$^+$,K$^+$, 2Cl$^-$ cotransporter, which both accumulate ions and osmotically obliged water and thus increase cell volume (Lang *et al.*, 1998a). Activation of I$_{CRAC}$, Ca^{2+} oscillations, and depolymerization of the actin filament network are prerequisites for the stimulation of cell proliferation (Dartsch *et al.*, 1995; Lang *et al.*, 1992, 2000c; Ritter *et al.*, 1997).

CD95 receptor triggering is paralleled by inhibition of I$_{CRAC}$ in Jurkat T lymphocytes (Dangel *et al.*, 2005; Lepple-Wienhues *et al.*, 1999). Inhibition of I$_{CRAC}$ prevents activation and proliferation of lymphocytes but does not necessarily lead to apoptotic cell death. At a later stage, CD95 stimulation may, in some cells, lead to a sustained increase of cytosolic Ca^{2+} activity, which has been shown to trigger apoptosis in a variety of nucleated cells

(Berridge *et al.*, 2000; Green and Reed, 1998; Liu *et al.*, 2005; Parekh and Penner, 1997; Parekh and Putney, 2005; Spassova *et al.*, 2004). Furthermore, Ca^{2+}-permeable cation channels trigger apoptosis-like suicidal death of erythrocytes (eryptosis) (Brand *et al.*, 2003; Lang *et al.*, 2002b, 2003b). Accordingly, eryptosis is elicited by exposure to the Ca^{2+} ionophore ionomycin (Berg *et al.*, 2001; Bratosin *et al.*, 2001; Daugas *et al.*, 2001; Lang *et al.*, 2002b, 2003b) and blunted in the nominal absence of Ca^{2+} (Lang *et al.*, 2003b). The Ca^{2+}-permeable erythrocyte cation channels are activated by osmotic shock (Huber *et al.*, 2001), oxidative stress (Duranton *et al.*, 2002), energy depletion (Lang *et al.*, 2003b), and infection with the malaria pathogen *Plasmodium falciparum* (Duranton *et al.*, 2003; Lang *et al.*, 2003b, 2004a). Energy depletion is presumably effective through impairment of GSH replenishment, thus weakening the antioxidative defense of the erythrocytes (Bilmen *et al.*, 2001; Mavelli *et al.*, 1984). The erythrocyte cation channels are inhibited by Cl^- and are activated by replacement of Cl^- with gluconate (Duranton *et al.*, 2002; Huber *et al.*, 2001). Similar or identical cation channels are activated by incubation of human erythrocytes in low ionic strength (Bernhardt *et al.*, 1991; Jones and Knauf, 1985; LaCelle and Rothsteto, 1966) or by depolarization (Bennekou, 1993; Christophersen and Bennekou, 1991; Kaestner *et al.*, 1999).

Increased cytosolic Ca^{2+} concentrations somehow trigger the scrambling of the erythrocyte cell membrane (Zhou *et al.*, 2002) with breakdown of phosphatidylserine asymmetry and phosphatidylserine exposure at the cell surface (Lang *et al.*, 2003b). The cation channels are activated by prostaglandin E_2, which is released upon osmotic shock (Lang *et al.*, 2005a). The cation channel blockers amiloride (Lang *et al.*, 2003b) and ethylisopropylamiloride (Lang *et al.*, 2003c) blunt the phosphatidylserine exposure following osmotic shock.

Cell volume-sensitive cation channels are similarly expressed in nucleated cells, such as airway epithelia cells (Chan *et al.*, 1992), vascular smooth muscle, colon carcinoma and neuroblastoma cells (Koch and Korbmacher, 1999), cortical collecting duct cells (Volk *et al.*, 1995), hepatocytes (Wehner *et al.*, 1995, 2000), mast cells (Cabado *et al.*, 1994), and macrophages (Gamper *et al.*, 2000). Cation channels activated by Cl^- removal are expressed in salivary and lung epithelial cells (Dinudom *et al.*, 1995; Marunaka *et al.*, 1994; Tohda *et al.*, 1994). Whether or not those channels participate in the stimulation of apoptosis remains elusive.

4. K^+ CHANNELS

Several K^+ channels participate in the regulation of cell proliferation (Patel and Lazdunski, 2004; Wang, 2004). Growth factors activate K^+ channels (Enomoto *et al.*, 1986; Faehling *et al.*, 2001; Lang *et al.*, 1991;

Liu *et al.*, 2001; O'Lague *et al.*, 1985; Sanders *et al.*, 1996; Wiecha *et al.*, 1998), and enhanced K^+ channel activity is observed in tumor cells (DeCoursey *et al.*, 1984; Mauro *et al.*, 1997; Nilius and Wohlrab, 1992; Pappas and Ritchie, 1998; Pappone and Ortiz-Miranda, 1993; Patel and Lazdunski, 2004; Skryma *et al.*, 1997; Strobl *et al.*, 1995; Wang, 2004; Zhou *et al.*, 2003). In ras oncogene-expressing cells, repetitive activation of Ca^{2+}-sensitive K^+ channels by oscillating cytosolic Ca^{2+} activity leads to oscillations of cell membrane potential (Lang *et al.*, 1991). Several K^+ channel inhibitors disrupt cell proliferation (for review, see Wang 2004). K^+ channel activation is apparently important for the early G1 phase of the cell cycle (Wang *et al.*, 1998; Wonderlin and Strobl, 1996). The maintenance of cell membrane potential by K^+ channels provides the electrical driving force for Ca^{2+} entry through I_{CRAC} (Parekh and Penner, 1997), which is required for stimulation of cell proliferation.

The role of K^+ channels in apoptosis is less obvious. In some cells, inhibition of K^+ channels participates in the stimulation of apoptosis (Bankers-Fulbright *et al.*, 1998; Chin *et al.*, 1997; Han *et al.*, 2004; Miki *et al.*, 1997; Pal *et al.*, 2004; Patel and Lazdunski, 2004), and activation of K^+ channels inhibits apoptosis (Jakob and Krieglstein, 1997; Lauritzen *et al.*, 1997). Along those lines, extensive neuronal cell death is observed in mice carrying a mutation of G-coupled inward rectifier K^+ channels (*Weaver* mice) (Harrison and Roffler-Tarlov, 1998; Migheli *et al.*, 1995, 1997; Murtomaki *et al.*, 1995; Oo *et al.*, 1996).

However, in other cells apoptosis is stimulated by activation of K^+ channels (Wei *et al.*, 2004; Yu *et al.*, 1997) and inhibited by increase of extracellular K^+ concentration (Colom *et al.*, 1998; Lang *et al.*, 2003e; Prehn *et al.*, 1997) or K^+ channel blockade (Gantner *et al.*, 1995; Lang *et al.*, 2003e). Cellular loss of K^+ apparently favors apoptosis in a wide variety of cells (Beauvais *et al.*, 1995; Benson *et al.*, 1996; Bortner and Cidlowski, 1999, 2004; Bortner *et al.*, 1997; Gomez-Angelats *et al.*, 2000; Hughes and Cidlowski, 1999; Hughes *et al.*, 1997; Maeno *et al.*, 2000; Montague *et al.*, 1999; Perez *et al.*, 2000; Yurinskaya *et al.*, 2005a,b). Moreover, activation of K^+ channels hyperpolarizes the cell membrane, thus increasing the electrical driving force for Cl^- exit. Depending on Cl^- channel activity, K^+ channel activity leads to cellular loss of KCl with osmotically obliged water and hence to apoptotic cell shrinkage (Lang *et al.*, 1998a).

In Jurkat lymphocytes, CD95 activation is followed within a few minutes by inhibition of Kv1.3 K^+ channels (Szabo *et al.*, 1996, 1997, 2004), the cell volume regulatory K^+ channel of those cells (Deutsch and Chen, 1993). CD95 triggering leads to tyrosine phosphorylation of the Kv1.3 channel protein (Gulbins *et al.*, 1997; Szabo *et al.*, 1996). Accordingly, CD95-induced inhibition of Kv1.3 requires Lck[56] (Gulbins *et al.*, 1997; Szabo *et al.*, 1996). The inhibitory effect of CD95 triggering is mimicked by the sphingomyelinase product ceramide, which similarly induces apoptosis

(Gulbins *et al.*, 1997). In other cells, Kv1.3 is similarly regulated by tyrosine phosphorylation (Holmes *et al.*, 1996). Moreover, Kv1.3 is upregulated by the serum and glucocorticoid-inducible kinase (Lang *et al.*, 2003a), which similarly inhibits apoptosis (Aoyama *et al.*, 2005). Following CD95 activation, the early inhibition of Kv1.3 is followed by late activation of Kv1.3 (Storey *et al.*, 2003). Early inhibition of Kv1.3 channels in CD95-activated cells may serve to prevent premature cell shrinkage which otherwise may interfere with signaling of apoptosis (Lang *et al.*, 1998a). The late activation of Kv1.3 channels during the execution phase of apoptosis supports apoptotic cell shrinkage (Storey *et al.*, 2003).

In suicidal erythrocytes, Ca^{2+}-sensitive K^+ channels (GARDOS channels) are activated by increased cytosolic Ca^{2+} activity (Brugnara *et al.*, 1993; Del Carlo *et al.*, 2002; Dunn, 1998; Gardos, 1958; Grygorczyk and Schwarz, 1983; Leinders *et al.*, 1992; Pellegrino and Pellegrini, 1998; Shindo *et al.*, 2000). Activation of GARDOS channels hyperpolarizes the cell membrane and, because of high erythrocyte Cl^- permeability, leads to parallel exit of K^+ and Cl^-. The cellular loss of KCl and osmotically obliged water leads to cell shrinkage, which in turn favors phosphatidylserine exposure (Lang *et al.*, 2003d). An increase of extracellular K^+ or a pharmacological inhibition of GARDOS channels blunts cell shrinkage and has a moderate inhibitory effect on phosphatidylserine scrambling following exposure to the Ca^{2+} ionophore ionomycin (Lang *et al.*, 2003d). Erythrocyte shrinkage stimulates formation of the platelet-activating factor, which in turn activates a sphingomyelinase (Lang *et al.*, 2005b). The ceramide generated by the sphingomyelinase then sensitizes the cell for the scrambling effect of Ca^{2+} (Lang *et al.*, 2004b, 2005b).

5. SWITCHING FROM CELL PROLIFERATION TO SUICIDAL CELL DEATH

The same or similar channels could participate in the stimulation of both cell proliferation and apoptosis. The effect of channel activation depends on further properties of the cell. For instance, it may depend on the activity of other channels. Activation of K^+ channels without parallel activity of electrogenic anion transporters or Cl^- channels, for instance, hyperpolarizes the cell membrane but does not shrink the cell (Lang *et al.*, 1998a). Moreover, activation of K^+ channels may increase Ca^{2+} entry and cytosolic Ca^{2+} activity only in the presence of active Ca^{2+} channels.

The effect may further depend on the temporal pattern of channel activation. The oscillating K^+ channel activity typical of proliferating cells (Lang *et al.*, 1991; Pandiella *et al.*, 1989) has different effects as sustained K^+ channel activation typical of apoptotic cells (Lang *et al.*, 2003d). Oscillations of

Ca^{2+} channel activity lead to fluctuations of cytosolic Ca^{2+} concentration, which depolymerize the cytoskeleton (Dartsch *et al.*, 1995; Lang *et al.*, 1992, 2000c; Ritter *et al.*, 1997), whereas permanent opening of Ca^{2+} channels leads to sustained increases of cytosolic Ca^{2+} activity, which may activate caspases (Whitfield *et al.*, 1995) or trigger scrambling of the cell membrane (Dekkers *et al.*, 2002; Woon *et al.*, 1999).

The outcome further depends on the amplitude of channel activity. The amplitude of TASK-3 K^+ channel activity during apoptosis is one order of magnitude higher than in tumor cells (Patel and Lazdunski, 2004; Wang, 2004), and the Ca^{2+} entry required for stimulation of mitogenic transcription factors may remain well below the Ca^{2+} entry required for the triggering of suicidal cell death (Whitfield *et al.*, 1995).

ACKNOWLEDGMENTS

The authors acknowledge the meticulous preparation of the manuscript by Jasmin Bühringer and Lejla Subasic. The work of the authors was supported by the Deutsche Forschungsgemeinschaft, Nr. La 315/4-3, La 315/6-1, Le 792/3-3, DFG Schwerpunkt Intrazelluläre Lebensformen La 315/11-1, and Bundesministerium für Bildung, Wissenschaft, Forschung und Technologie (Center for Interdisciplinary Clinical Research) 01 KS 9602. The study performed by A. A. Vereninov was supported by the Russian Foundation for Basic Research, projects RFFI No. 06-04 48060, RFFI-DFG 06-04 04000, as well as the joint German–Russian Grant DFG 436 RUS 113/488/0-2R.

REFERENCES

Adams, T. E., McKern, N. M., and Ward, C. W. (2004). Signalling by the type 1 insulin-like growth factor receptor: Interplay with the epidermal growth factor receptor. *Growth Factors* **22,** 89–95.

Albright, C. D., da Costa, K. A., Craciunescu, C. N., Klem, E., Mar, M. H., and Zeisel, S. H. (2005). Regulation of choline deficiency apoptosis by epidermal growth factor in CWSV-1 rat hepatocytes. *Cell. Physiol. Biochem.* **15,** 59–68.

Alisi, A., Demori, I., Spagnuolo, S., Pierantozzi, E., Fugassa, E., and Leoni, S. (2005). Thyroid status affects rat liver regeneration after partial hepatectomy by regulating cell cycle and apoptosis. *Cell. Physiol. Biochem.* **15,** 69–76.

Aoyama, T., Matsui, T., Novikov, M., Park, J., Hemmings, B., and Rosenzweig, A. (2005). Serum and glucocorticoid-responsive kinase-1 regulates cardiomyocyte survival and hypertrophic response. *Circulation* **111,** 1652–1659.

Bankers-Fulbright, J. L., Kephart, G. M., Loegering, D. A., Bradford, A. L., Okada, S., Kita, H., and Gleich, G. J. (1998). Sulfonylureas inhibit cytokine-induced eosinophil survival and activation. *J. Immunol.* **160,** 5546–5553.

Barvitenko, N. N., Adragna, N. C., and Weber, R. E. (2005). Erythrocyte signal transduction pathways, their oxygenation dependence and functional significance. *Cell. Physiol. Biochem.* **15,** 1–18.

Beauvais, F., Michel, L., and Dubertret, L. (1995). Human eosinophils in culture undergo a striking and rapid shrinkage during apoptosis: Role of K+ channels. *J. Leukocyte Biol.* **57,** 851–855.

Bennekou, P. (1993). The voltage-gated non-selective cation channel from human red cells is sensitive to acetylcholine. *Biochim. Biophys. Acta* **1147,** 165–167.

Benson, R. S., Heer, S., Dive, C., and Watson, A. J. (1996). Characterization of cell volume loss in CEM-C7A cells during dexamethasone-induced apoptosis. *Am. J. Physiol.* **270,** C1190–C1203.

Berg, C. P., Engels, I. H., Rothbart, A., Lauber, K., Renz, A., Schlosser, S. F., Schulze-Osthoff, K., and Wesselborg, S. (2001). Human mature red blood cells express caspase-3 and caspase-8, but are devoid of mitochondrial regulators of apoptosis. *Cell Death Differ.* **8,** 1197–1206.

Bernhardt, I., Hall, A. C., and Ellory, J. C. (1991). Effects of low ionic strength media on passive human red cell monovalent cation transport. *J. Physiol.* **434,** 489–506.

Berridge, M. J., Bootman, M. D., and Lipp, P. (1998). Calcium: A life and death signal. *Nature* **395,** 645–648.

Berridge, M. J., Bootman, M. D., and Roderick, H. L. (2003). Calcium signalling: Dynamics, homeostasis and remodelling. *Nat. Rev. Mol. Cell. Biol.* **4,** 517–529.

Berridge, M. J., Lipp, P., and Bootman, M. D. (2000). The versatility and universality of calcium signalling. *Nat. Rev. Mol. Cell. Biol.* **1,** 11–21.

Bikfalvi, A., Savona, C., Perollet, C., and Javerzat, S. (1998). New insights in the biology of fibroblast growth factor-2. *Angiogenesis* **1,** 155–173.

Bilmen, S., Aksu, T. A., Gumuslu, S., Korgun, D. K., and Canatan, D. (2001). Antioxidant capacity of G-6-PD-deficient erythrocytes. *Clin. Chim. Acta* **303,** 83–86.

Bortner, C. D., and Cidlowski, J. A. (1998). A necessary role for cell shrinkage in apoptosis. *Biochem. Pharmacol.* **56,** 1549–1559.

Bortner, C. D., and Cidlowski, J. A. (1999). Caspase independent/dependent regulation of K(+), cell shrinkage, and mitochondrial membrane potential during lymphocyte apoptosis. *J. Biol. Chem.* **274,** 21953–21962.

Bortner, C. D., and Cidlowski, J. A. (2004). The role of apoptotic volume decrease and ionic homeostasis in the activation and repression of apoptosis. *Pflüg. Arch.* **448,** 313–318.

Bortner, C. D., Hughes, F. M., Jr., and Cidlowski, J. A. (1997). A primary role for K+ and Na+ efflux in the activation of apoptosis. *J. Biol. Chem.* **272,** 32436–32442.

Brand, V. B., Sandu, C. D., Duranton, C., Tanneur, V., Lang, K. S., Huber, S. M., and Lang, F. (2003). Dependence of *Plasmodium falciparum in vitro* growth on the cation permeability of the human host erythrocyte. *Cell. Physiol. Biochem.* **13,** 347–356.

Bratosin, D., Leszczynski, S., Sartiaux, C., Fontaine, O., Descamps, J., Huart, J. J., Poplineau, J., Goudaliez, F., Aminoff, D., and Montreuil, J. (2001). Improved storage of erythrocytes by prior leukodepletion: Flow cytometric evaluation of stored erythrocytes. *Cytometry* **46,** 351–356.

Brugnara, C., de Franceschi, L., and Alper, S. L. (1993). Inhibition of Ca(2+)-dependent K+ transport and cell dehydration in sickle erythrocytes by clotrimazole and other imidazole derivatives. *J. Clin. Invest.* **92,** 520–526.

Cabado, A. G., Vieytes, M. R., and Botana, L. M. (1994). Effect of ion composition on the changes in membrane potential induced with several stimuli in rat mast cells. *J. Cell. Physiol.* **158,** 309–316.

Cariers, A., Reinehr, R., Fischer, R., Warskulat, U., and Haussinger, D. (2002). c-Jun-N-terminal kinase dependent membrane targeting of CD95 in rat hepatic stellate cells. *Cell. Physiol. Biochem.* **12,** 179–186.

Chan, H. C., Goldstein, J., and Nelson, D. J. (1992). Alternate pathways for chloride conductance activation in normal and cystic fibrosis airway epithelial cells. *Am. J. Physiol.* **262,** C1273–C1283.

Chin, L. S., Park, C. C., Zitnay, K. M., Sinha, M., DiPatri, A. J., Jr., Perillan, P., and Simard, J. M. (1997). 4-Aminopyridine causes apoptosis and blocks an outward rectifier K+ channel in malignant astrocytoma cell lines. *J. Neurosci. Res.* **48,** 122–127.

Christophersen, P., and Bennekou, P. (1991). Evidence for a voltage-gated, non-selective cation channel in the human red cell membrane. *Biochim. Biophys. Acta* **1065**, 103–106.

Colom, L. V., Diaz, M. E., Beers, D. R., Neely, A., Xie, W. J., and Appel, S. H. (1998). Role of potassium channels in amyloid-induced cell death. *J. Neurochem.* **70**, 1925–1934.

Dangel, G. R., Lang, F., and Lepple-Wienhues, A. (2005). Effect of sphingosine on Ca2+ entry and mitochondrial potential of Jurkat T cells: Interaction with Bcl2. *Cell. Physiol. Biochem.* **16**, 9–14.

Dartsch, P. C., Ritter, M., Gschwentner, M., Lang, H. J., and Lang, F. (1995). Effects of calcium channel blockers on NIH 3T3 fibroblasts expressing the Ha-ras oncogene. *Eur. J. Cell Biol.* **67**, 372–378.

Daugas, E., Cande, C., and Kroemer, G. (2001). Erythrocytes: Death of a mummy. *Cell Death Differ.* **8**, 1131–1133.

Davies, A. M. (2003). Regulation of neuronal survival and death by extracellular signals during development. *EMBO J.* **22**, 2537–2545.

DeCoursey, T. E., Chandy, K. G., Gupta, S., and Cahalan, M. D. (1984). Voltage-gated K+ channels in human T lymphocytes: A role in mitogenesis? *Nature* **307**, 465–468.

Dekkers, D. W., Comfurius, P., Bevers, E. M., and Zwaal, R. F. (2002). Comparison between Ca2+-induced scrambling of various fluorescently labelled lipid analogues in red blood cells. *Biochem. J.* **362**, 741–747.

Del Carlo, B., Pellegrini, M., and Pellegrino, M. (2002). Calmodulin antagonists do not inhibit IK(Ca) channels of human erythrocytes. *Biochim. Biophys. Acta* **1558**, 133–141.

Deutsch, C., and Chen, L. Q. (1993). Heterologous expression of specific K+ channels in T lymphocytes: Functional consequences for volume regulation. *Proc. Natl. Acad. Sci. USA* **90**, 10036–10040.

Dinudom, A., Komwatana, P., Young, J. A., and Cook, D. I. (1995). Control of the amiloride-sensitive Na+ current in mouse salivary ducts by intracellular anions is mediated by a G protein. *J. Physiol.* **487**(Pt. 3), 549–555.

Dunn, P. M. (1998). The action of blocking agents applied to the inner face of Ca(2+)-activated K+ channels from human erythrocytes. *J. Membr. Biol.* **165**, 133–143.

Duranton, C., Huber, S., Tanneur, V., Lang, K., Brand, V., Sandu, C., and Lang, F. (2003). Electrophysiological properties of the *Plasmodium falciparum*-induced cation conductance of human erythrocytes. *Cell. Physiol. Biochem.* **13**, 189–198.

Duranton, C., Huber, S. M., and Lang, F. (2002). Oxidation induces a Cl(−)-dependent cation conductance in human red blood cells. *J. Physiol.* **539**, 847–855.

Enomoto, K., Cossu, M. F., Edwards, C., and Oka, T. (1986). Induction of distinct types of spontaneous electrical activities in mammary epithelial cells by epidermal growth factor and insulin. *Proc. Natl. Acad. Sci. USA* **83**, 4754–4758.

Erdo, S., Michler, A., and Wolff, J. R. (1991). GABA accelerates excitotoxic cell death in cortical cultures: Protection by blockers of GABA-gated chloride channels. *Brain Res.* **542**, 254–258.

Faehling, M., Koch, E. D., Raithel, J., Trischler, G., and Waltenberger, J. (2001). Vascular endothelial growth factor-A activates Ca^{2+}-activated K+ channels in human endothelial cells in culture. *Int. J. Biochem. Cell Biol.* **33**, 337–346.

Fillon, S., Klingel, K., Warntges, S., Sauter, M., Gabrysch, S., Pestel, S., Tanneur, V., Waldegger, S., Zipfel, A., Viebahn, R., Haussinger, D., Broer, S., Kandolf, R., and Lang, F. (2002). Expression of the serine/threonine kinase hSGK1 in chronic viral hepatitis. *Cell. Physiol. Biochem.* **12**, 47–54.

Gamper, N., Huber, S. M., Badawi, K., and Lang, F. (2000). Cell volume-sensitive sodium channels upregulated by glucocorticoids in U937 macrophages. *Pflüg. Arch.* **441**, 281–286.

Gantner, F., Uhlig, S., and Wendel, A. (1995). Quinine inhibits release of tumor necrosis factor, apoptosis, necrosis and mortality in a murine model of septic liver failure. *Eur. J. Pharmacol.* **294**, 353–355.

Gardos, G. (1958). The function of calcium in the potassium permeability of human erythrocytes. *Biochim. Biophys. Acta* **30,** 653–654.

Gomez-Angelats, M., Bortner, C. D., and Cidlowski, J. A. (2000). Protein kinase C (PKC) inhibits fas receptor-induced apoptosis through modulation of the loss of K+ and cell shrinkage: A role for PKC upstream of caspases. *J. Biol. Chem.* **275,** 19609–19619.

Green, D. R., and Reed, J. C. (1998). Mitochondria and apoptosis. *Science* **281,** 1309–1312.

Grygorczyk, R., and Schwarz, W. (1983). Properties of the CA2+-activated K+ conductance of human red cells as revealed by the patch-clamp technique. *Cell Calcium* **4,** 499–510.

Gulbins, E., Jekle, A., Ferlinz, K., Grassme, H., and Lang, F. (2000). Physiology of apoptosis. *Am. J. Physiol. Renal Physiol.* **279,** F605–F615.

Gulbins, E., Szabo, I., Baltzer, K., and Lang, F. (1997). Ceramide-induced inhibition of T lymphocyte voltage-gated potassium channel is mediated by tyrosine kinases. *Proc. Natl. Acad. Sci. USA* **94,** 7661–7666.

Han, H., Wang, J., Zhang, Y., Long, H., Wang, H., Xu, D., and Wang, Z. (2004). HERG K channel conductance promotes H_2O_2-induced apoptosis in HEK293 cells: Cellular mechanisms. *Cell. Physiol. Biochem.* **14,** 121–134.

Harrison, S. M., and Roffler-Tarlov, S. K. (1998). Cell death during development of testis and cerebellum in the mutant mouse weaver. *Dev. Biol.* **195,** 174–186.

Holmes, T. C., Fadool, D. A., and Levitan, I. B. (1996). Tyrosine phosphorylation of the Kv1.3 potassium channel. *J. Neurosci.* **16,** 1581–1590.

Huber, S. M., Gamper, N., and Lang, F. (2001). Chloride conductance and volume-regulatory nonselective cation conductance in human red blood cell ghosts. *Pflüg. Arch.* **441,** 551–558.

Hughes, F. M., Jr., Bortner, C. D., Purdy, G. D., and Cidlowski, J. A. (1997). Intracellular K+ suppresses the activation of apoptosis in lymphocytes. *J. Biol. Chem.* **272,** 30567–30576.

Hughes, F. M., Jr., and Cidlowski, J. A. (1999). Potassium is a critical regulator of apoptotic enzymes *in vitro* and *in vivo*. *Adv. Enzyme Regul.* **39,** 157–171.

Jakob, R., and Krieglstein, J. (1997). Influence of flupirtine on a G-protein coupled inwardly rectifying potassium current in hippocampal neurones. *Br. J. Pharmacol.* **122,** 1333–1338.

Jiang, B., Hattori, N., Liu, B., Nakayama, Y., Kitagawa, K., and Inagaki, C. (2004). Suppression of cell proliferation with induction of p21 by Cl(−) channel blockers in human leukemic cells. *Eur. J. Pharmacol.* **488,** 27–34.

Jones, G. S., and Knauf, P. A. (1985). Mechanism of the increase in cation permeability of human erythrocytes in low-chloride media: Involvement of the anion transport protein capnophorin. *J. Gen. Physiol.* **86,** 721–738.

Kaestner, L., Bollensdorff, C., and Bernhardt, I. (1999). Non-selective voltage-activated cation channel in the human red blood cell membrane. *Biochim. Biophys. Acta* **1417,** 9–15.

Kitamura, H., Yamauchi, A., Sugiura, T., Matsuoka, Y., Horio, M., Tohyama, M., Shimada, S., Imai, E., and Hori, M. (1998). Inhibition of myo-inositol transport causes acute renal failure with selective medullary injury in the rat. *Kidney Int.* **53,** 146–153.

Koch, J., and Korbmacher, C. (1999). Osmotic shrinkage activates nonselective cation (NSC) channels in various cell types. *J. Membr. Biol.* **168,** 131–139.

LaCelle, P. L., and Rothsteto, A. (1966). The passive permeability of the red blood cell in cations. *J. Gen. Physiol.* **50,** 171–188.

Lang, F., Busch, G. L., Ritter, M., Volkl, H., Waldegger, S., Gulbins, E., and Haussinger, D. (1998a). Functional significance of cell volume regulatory mechanisms. *Physiol. Rev.* **78,** 247–306.

Lang, F., Friedrich, F., Kahn, E., Woll, E., Hammerer, M., Waldegger, S., Maly, K., and Grunicke, H. (1991). Bradykinin-induced oscillations of cell membrane potential in cells expressing the Ha-ras oncogene. *J. Biol. Chem.* **266,** 4938–4942.

Lang, F., Henke, G., Embark, H. M., Waldegger, S., Palmada, M., Bohmer, C., and Vallon, V. (2003a). Regulation of channels by the serum and glucocorticoid-inducible kinase: Implications for transport, excitability and cell proliferation. *Cell. Physiol. Biochem.* **13**, 41–50.

Lang, F., Lang, P. A., Lang, K. S., Brand, V., Tanneur, V., Duranton, C., Wieder, T., and Huber, S. M. (2004a). Channel-induced apoptosis of infected host cells-the case of malaria. *Pflüg. Arch.* **448**, 319–324.

Lang, F., Madlung, J., Bock, J., Lukewille, U., Kaltenbach, S., Lang, K. S., Belka, C., Wagner, C. A., Lang, H. J., Gulbins, E., and Lepple-Wienhues, A. (2000a). Inhibition of Jurkat-T-lymphocyte Na+/H+-exchanger by CD95(Fas/Apo-1)-receptor stimulation. *Pflüg. Arch.* **440**, 902–907.

Lang, F., Madlung, J., Siemen, D., Ellory, C., Lepple-Wienhues, A., and Gulbins, E. (2000b). The involvement of caspases in the CD95(Fas/Apo-1)- but not swelling-induced cellular taurine release from Jurkat T-lymphocytes. *Pflüg. Arch.* **440**, 93–99.

Lang, F., Madlung, J., Uhlemann, A. C., Risler, T., and Gulbins, E. (1998b). Cellular taurine release triggered by stimulation of the Fas(CD95) receptor in Jurkat lymphocytes. *Pflüg. Arch.* **436**, 377–383.

Lang, F., Ritter, M., Gamper, N., Huber, S., Fillon, S., Tanneur, V., Lepple-Wienhues, A., Szabo, I., and Gulbins, E. (2000c). Cell volume in the regulation of cell proliferation and apoptotic cell death. *Cell. Physiol. Biochem.* **10**, 417–428.

Lang, F., Szabo, I., Lepple-Wienhues, A., Siemen, D., and Gulbins, E. (1999). Physiology of receptor-mediated lymphocyte apoptosis. *News Physiol. Sci.* **14**, 194–200.

Lang, F., Waldegger, S., Woell, E., Ritter, M., Maly, K., and Grunicke, H. (1992). Effects of inhibitors and ion substitutions on oscillations of cell membrane potential in cells expressing the RAS oncogene. *Pflüg. Arch.* **421**, 416–424.

Lang, K. S., Duranton, C., Poehlmann, H., Myssina, S., Bauer, C., Lang, F., Wieder, T., and Huber, S. M. (2003b). Cation channels trigger apoptotic death of erythrocytes. *Cell Death Differ.* **10**, 249–256.

Lang, K. S., Fillon, S., Schneider, D., Rammensee, H. G., and Lang, F. (2002a). Stimulation of TNF alpha expression by hyperosmotic stress. *Pflüg. Arch.* **443**, 798–803.

Lang, K. S., Myssina, S., Brand, V., Sandu, C., Lang, P. A., Berchtold, S., Huber, S. M., Lang, F., and Wieder, T. (2004b). Involvement of ceramide in hyperosmotic shock-induced death of erythrocytes. *Cell Death Differ.* **11**, 231–243.

Lang, K. S., Myssina, S., Tanneur, V., Wieder, T., Huber, S. M., Lang, F., and Duranton, C. (2003c). Inhibition of erythrocyte cation channels and apoptosis by ethylisopropylamiloride. *Naunyn Schmiedebergs Arch. Pharmacol.* **367**, 391–396.

Lang, K. S., Roll, B., Myssina, S., Schittenhelm, M., Scheel-Walter, H. G., Kanz, L., Fritz, J., Lang, F., Huber, S. M., and Wieder, T. (2002b). Enhanced erythrocyte apoptosis in sickle cell anemia, thalassemia and glucose-6-phosphate dehydrogenase deficiency. *Cell. Physiol. Biochem.* **12**, 365–372.

Lang, P. A., Kaiser, S., Myssina, S., Wieder, T., Lang, F., and Huber, S. M. (2003d). Role of Ca2+-activated K+ channels in human erythrocyte apoptosis. *Am. J. Physiol. Cell. Physiol.* **285**, C1553–C1560.

Lang, P. A., Kempe, D. S., Myssina, S., Tanneur, V., Birka, C., Laufer, S., Lang, F., Wieder, T., and Huber, S. M. (2005a). PGE(2) in the regulation of programmed erythrocyte death. *Cell Death Differ.* **12**, 415–428.

Lang, P. A., Kempe, D. S., Tanneur, V., Eisele, K., Klarl, B. A., Myssina, S., Jendrossek, V., Ishii, S., Shimizu, T., Waidmann, M., Hessler, G., Huber, S. M., Lang, F., and Wieder, T. (2005b). Stimulation of erythrocyte ceramide formation by platelet-activating factor. *J. Cell. Sci.* **118**, 1233–1243.

Lang, P. A., Warskulat, U., Heller-Stilb, B., Huang, D. Y., Grenz, A., Myssina, S., Duszenko, M., Lang, F., Haussinger, D., Vallon, V., and Wieder, T. (2003e). Blunted apoptosis of erythrocytes from taurine transporter deficient mice. *Cell. Physiol. Biochem.* **13,** 337–346.

Lauritzen, I., De Weille, J. R., and Lazdunski, M. (1997). The potassium channel opener (−)-cromakalim prevents glutamate-induced cell death in hippocampal neurons. *J. Neurochem.* **69,** 1570–1579.

Leinders, T., van Kleef, R. G., and Vijverberg, H. P. (1992). Single Ca(2+)-activated K+ channels in human erythrocytes: Ca2+ dependence of opening frequency but not of open lifetimes. *Biochim. Biophys. Acta* **1112,** 67–74.

Lepple-Wienhues, A., Belka, C., Laun, T., Jekle, A., Walter, B., Wieland, U., Welz, M., Heil, L., Kun, J., Busch, G., Weller, M., Bamberg, M., Gulbins, E., and Lang, F. (1999). Stimulation of CD95 (Fas) blocks T lymphocyte calcium channels through sphingomyelinase and sphingolipids. *Proc. Natl. Acad. Sci. USA* **96,** 13795–13800.

Lepple-Wienhues, A., Szabo, I., Laun, T., Kaba, N. K., Gulbins, E., and Lang, F. (1998). The tyrosine kinase p56lck mediates activation of swelling-induced chloride channels in lymphocytes. *J. Cell Biol.* **141,** 281–286.

Lepple-Wienhues, A., Wieland, U., Laun, T., Heil, L., Stern, M., and Lang, F. (2001). A src-like kinase activates outwardly rectifying chloride channels in CFTR-defective lymphocytes. *FASEB J.* **15,** 927–931.

Liu, X. H., Kirschenbaum, A., Yu, K., Yao, S., and Levine, A. C. (2005). Cyclooxygenase-2 suppresses hypoxia-induced apoptosis via a combination of direct and indirect inhibition of p53 activity in a human prostate cancer cell line. *J. Biol. Chem.* **280,** 3817–3823.

Liu, X. M., Tao, M., Han, X. D., Fan, Q., and Lin, J. R. (2001). Gating kinetics of potassium channel and effects of nerve growth factors in PC12 cells analyzed with fractal model. *Acta Pharmacol. Sin* **22,** 103–110.

Long, H., Han, H., Yang, B., and Wang, Z. (2003). Opposite cell density-dependence between spontaneous and oxidative stress-induced apoptosis in mouse fibroblast L-cells. *Cell. Physiol. Biochem.* **13,** 401–414.

Maeno, E., Ishizaki, Y., Kanaseki, T., Hazama, A., and Okada, Y. (2000). Normotonic cell shrinkage because of disordered volume regulation is an early prerequisite to apoptosis. *Proc. Natl. Acad. Sci. USA* **97,** 9487–9492.

Marunaka, Y., Nakahari, T., and Tohda, H. (1994). Cytosolic (Cl-) regulates Na+ absorption in fetal alveolar epithelium?: Roles of cAMP and Cl- channels. *Jpn. J. Physiol.* **44** (Suppl. 2), S281–S288.

Mauro, T., Dixon, D. B., Komuves, L., Hanley, K., and Pappone, P. A. (1997). Keratinocyte K+ channels mediate Ca2+-induced differentiation. *J. Invest. Dermatol.* **108,** 864–870.

Mavelli, I., Ciriolo, M. R., Rossi, L., Meloni, T., Forteleoni, G., De Flora, A., Benatti, U., Morelli, A., and Rotilio, G. (1984). Favism: A hemolytic disease associated with increased superoxide dismutase and decreased glutathione peroxidase activities in red blood cells. *Eur. J. Biochem.* **139,** 13–18.

Michea, L., Ferguson, D. R., Peters, E. M., Andrews, P. M., Kirby, M. R., and Burg, M. B. (2000). Cell cycle delay and apoptosis are induced by high salt and urea in renal medullary cells. *Am. J. Physiol. Renal Physiol.* **278,** F209–F218.

Migheli, A., Attanasio, A., Lee, W. H., Bayer, S. A., and Ghetti, B. (1995). Detection of apoptosis in weaver cerebellum by electron microscopic *in situ* end-labeling of fragmented DNA. *Neurosci. Lett.* **199,** 53–56.

Migheli, A., Piva, R., Wei, J., Attanasio, A., Casolino, S., Hodes, M. E., Dlouhy, S. R., Bayer, S. A., and Ghetti, B. (1997). Diverse cell death pathways result from a single missense mutation in weaver mouse. *Am. J. Pathol.* **151,** 1629–1638.

Miki, T., Tashiro, F., Iwanaga, T., Nagashima, K., Yoshitomi, H., Aihara, H., Nitta, Y., Gonoi, T., Inagaki, N., Miyazaki, J., and Seino, S. (1997). Abnormalities of pancreatic islets by targeted expression of a dominant-negative KATP channel. *Proc. Natl. Acad. Sci. USA* **94,** 11969–11973.

Miller, G. W., and Schnellmann, R. G. (1993). Cytoprotection by inhibition of chloride channels: The mechanism of action of glycine and strychnine. *Life Sci.* **53,** 1211–1215.

Montague, J. W., Bortner, C. D., Hughes, F. M., Jr., and Cidlowski, J. A. (1999). A necessary role for reduced intracellular potassium during the DNA degradation phase of apoptosis. *Steroids* **64,** 563–569.

Moran, J., Hernandez-Pech, X., Merchant-Larios, H., and Pasantes-Morales, H. (2000). Release of taurine in apoptotic cerebellar granule neurons in culture. *Pflüg. Arch.* **439,** 271–277.

Murtomaki, S., Trenkner, E., Wright, J. M., Saksela, O., and Liesi, P. (1995). Increased proteolytic activity of the granule neurons may contribute to neuronal death in the weaver mouse cerebellum. *Dev. Biol.* **168,** 635–648.

Nilius, B., and Droogmans, G. (2001). Ion channels and their functional role in vascular endothelium. *Physiol. Rev.* **81,** 1415–1459.

Nilius, B., and Wohlrab, W. (1992). Potassium channels and regulation of proliferation of human melanoma cells. *J. Physiol.* **445,** 537–548.

O'Lague, P. H., Huttner, S. L., Vandenberg, C. A., Morrison-Graham, K., and Horn, R. (1985). Morphological properties and membrane channels of the growth cones induced in PC12 cells by nerve growth factor. *J. Neurosci. Res.* **13,** 301–321.

Okada, Y., Maeno, E., Shimizu, T., Manabe, K., Mori, S., and Nabekura, T. (2004). Dual roles of plasmalemmal chloride channels in induction of cell death. *Pflüg. Arch.* **448,** 287–295.

Oo, T. F., Blazeski, R., Harrison, S. M., Henchcliffe, C., Mason, C. A., Roffler-Tarlov, S. K., and Burke, R. E. (1996). Neuron death in the substantia nigra of weaver mouse occurs late in development and is not apoptotic. *J. Neurosci.* **16,** 6134–6145.

Pal, S., He, K., and Aizenman, E. (2004). Nitrosative stress and potassium channel-mediated neuronal apoptosis: Is zinc the link? *Pflüg. Arch.* **448,** 296–303.

Pandiella, A., Magni, M., Lovisolo, D., and Meldolesi, J. (1989). The effect of epidermal growth factor on membrane potential: Rapid hyperpolarization followed by persistent fluctuations. *J. Biol. Chem.* **264,** 12914–12921.

Pappas, C. A., and Ritchie, J. M. (1998). Effect of specific ion channel blockers on cultured Schwann cell proliferation. *Glia* **22,** 113–120.

Pappone, P. A., and Ortiz-Miranda, S. I. (1993). Blockers of voltage-gated K channels inhibit proliferation of cultured brown fat cells. *Am. J. Physiol.* **264,** C1014–C1019.

Parekh, A. B., and Penner, R. (1997). Store depletion and calcium influx. *Physiol. Rev.* **77,** 901–930.

Parekh, A. B., and Putney, J. W., Jr. (2005). Store-operated calcium channels. *Physiol. Rev.* **85,** 757–810.

Patel, A. J., and Lazdunski, M. (2004). The 2P-domain K+ channels: Role in apoptosis and tumorigenesis. *Pflüg. Arch.* **448,** 261–273.

Pellegrino, M., and Pellegrini, M. (1998). Modulation of Ca^{2+}-activated K^+ channels of human erythrocytes by endogenous cAMP-dependent protein kinase. *Pflüg. Arch.* **436,** 749–756.

Perez, G. I., Maravei, D. V., Trbovich, A. M., Cidlowski, J. A., Tilly, J. L., and Hughes, F. M., Jr. (2000). Identification of potassium-dependent and -independent components of the apoptotic machinery in mouse ovarian germ cells and granulosa cells. *Biol. Reprod.* **63,** 1358–1369.

Phipps, D. J., Branch, D. R., and Schlichter, L. C. (1996). Chloride-channel block inhibits T lymphocyte activation and signalling. *Cell Signal* **8,** 141–149.

Pozzi, S., Malferrari, G., Biunno, I., and Samaja, M. (2002). Low-flow ischemia and hypoxia stimulate apoptosis in perfused hearts independently of reperfusion. *Cell. Physiol. Biochem.* **12,** 39–46.

Prehn, J. H., Jordan, J., Ghadge, G. D., Preis, E., Galindo, M. F., Roos, R. P., Krieglstein, J., and Miller, R. J. (1997). Ca2+ and reactive oxygen species in staurosporine-induced neuronal apoptosis. *J. Neurochem.* **68,** 1679–1685.

Qian, D., and Weiss, A. (1997). T cell antigen receptor signal transduction. *Curr. Opin. Cell Biol.* **9,** 205–212.

Rice, L., and Alfrey, C. P. (2005). The negative regulation of red cell mass by neocytolysis: Physiologic and pathophysiologic manifestations. *Cell. Physiol. Biochem.* **15,** 245–250.

Ritter, M., Woll, E., Haller, T., Dartsch, P. C., Zwierzina, H., and Lang, F. (1997). Activation of Na+/H(+)-exchanger by transforming Ha-ras requires stimulated cellular calcium influx and is associated with rearrangement of the actin cytoskeleton. *Eur. J. Cell Biol.* **72,** 222–228.

Ritter, M., Woll, E., Waldegger, S., Haussinger, D., Lang, H. J., Scholz, W., Scholkens, B., and Lang, F. (1993). Cell shrinkage stimulates bradykinin-induced cell membrane potential oscillations in NIH 3T3 fibroblasts expressing the ras-oncogene. *Pflüg. Arch.* **423,** 221–224.

Rosette, C., and Karin, M. (1996). Ultraviolet light and osmotic stress: Activation of the JNK cascade through multiple growth factor and cytokine receptors. *Science* **274,** 1194–1197.

Rotoli, B. M., Uggeri, J., Dall'Asta, V., Visigalli, R., Barilli, A., Gatti, R., Orlandini, G., Gazzola, G. C., and Bussolati, O. (2005). Inhibition of glutamine synthetase triggers apoptosis in asparaginase-resistant cells. *Cell. Physiol. Biochem.* **15,** 281–292.

Rouzaire-Dubois, B., Milandri, J. B., Bostel, S., and Dubois, J. M. (2000). Control of cell proliferation by cell volume alterations in rat C6 glioma cells. *Pflüg. Arch.* **440,** 881–888.

Sanders, D. A., Fiddes, I., Thompson, D. M., Philpott, M. P., Westgate, G. E., and Kealey, T. (1996). In the absence of streptomycin, minoxidil potentiates the mitogenic effects of fetal calf serum, insulin-like growth factor 1, and platelet-derived growth factor on NIH 3T3 fibroblasts in a K+ channel-dependent fashion. *J. Invest. Dermatol.* **107,** 229–234.

Santella, L. (1998). The role of calcium in the cell cycle: Facts and hypotheses. *Biochem. Biophys. Res. Commun.* **244,** 317–324.

Santella, L., Kyozuka, K., De Riso, L., and Carafoli, E. (1998). Calcium, protease action, and the regulation of the cell cycle. *Cell Calcium* **23,** 123–130.

Shen, M. R., Droogmans, G., Eggermont, J., Voets, T., Ellory, J. C., and Nilius, B. (2000). Differential expression of volume-regulated anion channels during cell cycle progression of human cervical cancer cells. *J. Physiol.* **529**(Pt. 2), 385–394.

Shindo, M., Imai, Y., and Sohma, Y. (2000). A novel type of ATP block on a Ca(2+)-activated K(+) channel from bullfrog erythrocytes. *Biophys. J.* **79,** 287–297.

Shrode, L. D., Tapper, H., and Grinstein, S. (1997). Role of intracellular pH in proliferation, transformation, and apoptosis. *J. Bioenerg. Biomembr.* **29,** 393–399.

Skryma, R. N., Prevarskaya, N. B., Dufy-Barbe, L., Odessa, M. F., Audin, J., and Dufy, B. (1997). Potassium conductance in the androgen-sensitive prostate cancer cell line, LNCaP: Involvement in cell proliferation. *Prostate* **33,** 112–122.

Spassova, M. A., Soboloff, J., He, L. P., Hewavitharana, T., Xu, W., Venkatachalam, K., van Rossum, D. B., Patterson, R. L., and Gill, D. L. (2004). Calcium entry mediated by SOCs and TRP channels: Variations and enigma. *Biochim. Biophys. Acta* **1742,** 9–20.

Storey, N. M., Gomez-Angelats, M., Bortner, C. D., Armstrong, D. L., and Cidlowski, J. A. (2003). Stimulation of Kv1.3 potassium channels by death receptors during apoptosis in Jurkat T lymphocytes. *J. Biol. Chem.* **278,** 33319–33326.

Strobl, J. S., Wonderlin, W. F., and Flynn, D. C. (1995). Mitogenic signal transduction in human breast cancer cells. *Gen. Pharmacol.* **26,** 1643–1649.

Sturm, J. W., Zhang, H., Magdeburg, R., Hasenberg, T., Bonninghoff, R., Oulmi, J., Keese, M., and McCuskey, R. (2004). Altered apoptotic response and different liver structure during liver regeneration in FGF-2-deficient mice. *Cell. Physiol. Biochem.* **14,** 249–260.

Szabo, I., Gulbins, E., and Lang, F. (1997). Regulation of Kv1.3 during Fas-induced apoptosis. *Cell. Physiol. Biochem.* **7,** 148–158.

Szabo, I., Adams, C., and Gulbins, E. (2004). Ion channels and membrane rafts in apoptosis. *Pflüg. Arch.* **448,** 304–312.

Szabo, I., Gulbins, E., Apfel, H., Zhang, X., Barth, P., Busch, A. E., Schlottmann, K., Pongs, O., and Lang, F. (1996). Tyrosine phosphorylation-dependent suppression of a voltage-gated K+ channel in T lymphocytes upon Fas stimulation. *J. Biol. Chem.* **271,** 20465–20469.

Szabo, I., Lepple-Wienhues, A., Kaba, K. N., Zoratti, M., Gulbins, E., and Lang, F. (1998). Tyrosine kinase-dependent activation of a chloride channel in CD95-induced apoptosis in T lymphocytes. *Proc. Natl. Acad. Sci. USA* **95,** 6169–6174.

Takahashi, N., Wang, X., Tanabe, S., Uramoto, H., Jishage, K., Uchida, S., Sasaki, S., and Okada, Y. (2005). ClC-3-independent sensitivity of apoptosis to Cl- channel blockers in mouse cardiomyocytes. *Cell. Physiol. Biochem.* **15,** 263–270.

Tallquist, M., and Kazlauskas, A. (2004). PDGF signaling in cells and mice. *Cytokine Growth Factor Rev.* **15,** 205–213.

Teijeiro, R., Rios, R., Costoya, J. A., Castro, R., Bello, J. L., Devesa, J., and Arce, V. M. (2002). Activation of human somatostatin receptor 2 promotes apoptosis through a mechanism that is independent from induction of p53. *Cell. Physiol. Biochem.* **12,** 31–38.

Tohda, H., Foskett, J. K., O'Brodovich, H., and Marunaka, Y. (1994). Cl- regulation of a Ca(2+)-activated nonselective cation channel in beta-agonist-treated fetal distal lung epithelium. *Am. J. Physiol.* **266,** C104–C109.

Varela, D., Simon, F., Riveros, A., Jorgensen, F., and Stutzin, A. (2004). NAD(P)H oxidase-derived H_2O_2 signals chloride channel activation in cell volume regulation and cell proliferation. *J. Biol. Chem.* **279,** 13301–13304.

Volk, T., Fromter, E., and Korbmacher, C. (1995). Hypertonicity activates nonselective cation channels in mouse cortical collecting duct cells. *Proc. Natl. Acad. Sci. USA* **92,** 8478–8482.

Walsh, M. F., Thamilselvan, V., Grotelueschen, R., Farhana, L., and Basson, M. (2003). Absence of adhesion triggers differential FAK and SAPKp38 signals in SW620 human colon cancer cells that may inhibit adhesiveness and lead to cell death. *Cell. Physiol. Biochem.* **13,** 135–146.

Wang, G. L., Wang, X. R., Lin, M. J., He, H., Lan, X. J., and Guan, Y. Y. (2002). Deficiency in ClC-3 chloride channels prevents rat aortic smooth muscle cell proliferation. *Circ. Res.* **91,** E28–E32.

Wang, S., Melkoumian, Z., Woodfork, K. A., Cather, C., Davidson, A. G., Wonderlin, W. F., and Strobl, J. S. (1998). Evidence for an early G1 ionic event necessary for cell cycle progression and survival in the MCF-7 human breast carcinoma cell line. *J. Cell. Physiol.* **176,** 456–464.

Wang, Z. (2004). Roles of K+ channels in regulating tumour cell proliferation and apoptosis. *Pflüg. Arch.* **448,** 274–286.

Wehner, F., Böhmer, C., Heinzinger, H., van den, B. F., and Tinel, H. (2000). The hypertonicity-induced Na(+) conductance of rat hepatocytes: Physiological significance and molecular correlate. *Cell. Physiol. Biochem.* **10,** 335–340.

Wehner, F., Sauer, H., and Kinne, R. K. (1995). Hypertonic stress increases the Na+ conductance of rat hepatocytes in primary culture. *J. Gen. Physiol.* **105,** 507–535.

Wei, L., Xiao, A. Y., Jin, C., Yang, A., Lu, Z. Y., and Yu, S. P. (2004). Effects of chloride and potassium channel blockers on apoptotic cell shrinkage and apoptosis in cortical neurons. *Pflüg. Arch.* **448**, 325–334.

Wenzel, U., and Daniel, H. (2004). Early and late apoptosis events in human transformed and non-transformed colonocytes are independent on intracellular acidification. *Cell. Physiol. Biochem.* **14**, 65–76.

Whitfield, J. F., Bird, R. P., Chakravarthy, B. R., Isaacs, R. J., and Morley, P. (1995). Calcium-cell cycle regulator, differentiator, killer, chemopreventor, and maybe, tumor promoter. *J. Cell Biochem. Suppl.* **22**, 74–91.

Wiecha, J., Reineker, K., Reitmayer, M., Voisard, R., Hannekum, A., Mattfeldt, T., Waltenberger, J., and Hombach, V. (1998). Modulation of Ca2+-activated K+ channels in human vascular cells by insulin and basic fibroblast growth factor. *Growth Horm. IGF Res.* **8**, 175–181.

Wieder, T., Essmann, F., Prokop, A., Schmelz, K., Schulze-Osthoff, K., Beyaert, R., Dorken, B., and Daniel, P. T. (2001). Activation of caspase-8 in drug-induced apoptosis of B-lymphoid cells is independent of CD95/Fas receptor-ligand interaction and occurs downstream of caspase-3. *Blood* **97**, 1378–1387.

Wondergem, R., Gong, W., Monen, S. H., Dooley, S. N., Gonce, J. L., Conner, T. D., Houser, M., Ecay, T. W., and Ferslew, K. E. (2001). Blocking swelling-activated chloride current inhibits mouse liver cell proliferation. *J. Physiol.* **532**, 661–672.

Wonderlin, W. F., and Strobl, J. S. (1996). Potassium channels, proliferation and G1 progression. *J. Membr. Biol.* **154**, 91–107.

Woon, L. A., Holland, J. W., Kable, E. P., and Roufogalis, B. D. (1999). Ca2+ sensitivity of phospholipid scrambling in human red cell ghosts. *Cell Calcium* **25**, 313–320.

Yu, S. P., Yeh, C. H., Sensi, S. L., Gwag, B. J., Canzoniero, L. M., Farhangrazi, Z. S., Ying, H. S., Tian, M., Dugan, L. L., and Choi, D. W. (1997). Mediation of neuronal apoptosis by enhancement of outward potassium current. *Science* **278**, 114–117.

Yurinskaya, V. E., Goryachaya, T. S., Guzhova, T. V., Moshkov, A. V., Rozanov, Y. M., Sakuta, G. A., Shirokova, A. V., Shumilina, E. V., Vassilieva, I. O., Lang, F., and Vereninov, A. A. (2005a). Potassium and sodium balance in U937 cells during apoptosis with and without cell shrinkage. *Cell. Physiol. Biochem.* **16**, 155–162.

Yurinskaya, V. E., Moshkov, A. V., Rozanov, Y.M, Shirokova, A. V., Vassilieva, I. O., Shumilina, E. V., Lang, F., Volgareva, A. A., and Vereninov, A. A. (2005b). Thymocyte K+, Na+ and water balance during dexamethasone and etoposide induced apoptosis. *Cell. Physiol. Biochem.* **16**, 15–22.

Zhou, Q., Kwan, H. Y., Chan, H. C., Jiang, J. L., Tam, S. C., and Yao, X. (2003). Blockage of voltage-gated K+ channels inhibits adhesion and proliferation of hepatocarcinoma cells. *Int. J. Mol. Med.* **11**, 261–266.

Zhou, Q., Zhao, J., Wiedmer, T., and Sims, P. J. (2002). Normal hemostasis but defective hematopoietic response to growth factors in mice deficient in phospholipid scramblase 1. *Blood* **99**, 4030–4038.

CHAPTER TWELVE

ACTIN CYTOSKELETON ARCHITECTURE AND SIGNALING IN OSMOSENSING

Evangelia A. Papakonstanti *and* Christos Stournaras

Contents

Abstract

Since the early days of cell volume regulation research, the role of actin cytoskeleton organization and rearrangement has attracted specific interest. Rapid modifications in actin dynamics and architecture have been described. They were shown to regulate cell volume changes, as well as regulatory volume decrease in a large variety of cell types, including hepatocytes, lymphocytes,

Department of Biochemistry, University of Crete Medical School, Heraklion, Greece

Methods in Enzymology, Volume 428
ISSN 0076-6879, DOI: 10.1016/S0076-6879(07)28012-7

fibroblasts, myocytes, and various tumor cells. Using microscopic and biochemical analyses, modifications of actin organization and polymerization dynamics were studied. This chapter summarizes the molecular approaches applied so far for the quantitative assessment of actin cytoskeleton dynamics in the various cell types. It demonstrates that rapid modifications of actin cytoskeleton dynamics regulated by specific signaling pathways play a functional role in cell volume regulation. It is concluded that studying actin polymerization dynamics and signaling represents a challenging tool for the understanding of osmosensing and osmosignaling regulation in cellular physiology.

1. INTRODUCTION

Actin cytoskeleton is a dynamic cellular structure known to regulate many aspects of cell physiology. Various effectors actively modulate actin architecture governed by specific signaling cascades, including both nongenomic and transcriptional pathways (Papakonstanti and Stournaras, 2002, 2004; Papakonstanti et al., 2003; Rivera et al., 2006; Theriot, 1994; Vardouli et al., 2005). The signaling transducers generate rapid and long-term modifications of actin polymerization dynamics and microfilament organization and regulate various cell functions (Gourlay and Ayscough, 2005; Papakonstanti and Stournaras, 2004; Pollard and Borisy, 2003; Raftopoulou and Hall, 2004; Thomas et al., 2006; Yamazaki et al., 2005). These reports provided convincing evidence that actin cytoskeleton dynamics—in addition to their structural and mechanical properties—may function as a "sensory system" receiving extracellular signals and mediating cell responses, implying a functional role for the actin cytoskeleton dynamics in cellular physiology (for review, see Stournaras, 2006).

Many animal cells have the ability to regulate their volume in response to alterations of environmental osmotic conditions. Regulation of cell volume is a critical response for cell metabolism and survival. Exposure to a hypotonic milieu induces initial swelling followed by slow compensatory cell shrinkage through activation of various transport systems that regulate the efflux of intracellular solutes. On the contrary, exposure to a hypertonic environment stimulates activation of uptake systems, allowing osmolyte influx. These regulatory processes are termed regulatory volume decrease (RVD) and regulatory volume increase (RVI), respectively (for review, see Okada and Maeno, 2001). Various effectors and uptake systems, including plasma membrane ion channels and transporters, contribute to the overall regulation of cell volume (Lang et al., 1998; Okada, 2004; Wehner et al., 2003). Although a large amount of experimental results has accumulated, the nature of the volume sensor and the signaling pathways that cells are using to adjust their volume are still not fully understood. In addition, a definition of unambiguously common sensing and signaling mechanisms for cell volume regulation is not evident (for review, see Jakab *et al.*, 2002).

The involvement of actin cytoskeleton dynamics in cell volume control was studied in detail in the past (for reviews, see Moustakas *et al.*, 1998; Pedersen *et al.*, 2001). In response to changes in osmolarity, modulation of actin organization and polymerization dynamics was shown in several cell types (Bibby and McCulloch, 1994; Ebner *et al.*, 2005; Hallows *et al.*, 1991, 1996; Mountain *et al.*, 1998; Theodoropoulos *et al.*, 1992), while the regulatory role of actin reorganization during RVD and RVI processes was analyzed as well in various cell models (Ebner *et al.*, 2005; Liu *et al.*, 2006; Pedersen *et al.*, 1999). Taken together, these reports provide evidence that rapid modifications of actin cytoskeleton dynamics may be considered a potential volume sensor mechanism (Jakab *et al.*, 2002; Papakonstanti *et al.*, 2000b; Wehner *et al.*, 2003). This chapter summarizes the role of actin cytoskeleton organization in osmosensing and osmosignaling and discusses in detail techniques aimed at evaluating quantitatively actin polymerization dynamics in cell volume regulation.

2. Morphological Analysis of Actin Cytoskeleton During Cell Volume Changes

The implication of microfilament reorganization in cell volume regulation was initially studied by applying qualitative microscopic analysis (immunofluorescence and confocal laser-scanning microscopy). In cells exposed to hypotonic media, the majority of microscopic studies reported actin cytoskeleton disorganization and loss of microfilamentous structures such as stress fibers and formation of submembranous F-actin aggregations (Cornet *et al.*, 1994; Dartsch *et al.*, 1994; Hallows *et al.*, 1991; Moran *et al.*, 1996; Schwiebert *et al.*, 1994). In specific cell types, such as astrocytes and pheochromocytoma PC12 cells, the reorganization of actin was a transient phenomenon (Cornet *et al.*, 1994; Moran *et al.*, 1996), whereas in renal epithelial OK cells, actin microfilament reorganization could not be detected by fluorescence microscopy (Dartsch *et al.*, 1994). From these reports it became evident that cell volume changes result in reorganization of F-actin structures. It should be noted that although microscopic analysis provides an overall qualitative representation of cellular morphology, it could by no means provide specific insights on the dynamics of actin cytoskeleton organization. In addition, in cell systems expressing feeble microfilamentous structures, such as leukocytes, hepatocytes, or tumor cells, the evaluation of actin reorganization by fluorescence microscopy is problematical. Despite those limitations, however, microscopic approaches established a pivotal role of F-actin reorganization for the adjustment of cell volume (Hallows *et al.*, 1991; Schwiebert *et al.*, 1994).

An indirect approach in monitoring the dynamic state of actin polymerization during cell volume regulation was addressed by microscopic analysis (immunofluorescence and confocal laser-scanning microscopy) using various microfilament-disrupting agents. The existing literature reports that in various cell types, such as leukocytes, myocytes, Ehrlich ascites tumor cells, HEK 293 cells, and hepatocytes, cytochalasins inhibit RVD and RVI, indicating that an intact cytoskeleton is required for the mediation of cellular signals in response to cell volume alterations (Downey *et al.*, 1995; Ebner *et al.*, 2005; Hall *et al.*, 1997; Jorgensen *et al.*, 2003; Pedersen *et al.*, 1999). It has been reported that involvement of the actin cytoskeleton in regulating RVD was directly verified in salivary gland cells (Liux *et al.*, 2006). Indeed, by using N terminus-deleted aquaporin 5, the osmosensing transient receptor potential vanalloid 4 and RVD were suppressed and this effect was inhibited by actin depolymerization. However, cytochalasin-independent effects on regulated volume decrease have also been reported (Downey *et al.*, 1995; Hallows *et al.*, 1996; Moran *et al.*, 1996), indicating that changes in the actin cytoskeleton are quite diverse in different cell types. Thus, evaluation of these results should be considered with caution. Although cytochalasins or phalloidins are well-studied drugs affecting actin organization and polymerization, their side effects on other aspects of cell physiology cannot be excluded. Complementary analysis of actin polymerization dynamics by quantitative biochemical analysis is needed to provide molecular insights into the regulatory role of the actin cytoskeleton in cell volume regulation.

3. QUANTITATIVE BIOCHEMICAL ANALYSIS OF ACTIN CYTOSKELETON DYNAMICS DURING CELL VOLUME CHANGES

A much more detailed analysis of microfilament reorganization during the different phases of cell volume changes became possible by quantitative biochemical measurements of intracellular actin polymerization equilibrium, including assessment of cellular monomeric and polymerized actin levels using various techniques. In an initial study using the DNase I inhibition assay to assess the intracellular monomeric and total actin content, we determined actin polymerization dynamics in primary cultures of rat hepatocytes after hypotonic exposure (Theodoropoulos *et al.*, 1992). Detailed analysis revealed rapid decrease of the monomeric to total actin ratio (G-/T-actin) upon induction of dramatic cell volume alterations. Interestingly, during the initial phase of cell swelling the total actin content of cells treated with hypotonic (or hypertonic) media remained unchanged, indicating a net actin polymerization. These findings were confirmed by applying an independent method for the quantitative F-actin determination in hepatocytes by using fluorimetry of detergent cell extracts after binding to rhodamine–phalloidin, which monitors polymerized F-actin. In line with these observations, Henson *et al.* (1997) reported that quantitative Western analysis of total F-actin in dogfish shark rectal gland cells revealed stable actin content in cells undergoing controlled cell volume changes. Using the quantitative phalloidin assay for F-actin determinations, Pedersen *et al.* (1999) reported a very fast decrease in cellular F-actin content in Ehrlich ascites tumor cells upon hypotonic treatment. In HL-60 monocytes the cellular F-actin level increases during hypotonic exposure, and evidence was provided for the existence of cytoplasmic subpopulations of F-actin that exhibit a differential response to volume changes (Hallows *et al.*, 1996). In this report, Western analysis of detergent soluble and insoluble cell fractions monitored the cellular soluble and insoluble F-actin levels. In a more recent study the assessment of the ratio of cortical to cytoplasmic F-actin provided additional evidence for the crucial role of actin cytoskeleton in cellular responses to anisotonic conditions (Ebner *et al.*, 2005). Thus, the application of detailed analysis of actin polymerization dynamics using specific methodologies emphasizes the importance of dynamic changes in the organization of actin microfilaments in cell volume regulation. These biochemical approaches, in combination with microscopic analysis, are important tools when addressing insights of the molecular mechanisms of actin cytoskeleton involvement in regulating osmosensing.

4. SIGNALING PATHWAYS LINKING ACTIN REORGANIZATION AND CELL VOLUME REGULATION

Although it is widely accepted that actin polymerization is a primary receiver of cell volume changes, mechanisms linking actin cytoskeleton reorganization in response to cell volume regulation are still not fully understood. Until the present time, the actions of specific membrane channels and transporters are the best-studied regulatory circuits associated with cell volume regulation and actin reorganization (Cantiello, 1997; Dartsch *et al.*, 1995; Jorgensen *et al.*, 2003; Schwartz *et al.*, 1997; Schwiebert *et al.*, 1994; Tilly *et al.*, 1996). Key signaling molecules, such as actin-binding protein 280 (Cantiello *et al.*, 1997), Rho GTPase, focal adhesion kinase (FAK), and phosphoinositol-3-kinase (Okada, 1997; Tilly *et al.*, 1996), were shown to mediate actin reorganization upon modulation of membrane ion channel activity. In another study, tumor necrosis factor α-induced cell volume changes were associated with actin polymerization in glomerular epithelial cells (Koukouritaki *et al.*, 1999). This effect was regulated by vinculin, FAK, and paxillin activation via phosphorylation, supporting the idea that signaling molecules operating at focal adhesions may trigger actin cytoskeleton responses to cell volume regulation. In a more recent study, a novel scaffold protein termed OSM (osmosensing scaffold for MEKK3) that binds to actin, the small GTPase Rac, and the upstream kinases MEKK3 and MKK3 complex was identified. It was demonstrated that this MEKK3 scaffold protein complex is recruited to dynamic actin structures and that this scaffold protein complex is required for p38 activation in response to hyperosmolarity (Uhlik *et al.*, 2003). The role of small GTPase activation in volume regulation and actin reorganization was addressed in Ehrlich ascites tumor cells (Klausen *et al.*, 2006). It was shown that the mechanism controlling the volume–regulated anion current was regulated by a signaling cascade involving Rho, Rho kinase, phosphatidylinositol (4,5)-bisphosphate activation, and actin reorganization. From these reports it is evident that activation of signaling cascades that operate at focal adhesions and/or membrane ion channels and receptors appear as strong candidates that control the actin cytoskeleton response to cell volume regulation. Reorganization of actin architecture participates in the molecular "sensing" of osmotic changes in the cell environment, implying a major functional role of actin dynamics and signaling in cellular physiology. Thus, assessing actin dynamics modifications and elucidating the signal transducers regulating these changes may determine the efforts needed for understanding osmoregulatory cell physiology.

5. QUANTITATIVE ASSESSMENT OF CELLULAR ACTIN CYTOSKELETON DYNAMICS

5.1. Overview

From the reports presented so far, it is evident that quantitative biochemical approaches can measure subtle changes in the intracellular actin monomer polymer equilibrium, corresponding even to local actin reorganization events, not easily detected by microscopic analysis. Those approaches became necessary in various cell types expressing feeble actin structures in which microscopic analysis failed to provide reliable information on actin reorganization. The example shown in Fig. 12.1 demonstrates rapid and transient actin cytoskeleton reorganization induced by transforming growth factor (TGF)-β in A375 melanoma cells using microscopic (A) and biochemical analyses (B) (Vardouli and Stournaras, unpublished result). Although the morphological observation indicates actin restructuring and disappearance of submembranous microfilament structures after 15 and 30 min (Fig. 12.1A, b and c), it is unclear whether this is because of net actin depolymerization. This could be verified

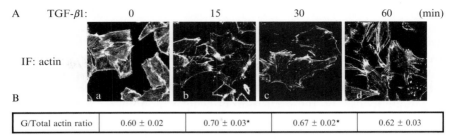

G/Total actin ratio	0.60 ± 0.02	$0.70 \pm 0.03^*$	$0.67 \pm 0.02^*$	0.62 ± 0.03

Figure 12.1 Actin redistribution and depolymerization in response to TGF-β in A375 melanoma cells. (A) Cells were incubated with TGF-β for 15 (b), 30 (c), or 60 (d) min and then the redistribution of filamentous actin was determined with rhodamine–phalloidin staining by immunofluorescence microscopy. (a) Untreated cells. (B) Cells were incubated for the times indicated in A with 10 ng/ml TGF-β and then monomeric (G) and total (T) actin levels were measured according to the DNase I inhibition assay. Data are presented as G/T actin ratio (mean $\pm$ SEM from five separate experiments). $^*p < 0.05$.

by applying quantitative molecular assessment of cellular monomeric and total actin content (DNase I inhibition assay, see later) showing increased G-/T-actin ratios after 15 and 30 min of incubation, which documents net actin depolymerization upon TGF-β treatment. Note that the G-/T-actin ratio returns to control levels after 60 min, in line with the similar morphological actin patterns shown in Fig. 12.1A (a and d).

The following sections summarize the molecular approaches applied so far for the quantitative analysis of actin cytoskeleton dynamics in various cell types. In particular we will present specific techniques aimed at determining the intracellular monomeric, filamentous, and total actin content assessed by quantitative biochemical determinations of actin dynamics during different phases of cell volume changes. These methodologies include (1) the DNase I inhibition assay, which monitors intracellular monomeric (G-) and total (T-) actin content; (2) fluorescence measurements of phalloidin-labeled detergent cell extracts, which monitor filamentous (F-) actin content; and (3) immunoblot analysis of detergent insoluble cytoskeletal pellets and soluble supernatants using specific actin antibodies to determine monomeric and polymerized actin levels. As pointed out later, a combination of these techniques is often required to provide an intergraded description of what happens during changes of actin polymerization dynamics.

6. QUANTIFICATION OF CELLULAR MONOMERIC AND TOTAL ACTIN USING THE DNASE I INHIBITION ASSAY

Since the early days of actin cytoskeleton research monomeric (G-) actin was shown to be a specific inhibitor of DNase I. Selective assays for monomeric and filamentous actin determinations were proposed based on the inhibitory activity of G–actin (Blikstad *et al.*, 1978). This method was widely used in the past to assess quantitatively the rapid modifications of actin cytoskeleton dynamics in response to extracellular signals, including osmosensing (Koukouritaki *et al.*, 1999; Papakonstanti and Stournaras, 2002, 2004; Papakonstanti *et al.*, 1996, 2000a; Stournaras *et al.*, 1996; Theodoropoulos *et al.*, 1992). It should be noted, however, that this technique focuses on the evaluation of cellular monomeric and total actin content, without providing reliable calculation of filamentous actin levels. Accordingly, when necessary it should be combined with direct F–actin measurements (see following methods).

7. DNase I Inhibition Assay Protocol

1. Cells (usually 5×10^6 in 75-cm^2 flasks depending on cell type), appropriately treated, are washed three times with ice-cold phosphate-buffered saline (PBS) and suspended in 300 μl of lysis buffer containing 10 mM K$_2$HPO$_4$, 100 mM NaF, 50 mM KCl, 2 mM MgCl$_2$, 1 mM EGTA, 0.2 mM dithiothreitol (DTT), 0.5% Triton X-100, and 1 M sucrose, pH 7.0.

2. For determination of the G–actin content, 10 μl of the lysate is added to the assay mixture containing 10 μl of DNase I solution (0.1 mg/ml DNase I in 50 mM Tris/HCl, 10 mM phenylmethylsulfonyl fluoride [PMSF], 0.5 mM CaCl$_2$, pH 7.5) and 1 ml of DNA solution (40 μg/ml DNA in 100 mM Tris/HCl, 4 mM MgSO$_4$, 1.8 mM CaCl$_2$, pH 7.5). Measurements of G–actin must start immediately after lysis of the cells and should be completed during the first 10 min or earlier, as depolymerization of F–actin occurs, giving false results.

3. The DNase I activity is monitored continuously in a double-beam spectrophotometer at 260 nm. In parallel with the measurement of DNase I inhibition by cellular actin, a control measurement of DNase I activity must be done using 10 μl of the lysis buffer instead of cell lysate. The value that will be obtained represents 100% of DNase I activity and will be used for calculation of DNase I inhibition by the monomeric actin in cell samples.

4. The actin content in the sample is measured by reference to a standard curve for the inhibition of DNase I activity, prepared with purified rabbit muscle actin. Note that reliable values are obtained entirely over the linear range of DNase I inhibition (usually the range between 25 and 70% of the inhibition of DNase I activity).

5. To measure total actin, aliquots of the lysed cell suspension are diluted two to three times with lysis buffer (see step 1) and then incubated for 10 min with an equal volume of guanidine/HCl buffer (1.5 M guanidine/HCl,

1 M sodium acetate, 1 mM $CaCl_2$, 1 mM ATP, 20 mM Tris/HCl, pH 7.5) to depolymerize F-actin to monomeric G-actin.

6. Both G- and T-actin contents can be related to the total protein amount of cell lysates.

8. QUANTIFICATION OF FILAMENTOUS ACTIN USING RHODAMINE–PHALLOIDIN FLUORESCENCE MEASUREMENTS OF ACTIN IN DETERGENT CELL EXTRACTS

As indicated earlier, the G-actin-dependent DNase I inhibition assay does not permit direct quantification of polymeric actin levels. F-actin can be simply calculated from the difference between total and monomeric actin content; however, this estimation is not precise, as it cannot differentiate between filaments (short or long) and actin aggregates. The introduction of methods for quantitative fluorescence measurements of phalloidin- or rhodamine–phalloidin-labeled detergent cell extracts offered objective techniques for the assessment of total intracellular F-actin content (Cable et al., 1995; Wu et al., 1992). These methods, with minor modifications, were widely used in the past for F-actin determinations in response to osmosensing and as complementary techniques to the DNase I inhibition assay for definition of the G-/F-actin ratio (Ebner et al., 2005; Hallows et al., 1996; Henson et al., 1997; Papakonstanti et al., 2000a; Pedersen et al., 1999; Stournaras et al., 1996). The following section presents a modified version (Stournaras et al., 1996) of the widely used protocol described by Wu et al. (1992) and Cable et al. (1995).

9. FILAMENTOUS (F-) ACTIN QUANTIFICATION PROTOCOL

1. Cells grown in cell culture dishes (usually 24-well plates) and treated appropriately are fixed by adding 0.3 ml of formaldehyde (3.7% in PBS), followed by a 15-min incubation at room temperature.
2. Cells are permeabilized by adding 0.3 ml of Triton X-100 (0.2% in PBS) for 5 min at room temperature.
3. After adding 0.3 ml of the labeling solution (rhodamine–phalloidin, 1.5 μM in PBS) to the permeabilized cells, the cells are incubated for 30 min at room temperature in the dark.
4. Cells are washed three times with 1 ml PBS and scraped off into 0.5 ml of 0.1 M NaOH.

5. Fluorescence of the samples is measured in an appropriate fluorimeter using excitation and emission wavelengths of 550 and 580 nm, respectively. The relative F-actin content is calculated as the ratio of fluorescence units per milligram total protein.

10. QUANTITATIVE IMMUNOBLOT ANALYSIS OF TRITON X-100 INSOLUBLE CYTOSKELETAL PELLETS AND CORRESPONDING SUPERNATANTS

This approach addresses the quantification of three distinct cellular actin cytoskeleton fractions—the soluble G-actin, short actin filaments, and the microfilamentous network—and was initially reported by Golenhofen *et al.* (1995). The specificity of this technique focuses on the capacity to separate short and long actin filaments, which may be important for the understanding of differential subpopulations of actin filaments that may exhibit differential response to volume changes (Hallows *et al.*, 1996). This method, with minor adaptations, is widely used in the literature for soluble/insoluble actin ratio determinations in various cell models triggered by several stimuli, including anisotonicity (Ebner *et al.*, 2005; Hallows *et al.*, 1996; Papakonstanti and Stournaras, 2002, 2004; Vardouli *et al.*, 2005). The next section presents the widely used protocol of Golenhofen *et al.* (1995), modified accurately to address the assessment of either (a) three fractions: G-actin and short and long actin filaments (protocol I; Papakonstanti *et al.*, 2000a), or (b) two fractions: monomeric G-actin and total filamentous F-actin (protocol II; Papakonstanti and Stournaras, 2002).

11. TRITON SOLUBLE/INSOLUBLE ACTIN RATIO DETERMINATION

11.1. Protocol I

1. Cells are incubated for 20 min at $4°$ in 1 ml of cytoskeleton extraction buffer consisting of 0.5% Triton X-100, 10 mM EGTA, 40 mM KCl, 5 μg/ml leupeptin, 1 μg/ml aprotinin, 1 mM PMSF, and 10 mM imidazole, pH 7.15, on ice.
2. Cell extracts are centrifuged for 4 min at 16,000g, and the resulting low-speed pellet (LSP; corresponding to the microfilamentous network) is dissolved in a Tris/SDS buffer consisting of 0.625 M Tris/HCl, pH 7.4, 2% SDS, and 10% glycerol.

3. The remaining supernatant is centrifuged for 2.5 h at 100,000*g* to obtain the high-speed pellet (HSP; corresponding to short actin filaments), which is dissolved in the same Tris/HCl buffer as in step 2.

4. Proteins in the remaining high-speed supernatant (HSS; corresponding to the monomeric G-actin) are precipitated by mixing with 4 volumes of ice-cold acetone and subsequently dissolved in the Tris/HCL buffer as in step 2.

5. The three fractions (LSP, HSP, and HSS) are subjected to SDS–PAGE and Western blotting analysis using monoclonal antiactin antibodies.

11.2. Protocol II

1. Cells are incubated in 500 μl of Triton-extraction buffer consisting of 0.3% Triton X-100, 5 mM Tris, pH 7.4, 2 mM EGTA, 300 mM sucrose, 2 μM phalloidin, 1 mM PMSF, 10 μg/ml leupeptin, 20 μg/ml aprotinin, 1 mM sodium orthovanadate, and 50 mM NaF for 5 min on ice.

2. The supernatant containing the soluble proteins is removed by aspiration from the Triton insoluble fraction, which remained on the plate.

3. The Triton insoluble pellet is scraped from the plate directly into 500 μl of RIPA buffer consisting of 50 mM Tris/HCl, pH 7.4, 1% Triton X-100, 1% sodium deoxycholate, 0.1% SDS, 0.15 M NaCl, 1 mM EDTA, 1 mM DTT, and 1 mM sodium orthovanadate.

4. Any remaining insoluble material is removed by centrifugation.

5. Equal volumes of each fraction are subjected to SDS–PAGE and Western blotting analysis using monoclonal antiactin antibodies.

6. Alternatively, Triton soluble proteins (step 2) are precipitated with an equal volume of 0.6% perchloric acid (PCA) and the Triton insoluble pellet (step 3) is scraped from the plate directly into 1 ml 3% PCA. Both cellular fractions are centrifuged at 12,000 rpm for 5 min, and the resulting pellets are dissolved in 0.1 N NaOH. Equal volumes of each are subjected SDS–PAGE and Western blotting analysis using monoclonal antiactin antibodies.

REFERENCES

Bibby, K. J., and McCulloch, C. A. (1994). Regulation of cell volume and [Ca2+]i in attached human fibroblasts responding to anisosmotic buffers. *Am. J. Physiol.* **266,** 1639–1649.

Blikstad, I., Markey, F., Carrison, L., Persson, T., and Lindberg, U. (1978). Selective assay of monomeric and filamentous actin in cell extras, using inhibition of deoxyribonuclease I. *Cell* **15,** 935–943.

Cable, H. C., El-Mansoury, A, and Morgan, N. J. (1995). Activation of alpha-2-adreno-ceptors results in an increase in F-actin formation in HIT-T15 pancreatic B-cells. *Biochem. J.* **307,** 169–174.

Cantiello, H. F. (1997). Role of actin filament organization in cell volume and ion channel regulation. *J. Exp. Zool.* **279,** 425–435.

Cornet, M., Isobe, Y., and Lemanski, L. F. (1994). Effects of anisosmotic conditions on the cytoskeletal architecture of cultured PC12 cells. *J. Morphol.* **222,** 269–286.

Dartsch, P. C., Kolb, H. A., Beckmann, M., and Lang, F. (1994). Morphological alterations and cytoskeletal reorganization in opossum kidney (OK) cells during osmotic swelling and volume regulation. *Histochemistry* **102,** 69–75.

Dartsch, P. C., Ritter, M., Gschwentner, M., Lang, H. J., and Lang, F. (1995). Effects of calcium channel blockers on NIH 3T3 fibroblasts expressing the Ha-ras oncogene. *Eur. J. Cell Biol.* **67,** 372–378.

Downey, G. P., Grinstein, S., Sue, A. Q., Czaban, B., and Chan, C. K. (1995). Volume regulation in leukocytes: Requirement for an intact cytoskeleton. *J. Cell Physiol.* **163,** 96–104.

Ebner, H. L., Cordas, A., Pafundo, D. E., Schwarzbaum, P. J., Pelster, B., and Krumschnabel, G. (2005). Importance of cytoskeletal elements in volume regulatory responses of trout hepatocytes. *Am. J. Physiol. Regul. Integr. Comp. Physiol.* **289,** 887–890.

Gourlay, C. W., and Ayscough, K. R. (2005). The actin cytoskeleton: A key regulator of apoptosis and ageing? *Nat. Rev. Mol. Cell Biol.* **6,** 583–589.

Golenhofen, N., Doctor, R. B., Bacallao, R., and Mandel, L. J. (1995). Actin and villin compartmentation during ATP depletion and recovery in renal cultured clls. *Kidney Int.* **48,** 1837–1845.

Hall, S. K., Zhang, J., and Lieberman, M. (1997). An early transient current is associated with hyposmotic swelling and volume regulation in embryonic chick cardiac myocytes. *Exp. Physiol.* **82,** 43–54.

Hallows, K. R., Law, F. Y., Packman, C. H., and Knauff, P. A. (1996). Changes in cytoskeletal actin content, F-actin distribution, and surface morphology during HL-60 cell volume regulation. *J. Cell Physiol.* **167,** 60–71.

Hallows, K. R., Packman, C. H., and Knauf, P. A. (1991). Acute cell volume changes in anisotonic media affect F–actin content of HL-60 cells. *Am. J. Physiol.* **261,** 1154–1161.

Henson, J. H., Roesener, C. D., Gaetano, C. J., Mendola, R. J., Forrest, J. N. Jr., Holy, J., and Kleinzeller, A. (1997). Confocal microscopic observation of cytoskeletal reorganizations in cultured shark rectal gland cells following treatement with hypotonic shock and high external K+. *J. Exp.Zool.* **279,** 415–424.

Jakab, M., Furst, J., Gschwentner, M., Botta, G., Garavaglia, M. L., Bazzini, C., Rodighiero, S., Meyer, G., Eichmueller, S., Woll, E., Chwatal, S., Ritter, M., *et al.* (2002). Mechanisms sensing and modulating signals arising from cell swelling. *Cell Physiol. Biochem.* **12,** 235–258.

Jorgensen, N. K., Pedersen, S. F., Rasmussen, H. B., Grunnet, M., Klaerke, D. A., and Olesen, S. P. (2003). Cell swelling activates cloned Ca(2+)-activated K(+) channels: A role for the F–actin cytoskeleton. *Biochim. Biophys. Acta* **1615,** 115–125.

Klausen, T. K., Hougaard, C., Hoffmann, E. K., and Pedersen, S. F. (2006). Cholesterol modulates the volume-regulated anion current in Ehrlich-Lettre ascites cells via effects on Rho and F-actin. *Am. J. Physiol. Cell Physiol.* **291,** 757–771.

Koukouritaki, S. B., Vardaki, E. A., Papakonstanti, E., Lianos, E., Stournaras, C., and Emmanouel, D. S. (1999). TNFa induces actin cytoskeleton reorganization in glomerular epithelial cells involving tyrosine phosphorylation of paxillin and focal adhesion kinase. *Mol. Med.* **5,** 382–392.

Lang, F., Busch, G. L., Ritter, M., Volkl, H., Waldegger, S., Gulbins, E., and Häussinger, D. (1998). Functional significance of cell volume regulatory mechanisms. *Physiol. Rev.* **78,** 247–306.

Liu, X., Bandyopadhyay, B., Nakamoto, T., Singh, B., Liedtke, W., Melvin, J. E., and Ambudkar, I. (2006). A role for AQP5 in activation of TRPV4 by hypotonicity:

Concerted involvement of AQP5 and TRPV4 in regulation of cell volume recovery. *J. Biol. Chem.* **281,** 15485–15495.

Moran, J., Sabanero, M., Meza, I., and Pasantes-Morales, H. (1996). Changes of actin cytoskeleton during swelling and regulatory volume decrease in cultured astrocytes. *Am. J. Physiol.* **271,** 1901–1907.

Mountain, I., Waelkens, E., Missiaen, L., and Van Driessche, W. (1998). Changes in actin cytoskeleton during volume regulation in C6 glial cells. *Eur. J. Cell Biol.* **77,** 196–204.

Moustakas, A., Theodoropoulos, P. A., Gravanis, A., Häussinger, D., and Stournaras, C. (1998). The cytoskeleton in cell volume regulation. *Contrib. Nephrol. Basel Karger* **123,** 121–134.

Okada, Y. (1997). Volume expansion-sensing outward-rectifier Cl channel: Fresh start to the molecular indentity and volume sensor. *Am. J. Physiol.* **273,** 755–789.

Okada, Y. (2004). Ion channels and transporters involved in cell volume regulation and sensor mechanisms. *Cell Biochem. Biophys.* **41,** 233–258.

Okada, Y., and Maeno, E. (2001). Apoptosis, cell volume regulation and volume-regulatory chloride channels. *Comp. Biochem. Physiol. A Mol. Integr. Physiol.* **130,** 377–383.

Papakonstanti, E., Emmanouel, D. S., Gravanis, A., and Stournaras, C. (1996). Na$^+$/Pi cotransport alters rapidly cytoskeletal protein polymerization dynamics in opossum kidney cells. *Biochem. J.* **315,** 241–247.

Papakonstanti, E. A., Emmanouel, D. S., Gravanis, A., and Stournaras, C. (2000a). PLC-γ1 signaling pathway and villin activation are involved in actin cytoskeletal reorganization induced by Na+/Pi cotransport up-regulation. *Mol. Med.* **6,** 303–318.

Papakonstanti, E. A., Kampa, M., Castanas, E., and Stournaras, C. (2003). A rapid, non-genomic, signaling pathway regulates the actin reorganization and PSA secretion induced by membrane testosterone receptors' activation. *Mol. Endocrinol.* **17,** 870–881.

Papakonstanti, E. A., and Stournaras, C. (2002). Association of PI-3 kinase with PAK1 leads to direct phosphorylation of actin and cytoskeletal reorganization. *Mol. Biol. Cell* **13,** 2946–2962.

Papakonstanti, E. A., and Stournaras, C. (2004). TNFa promotes survival of proximal tubular epithelial cells via Cdc42-induced PLC- gamma 1 activation and actin filament redistribution. *Mol. Biol. Cell* **15,** 1273–1286.

Papakonstanti, E. A., Vardaki, E. A., and Stournaras, C. (2000b). Actin cytoskeleton: A signaling sensor in cell volume regulation. *Cell Physiol. Biochem.* **10,** 257–264.

Pedersen, S. F., Hoffman, E. K., and Mills, J. W. (2001). The cytoskeleton and cell volume regulation. *Comp. Biochem. Physiol. A Mol. Integr. Physiol.* **130,** 385–399.

Pedersen, S. F., Mills, J. W., and Hoffmann, E. K. (1999). Role of the F–actin cytoskeleton in the RVD and RVI processes in Ehrlich ascites tumor cells. *Exp. Cell Res.* **252,** 63–74.

Pollard, T. D., and Borisy, G. G. (2003). Cellular motility driven by assembly and disassembly of actin filaments. *Cell* **112,** 453–465.

Raftopoulou, M., and Hall, A. (2004). Cell migration: Rho GTPases lead the way. *Dev. Biol.* **265,** 23–32.

Rivera, G. M., Antoku, S., Gelkop, S., Shin, N. Y., Hanks, S. K., Pawson, T., and Mayer, B. J. (2006). Requirement of Nck adaptors for actin dynamics and cell migration stimulated by platelet-derived growth factor B. *Proc. Natl. Acad. Sci. USA* **103,** 9536–9541.

Schwartz, R. S., Rybicki, A. C., and Nagel, R. L. (1997). Molecular cloning and expression of a chloride channel-associated protein plCln in human young red blood cells: Association with actin. *Biochem. J.* **327,** 609–616.

Schwiebert, E. M., Mills, J. W., and Stanton, B. A. (1994). Actin-based cytoskeleton regulates a chloride channel and cell volume in a renal cortical collecting duct cell line. *J. Biol. Chem.* **269,** 7081–7089.

Stournaras, C. (2006). The functional role of actin cytoskeleton dynamics and signaling in early cell responses. *Adv. Mol. Cell Biol.* **37**, 181–200.

Stournaras, C., Stiakaki, E., Koukouritaki, S., Theodoropoulos, P. A., Kalmanti, M., and Gravanis, A. (1996). Altered actin polymerization dynamics in various malignant cell types: Evidence for differential sensitivity to cytochalasin B. *Biochem. Pharmacol.* **52**, 1339–1346.

Theodoropoulos, P., Stournaras, C., Stoll, B., Markogiannakis, E., Lang, F., Gravanis, A., and Häussinger, D. (1992). Hepatocytes swelling leads to rapid decrease of the G-/total actin ratio and increases actin mRNA levels. *FEBS Lett.* **311**, 241–245.

Theriot, J. A. (1994). Regulation of the actin cytoskeleton in living cells. *Semin. Cell Biol.* **5**, 193–199.

Thomas, S. G., Huang, S., Li, S., Staiger, C. J., and Franklin-Tong, V. E. (2006). Actin depolymerization is sufficient to induce programmed cell death in self-incompatible pollen. *J. Cell Biol.* **174**, 221–229.

Tilly, B. C., Edixhoven, M. J., Tertoolen, L. G., Morii, N., Saitoh, Y., Narumiya, S., and de Jonge, H. R. (1996). Activation of the osmo-sensitive chloride conductance involves p21rho and is accompanied by a transient reorganization of the F-actin cytoskeleton. *Mol. Biol. Cell* **7**, 1419–1427.

Uhlik, M. T., Abell, A. N., Johnson, N. L., Sun, W., Cuevas, B. D., Lobel-Rice, K. E., Horne, E. A., Dell'Aqua, M. L., and Johnson, G. L. (2003). Rac-MEKK3 scaffolding for p38 MAPK activation during hyperosmotic shock. *Nat. Cell Biol.* **5**, 1104–1110.

Vardouli, L., Moustakas, A., and Stournaras, C. (2005). LIM-kinase 2 and cofilin phosphorylation mediate actin cytoskeleton reorganization induced by transforming growth factor-beta. *J. Biol. Chem.* **280**, 11448–11457.

Wehner, F., Olsen, H., Tinel, H., Kinne-Saffran, E., and Kinne, R. K. H. (2003). Cell volume regulation: Osmolytes, osmolyte transport, and signal transduction. *Rev. Physiol. Biochem. Pharmacol.* **148**, 1–80.

Wu, Y. N., Yang, Y. C., and Wagner, P. D. (1992). Modification of chromaffin cells with pertussis toxin or N-ethylmaleimide lowers cytoskeletal F-actin and enhances Ca(2+)-dependent secretion. *J. Biol. Chem.* **267**, 8396–8403.

Yamazaki, D., Kurisu, S., and Takenawa, T. (2005). Regulation of cancer cell motility through actin reorganization. *Cancer Sci.* **96**, 379–386.

CHAPTER THIRTEEN

OSMOTIC STRESS AND DNA DAMAGE

Natalia I. Dmitrieva *and* Maurice B. Burg

Contents

Abstract

Mammalian renal inner medullary cells are normally exposed to extremely high NaCl concentrations. The interstitial NaCl concentration in parts of a normal renal medulla can be 500 mM or more, depending on the species. Remarkably, under these normal conditions, the high NaCl causes DNA damage, yet the cells survive and function both in cell culture and *in vivo*. Both in cell culture and *in vivo* the breaks are repaired rapidly if the NaCl concentration is lowered. This chapter describes two methods used to detect and study the DNA damage induced by osmotic stress: comet assay or single cell electrophoresis and TUNEL assay or *in situ* labeling of 3′-OH ends of DNA strands. This chapter also discusses how specifics of the protocols influence the conclusions about types of DNA damage and what the limitations of these methods are for detecting different types of DNA damage.

Laboratory of Kidney and Electrolyte Metabolism, National Heart, Lung, and Blood Institute, National Institutes of Health, Bethesda, Maryland

Methods in Enzymology, Volume 428
ISSN 0076-6879, DOI: 10.1016/S0076-6879(07)28013-9

1. INTRODUCTION

Studies have shown that osmotic stress induces DNA damage. The first indication of genotoxicity from high NaCl was that there is a transient cell cycle arrest after acute elevation of NaCl in cell culture (Kultz *et al.*, 1998). This arrest is very similar to that induced by the classical genotoxic agents that induce DNA breaks, such as ionizing and ultraviolet (UV) radiation. In the case of ionizing or UV irradiation, cell cycle arrest occurs within minutes, coincident with activation of DNA repair. The cell cycle begins again only after DNA repair is complete. If the DNA damage is so severe that it cannot be repaired, apoptosis occurs. The response of cells to high NaCl is remarkably similar, namely the high NaCl-induced arrest activates quickly, is maintained for several hours, and then the cells begin to proliferate again (Dmitrieva *et al.*, 2001, 2002). If NaCl is raised too much, apoptosis occurs (Santos *et al.*, 1998). Further studies revealed the actual occurrence of increased DNA breaks (Kultz and Chakravarty, 2001). Surprisingly, even after the cells adapt and reenter the cell cycle, numerous DNA breaks persist (Dmitrieva *et al.*, 2004). Even more surprising, numerous DNA breaks are present in normal cells in the mouse renal inner medulla, where NaCl is always high as part of the urinary concentrating mechanism (Dmitrieva *et al.*, 2004), as well as in cells of marine invertebrates from diverse phyla while they are exposed to seawater, which normally contains high NaCl (Dmitrieva *et al.*, 2006). To study these genotoxic effects of osmotic stress, we have used the methods generally used in studies of DNA damage, repair, and mutagenesis. This chapter describes methods used to detect and study the DNA damage induced by osmotic stress.

2. DETECTION OF DNA DAMAGE BY COMET ASSAY

2.1. Overview of the technique

The comet assay is a gel electrophoresis method used to visualize and measure DNA strand breaks in individual cells, using microscopy. In the simplest form, cells are embedded in agarose on a microscope slide, immersed in a lysis solution, and then exposed to an electric field to attract negatively charged fragments of DNA toward the anode. After electrophoresis, the DNA is stained using a fluorescent dye and viewed using a fluorescence microscope. Broken DNA migrates out of the nucleus in the electric field, and the preparation then resembles a "comet" with a brightly fluorescent head (the nucleus) and a "tail" (fragmented DNA). The relative size of the tail is an index of DNA damage. The comet assay can measure

DNA single-strand breaks, DNA double-strand breaks, damage to DNA bases, DNA interstrand cross-links, and apoptotic fragments. Reviews and protocols for the comet assay have been published elsewhere (Collins, 2002; Olive, 2002). This chapter describes modifications of the comet assay used to detect DNA damage induced by osmotic stress. Also, this chapter discusses how specifics of the protocols influence the conclusions about types of DNA damage and what the limitations of this method are for detecting different types of DNA damage. The methods used are mostly based on a commercially available comet assay kit from Trevigen Inc. (Gaithersburg, MD).

2.2. Alkaline comet assay

2.2.1. Preparation of cells for comet assay

Cell growth and exposure to osmotic stress   Because the comet assay requires very few cells, any size dish can be used from which cells can conveniently be scraped. On the day of the analysis it is critical to avoid exposure to ultraviolet (UV) irradiation, as this in itself can introduce DNA damage (Bradley *et al.*, 1978). We routinely handle cells with the room lights off, and we cover the preparations with aluminum foil during all subsequent manipulations. To induce osmotic stress, we replace the culture medium with one of the desired osmolality. We find it best to preequilibrate the replacement medium in an incubator for several hours in order to avoid changes in temperature and pH.

Single cell preparations from adherent cultures   Wash cells with ice-cold, Ca^{2+}- and Mg^{2+}-free phosphate-buffered saline (PBS), scrape them from the plate, and transfer them in that PBS in centrifuge tubes kept on ice in order to minimize endonuclease activity. Scraping the cells from the plate may disrupt some plasma membranes, but that is unimportant because the comet assay requires only that nuclei are initially intact.

Single cell preparation from kidney cortex and inner medulla   Place a small piece of cortex or inner medulla on a glass slide on ice, and disrupt the tissue in a small amount of the ice-cold, Ca^{2+}- and Mg^{2+}-free PBS using scalpels and needles. Keep the cell suspensions ice cold during subsequent manipulations.

2.2.2. Mixing the cells with agarose

Prepare 1% low melting point agarose in PBS in advance by heating it in a 90 to 100° water bath until the agarose is molten. Cool the agarose to 42°. It will remain molten at 42° indefinitely. Mix 1 part of cells with 10 parts of molten agarose (at 42°), and immediately pipette 75 μl of the mixture onto a pre-coated slide. Precoated slides can be purchased (Trevigen Inc.) or prepared by the investigator using a published protocol (Olive, 2002; Singh *et al.*, 1988). Keep the slides horizontal at 4° for 5 to 10 min to allow the agarose to gel. Because the DNA is not fixed, and fluorescent lights in combination with alkali treatment can create DNA breaks (Bradley *et al.*, 1978), continue to protect cells from light until the electrophoresis is complete.

2.2.3. Cell lysis and DNA denaturation

After the agarose has gelled, submerse each slide horizontally in prechilled lysis solution (2.5 *M* sodium chloride, 100 m*M* EDTA, pH 10, 10 m*M* Tris base, 1% sodium lauryl sarcosinate, 1% Triton X-100) and leave this preparation on ice or at 4° for 30 to 60 min. It is important that the duration of lysis be consistent. Then, tap off excess buffer and submerse the slides in freshly prepared alkaline solution (0.3 *M* NaOH, 1 m*M* EDTA, pH > 13) for 1 h at room temperature.

2.2.4. Electrophoresis and DNA staining

Transfer slides from the alkaline solution to a horizontal electrophoresis apparatus setup at 4°. Fill the electrophoresis tank with the alkaline solution until it just covers the samples. Set the voltage at about 1 V/cm. Add or remove buffer until the current is approximately 300 mA and electrophorese for 15 to 60 min. The optimal voltage and duration of electrophoresis depend on the particular type of cell and particular experiment. We routinely test durations of 15, 30, and 45 min when we start working with a new cell type or a new experimental condition and choose the time that provides the best image. After electrophoresis, tap off any excess solution and then immerse each slide in 70% ethanol for 5 min. Air dry the samples. They can be stored in this condition at room temperature until they are examined. Stain the DNA by covering the dried agarose with SYBR Green DNA. Stain for 15 min. SYBR Green is a component of the comet assay kit from Trevigen Inc., but also is available commercially elsewhere. The DNA can also be stained with propidium iodide (2.5 µg/ml for 15 to 30 min).

2.2.5. Comet analysis

Photograph images of the comets using a fluorescence microscope. It is essential to not saturate any part of the image. Although it is tempting to increase the camera gain or integration time in order to visualize the comet tail better, if that results in saturating the head, calculation of the percentage of DNA in the tail will be inaccurate. Important features of the analysis include total image intensity (DNA content), comet tail length, head diameter/tail length, percentage of DNA in the comet tail, and tail moment (Olive *et al.*, 1990). We prefer to calculate the percentage of total cellular DNA that is in the comet tail. We use Scion Image for Windows. This software is available free at http://www.scioncorp.com/pages/scion_image_windows.htm.

Free Macintosh software for image analysis is available at http://rsb.info.nih.gov/nih-image (Helma and Uhl, 2000).

2.3. Measurements of oxidative base damage by the comet assay

In addition to DNA strand breaks, the comet assay can also be used to measure oxidative base damage (Collins *et al.*, 1993). This entails the additional step of digesting the DNA with an enzyme that recognizes oxidized bases and breaks the DNA in their vicinity. Thus, endonuclease III is used to detect oxidized pyrimidines, and human 8-oxoguanine DNA glycosylase (hOGG1) or formamidopyrimidine DNA glycosylase is used to detect 8-oxoguanine, the major product of purine oxidation (Collins *et al.*, 1993, 1996). Cleaving the DNA at oxidized bases increases the number of DNA breaks and, consequently, the intensity of the comet tail. Using this approach we detected oxidative DNA damage after treatment of cells with high urea (Zhang *et al.*, 2004). We used a commercially available assay kit, hOGG1 FLARE (Trevigen Inc.). The protocol for this assay is the same as for the alkaline comet assay described earlier with the additional step of treatment with the hOGG1 enzyme after cells are lysed.

2.4. Detection of DNA double-strand breaks by the comet assay

The alkaline comet assay, described earlier, which includes incubation at high pH before and during electrophoresis, is the most common variant now employed. The alkaline comet assay identifies both single- and double-strand breaks because the two strands of DNA are uncoiled at the breaks by alkaline denaturation. The idea came about that if the assay is performed in neutral buffers (pH < 10), it detects only double-strand breaks because the DNA is not unwound. However, it now seems likely that both single- and double-strand breaks are identified by the comet assay at neutral pH (Collins *et al.*, 1997; Olive *et al.*, 1991). The reasoning is as follows. Treatment of cells with detergents and/or high NaCl removes membranes, cytoplasm, and nucleoplasm and disrupts nucleosomes (histones are solubilized by high salt or ionic detergents, such as SDS). The nucleoid that is left consists of nuclear matrix and DNA that is negatively supercoiled consequent to the turns that the double helix had made around the now absent histones of the nucleosome. Cook *et al.* (1976) first recognized that DNA attached to the matrix is looped rather than linear. If the negative supercoiling is relaxed by the intercalating agent ethidium bromide, the loops expand out of the nucleoid core to form a "halo." A similar result occurs when ionizing radiation relaxes the loops, with one single-strand break being sufficient for relaxation of the supercoiling in a loop. During electrophoresis the halo of relaxed loops is pulled into the comet tail. The number of relaxed loops in the tail (or tail intensity) is then an index of the total number of DNA breaks, both single and double stranded. As is predicted by this model, there is no difference between neutral and alkaline comet assays in respect to lowest levels of DNA damage that they can detect (Collins *et al.*, 1997). Also the model fits the observation that, with increasing amount of damage, tail intensity rather than length increases, because tail length is determined primarily by the length of the loops (Collins *et al.*, 1997).

Following the realization that the neutral comet assay in its original form (Ostling and Johanson, 1984) does not distinguish between single- and double-strand breaks, a special modification was devised to make the distinction (Olive *et al.*, 1991). In the modified method the lysed cells are additionally treated in the agarose at 50° with proteinase K. The idea is to digest and disrupt the nuclear matrix to which DNA is attached in order to destroy the "loop" structure of the DNA and facilitate the release of double-stranded pieces of DNA. The detailed protocol for the neutral comet assay that specifically detects double-strand breaks is published elsewhere (Olive, 2002; Olive *et al.*, 1991). The general procedure is the same as for the alkaline comet assay described earlier. Differences are in the conditions of cell lysis and the composition of the electrophoresis buffer. Lysis is at 50° for 4 h in a buffer containing 30 mM EDTA and 0.5% SDS at pH 8.0. Proteinase K (0.5 mg/ml) is added if the duration of lysis is less than 4 h. After lysis, the slides are rinsed in electrophoresis buffer (TBE buffer) for 4 h or overnight and then transferred to an electrophoresis chamber filled with TBE buffer (90 mM Tris, 90 mM boric acid, 2 mM EDTA, pH 8.5).

3. DETECTION OF DNA BREAKS BY *IN SITU* END LABELING

3.1. Overview of techniques

Polymerase-based DNA end-labeling assays have been used for decades to detect DNA breaks. The basic concept is that labeled nucleotides are added to gaps or nicks in one of the DNA strands and to recessing strands at the double-strand break. A reaction mixture containing labeled nucleotides and appropriate enzymes is applied to permeabilized tissue culture cells or tissue sections. The enzymes that are used can catalyze polymerization of added nucleotides without a single-strand DNA template. A variety of methods are used to detect the labels; detailed descriptions of them are published elsewhere (Didenko, 2002). Terminal transferase, *Escherichia coli* DNA polymerase (Pol I), and the "Klenow fragment" are the polymerases used most often in the end-labeling reactions. The different modes of action of these enzymes can provide information about the nature of DNA breaks.

Pol I binds to DNA at a single-strand break ("nick") with a 3'-OH terminus. Replication starts by covalent extension of the 3'-OH terminus. The enzyme has concurrent 5'-3' nuclease and polymerase activity. A primer strand is constantly uncovered by the nuclease activity by hydrolysis at the 5' side of the nick, while the polymerase activity catalyses the addition of nucleotides to the 3' end. The net result of simultaneous hydrolysis and synthesis is translation of the nick along the DNA duplex in the 5'-3' direction, incorporating labeled nucleotides (Kelly *et al.*, 1970; Manoharan *et al.*, 1987; Nose and Okamoto, 1983; Snyder and Matheson, 1985). Cleavage of Pol I by subtilisin creates fragments of 76 and 34 kDa. The 76-kDa fragment has polymerizing but not 5'→3' nuclease activity. Following its discovery by Klenow and Henningsen (1970), the large fragment has been called the "Klenow fragment." In order to polymerize the addition of nucleotides to the 3'-OH end of DNA, the Klenow fragment requires a single-stranded DNA template and a DNA or RNA primer with a 3'-OH terminus. The ability of the Klenow fragment to label DNA with a 5' overhang is used to detect DNA breaks *in situ* (reviewed in Wood, 2002). Terminal deoxynucleotidyl transferase (TdT) is another enzyme often used for DNA end labeling (reviewed in Loo, 2002; Walker *et al.*, 2002). TdT catalyzes the addition of deoxyribonucleotides to the 3'-OH end of DNA strands without any template or primer. This contrasts with most other enzymes that incorporate nucleotides into duplex DNA, as the others generally require a string of nucleotides on the opposite strand, providing a template that directs the type of nucleotide to be added. Other enzymes require a mixture of all four nucleotides, but TdT can create a homopolymer containing only one type of nucleotide. Typically, deoxyuridine triphosphate (dUTP) is used. TdT labels a variety of 3'-OH DNA ends, including double-stranded ends that are blunt, double strands that are protruding or recessed with a 3'-hydroxyl single-stranded end, and single-strand nicks. The DNA end-labeling technique that utilizes TdT is called TdT-mediated dUTP-biotin nick end labeling (TUNEL). We will now describe the modification of the TUNEL method used to detect DNA breaks associated with high NaCl in the mouse renal inner medulla (Dmitrieva *et al.*, 2004), in *Caenorhabditis elegans* (Dmitrieva *et al.*, 2005), and in marine invertebrates (Dmitrieva *et al.*, 2006).

3.2. Labeling of DNA breaks in paraffin-embedded tissue sections by the TUNEL assay

Several kits for labeling DNA strand breaks are available commercially. They all use terminal transferase to label 3'-OH DNA ends (TUNEL staining) and are advertised as kits for detecting apoptotic cell death. However, as explained earlier, they can be used for detection of any type of DNA damage. In our experiments we used the TdT-FragEL DNA fragmentation detection kit (EMD Biosciences, San Diego, CA). This kit uses BrdUTP to label 3′-OH termini of DNA breaks. After labeling, the BrdUTP is detected by immunocytochemistry, using the anti–BrdU antibody.

1. Fixed samples embedded in paraffin must be deparaffinized and hydrated prior to staining. To deparaffinize sections, bathe slides twice for 5 min in xylene. Then, hydrate the sections by bathing the slides in serial dilutions of ethanol (5 min each: twice in 100% ethanol, then once each in 90% ethanol, 80% ethanol, and 70% ethanol). Finally, bathe the slides three times for 5 min in changes of PBS.

2. Permeabilize the sections for 15 min in proteinase K solution (20 μg/ml) in a humid chamber at room temperature. Then, bathe the slides three times for 5 min in changes of PBS.

3. Inactivate endogenous peroxidases by covering the slides with 3% hydrogen peroxide in PBS for 10 min at room temperature. Then, bathe the slides three times for 5 min in changes of PBS.

4. Cover sections on the slides with TdT equilibration buffer (25 mM Tris-HCl, pH 6.6; 200 mM potassium cacodylate; 5 mM CoCl$_2$; 0.25 mg/ml bovine serum albumin [BSA]) for 10 to 30 min at room temperature.

5. Remove the TdT equilibration buffer and cover the sections with TdT reaction buffer (TdT equilibration buffer containing 0.5 U/μl of TdT enzyme and 80 μM of BrdUTP). TdT enzyme and 5× reaction buffer can be purchased from Roche Applied Science (Indianapolis, IN). Incubate the slides in a humidified chamber for 1.5 h at 37°. To prevent evaporation, cover the specimen with a piece of Parafilm.

6. Bathe the slides for 5 min three times in changes of PBS and then block with 2% BSA in PBS solution for 30 min at room temperature.

7. Cover sections with biotin–conjugated anti–BrdU monoclonal antibody, diluted in PBS containing 2% BSA and 0.1% Triton X-100. Incubate for 1 h. Bathe the slides for 5 min three times in changes of PBS.

8. Cover sections for 1 h with peroxidase-conjugated strepavidin in PBS that contains 2% BSA and 0.1% Triton X-100. Bathe the slides for 5 min three times in changes of PBS.

9. Stain the sections with 3,3′-diaminobenzidine tetrahydrochloride and watch with a microscope until a brown color appears (2 to 15 min). Rinse slides with water.

10. Dehydrate the slides with two washes of 100% ethanol and then with two washes of xylene. Mount and examine the specimens using bright-field optics.

Note: A kit for BrdU immunostaining is available from Zymed and can be used to perform steps 7 to 10 of the protocol just given.

3.3. Apoptotic vs nonapoptotic DNA breaks: technical considerations for increasing sensitivity to detect nonapoptotic DNA breaks

The TUNEL assay is widely used to detect apoptosis, implying a specificity for apoptosis that does not exist. Many companies provide "apoptosis detection" kits that label DNA breaks using the TdT enzyme. However, detection of DNA breaks with these kits is not a specific indicator of apoptosis. The idea of using polymerases to label radiation-induced DNA strand breaks originated in the early 1970s (Modak and Price, 1971), predating the idea of apoptosis. The technique has been widely used to measure DNA damage from genotoxic agents and the repair thereof (Manoharan et al., 1987; Nose and Okamoto, 1983; Snyder and Matheson, 1985). After the discovery of apoptosis, a variety of the existing DNA end-labeling assay was used to detect it, and the procedure was renamed the TUNEL assay (Gavrieli et al., 1992). During the late stages of apoptosis, nucleases are activated that cleave DNA. This produces double-strand breaks that are labeled in reactions catalyzed by the TdT enzyme. Since the initial publication of the TUNEL method, it has been emphasized that the assay is not specific for apoptosis because DNA fragments with 3'-OH ends can be produced without apoptosis (Charriaut-Marlangue and Ben-Ari, 1995). Because the method labels any free 3'-OH end in DNA, other forms of damage to DNA, apart from apoptosis, produce positive signals. For example, the TUNEL assay detects DNA breaks associated with necrotic cell death (Ansari et al., 1993; Nishiyama et al., 1996). Also, it detects active DNA repair (Kanoh et al., 1999). Therefore, TUNEL staining is a general method for the detecting of DNA breaks, one of several in situ DNA end-labeling techniques. In order to establish whether the breaks are due to apoptosis, additional criteria should be used, such as morphological examination or detection of activated caspase-3.

Although the TdT enzyme does not distinguish between apoptotic- and nonapoptotic-induced DNA 3'-OH ends, the number of DNA strand breaks produced by apoptosis is so high that the intensity of TUNEL labeling produced by apoptosis exceeds at least by one and often two orders of magnitude the labeling of DNA breaks induced by DNA-damaging agents in the absence of apoptosis. That means that nonapoptotic DNA breaks may be overlooked in TUNEL assays calibrated to measure apoptosis. However, several technical modifications can increase its sensitivity enough to detect a relatively small number of DNA breaks in nonapoptotic cells. Many factors influence the intensity of labeling of DNA breaks (reviewed in van Dierendonck, 2002). The main point is accessibility of the DNA ends to the labeling reagents. This is impaired by proteins surrounding the DNA, especially when there is fixative-induced cross-linking. Accessibility is improved by treating the tissue sections with proteases, such as proteinase K. However, because excess protein digestion may result in DNA extraction, the reaction must be calibrated (Migheli *et al.*, 1994). Another factor that can influence the accessibility of reagents to DNA ends is the duration of fixation. Prolonged fixation can lead to irreversible cross-linking between different DNA strands and between DNA and proteins, making the DNA ends inaccessible (Davison *et al.*, 1995). Heating can be used to break formaldehyde-induced cross-links and thus enhance detection of DNA breaks. Lucassen *et al.* (1995) reported that suppression of end labeling by excessive fixation can be reversed by microwave heating (up to 90 to 100°) of fixed tissue sections in citrate buffer. The best results occur at low pH (pH 3). However, the same method applied to lightly fixed material results in strong nonspecific labeling (Lucassen *et al.*, 1995). Another factor that influences the sensitivity of DNA end labeling is the type of nucleotides used in the labeling reaction. The intensity of labeling detected by BrdUTP immunochemistry is nearly four times that obtained using biotin-conjugated dUTP, twice that using digoxygenin-conjugated dUTP, and over eight times that using direct labeling with fluorochrome (fluorescein or BODIPY)-conjugated deoxynucleotides (Li and Darzynkiewicz, 1995). The greater labeling with BrdUTP may reflect more efficient incorporation of this nucleotide by terminal transferase because of its smaller size than nucleotides with bulky fluorochrome conjugates. Thus, many factors can influence sensitivity of the TUNEL assay and can create both false-positive and false-negative results. Optimization by manipulating these factors allows specific detection of a relatively low, but still important, number of DNA breaks caused by genotoxic agents.

REFERENCES

Ansari, B., Coates, P. J., Greenstein, B. D., and Hall, P. A. (1993). *In situ* end-labeling detects DNA strand breaks in apoptosis and other physiological and pathological states. *J. Pathol.* **170,** 1–8.

Bradley, M. O., Erickson, L. C., and Kohn, K. W. (1978). Non-enzymatic DNA strand breaks induced in mammalian cells by fluorescent light. *Biochim. Biophys. Acta* **520,** 11–20.

Charriaut-Marlangue, C., and Ben-Ari, Y. (1995). A cautionary note on the use of the TUNEL stain to determine apoptosis. *Neuroreport* **7,** 61–64.

Collins, A. R. (2002). The comet assay: Principles, applications, and limitations. *In* "In Situ Detection of DNA Damage: Methods and Protocols" (V. V. Didenko, ed.), Vol. 203, pp. 163–177. Humana Press, Totowa, NJ.

Collins, A. R., Dobson, V. L., Dusinska, M., Kennedy, G., and Stetina, R. (1997). The comet assay: What can it really tell us? *Mutat. Res.* **375,** 183–193.

Collins, A. R., Dusinska, M., Gedik, C. M., and Stetina, R. (1996). Oxidative damage to DNA: Do we have a reliable biomarker? *Environ. Health Perspect.* **104**(Suppl. 3), 465–469.

Collins, A. R., Duthie, S. J., and Dobson, V. L. (1993). Direct enzymic detection of endogenous oxidative base damage in human lymphocyte DNA. *Carcinogenesis* **14,** 1733–1735.

Cook, P. R., Brazell, I. A., and Jost, E. (1976). Characterization of nuclear structures containing superhelical DNA. *J. Cell Sci.* **22,** 303–324.

Davison, F. D., Groves, M., and Scaravilli, F. (1995). The effects of formalin fixation on the detection of apoptosis in human brain by *in situ* end-labelling of DNA. *Histochem. J.* **27,** 983–988.

Didenko, V. V. (2002). "*In Situ* Detection of DNA damage: Methods and Protocols." Humana Press, Totowa, NJ.

Dmitrieva, N., Michea, L., and Burg, M. (2001). p53 Protects renal inner medullary cells from hypertonic stress by restricting DNA replication. *Am. J. Physiol. Renal Physiol.* **281,** F522–F530.

Dmitrieva, N. I., Bulavin, D. V., Fornace, A. J., Jr., and Burg, M. B. (2002). Rapid activation of G2/M checkpoint after hypertonic stress in renal inner medullary epithelial (IME) cells is protective and requires p38 kinase. *Proc. Natl. Acad. Sci. USA* **99,** 184–189.

Dmitrieva, N. I., Cai, Q., and Burg, M. B. (2004). Cells Adapted to high NaCl have many DNA breaks and impaired DNA repair both in cell culture and *in vivo. Proc. Natl. Acad. Sci. USA* **101,** 2317–2322.

Dmitrieva, N. I., Celeste, A., Nussenzweig, A., and Burg, M. B. (2005). Ku86 preserves chromatin integrity in cells adapted to high NaCl. *Proc. Natl. Acad. Sci. USA* **102,** 10730–10735.

Dmitrieva, N. I., Ferraris, J. D., Norenburg, J. L., and Burg, M. B. (2006). The saltiness of the sea breaks DNA in marine invertebrates: Possible implications for animal evolution. *Cell Cycle* **5,** 1320–1323.

Gavrieli, Y., Sherman, Y., and Ben-Sasson, S. A. (1992). Identification of programmed cell death *in situ* via specific labeling of nuclear DNA fragmentation. *J. Cell Biol.* **119,** 493–501.

Helma, C., and Uhl, M. (2000). A public domain image-analysis program for the single-cell gel-electrophoresis (comet) assay. *Mutat. Res.* **466,** 9–15.

Kanoh, M., Takemura, G., Misao, J., Hayakawa, Y., Aoyama, T., Nishigaki, K., Noda, T., Fujiwara, T., Fukuda, K., Minatoguchi, S., and Fujiwara, H. (1999). Significance of myocytes with positive DNA *in situ* nick end-labeling (TUNEL) in hearts with dilated cardiomyopathy: Not apoptosis but DNA repair. *Circulation* **99,** 2757–2764.

Kelly, R. B., Cozzarelli, N. R., Deutscher, M. P., Lehman, I. R., and Kornberg, A. (1970). Enzymatic synthesis of deoxyribonucleic acid. XXXII. Replication of duplex deoxyribonucleic acid by polymerase at a single strand break. *J. Biol. Chem.* **245,** 39–45.

Klenow, H., and Henningsen, I. (1970). Selective elimination of the exonuclease activity of the deoxyribonucleic acid polymerase from *Escherichia coli* B by limited proteolysis. *Proc. Natl. Acad. Sci. USA* **65,** 168–175.

Kultz, D., and Chakravarty, D. (2001). Hyperosmolality in the form of elevated NaCl but not urea causes DNA damage in murine kidney cells. *Proc. Natl. Acad. Sci. USA* **98,** 1999–2004.

Kultz, D., Madhany, S., and Burg, M. B. (1998). Hyperosmolality causes growth arrest of murine kidney cells: Induction of GADD45 and GADD153 by osmosensing via stress-activated protein kinase 2. *J. Biol. Chem.* **273,** 13645–13651.

Li, X., and Darzynkiewicz, Z. (1995). Labelling DNA strand breaks with BrdUTP: Detection of apoptosis and cell proliferation. *Cell Prolif.* **28,** 571–579.

Loo, D. T. (2002). TUNEL assay: An overview of techniques. *In* "*In Situ* Detection of DNA Damage: Methods and Protocols" (V. V. Didenko, ed.), Vol. 203, pp. 21–30. Humana Press, Totowa, NJ.

Lucassen, P. J., Chung, W. C., Vermeulen, J. P., Van Lookeren, C. M., van Dierendonck, J. H., and Swaab, D. F. (1995). Microwave-enhanced *in situ* end-labeling of fragmented DNA: Parametric studies in relation to postmortem delay and fixation of rat and human brain. *J. Histochem. Cytochem.* **43,** 1163–1171.

Manoharan, K., Kinder, D., and Banerjee, M. R. (1987). DMBA induced DNA damage and repair in mammary epithelial cells *in vitro* measured by a nick translation assay. *Cancer Biochem. Biophys.* **9**, 127–132.

Migheli, A., Cavalla, P., Marino, S., and Schiffer, D. (1994). A study of apoptosis in normal and pathologic nervous tissue after *in situ* end-labeling of DNA strand breaks. *J. Neuropathol. Exp. Neurol.* **53**, 606–616.

Modak, S. P., and Price, G. B. (1971). Exogenous DNA polymerase-catalysed incorporation of deoxyribonucleotide monophosphates in nuclei of fixed mouse-brain cells. *Exp. Cell Res.* **65**, 289–296.

Nishiyama, K., Kwak, S., Takekoshi, S., Watanabe, K., and Kanazawa, I. (1996). *In situ* nick end-labeling detects necrosis of hippocampal pyramidal cells induced by kainic acid. *Neurosci. Lett.* **212**, 139–142.

Nose, K., and Okamoto, H. (1983). Detection of carcinogen-induced DNA breaks by nick translation in permeable cells. *Biochem. Biophys. Res. Commun.* **111**, 383–389.

Olive, P. L. (2002). The comet assay: An overview of techniques. In "*In Situ* Detection of DNA Damage: Methods and Protocols" (V. V. Didenko, ed.), Vol. 203, pp. 179–194. Humana Press, Totowa, NJ.

Olive, P. L., Banath, J. P., and Durand, R. E. (1990). Heterogeneity in radiation-induced DNA damage and repair in tumor and normal cells measured using the "comet" assay. *Radiat. Res.* **122**, 86–94.

Olive, P. L., Wlodek, D., and Banath, J. P. (1991). DNA double-strand breaks measured in individual cells subjected to gel electrophoresis. *Cancer Res.* **51**, 4671–4676.

Ostling, O., and Johanson, K. J. (1984). Microelectrophoretic study of radiation-induced DNA damages in individual mammalian cells. *Biochem. Biophys. Res. Commun.* **123**, 291–298.

Santos, B. C., Chevaile, A., Hebert, M. J., Zagajeski, J., and Gullans, S. R. (1998). A combination of NaCl and urea enhances survival of IMCD cells to hyperosmolality. *Am. J. Physiol.* **274**, F1167–F1173.

Singh, N. P., McCoy, M. T., Tice, R. R., and Schneider, E. L. (1988). A simple technique for quantitation of low levels of DNA damage in individual cells. *Exp. Cell Res.* **175**, 184–191.

Snyder, R. D., and Matheson, D. W. (1985). Nick translation: A new assay for monitoring DNA damage and repair in cultured human fibroblasts. *Environ. Mutagen.* **7**, 267–279.

van Dierendonck, J. H. (2002). DNA damage detection using DNA polymerase I or its Klenov fragment. In "*In Situ* Detection of DNA Damage: Methods and Protocols" (V. V. Didenko, ed.), Vol. 203, pp. 81–108. Humana Press, Totowa, NJ.

Walker, P. R., Carson, C., Leblanc, J., and Sikorska, M. (2002). Labeling DNA damage with terminal transferase: Applicability, specificity, and limitations. In "*In Situ* Detection of DNA Damage: Methods and Protocols" (V. V. Didenko, ed.), Vol. 203, pp. 3–19. Humana Press, Totowa, NJ.

Wood, K. A. (2002). Labeling DNA breaks *in situ* by Klenow enzyme. In "*In Situ* Detection of DNA Damage: Methods and Protocols" (V. V. Didenko, ed.), Vol. 203, pp. 109–119. Humana Press, Totowa, NJ.

Zhang, Z., Dmitrieva, N. I., Park, J. H., Levine, R. L., and Burg, M. B. (2004). High urea and NaCl carbonylate proteins in renal cells in culture and *in vivo*, and high urea causes 8-oxoguanine lesions in their DNA. *Proc. Natl. Acad. Sci. USA* **101**, 9491–9496.

CHAPTER FOURTEEN

Transcriptional Activator TonE-Binding Protein in Cellular Protection and Differentiation

Jeong Ah Kim, Un Sil Jeon, Min Seong Kwon, Sun Woo Lim, *and* H. Moo Kwon

Contents

Abstract

The TonE-binding protein (TonEBP) is a transcriptional activator in the Rel family that includes NFκB and NFAT. TonEBP is critical for the development and function of the renal medulla, which is a major regulator of water homeostasis. TonEBP is also implicated in diabetic nephropathy and inflammation. Established methods for biochemical and histochemical detection and functional

Department of Medicine, University of Maryland, Baltimore, Maryland

Methods in Enzymology, Volume 428
ISSN 0076-6879, DOI: 10.1016/S0076-6879(07)28014-0

analysis of TonEBP, including identification of novel TonEBP target genes, are described for those who are interested in investigating function and regulation of TonEBP.

1. INTRODUCTION

The TonE-binding protein (TonEBP) was cloned (Miyakawa *et al.*, 1999) as the DNA-binding protein for TonE (tonicity-responsive enhancer), a DNA sequence of 11 bp that mediates the stimulation of transcription in response to hypertonicity (Miyakawa *et al.*, 1998). Hypertonicity is defined as an increase in effective osmolality. The DNA-binding domain of TonEBP shares structural homology with the Rel family of proteins, NFκB and NFAT (Stroud *et al.*, 2002). The Rel proteins are present in vertebrates and invertebrates, but not in lower animals or plants (Graef *et al.*, 2001). Like other Rel proteins, TonEBP is a transcriptional stimulator.

Although TonEBP is expressed in many tissues during development and in adult animals (Maouyo *et al.*, 2002; Trama *et al.*, 2000), its function is best defined in the renal medulla related to osmotic regulation. Several functions of TonEBP have been defined in the renal medulla. First, TonEBP protects renal medullary cells from the deleterious effects of hypertonicity, which exists in the form of hyperosmotic salt concentration (reviewed in Jeon *et al.*, 2006). This is achieved by enhancing the expression of plasma membrane transporters and biosynthetic enzymes for organic osmolytes: sodium/*myo*-inositol cotransporter (SMIT1), sodium/chloride/betaine cotransporter (BGT1), sodium/chloride/taurine cotransporter (TauT), aldose reductase (AR, for the production of sorbitol), and a recently identified phospholipase named NTE for the production of glycerophosphorylcholine. Second, TonEBP directly stimulates transcription of heat shock protein 70 (HSP70) in a manner independent of heat (Woo *et al.*, 2002). HSP70 protects renal cells from the harmful effects of high urea concentration in the medulla (Neuhofer *et al.*, 2001). Third, TonEBP contributes to the urinary concentration independent of vasopressin by directly stimulating transcription of aquaporin 2 (Hasler *et al.*, 2006) and UT-A urea transporters (Nakayama *et al.*, 2000). Fourth, TonEBP is involved in postnatal development of the renal medulla (Lee *et al.*, 2007). During development of the kidney, expression of TonEBP temporally follows expression of the sodium/potassium/chloride cotransporter type 2 that drives medullary hypertonicity (Han *et al.*, 2004; Lee *et al.*, 2007). Thus, hypertonicity generated by active sodium transport is an important local signal for development of the renal medulla, for which TonEBP is a major mediator. Genetically modified mice with a deficiency in TonEBP, such as TonEBP-/- mice or transgenic mice expressing a dominant–negative form of TonEBP, display severe hypotrophy in the renal medulla because of the failure of cells to adapt to hyperosmolality and life-threatening volume depletion because of a defective urinary concentrating ability (Lam *et al.*, 2004; Lopez-Rodriguez *et al.*, 2004).

In cultured cells, TonEBP is regulated by ambient tonicity in a bidirectional manner: inhibited by hypotonicity and stimulated by hypertonicity (Woo *et al.*, 2000a). Regulation of TonEBP involves at least three pathways: nucleocytoplasmic trafficking (Dahl *et al.*, 2001; Woo *et al.*, 2000b), transactivation (Lee *et al.*, 2003), and abundance (Woo *et al.*, 2000a). Although phosphorylation of TonEBP is increased dramatically in response to hypertonicity (Dahl *et al.*, 2001), the role of phosphorylation has not been defined in terms of amino acid residues of TonEBP involved and identity of kinases (reviewed in Jeon *et al.*, 2006).

In the brain, TonEBP is expressed exclusively in neurons (Loyher *et al.*, 2004; Maallem *et al.*, 2006b). Hypernatremia causes an immediate increase in the nuclear TonEBP and a subsequent increase in AR mRNA, indicating that TonEBP controls cellular accumulation of organic osmolytes in the brain (Maallem *et al.*, 2006a). In the thymus, TonEBP is expressed abundantly (Trama *et al.*, 2000). Studies using genetically modified mice demonstrate that TonEBP is involved in the proliferation of T cells and immune response (Go *et al.*, 2004; Trama *et al.*, 2002). It has been demonstrated that increased activity of TonEBP is associated with the development of nephropathy among type 1 diabetics (Yang *et al.*, 2006). These studies illustrate the importance of TonEBP in osmotic regulation involving the kidney and brain, as well as disease processes such as glomerular sclerosis and inflammation.

This chapter describes laboratory methods for detection, assays for activity, and biochemical analysis of TonEBP. Because much is to be learned about the function of TonEBP, this chapter also discusses how to identify genes regulated by TonEBP.

2. IMMUNOLOGICAL DETECTION OF TonEBP

Like many other transcription factors, TonEBP is in low abundance in cultured cells and tissues. This is probably why many of the commercially available antibodies do not yield positive results. We find that rabbit polyclonal antibodies raised against the N-terminal 472 amino acids (Miyakawa *et al.*, 1999) and the C-terminal 300 amino acids (unpublished) work well in many procedures. It should be noted that TonEBP is extremely prone to proteolysis, probably because much of the molecule is unstructured or disordered. Extra care should be used to minimize proteolytic degradation, especially for quantitative analysis.

2.1. Immunoblotting

Cultured cells are washed three times in cold phosphate-buffered saline (PBS) and then lysed with 1 ml Triton X-100 extraction buffer (10 mM Tris, pH 7.4, 150 mM NaCl, 1 mM EDTA, 1 mM EGTA, 1% Triton X-100, 2 mM phenylmethylsulfonyl fluoride [PMSF], 20 μg/ml aprotinin, 12.5 μg/ml pepstatin, and 12.5 μg/ml leupeptin) for 1 h at 4°. Cell lysates are centrifuged for 10 min at 12,000g, and protein concentrations of super-natant are determined using a bicinchoninic acid (BCA)-based assay kit (Pierce, Rockford, IL). Cell lysates are diluted 1:1 with 2× SDS sample buffer (5% glycerol, 100 mM dithiothreitol, 2% SDS, 0.01% bromphenol blue, and 125 mM Tris, pH 6.8) and subjected to electrophoresis on 8% SDS–polyacrylamide gels. The separated proteins are transferred to a poly-vinylidene fluoride membrane (Fisher Immobilon-P IPVH00010) using a Bio-Rad Mini Protean II apparatus. Membranes are blocked for 1 h at room temperature with Tris-buffered saline (TBS; 150 mM NaCl and 50 mM Tris, pH 7.5) containing 0.1% Tween 20 (TBST) and 5% dry milk. Blots are incubated for 1 h at room temperature with anti–TonEBP antibodies diluted 1:2500 in TBST and washed three times in TBST for 10 min. Blots are then incubated an additional hour with horseradish peroxidase-conjugated antirabbit immunoglobulin diluted 1:5000. Proteins are detected using chemiluminescence using a commercial kit (Amersham).

For tissues such as kidney and liver, the organs should be chilled immediately by perfusion with or immersion in ice-cold PBS. Tissues are then homogenized promptly in hot tissue lysis buffer (1% SDS, 1 mM NaVO$_4$, and 10 mM Tris-Cl, pH 7.5) at full speed using a Polytron (Brinkmann) or a blender. For kidney samples, we use 4 ml of tissue lysis buffer for 0.12 g of tissue. The homogenate is mixed with 1 ml of 5× SDS sample buffer and boiled. Samples are processed for immunoblotting as described earlier or stored frozen with minimal thawing.

2.2. Single immunoprecipitation

This is useful for detecting posttranslational modifications such as phos-phorylation. Phosphorylation can be measured by immunoblotting the immunoprecipitate using antiphosphorylated serine, threonine, or tyrosin antibody or autoradiography when TonEBP is immunoprecipitated from cells labeled metabolically with [^{32}P]phosphate (Dahl *et al.*, 2001; Lee *et al.*, 2003).

Cells grown on a 100-mm dish are washed with ice-cold PBS and lysed with 0.1 ml of SDS extraction buffer (1% SDS, 50 mM Tris, pH 7.4, 5 mM EDTA, 2 mM PMSF, 20 μg/ml aprotinin, 12.5 μg/ml pepstatin, and 12.5 μg/ml leupeptin) at 4°. The lysate is boiled for 10 min and mixed with 0.9 ml Triton X-100 extraction buffer (10 mM Tris, pH 7.4, 150 mM NaCl, 1 mM EDTA, 1 mM EGTA, 1% Trition X-100, 2 mM PMSF, 20 μg/ml aprotinin, 12.5 μg/ml pepstatin, and 12.5 μg/ml leupeptin). The mixture is passed 10 times through a 25-gauge needle attached to a 1-ml syringe and then incubated on ice for 10 min. Protein concentrations of lysate are determined using a BCA-based assay kit (Pierce). For immunoprecipitation, 2 mg of protein is incubated overnight at 4° with 10 μl anti-TonEBP serum followed by incubation with 30 μl of 50% protein A–Sepharose 4B beads (Sigma) for 4 h at 4° with constant agitation. Immune complexes are centrifuged for 1 min at 3000 rpm and are then washed three times in Triton X-100 extraction buffer. Immune complexes are treated with 30 μl 2× SDS sample buffer after washing.

2.3. Coimmunoprecipitation

This is useful to identify proteins associated with TonEBP (Colla et al., 2006) or to detect the TonEBP dimer (Lee et al., 2002). Cells grown on a 100-mm dish are washed and then lysed with 1 ml Triton X-100 extraction buffer for 1 h at 4°. Cell lysates are centrifuged for 10 min at 12,000g, and protein concentrations of supernatant are determined. Lysates are either analyzed directly by immunoblotting or used for coimmunoprecipitation. For coimmunoprecipitation, 2 mg of total protein is processed as described earlier and analyzed with immunoblotting for associated proteins.

2.4. Immunofluorescence detection in cultured cells

Cells grown on 0.2% gelatin–coated coverslips are washed with PBS and fixed for 10 min at room temperature in 3.5% paraformaldehyde in PBS. After washing with PBS, cells are permeabilized with 0.1% Triton X-100 in PBS for 10 min and washed with PBS again. Cells are incubated for 60 min with anti-TonEBP antibody diluted 1:200. After three washes with PBS, cells are incubated for 60 min with Cy3-conjugated donkey antirabbit immunoglobulin (Jackson ImmunoResearch Laboratory; dilute 1:100) and then washed. In order to reduce background, 2 μl of Cy3-conjugated secondary antibodies is incubated with 78 μl donkey serums for 30 min at 4° and then centrifuged for 30 min at 12,000g, 4°. One hundred twenty microliters of PBS is added to the supernatant. The supernatant is now ready to use for immunofluorescence staining. Cells are mounted with FluorSave mounting regent (Calbiochem). Immunofluorescence is analyzed by standard microscopy for epifluorescence and images are captured by a digital camera.

2.5. Immunohistochemical detection of TonEBP in the kidney by preembedding technique

We have used the preembedding technique successfully for immunohistochemical analysis of TonEBP in kidneys from rat and mouse (Cha et al., 2001; Lee et al., 2007). For this, proper tissue preservation is essential.

Animals are anesthetized with ketamine and perfused with PBS via the abdominal aorta for rat or via the left ventricle of the heart for mouse to rinse out blood from the kidneys. This is followed by perfusion with periodatelysine–2% paraformaldehyde (PLP) solution for 10 min. Kidneys are removed and cut into sagittal slices of 1 to 2 mm thickness and postfixed overnight in PLP solution at 4°. Fixed slices are cut along the sagittal plane on a Vibratome1000Plus (Ted Pella Inc., Redding, CA) at a thickness of 50 μm and processed for immunohistochemistry.

The 50-μm-thick Vibratome sections are washed three times in PBS containing 50 mM NH$_4$Cl. They are then washed six times for 30 min each with PBS containing 1% BSA, 0.05% saponin, and 0.2% gelatin (solution B) on ice. Tissue sections are incubated overnight at 4° with the TonEBP antibody diluted 1:3000 to 10,000 in 1% BSA in PBS (solution A). After several washes with PBS containing 0.1% BSA, 0.05% saponin, and 0.2% gelatin (solution C), tissue sections are incubated for 10 min with PBS containing 10% methanol and 3% H$_2$O$_2$ and then rinsed several times with PBS until attached bubbles on tissue are removed. Tissue sections are incubated for 2 h in a 1:100 dilution of a peroxidase-conjugated AffiniPure F(ab') fragment donkey antirabbit IgG (Jackson ImmunoResearch Laboratories) in solution A at room temperature. Tissue sections are then rinsed, first in solution C and subsequently in 50 mM Tris-HCl, pH 7.6 (TB). For the detection of horseradish peroxidase, sections are incubated in 0.1% 3,3′-diaminobenzidine in the TB for 5 min. The reaction is stopped with a 10-min incubation in 0.01% H$_2$O$_2$. After being washed in the TB, sections are dehydrated in a graded series of ethanol and acetone for 10 min each. For embedding, tissue sections are incubated with a graded series of Epon 812 mixture (diluted acetone) for 1 h each (Polysciences, Warrington, CA) and then embedded with the pure Epon 812 mixture, from which microbubbles should be removed by vacuum. Embeddings are incubated for 3 to 4 days at 60° for 3 to 4 days for polymerization. They can be examined as 50-μm-thick sections or after 2-μm-thin sections are made.

3. FUNCTIONAL ANALYSES OF TonEBP

3.1. TonEBP-driven luciferase reporter gene

TonEBP is one of the easiest transcription factors to measure the activity of. This is because (i) cultured cells express TonEBP with few exceptions, (ii) the activity of TonEBP is robust, and (iii) stimulus is simple and easy— the addition of 50 to 100 mM NaCl to the culture medium.

A TonEBP reporter can be made using commercial plasmids such as the pGL3-promoter (Promega). This plasmid has the photinus luciferase gene under the control of the SV40 minimal promoter, which is not affected by hypertonicity. To make a TonEBP reporter, clone two or more repeats of the TonE sequence upstream of the SV promoter using the multiple cloning sites. We have made two to four repeats of the following sequence: CTTGGT-GGAAAATTACCGCTGGT (TonE is italicized) (Miyakawa et al., 1998). Higher repeats of TonE increase the tonicity responsiveness: fold increase in expression of luciferase in response to hypertonicity. The TonEBP reporter is cotransfected with a reference reporter such as pRL–CMV (Promega) where the renilla luciferase gene is driven by the cytomegalovirus promoter.

For a given cell line, amounts of plasmids for TonEBP-luciferase and pRL–CMV should be titrated. Commercial tranfection agents such as Lipofectamine 2000 (Invitrogen) or electroporation with a Gene Pulser (Bio-Rad) can be used. After 16 h of recovery, cells can be stimulated by adding 50 to 100 mM NaCl to the medium and analyzed 4 to 24 h later. We observe reliable stimulation after 4 h in COS7 and HeLa cells. We use a commercial kit (Dual-Luciferase Assay System, Promega) to measure the photinus and renilla luciferases separately. Renilla luciferase is used to estimate transfection efficiency.

Because luciferase is highly sensitive, care should be taken to titrate the amount of plasmid DNA, time of hypertonicity treatment, and the amount of lysates used for the assays of luciferases so that the assays are performed in the linear range. One should be mindful that the expression of luciferase can be dissociated from the activity of TonEBP. For example, we find that some compounds inhibit expression of the TonEBP-luciferase even though they do not affect expression of TonEBP target genes such as SMIT1 or AR.

3.2. Overexpression of TonEBP and DN-TonEBP

Overexpression of full-length TonEBP enhances the activity of TonEBP as measured by expression of a TonEBP reporter or abundance of SMIT1 mRNA (Hasler et al., 2006; Lee et al., 2003), as expected. However, TonEBP can be specifically inhibited by expression of the dominant negative form of TonEBP (DN-TonEBP) (Hasler et al., 2006; Miyakawa et al., 1999). A truncated TonEBP containing DNA-binding and dimerization domains but not transactivation domains is an effective DN-TonEBP. Full-length or DN-TonEBP is cloned in a commercial expression vector such as pcDNA3 (Invitrogen). The cDNA can be tagged in frame with FLAG, myc, or GFP. Efficiency of expression can be measured by immunoblotting or immunohistochemistry as described earlier.

3.3. RNA interference of TonEBP

Knockdown of TonEBP using small interfering RNA (siRNA) is a powerful tool for loss-of-function analysis (Na et al., 2003). siRNA for TonEBP can be purchased from commercial sources such as Ambion using the NCBI locus number (NM_006599 for human TonEBP). We found that two transfections of siRNA are required for reliable and significant knockdown of TonEBP. Control siRNA such as siCONTROL nontargeting siRNA #1 (Dharmacon, Inc., Lafayette, CO) may also be obtained. The following procedure is used for HeLa cells, but the procedure can be adapted easily for other cells.

We typically purchase 5 nmol of siRNA, which is stored frozen in a solution of 100 μM. The day before transfection, HeLa cells are seeded at 5×10^4/well in 12-well plates in Dulbecco's modified Eagle's medium with 10% serum. Transfection of siRNA is carried out using Oligofectamine (Invitrogen) with 100 nM siRNA in final culture medium. We found that knockdown of TonEBP is maximal at 100 nM. Sixteen to 24 h later, the transfection is repeated. The day after the second transfection, cells are switched to fresh medium without antibiotics and allowed to recover for 1 to 2 days. Cells can be analyzed by immunoblotting for the efficiency of knockdown. TonEBP-luciferase can be transfected at this point and analyzed as described earlier. Alternatively, cells are treated with hypertonic medium, and mRNA abundance of TonEBP target genes can be analyzed using quantitative polymerase chain reaction (PCR) or ribonuclease protection assay (RPA).

3.4. Electrophoretic mobility shift assay (EMSA)

Chromatin immunoprecipitation (ChIP) is the method of choice for directly detecting the binding of transcription factors to the chromatin *in situ*. However, ChIP is unlikely to work for TonEBP because TonEBP encircles DNA (Stroud *et al.*, 2002) and immunoprecipitated TonEBP is bound nonspecifically to DNA fragments (Colla *et al.*, 2006). However, *in vivo* footprinting has been used successfully to detect binding to the promoters *in situ* (Miyakawa *et al.*, 1998).

Because TonEBP is in low abundance in many tissues and cultured cells, EMSA is the method of choice for the quantification of TonEBP due to high sensitivity and reliability. Using this method, increased activity of TonEBP in peripheral blood mononuclear cells has been found in type 1 diabetics with nephropathy compared to those without nephropathy (Yang *et al.*, 2006). Because TonEBP has an extraordinarily large hydroscopic size, a porous gel is required for clear detection of TonEBP:DNA bands.

Nuclear extracts are prepared as follows. Cells cultured in isotonic or hypertonic medium are chilled on ice and washed with isotonic or hypertonic PBS to remove floating cells. The nuclear fraction is prepared using the NE-PER nuclear cytoplasmic extraction kit (Pierce). To prepare probes for EMSA, single-stranded oligonucleotides (sense: 5'-CTTGG*TGGAA-AATTA*CCGCTGGT-3', antisense: 5'-ACCAGC*GGTAATTTTCCACC*-AAG-3') are synthesized and purified using commercial sources. To obtain a double-stranded probe, 200 pmol of each complementary oligonucleotide is annealed in 100 μl of annealing buffer (150 mM NaCl, 10 mM MgCl$_2$, and 50 mM Tris, pH 7.9) by boiling for 5 min and slow cooling at room temperature. Annealed TonE oligonucleotides are end labeled by T4 polynucleotide kinase. Four picomoles of annealed TonE oligonucleotide is mixed with 50 μCi of [γ-^{32}P]ATP (3000 Ci/mmol, Amersham CA) and 20 U of T4 polynucleotide kinase (NEB) in 25 μl of PNK buffer supplied by the vendor. The mixture is incubated at 37° for 1 h. After the addition of 25 μl of STE (0.1 M NaCl in TE), the labeled oligolucleotide is purified by a Quick Spin Sephadex G-50 column (Roche). The purified probe is diluted with STE.

DNA-binding reactions are set up with 10 μg nuclear protein as follows: the nuclear extract is preincubated for 10 min with 1 μg of poly(dA·dT) DNA in 20 μl containing 20 mM HEPES, pH 7.9, 50 mM KCl, 5 mM MgCl$_2$, 0.1 mM EDTA, 1 mM dithiothreitol, and 5% (v/v) glycerol. After the addition of 10 fmol ^{32}P-labeled hTonE, the reaction is incubated for 20 min at room temperature. The mixture is electrophoresed for 4 h on a 4% polyacrylamide gel (79:1 = arylamide:bisacrylamide) in 45 mM Tris, 45 mM boric acid, and 1 mM EDTA with a constant voltage of 150 V. Radioactivity of TonEBP bands is visualized and quantified using a phosphoimager (Molecular Dynamics) after the gel is dried.

The specificity of the TonEBP band can be tested by specific competition with an unlabeled TonE oligonucleotide. The competing oligonucleotide is added at the preincubation step. Because the binding constant is about 10 nM (Miyakawa et al., 1998; Stroud et al., 2002), more than 50 nM is required for >90% inhibition. TonEBP antiserum can be added to cause TonEBP bands to supershift (Miyakawa et al., 1999).

4. How to Identify TonEBP Target Genes

As of this writing, nine genes are known to be stimulated transcriptionally by TonEBP: SMIT1, BGT1, TauT, AR, NTE, HSP70, AQP2, UT-A, and TNFα. Unpublished work from the authors' laboratory and others indicates that there are more genes regulated by TonEBP, that is, TonEBP target genes. Identification of all the TonEBP target genes is essential to understanding the biology of TonEBP and osmotic regulation. This section describes strategies used to identify novel TonEBP target genes.

The first step is to identify genes whose mRNA abundance is increased in response to hypertonicity. Cultured cells can be treated with hypertonic medium and induced mRNA can be searched using functional genomic analyses such as cDNA microarray analysis or targeted gene analysis using Northern analysis, RPA, or quantitative PCR. In the kidney, increased mRNA abundance in response to water deprivation is a good place to start. Likewise, in the brain, mRNA induction in response to systemic hypernatremia is an excellent clue. For those candidate genes identified in organs or tissues, it is desirable to find a cell culture model in which the candidate gene is induced by hypertonicity (an increase in effective osmolality) but not by hyperosmolality made by ineffective osmolytes such as urea or ethanol. This is because TonEBP is induced by hypertonicity but not by hyperosmolality per se. One should also be aware that switching to extreme hypertonicity, for example, from isotonic medium to 600 mOsm/kg or higher, results in cell death. However, most cells tolerate a switch to mild hypertonicity of up to 500 mOsm/kg, which can be made by the addition of 100 to 120 mM NaCl to isotonic medium. In addition, one should remember that the kinetics of mRNA induction varies: SMIT1 mRNA induces fast (within 6 h), whereas AR mRNA takes longer (more than 6 h).

Cellular evidence that TonEBP is involved in the induction of a candidate gene in response to hypertonicity can be obtained using cultured cells. In general, loss of function is more convincing. Effects of TonEBP knockdown using specific siRNA (see earlier discussion) on induction can be examined. Alternatively, cell lines derived from TonEBP-/- animals can be used. If cells can be transfected efficiently, effects of transient overexpression of TonEBP can be examined. If necessary, inhibitory effects due to expression of DN-TonEBP can be demonstrated as well. These experiments will provide a functional relationship between TonEBP and induction of a candidate gene in response to hypertonicity.

Direct action of TonEBP on the promoter of the candidate gene can be established in a variety of ways. As discussed previously, we doubt that ChIP will yield satisfactory results for TonEBP. Instead, *in vivo* footprinting analysis can be performed to detect *in situ* binding to the suspected TonEBP-binding sites in response to hypertonicity (Miyakawa *et al.*, 1998). Finally, the promoter can be analyzed using a reporter gene. All of these techniques require definitive identification of the start of the gene. Start sites for transcription as annotated by NCBI should be examined carefully. If there are solid studies on definition of the transcription start site published, one can use the information. For the vast majority of genes, this information is lacking. Sometimes the NCBI annotation is misleading. For example, in the current annotation of the human SMIT1 gene (SLC5A3) in the NCBI, the first exon is not recognized: the second exon is annotated as the only transcriptional unit. Although we reported that the first intron is ≈25 kB (Mallee *et al.*, 1997), this information is not recognized in the NCBI annotation. This illustrates that the transcription start site should be mapped to correctly identify the true promoter of a gene. We next describe primer extension analysis to map the transcription start site and construction of the promoter–reporter.

4.1. Primer extension analysis

High-quality RNA containing abundant mRNA of the target gene is essential. This can be obtained from cells cultured with the optimal hypertonic condition. One or two deoxyoligonucleotides (primers) of 25 to 40 bases complementary to the suspected first exon 50 to 100 bases downstream of the 5′ end are made. The primers are end labeled with T4 polynucleotide kinase and purified as described previously for the EMSA probe. The radiolabeled primers are annealed to 10 μg of total RNA or 1 μg of poly(A) RNA in 250 mM KCl, 2 mM EDTA, and 40 mM Tris-Cl, pH 8.0, in a volume of 20 μl. The tube is heated to 95° for 10 min and to 60° for 60 min before being placed on ice. Reverse transcription is performed using Superscript III reverse transcriptase (Invitrogen) using the supplier's instructions except that the reaction is carried out for 1 h at 50° to disrupt secondary structures in mRNA. The extended product is precipitated at −80° after the addition of 0.3 M sodium acetate, 10 mM MgCl$_2$, and 3 volumes of ethanol. The pellet will be analyzed with a standard DNA sequencing gel containing 8 M urea. A radioactive size marker can be produced by sequencing a plasmid using a commercial kit (USB).

Typically, the longest extension products form a few discrete bands. In this case, we use the longest band to estimate the site of transcription start. Because the intensity of primer extension products is quantitative, we compare RNA from isotonic and hypertonic cells and look for bands with higher intensity in hypertonicity. One can confirm the results by RPA using a probe designed from a genomic DNA sequence around the start site. Sometimes we find that the extension products are much longer than anticipated. In this case, we perform 5′ rapid amplification of cDNA ends using commercially available kits (Stratagene) to get closer to the true start sites.

4.2. Construction of promoter–reporter

The luciferase assay is a very sensitive and convenient way to examine the transcriptional activity of a gene. Typically, a promoter region containing 2 to 3 kb of a 5′ flanking sequence and part of the first exon is cloned in a commercial vector PGL3-basic vector (Promega) to obtain a photinus luciferase gene driven by the promoter of the target gene. PCR amplification of the genomic DNA is an easy way to obtain the promoter fragment. The promoter can be analyzed as described earlier for the TonEBP-driven luciferase reporter gene. A clear response to hypertonicity suggests the involvement of TonEBP. The next step is to make mutant versions of the promoter–reporter with inactive TonE. This is best achieved by mutating GG to AA in candidate TonEBP sites: TGGAAAxxYnY (Y is C or T). Loss of the response to

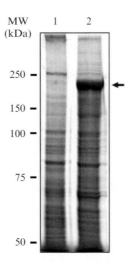

```
   1 mpsdfislls adldlespks lysresvvdl lpkelqlpps retsvasmsq tsggeagspp
  61 pavvaadass apssssmgga cssfttsssp tiystsvtds kamqvescss avqvsnrgvs
 121 ekgltsntvq ghpstpkrht vlyisppped lldnsrmscq degcgleseq scsmwmedsp
 181 snfsnmstss yndntevprk srkrnpkqrp gvkrrdcees nmdifdadsa kaphyvlsql
 241 ttdnkgnska gngtlenqkg tgvkkspmlc qqypvksegk elkivvqpet qhrarylteg
 301 srgsvkdrtq qqfptvkleg hnepvvlqvf vgndsgrvkp hqfyqacrvt grnttpckev
 361 diegttviev gldpsnnmtl avdcvgilkl rnadvearig iagskkkstr arlvfrvnim
 421 rkdgstltlq tpsspilctq pagvpeilkk slhscsvkqe eevfliqknf lkgtkvifqe
 481 nvsdenswks eaeidmelfh qnhlivkvpp yhdqhitlpv svqiyvvtna qrshdvqpft
 541 ytpdpaaaga lnvnvkkeis sparpcsfee amkamkttgc nldkvniipn almtplipss
 601 miksedvtpm evtaekrsst ifkttksvgs tqqtlenisn iagngsfssp ssshlpesene
 661 kqqqiqpkay npetlttiqt qdisqpgtfp avsassqlpn sdallqqatq fqtretqsre
 721 ilqsdgtvvn lsqlteasqq qqqsplqeqa qtlqqqissn ifpspnsvsq lqntiqqlqa
 781 gsftgstasg ssgsvdlvqq vleaqqqlss vlfsapdgne nvqeqlsadi fqqvsqiqsg
 841 vspgmfsste ptvhtrpdnl lpgraesvhp qsentlsnqq qqqqqqqqvm essaamvmem
 901 qqsicqaaaq iqselfpsta sangnlqqsp vyqqtshmms alstnedmqm qcelfsssppa
 961 vsgnetsttt tqqvatpgtt mfqtsssgdg eetgtqakqi qnsvfqtmvq mqhsgdnqpq
1021 vnlfsstksm msvqnsgtqq qgnglfqqgn emmslqsgnf lqqsshsqaq lfhpqnpiad
1081 aqnlsqetqg slfhspnpiv hsqtsttsse qmqppmfhsq stiavlqgss vpqdqqstni
1141 flsqspmnnl qtntvaqeaf faapnsispl qstsnseqqa afqqqapish iqtpmlsqeq
1201 aqppqqglfq pqvalgslpp npmpqsqqgt mfqsqhsiva mqsnspsqeq qqqqqqqqqq
1261 qqqqqqsilf snqntmatma spkqpppnmi fnpnqnpman qeqqnqsifh qqsnmapmnq
1321 eqqpmqfqsq stvvslqnpg ptqsessqtp lfhsspqiql vqgspssqeq qvtlflspas
1381 msalqtsinq qdmqqsplys pqnnmpgiqg atsspqpqat lfhntaggtm nqlqnspgss
1441 qqtsgmflfg iqnncsqllt sgpatlpdql maisqpgqpq negqppvttl lsqqqmpensp
1501 lassintnqn iekidlllvsl qnqgnnltgs f
```

Figure 14.1 Purification of TonEBP complex. (Top) Coomassie blue staining of proteins isolated using a streptavidin affinity column from a cell line expressing TAP (lane 1) or TAP-TonEBP (lane 2). The arrow denotes the TonEBP band, which was analyzed by mass spectrometry after trypsin digestion. Note that there are additional unique bands in lane 2. Positions of size markers are shown on the left. (Bottom) The amino acid sequence of human TonEBP (NCBI Locus number NP.006590) with peptides detected by MS/MS underlined.

hypertonicity in the mutant promoter–reporter provides a functional localization of TonEBP-binding sites for the gene.

5. Biochemical Analysis of TonEBP

Gel filtration analysis revealed that TonEBP has a large hydroscopic size, perhaps $\approx$2000 kDa, that is, more than 10 times its calculated molecular mass (Miyakawa *et al.*, 1999). Both a disordered nature and associated proteins might contribute to the extraordinary size. We have been interested in the immunoprecipitation of TonEBP in the hope that we would be able to copurify its associated proteins. Because TonEBP is in low abundance in most cultured cells, we have attempted to make cell lines overexpressing TonEBP. The classical transfection method turned out to be quite inefficient in yielding colonies expressing full-length TonEBP. We suspect that the repeat sequences within TonEBP, especially those encoding glutamine repeats, might interfere with genomic integration with an intact open reading frame. We have overcome this by using the Flp-In system (Invitrogen), which uses the FLP recombinase for genomic integration. We added tandem affinity purification (TAP) tags at the N terminus or C terminus of TonEBP for efficient and minimally disruptive affinity purification to maximize the chance of copurifying associated proteins. TAP tags consisted of a streptavidin-binding peptide (SBP: MDEKTTGWRGGHVVEGLAGE-LEQLRARLEHHPQGQREPSGG CKLG) and a calmodulin-binding peptide (CBP: KRRWKKNFIAVSAANRFKKISSSGAL). While SBP worked well, we could not get the CBP affinity purification to work. Nonetheless, we have been able to purify a biochemical quantity of TonEBP from Flp cell lines expressing TAP-TonEBP using the SBP affinity chromatography with a commercial kit (Interplay TAP purification kit, Stratagene) (Fig. 14.1). Analysis of the TonEBP band digested with trypsin using mass spectrometry resulted in the identification of 11 peptides, which covers ~10% of TonEBP (Fig. 14.1). This system can be used for the identification of associated proteins and analysis of posttranslational modifications.

ACKNOWLEDGMENTS

The work is supported by NIH Grants DK42479 and DK61677 to HMK. JAK is supported by a fellowship from the National Kidney Foundation.

REFERENCES

Cha, J. H., Woo, S. K., Han, K. H., Kim, Y. H., Hander, J. S., Kim, J., and Kwon, H. M. (2001). Hydration status affects nuclear distribution of transcription factor tonicity responsive enhancer binding protein in rat kidney. *J. Am. Soc. Nephrol.* **12**, 2221–2230.

Colla, E., Lee, S. D., Sheen, M. R., Woo, S. K., and Kwon, H. M. (2006). TonEBP is inhibited by RNA helicase A via interaction involving the E'F loop. *Biochem. J.* **393**, 411–419.

Dahl, S. C., Handler, J. S., and Kwon, H. M. (2001). Hypertonicity-induced phosphorylation and nuclear localization of the transcription factor TonEBP. *Am. J. Physiol.* **280**, C248–C253.

Go, W. Y., Liu, X., Roti, M. A., Liu, F., and Ho, S. N. (2004). NFAT5/TonEBP mutant mice define osmotic stress as a critical feature of the lymphoid microenvironment. *Proc. Natl. Acad. Sci. USA* **101**, 10673–10678.

Graef, I. A., Gastler, J. M., Francke, U., and Crabtree, G. R. (2001). Evolutionary relationships among Rel domains indicate functional diversification by recombination. *Proc. Natl. Acad. Sci. USA* **98**, 5740–5745.

Han, K. H., Woo, S. K., Kim, W. Y., Park, S. H., Cha, J. H., Kim, J., and Kwon, H. M. (2004). Maturation of TonEBP expression in developing rat kidney. *Am. J. Physiol.* **287**, F878–F885.

Hasler, U., Jeon, U. S., Kim, J. A., Mordasini, D., Kwon, H. M., Ferille, E., and Martin, P. Y. (2006). TonEBP is an essential regulator of aquaporin-2 expression in renal collecting duct principal cells. *J. Am. Soc. Nephrol.* **17**, 1521–1531.

Jeon, U. S., Kim, J. A., Sheen, M. R., and Kwon, H. M. (2006). How tonicity regulates genes: Story of TonEBP transcriptional activator. *Acta Physiol.* **187**, 241–247.

Lam, A. K. M., Ko, B. C. B., Tam, S., Morris, R., Yang, J. Y., Chung, S. K., and Chung, S. S. M. (2004). OREBP is an essential regulator of the urine concentration mechanism. *J. Biol. Chem.* **279**, 48048–48054.

Lee, H. W., Kim, W. Y., Song, H. K., Yang, C. W., Han, K. H., Kwon, H. M., and Kim, J. (2007). Sequential expression of NKCC2, TonEBP, aldose reductase, and urea transporter-A in developing mouse kidney. *Am. J. Physiol.* **292**, F269–F277.

Lee, S. D., Colla, E., Sheen, M. R., Na, K. Y., and Kwon, H. M. (2003). Multiple domains of TonEBP cooperate to stimulate transcription in response to hypertonicity. *J. Biol. Chem.* **278**, 47571–47577.

Lee, S. D., Woo, S. K., and Kwon, H. M. (2002). Dimerization is required for phosphorylation and DNA binding of TonEBP/NFAT5. *Biochem. Biophys. Res. Commun.* **294**, 968–975.

Lopez-Rodriguez, C., Antos, C. L., Shelton, J. M., Richardson, J. A., Lin, F., Novobrantseva, T. I., Bronson, R. T., Igarashi, P., Rao, A., and Olson, E. N. (2004). Loss of NFAT5 results in renal atrophy and lack of tonicity-responsive gene expression. *Proc. Natl. Acad. Sci. USA* **101**, 2392–2397.

Loyher, M. L., Mutin, M., Woo, S. K., Kwon, H. M., and Tappaz, M. L. (2004). Transcription factor TonEBP which transactivates osmoprotective genes is expressed and upregulated following acute systemic hypertonicity in neurons in brain. *Neuroscience* **124**, 89–104.

Maallem, S., Berod, A., Mutin, M., Kwon, H. M., and Tappaz, M. L. (2006a). Large discrepancies in cellular distribution of osmoprotective genes and their regulatory transcription factor TonEBP in rat brain. *Neuroscience* **142**, 355–368.

Maallem, S., Mutin, M., Kwon, H. M., and Tappaz, M. L. (2006b). Differential cellular distribution of tonicity-induced expression of transcription factor TonEBP in the rat brain following prolonged systemic hypertonicity. *Neuroscience* **137**, 51–71.

Mallee, J. J., Atta, M. G., Lonica, V., Rim, J. S., Kwon, H. M., Lucente, A. D., Wang, Y., and Berry, G. T. (1997). The structural organization of the human Na/myo-inositol cotransporter (SLC5A3) gene and characterization of the promoter. *Genomics* **46,** 459–465.

Maouyo, D., Kim, J. Y., Lee, S. D., Wu, Y., Woo, S. K., and Kwon, H. M. (2002). Mouse TonEBP-NFAT5: Expression in early development and alternative splicing. *Am. J. Physiol.* **282,** F802–F809.

Miyakawa, H., Woo, S. K., Chen, C., Dahl, S. C., Handler, J. S., and Kwon, H. M. (1998). *Cis-* and *trans*-acting factors regulating transcription of the BGT1 gene in response to hypertonicity. *Am. J. Physiol.* **274,** F753–F761.

Miyakawa, H., Woo, S. K., Dahl, S. C., Handler, J. S., and Kwon, H. M. (1999). Tonicity-responsive enhancer binding protein, a Rel-like protein that stimulates transcription in response to hypertonicity. *Proc. Natl. Acad. Sci. USA* **96,** 2538–2542.

Na, K. Y., Woo, S. K., Lee, S. D., and Kwon, H. M. (2003). Silencing of TonEBP/NFAT5 transcriptional activator by RNA interference. *J. Am. Soc. Nephrol.* **14,** 283–288.

Nakayama, Y., Peng, T., Sands, J. M., and Bagnasco, S. M. (2000). The TonE/TonEBP pathway mediates tonicity-responsive regulation of UT-A urea transporter expression. *J. Biol. Chem.* **275,** 38275–38280.

Neuhofer, W., Lagmayr, K., Fraek, M. L., and Beck, F.-X. (2001). Regulated overexpression of heat shock protein 72 protects MDCK cells from the detrimental effects of high urea concentrations. *J. Am. Soc. Nephrol.* **12,** 2565–2571.

Stroud, J. C., Lopez-Rodriguez, C., Rao, A., and Chen, L. (2002). Structure of a TonEBP-DNA complex reveals DNA encircled by a transcription factor. *Nat. Struct. Biol.* **9,** 90–94.

Trama, J., Go, W. Y., and Ho, S. N. (2002). The osmoprotective function of the NFAT5 transcription factor in T cell development and activation. *J. Immunol.* **169,** 5477–5488.

Trama, J., Lu, Q., Hawley, R. G., and Ho, S. N. (2000). The NFAT-related NFATL1 is induced upon T cell activation in a calcineurin-dependent manner. *J. Immunol.* **165,** 4884–4894.

Woo, S. K., Dahl, S. C., Handker, J. S., and Kwon, H. M. (2000a). Bidirectional regulation of tonicity-responsive enhancer binding protein in response to changes in tonicity. *Am. J. Physiol.* **278,** F1006–F1012.

Woo, S. K., Lee, S. D., Na, K. Y., Park, W. K., and Kwon, H. M. (2002). TonEBP/NFAT5 stimulates transcription of HSP70 in response to hypertonicity. *Mol. Cell. Biol.* **22,** 5753–5760.

Woo, S. K., Maouyo, D., Handler, J. S., and Kwon, H. M. (2000b). Nuclear redistribution of tonicity-responsive enhancer binding protein requires proteasome activity. *Am. J. Physiol.* **278,** C323–C330.

Yang, B., Hodgkinson, A. D., Oates, P. J., Kwon, H. M., Millward, B. A., and Demaine, A. G. (2006). Elevated activity of transcription factor nuclear factor of activated T-cells 5 (NFAT5) and diabetic nephropahy. *Diabetes* **55,** 1450–1455.

CHAPTER FIFTEEN

DESICCATION RESPONSE OF MAMMALIAN CELLS: ANHYDROSIGNALING

Zebo Huang[*,†] *and* Alan Tunnacliffe[†]

Contents

Abstract

Dehydration through evaporation, or air drying, is expected to have both similarities and differences to osmostress. Both stresses involve water loss, but the degree of dehydration will ultimately be more severe during desiccation. Despite the severity of desiccation stress, there are examples of organisms that can survive almost complete water loss, including resurrection plants and plant seeds, certain invertebrates among the nematodes, brine shrimps, tardigrades and bdelloid rotifers, and many microorganisms, including bakers' yeast. During desiccation, these organisms enter a state of suspended animation, a process known as anhydrobiosis ("life without water"). For other organisms, desiccation is lethal, but there is considerable interest in using what is known about anhydrobiosis to confer desiccation tolerance on sensitive cell types, such as mammalian cells. Success with this approach, which we have termed anhydrobiotic engineering, will require a more complete knowledge of the mechanisms of desiccation tolerance and the sensing and response of nontolerant organisms to extreme dehydration. With this goal in mind, we have attempted to characterize the response of human tissue culture cells to desiccation and to compare this response with osmotic upshift. This chapter

* College of Pharmacy, Wuhan University, Wuhan, China
† Institute of Biotechnology, University of Cambridge, Cambridge, United Kingdom

Methods in Enzymology, Volume 428
ISSN 0076-6879, DOI: 10.1016/S0076-6879(07)28015-2

describes some of the methods used to begin to uncover the response to evaporative water loss in human cell cultures.

 1. INTRODUCTION

Water is essential to all life forms on Earth, but certain organisms (i.e., anhydrobiotes) can survive almost complete desiccation—the water left in a dry, anhydrobiotic cell is inadequate to hydrate its components to a functional level (Clegg, 2001; Potts, 1994; Tunnacliffe and Lapinski, 2003). As a survival strategy in nature, successful anhydrobiosis enables organisms in a quiescent state to withstand not only extended desiccation, but also a variety of other harsh environmental conditions, including both chemical and physical extremes (e.g., salinity, temperature, pressure, radiation). This exceptional biostability has provoked increasing interest in developing biotechnologies from its underpinning mechanisms to improve the stability of biomaterials such as vaccines, cells, and tissues for biomedical applications in biosensors, tissue engineering, cell banks, etc. (Bloom *et al.*, 2001; Tunnacliffe *et al.*, 2001). To this end, mammalian cell cultures are the model of choice for anhydrobiotic engineering, which aims to confer desiccation tolerance on otherwise sensitive cells or organisms (Tunnacliffe *et al.*, 2001). In addition, anhydrobiotic organisms can remain viable in the quiescent dry state for long periods of time; provided that "full" anhydrobiosis is achieved, their life span may virtually be extended dramatically, if the inactive anhydrobiotic form ("suspended animation") is counted as "life" (Huang and Tunnacliffe, 2006). Therefore, the ability to perform anhydrobiosis might also have important implications in aging studies, particularly at the molecular level.

At least two groups have previously claimed to have achieved anhydrobiotic engineering of mammalian cells (Bloom *et al.*, 2001; Guo *et al.*, 2000), although this has been disputed in the latter case (García de Castro *et al.*, 2000) and qualified markedly in the former case (Jack *et al.*, 2006; Potts *et al.*, 2005). Although some improvements to desiccation tolerance can be achieved using sugar, protein, or amphiphile protectants (Acker *et al.*, 2002; Chen *et al.*, 2001; García de Castro and Tunnacliffe, 2000; Jamil *et al.*, 2005; Ma *et al.*, 2005; Walton, 2005), the current consensus seems to be that significant further progress is necessary before anhydrobiotic engineering can be performed successfully on proliferation-competent mammalian cells. This will undoubtedly require a deeper understanding of not only how anhydrobiotic organisms withstand extreme water loss, but also how mammalian cells themselves perceive and respond to this stress.

Identification of desiccation-responsive genes in both anhydrobiotic and nonanhydrobiotic organisms may facilitate the definition of key adaptations and thus provide an anhydrobiotic gene set for potential use in anhydrobiotic engineering. Using cDNA microarray and quantitative real-time polymerase chain reaction (PCR) analyses, we have shown that a number of human genes, including early stress response and zinc finger transcription factor genes, are considerably induced by desiccation stress in the cell cultures of a human embryonic kidney cell line (T-REx 293) and that these genes are also highly induced by hyperosmotic stress (Huang and Tunnacliffe, 2005). Interestingly, however, three genes, which are induced later in the osmotic stress response, that is, *AR* (aldose reductase), *BGT1* (betaine-γ-amino-n-butyric acid transporter), and *SMIT* (sodium-dependent *myo*-inositol transporter), are not significantly induced by desiccation (Huang and Tunnacliffe, 2004). The difference between the responses of these genes to the two closely related dehydration stresses is intriguing. For example, *AR*, *BGT1*, and *SMIT*, which encode proteins governing organic osmolyte accumulation to mitigate hyperosmotic water loss, are reported to be induced by changes of intracellular ionic strength and cell volume as a consequence of osmotic efflux of water, and these changes should be similarly affected in drying cells because of water evaporation, yet the response is different. Although osmotic stress has been studied extensively, the response of mammalian cells to desiccation has yet to be fully explored (Ferraris and Burg, 2004; Huang and Tunnacliffe, 2004).

Like other stress response cascades, the molecular response of cells to desiccation may also be divided into signal transduction, gene activation/suppression, and biochemical modulation, leading to survival (anhydrobiotes) or death (nonanhydrobiotes) under persistent stress. We have demonstrated previously that human cell cultures are capable of responding to desiccation via rapid activation of c-Jun N-terminal kinases (JNKs) and p38 mitogen-activated protein kinases (MAPKs) (Huang and Tunnacliffe, 2004), and the involvement of JNKs and NF-κB in the human cell desiccation response has also been confirmed (Jack *et al.*, 2006). There are three main mammalian MAPK families, that is, extracellular signal–regulated kinases (ERKs), JNKs, and p38 MAPKs. The phylogenetically conserved MAPK signal transduction networks are widespread in eukaryotes and have important physiological functions in the regulation of cell metabolism. In response to pathophysiological stimuli and environmental stresses, the MAPK pathways are activated to modulate gene expression and cell proliferation, as well as cell survival and death (Chang and Karin, 2001; Kyriakis and Avruch, 2001). For example, regulation of gene expression is one of the most important functions of MAPK cascades in osmotic stress (Burg *et al.*, 1997; Sheikh-Hamad and Gustin, 2004). We have also found that the induction of a number of genes, including early growth responses 1 and 3 (*EGR1* and *EGR3*), snail homolog 1 (*SNAI1*), dexamethasone-induced RAS 1 (*RASD1*), and growth arrest and DNA damage-inducible gene (*GADD45B*), by desiccation stress in human cell cultures is related to ERK activation (Z. Huang and A. Tunnacliffe, in preparation). Therefore, the response of human cells to desiccation is distinct from but also overlapping with that of osmotic stress as seen at protein phosphorylation, as well as at gene induction, levels (Huang and Tunnacliffe, 2004, 2005). Nevertheless, despite some fundamental differences, both desiccation and osmotic stress lead to an efflux of intracellular water and consequent concentration of inorganic salts, crowding of macromolecules, and potential damage to macromolecular structure and function, as well as mechanical damage because of cell shrinkage and distortion (Potts, 1994). Therefore, studies on the signals arising from desiccation in mammalian cells may also provide insight into the extensively studied area of signals arising from osmotic stress (Ferraris and Burg, 2004).

In summary, the long-term preservation of mammalian cells by desiccation is one of the ultimate goals of anhydrobiotic engineering. However, the molecular mechanisms underlying successful anhydrobiosis of desiccation-tolerant organisms and underlying the adaptation failure of desiccation-sensitive organisms are still lacking. Therefore, further studies are needed to reveal the molecular response of both anhydrobiotes and nonanhydrobiotes in order to fully explore the potentials of the molecular mechanism in medicine and research.

2. Cell Cultures

The human embryonic kidney cell line T-REx 293 (Invitrogen) was chosen as it can be used with an inducible expression system for subsequent experiments involving candidate desiccation protection genes. The cell line was originally derived from HEK 293, whose response to hyperosmotic stress has been studied previously (e.g., Ferraris et al., 2002). The relative survival rate of T-REx 293 cells dried at 98% relative humidity (RH) and 37° is ≈50% after drying for 12 h, ≈30% after drying for 24 h, and ≈4% after drying for 48 h (Huang and Tunnacliffe, 2004). Other cell lines may be used but the survival time and rate may vary greatly.

T-REx 293 cells are maintained routinely for up to 50 passages at 37° in an atmosphere of 5% CO_2 and 95% air in a 75-cm^2 tissue culture flask containing 25 ml of high glucose (4500 mg/liter) Dulbecco's modified Eagle's medium (Sigma) supplemented with 10% fetal bovine serum (FBS), 2 mM L-glutamine, 100 units/ml penicillin, 100 μg/ml streptomycin (all Sigma), and 5 μg/ml blasticidin (Invitrogen). For quantitative real-time PCR samples, cells are grown in 0.5 ml of medium in multidishes (1.9 cm^2/well, flat bottom; Nunc). For Western blot analysis, cells are grown in 2.5 ml of medium in multidishes (9.6 cm^2/well, flat bottom; Nunc) or air-vent culture dishes of similar size. All treatments are performed on the same number of cells in an identical fashion in each experiment.

3. DESICCATION

Cells can be dried in a number of ways, including air drying, vacuum drying, and freeze drying, depending on the experiments (e.g., Baust *et al.*, 2004; Yu *et al.*, 2004). For Western blot and real-time PCR analyses, T-REx 293 cells are grown to near confluence prior to desiccation treatment. Then the culture medium is removed carefully but completely from the cells by two sequential pipetting operations, with a brief draining period in between when the dish is tilted on its side. Cells in the air-vent (multi) dishes are then placed in humidity chambers and dried for the needed times. The humidity in the chambers is maintained by saturated $BaCl_2$ or saturated $CuSO_4$ in a Stewart air-tight polypropylene box of suitable size at appropriate temperature, providing 90 or 98% RH, respectively, which can be confirmed at room temperature using a thermohygrometer (Cole-Parmer). Although water content is not determined directly in the dried cells, the controlled drying procedure used is assumed to result in similar residual water contents from experiment to experiment for specific drying times.

4. CELL VIABILITY TEST

Live cells after drying can be counted using Trypan blue or other live/dead staining methods. Viable cells can also be determined using the CellTiter 96 AQueous One Solution Cell Proliferation Assay (Promega) according to the manufacturer's instruction as described (Huang and Tunnacliffe, 2004). Diluted cell suspensions are passaged into 96-well plates (100 μl/well) and incubated at 37° until near confluence. The cells are then dried at 37° as described earlier in a humidity chamber for specific times and rehydrated for 2 h in 100 μl of medium at 37°. After the addition of 20 μl of MTS reagent, the plate is incubated at 37° and 5% CO_2 for 3 h. Absorbance is measured at 490 nm using a 96-well plate reader and corrected by subtraction of background absorbance (same amount of medium without cells). The viability of cells dried for various times is assessed through the effect of intracellular dehydrogenases on the MTS tetrazolium compound. Relative cell survival rate using this assay is expressed as the percentage of absorbance of dehydrated cells relative to that of fully hydrated control cells (Huang and Tunnacliffe, 2004).

5. WESTERN BLOTTING

Western blot analysis of total cell lysates is performed using antibodies able to recognize either activated or total MAPKs following the manufacturer's instructions (Huang and Tunnacliffe, 2004). T-REx 293 cells are grown and dried as described previously. In order to reduce basal level phosphorylation of ERKs, near confluent cells are incubated overnight in FBS-free medium prior to desiccation or lysis (Z. Huang and A. Tunnacliffe, in preparation).

Protein samples for Western blotting are prepared by direct lysis of cells in 9.6-cm^2 wells with equal numbers of starting cells. After complete and careful removal of medium by aspiration at the time points of desiccation as described earlier, 100 μl of 1× SDS sample buffer (62.5 mM Tris-HCl, pH 6.8, 2% [w/v] SDS, 10% glycerol, 50 mM dithiothreitol, 0.01% [w/v] bromphenol blue) is added directly to the cell monolayer, and cells are immediately scraped and transferred to a cold microfuge tube on ice. The cell lysates are then sonicated in a sonication bath to reduce viscosity. The sonication is performed at 4° for 10 to 15 s and repeated after 30 s on ice until sample viscosity is reduced. Then the samples are boiled for 5 min and centrifuged at 4° for 15 min at 15,000g. Equal volumes of the supernatants are used for immunoblotting. The lysates can also be stored at −70° until use.

Up to 20 μl of each lysate is loaded onto an SDS–PAGE gel (10 × 10 cm), and the gel-resolved proteins are transferred to nitrocellulose membrane (0.45-μm pore size; Bio-Rad) using a Trans-Blot SD semidry electrophoretic transfer cell (Bio-Rad). Equal loading on SDS–PAGE can be demonstrated by staining gels with Coomassie brilliant blue after transfer and by staining Western blots with Ponceau S solution (which is reversible, i.e., the membrane can be destained) if necessary. The membrane is washed with 25 ml of Tris-buffered saline (TBS; pH 7.6) and blocked with 50 ml of 5% skimmed milk in TBS/T (0.1% Tween-20 in TBS) at room temperature for 1 h. The membrane is washed with TBS/T (25 ml, 3 × 5 min) and incubated overnight at 4° with 10 ml of the primary antibody dilution, which is diluted at 1:1000 (or according to the manufacturer's suggestion) with 5% (w/v) bovine serum albumin (for polyclonal antibodies) or 5% (w/v) nonfat dry milk (for monoclonal antibodies) in TBS/T. After overnight incubation, the membrane is washed again with TBS/T (25 ml, 3 × 5 min) and then incubated at room temperature for 1 h with horseradish peroxidase-conjugated secondary antibody, which is diluted at 1:1000 (or according to the manufacturer's suggestion) with 1% skimmed milk in TBS/T. After washing the membrane with TBS/T (25 ml, 3 × 5 min), antibody binding is detected using either the ECL Western blotting analysis system (Amersham Biosciences) or the Phototope-HRP Western detection kit (Cell Signaling Technology). After further washing with TBS/T (25 ml, 4 × 5 min), the antibody can be removed from the membrane by incubation in stripping buffer (63 mM Tris, 2% [w/v] SDS, and 0.7% 2-mercaptoethanol) at 50° for 30 min. After washing again with TBS/T (25 ml, 10 × 5 min), the membrane can be reprobed as described earlier for another antibody.

The stripping and reprobing procedure not only saves time and materials, but also greatly facilitates the comparison of the phosphoproteins and their total protein counterparts by eliminating the need to compare separate membrane blots. Because some antigens may be lost with each stripping/reprobing cycle, it is advisable to probe the antigens expected to be in the least amounts first (e.g., phospho-MAPKs vs total MAPKs). It is also advisable to put membranes into the washing buffer immediately after the final X-ray film exposure. Conditions can be optimized to perform up to five rounds of stripping and reprobing of the same blot.

6. RELATIVE QUANTIFICATION OF GENE EXPRESSION

Relative quantification of gene expression may be analyzed by quantitative real-time PCR as described (Huang and Tunnacliffe, 2004, 2005). Total RNA is prepared using the Cells-to-cDNA II kit (Ambion) following the instruction manual. T-REx 293 cells are grown in 0.5 ml of medium in multidishes (1.9 cm^2/well) as described earlier. After complete removal of medium by aspiration from the 1.9-cm^2 wells or at the time points of desiccation, 100 μl of ice-cold cell lysis II buffer is added directly to the cell layer. The cells are immediately scraped and pipetted quickly to a cold 1.5-ml microfuge tube in ice. The lysates are processed according to instruction until just before the reverse transcription step, i.e., 75° for 10 min, DNase I digestion at 37° for 15 min, and 75° for 5 min. RNA preparations can be stored at −20° for a short period or at −80° for a relatively longer time. Reverse transcription is also performed using the kit according to instructions with the random decamers provided and 5 μl of cell lysate (RNA). Resultant cDNA can be diluted 10 to 20 times with water and stored at −20° for a short time.

Relative quantification of cDNA by real-time PCR is performed using a Rotor-Gene real-time cycler (Corbett Research) and QuantiTect SYBR Green PCR kit (Qiagen). The critical threshold values are used to calculate the relative amounts of cDNA according to the delta-delta method (Pfaffl, 2001). *ACTB* (β-actin) and *GAPDH* (glyceraldehyde 3-phosphate dehydrogenase) are used as reference gene transcripts. Whenever appropriate the primers for target genes are designed to span exon–exon junctions to avoid amplification from potential contaminating genomic DNA (although RNA preparations are digested with DNase I). Primers are also chosen so that the size of the resulting amplicons is 50 to 150 bp. Specific amplification of transcripts is verified by gel electrophoresis, where only one DNA fragment should be observed, and by melt curve analysis of the real-time PCR products, where a single peak should be seen. Typical primer sequences are given in Huang and Tunnacliffe (2004, 2005).

ACKNOWLEDGMENT

This work was supported by Grant 8/C17391 from the Biotechnology and Biological Sciences Research Council of the UK.

REFERENCES

Acker, J. P., Fowler, A., Lauman, B., Cheley, S., and Toner, M. (2002). Survival of desiccated mammalian cells: Beneficial effects of isotonic media. *Cell Preserv. Technol.* **1,** 129–140.

Baust, J. M., Fowler, A., and Toner, M. (2004). Induction of apoptosis in response to anhydrobiotic conditions in mammalian cells. *Cell Preserv. Technol.* **2,** 248–259.

Bloom, F. R., Price, P., Lao, G. F., Xia, J. L., Crowe, J. H., Battista, J. R., Helm, R. F., Slaughter, S., and Potts, M. (2001). Engineering mammalian cells for solid-state sensor applications. *Biosens. Bioelectron.* **16,** 603–608.

Burg, M. B., Kwon, E. D., and Kültz, D. (1997). Regulation of gene expression by hypertonicity. *Annu. Rev. Physiol.* **59,** 437–455.

Chang, L., and Karin, M. (2001). Mammalian MAP kinase signalling cascades. *Nature* **410,** 37–40.

Chen, T., Acker, J. P., Eroglu, A., Cheley, S., Bayley, H., Fowler, A., and Toner, M. L. (2001). Beneficial effect of intracellular trehalose on the membrane integrity of dried mammalian cells. *Cryobiology* **43,** 168–181.

Clegg, J. S. (2001). Cryptobiosis: A peculiar state of biological organization. *Comp. Biochem. Physiol. B* **128,** 613–624.

Ferraris, J. D., and Burg, M. B. (2004). Drying and salting send different messages. *J. Physiol.* **558,** 3.

Ferraris, J. D., Williams, C. K., Persaud, P., Zhang, Z., Chen, Y., and Burg, M. B. (2002). Activity of the TonEBP/OREBP transactivation domain varies directly with extracellular NaCl concentration. *Proc. Natl. Acad. Sci. USA* **99,** 739–744.

García de Castro, A., Lapinski, J., and Tunnacliffe, A. (2000). Anhydrobiotic engineering. *Nat. Biotechnol.* **18,** 473.

García de Castro, A., and Tunnacliffe, A. (2000). Intracellular trehalose improves osmotolerance but not desiccation tolerance in mammalian cells. *FEBS Lett.* **487,** 199–202.

Guo, N., Puhlev, I., Brown, D. R., Mansbridge, J., and Levine, F. (2000). Trehalose expression confers desiccation tolerance on human cells. *Nat. Biotechnol.* **18,** 168–171.

Huang, Z., and Tunnacliffe, A. (2004). Response of human cells to desiccation: Comparison with hyperosmotic stress response. *J. Physiol.* **558,** 181–191.

Huang, Z., and Tunnacliffe, A. (2005). Gene induction by desiccation stress in human cell cultures. *FEBS Lett.* **579,** 4973–4977.

Huang, Z., and Tunnacliffe, A. (2006). Cryptobiosis, aging and cancer: Yin-yang balancing of signaling networks. *Rejuvenation Res.* **9,** 292–296.

Jack, G. D., Mead, E. A., Garst, J. F., Cabrera, M. C., Desantis, A. M., Slaughter, S. M., Jervis, J., Brooks, A. I., Potts, M., and Helm, R. F. (2006). Long term metabolic arrest and recovery of HEK293 spheroids involves NF-kappaB signaling and sustained JNK activation. *J. Cell Physiol.* **206,** 526–536.

Jamil, K., Crowe, J. H., Tablin, F., and Oliver, A. E. (2005). Arbutin enhances recovery and osteogenic differentiation in dried and rehydrated human mesenchymal stem cells. *Cell Preserv. Technol.* **3,** 244–255.

Kyriakis, J. M., and Avruch, J. (2001). Mammalian mitogen-activated protein kinase signal transduction pathways activated by stress and inflammation. *Physiol. Rev.* **81,** 807–869.

Ma, X., Jamil, K., MacRae, T. H., Clegg, J. S., Russell, J. M., Villeneuve, T. S., Euloth, M., Sun, Y., Crowe, J. H., Tablin, F., and Oliver, A. E. (2005). A small stress protein acts synergistically with trehalose to confer desiccation tolerance on mammalian cells. *Cryobiology* **51,** 15–28.

Pfaffl, M. W. (2001). A new mathematical model for relative quantification in real-time RT-PCR. *Nucleic Acids Res.* **29,** 2002–2007.

Potts, M. (1994). Desiccation tolerance of prokaryotes. *Microbiol. Rev.* **58,** 755–805.

Potts, M., Slaughter, S. M., Hunneke, F.-U., Garst, J. F., and Helm, R. F. (2005). Desiccation tolerance of prokaryotes: Application of principles to human cells. *Integr. Comp. Biol.* **45,** 800–809.

Sheikh-Hamad, D., and Gustin, M. C. (2004). MAP kinases and the adaptive response to hypertonicity: Functional preservation from yeast to mammals. *Am. J. Physiol. Renal Physiol.* **287,** F1102–F1110.

Tunnacliffe, A., and Lapinski, J. (2003). Resurrecting Van Leeuwenhoek's rotifers: A reappraisal of the role of disaccharides in anhydrobiosis. *Phil. Trans. R. Soc. Lond. B* **358,** 1755–1771.

Tunnacliffe, A., García de Castro, A., and Manzanera, M. (2001). Anhydrobiotic engineering of bacterial and mammalian cells: Is intracellular trehalose sufficient? *Cryobiology* **43,** 124–132.

Walton, L. J. (2005). "The Role of the LEA Proteins in Anhydrobiosis." Ph.D. thesis, University of Cambridge.

Yu, J. P., Liu, J. H., Pu, L. Q., Cui, X. D., Wang, C. Z., Ouyang, S. L., and Gao, D. Y. (2004). Freeze-drying of human red blood cells: Influence of carbohydrates and their concentrations. *Cell Preserv. Technol.* **2,** 270–275.

CHAPTER SIXTEEN

TONICITY-REGULATED GENE EXPRESSION

Joan D. Ferraris *and* Maurice B. Burg

Contents

Laboratory of Kidney and Electrolyte Metabolism, National Heart Lung Blood Institute, National Institutes of Health, Department of Health and Human Services, Bethesda, Maryland

Methods in Enzymology, Volume 428

ISSN 0076-6879, DOI: 10.1016/S0076-6879(07)28016-4

Abstract

Hypertonicity activates several different transcription factors, including TonEBP/OREBP, that in turn increase transcription of numerous genes. Hypertonicity elevates TonEBP/OREBP transcriptional activity by moving it into the nucleus, where it binds to its cognate DNA element (ORE), and by increasing its transactivational activity. This chapter presents protocols for measuring the transcriptional activity of TonEBP/OREBP and determining its subcellular localization, its binding to OREs, and activity of its transactivation domain.

1. INTRODUCTION

Numerous transcription factors are regulated by tonicity, as follows. (1) c-*fos* and c-*jun* are immediate-early genes (IEGs) and zinc finger transcription factors. They are members of the activator protein-1 family that regulates diverse processes, including cell proliferation, embryonic development, vascular smooth muscle cell growth, and apoptosis. Hypertonicity increases transcription of c-*fos* and c-*jun* in cultured cells (Cohen *et al.*, 1991; Wiese *et al.*, 1998; Wollnik *et al.*, 1993), phosphorylation of c–jun, and nuclear abundance of the phosphorylated, active form of c–jun (Clerk and Sugden, 1997). Hypertonicity in the form of high NaCl also increases the DNA-binding activity of AP-1 proteins, c-Fos/Fra, and c–jun *in vivo* (Borsook *et al.*, 1994; Wiese *et al.*, 1998; Ying *et al.*, 1996). (2) *Egr-1*, *Egr-3*, and *SNAI1* are other IEGs and zinc finger transcription factors. Hypertonicity increases their mRNA abundance (Cohen *et al.*, 1991; Huang and Tunnacliffe, 2005). Egr-1 and Egr-3 are involved in cell growth and survival but also have roles in inducing apoptosis (Thiel and Cibelli, 2002; Xi and Kersh, 2004). SNAI1, however, blocks cell cycle progression and is anti-apoptotic (Vega *et al.*, 2004). (3) p53 is a stress–induced tumor suppressor. High NaCl increases its protein abundance, phosphorylation, and activity, which contribute to cell survival (Dmitrieva *et al.*, 2000). Hypertonicity also elevates mRNA abundance and protein stability of p73β, which is a functional homologue of p53 (Lin *et al.*, 2004). (4) Pax2 is a paired box transcription factor necessary for kidney development (Torban and Goodyer, 1998). Hypertonicity increases its mRNA abundance, protein, and transcriptional activity (Cai *et al.*, 2005a). Induction of Pax2 expression protects cultured innermedullary cells by inhibiting hypertonicity-induced apoptosis (Cai *et al.*, 2005a). The mRNA abundance of another paired-box transcription factor, Pax8, also increases with tonicity but its function is undetermined in this regard (Cai *et al.*, 2005a). (5) HSF-1 is a heat shock transcription factor. A change in tonicity, in either direction, results in both HSF-1 trimerization and nuclear translocation (Caruccio *et al.*, 1997). (6) Signal transducers and activators of transcription (STATs) are important for the intracellular response to numerous cytokines, growth, and differentiation factors (Gatsios et al., 1998). STATs are regulated by Janus tyrosine kinases (Jaks). Hypertonicity increases tyrosine phosphorylation of Jak1, Jak2, and Tyk2, as well as phosphorylation of STAT1 and STAT3, and causes nuclear translocation of STAT3 (Gatsios et al., 1998). Tonicity-dependent activation of STAT1 appears to be mediated by p38 and its upstream activator MKK6 (Bode et al., 1999). Hypertonicity-induced phosphorylation of Jak2/STAT3 may lead to activation of the type 1 sodium hydrogen exchanger (NHE-1) (Garnovskaya et al., 2003). (7) NF-κB is a Rel family transcription factor that activates an inflammatory cascade, leading to interleukin (IL)-8 production. Hypertonicity elevates both NF-κB DNA binding and transcriptional activity in intestinal cells, increasing IL-8 production (Nemeth et al., 2002).

Another Rel family transcription factor, TonEBP/OREBP/NFAT5 (tonicity-responsive enhancer-/osmotic response element-binding protein/nuclear factor of activated T cells) (Ko et al., 2000; Lopez-Rodriguez et al., 1999; Miyakawa et al., 1999), is particularly well studied for its effects on tonicity-regulated gene expression. This chapter describes methods used to study how tonicity affects its activity. We start with a brief overview of the osmoadaptive genes that TonEBP/OREBP transactivates, followed by a description of the mechanisms involved in its tonicity-dependent activation. We concentrate on three aspects of TonEBP/OREBP activity that are common to all transcription factors, namely nuclear localization, DNA binding, and transactivational activity. We emphasize each of these aspects in the overview and in the protocols that follow.

1.1. Genes that TonEBP/OREBP transactivates in response to increased tonicity

Cells initially respond to hypertonicity by regulatory volume increase in which transporters convey inorganic ions into the cells, followed by an osmotic influx of water that restores cell volume. Although cell volume is corrected, intracellular ionic strength remains elevated, and high ionic strength perturbs the structure and function of cellular macromolecules (Yancey et al., 1982). Cells compensate by replacing excess inorganic ions with compatible organic osmolytes, which allows maintenance of cell volume along with less perturbation of macromolecules. Tonicity is normally higher in the renal inner medulla than elsewhere in the body. The principal organic osmolytes in cells of the renal medulla are sorbitol, glycine betaine (betaine), myo-inositol (inositol), taurine, and glycerophosphocholine (GPC) (Burg et al., 1997). Several enzymes and transporters are responsible for accumulation of these osmolytes. Aldose reductase (AR) converts glucose to sorbitol. The betaine/γ-aminobutyric acid transporter, the sodium inositol cotransporter, and the taurine transporter, respectively, mediate the cellular accumulation of betaine, inositol, and taurine (Burg *et al.*, 1997). The phospholipase, neuropathy target esterase, catalyzes the synthesis of GPC from phosphatidylcholine (Gallazzini *et al.*, 2006). Hypertonicity causes each of these proteins to increase in abundance by increasing the transcription of its gene (Burg *et al.*, 1997; Gallazzini *et al.*, 2006). The increase results from activation of the transcription factor, TonEBP/OREBP.

Other genes that are central to osmoprotection and urinary concentration are also transactivated by TonEBP/OREBP. These genes code for heat shock protein 70 (Woo *et al.*, 2002), vasopressin-activated urea transporters [UT-A1 (Nakayama *et al.*, 2000) and possibly UT-A2 (Lam *et al.*, 2004)], as well as aquaporin 2 (AQP2)(Hasler *et al.*, 2006). TonEBP/OREBP activates its target genes by binding to one or more consensus DNA elements in the regulatory regions of these genes. The TonEBP/OREBP-binding sites are called osmotic response elements (OREs) (Ferraris *et al.*, 1994; Ko *et al.*, 1997) or tonicity-responsive enhancers (TonEs) (Takenaka *et al.*, 1994).

1.2. Mechanisms of tonicity-dependent activation of TonEBP/OREBP

Hypertonicity increases TonEBP/OREBP mRNA abundance in various cell lines in culture (Cai *et al.*, 2005b; Ko *et al.*, 2000; Woo *et al.*, 2000) because of a transient ($\approx$6 h) increase in its stability. The tonicity-dependent increase in stability is mediated by elements within its $5'$-UTR (Cai *et al.*, 2005b). TonEBP/OREBP protein abundance increases secondary to the increase in its mRNA. Increased TonEBP/OREBP protein, coupled with tonicity-dependent nuclear translocation, results in more TonEBP/OREBP in the nucleus where it binds to OREs and transactivates its target genes.

Serine and tyrosine residues in TonEBP/OREBP become phosphorylated within 30 min of increasing NaCl, coincident with movement of TonEBP/OREBP into the nucleus (Dahl *et al.*, 2001). Transcription factors commonly are regulated by single or multisite phosphorylation (Cohen, 2000; Holmberg *et al.*, 2002), affecting their subcellular localization (Cyert, 2001). Tonicity-dependent phosphorylation also apparently affects TonEBP/OREBP transactivational activity. Phosphorylation of other transcription factors regulates transactivation (1) as a requirement for maximal transcriptional activity, (2) by stabilizing homodimer–DNA complexes (Groner *et al.*, 2000; Schmitz *et al.*, 2001), or (3) by recruiting coactivator proteins that contact the general transcriptional machinery (De Cesare *et al.*, 1999). High NaCl increases phosphorylation of the part of TonEBP/OREBP (amino acids 548 to 1531) that contains its high NaCl-activated transactivation domain (TAD) (Ferraris *et al.*, 2002b). Kinase inhibitors reduce the transactivational activity of TonEBP/OREBP at high NaCl (Ferraris *et al.*, 2002b), which could indicate direct inhibition of phosphorylation of TonEBP/OREBP itself or an indirect effect on a signaling partner or a transcription cofactor. An indirect effect is favored by the failure to find tonicity-dependent change in net phosphorylation of the TonEBP/OREBP TAD (Lee *et al.*, 2003). A direct effect is favored by the inhibition caused by site-directed mutation of putative phosphorylation sites in TonEBP/OREBP (Irarrazabal *et al.*, 2004). TonEBP/OREBP has many potential phosphorylation sites, namely 216 serines, 15 tyrosines, and 111 threonines. Defining which amino acids become phosphorylated and the functional consequences is a subject of ongoing research.

Transcription factors must enter the nucleus to bind to consensus DNA elements. Proteins the size of TonEBP/OREBP are too large (>50 kDa) to pass through nuclear pores by diffusion and, therefore, require active transport through the pores, mediated by importins (Cyert, 2001). Nuclear localization sequences (NLSs) mediate binding to importins. Similarly, because large proteins are unable to exit the nucleus by diffusion through the nuclear pores, nuclear export sequences (NESs) are required to mediate binding to exportins (Cyert, 2001). Several microscopy-based methods have been used to determine subcellular localization of TonEBP/OREBP, including immunofluorescence of fixed cells, using antibodies, against native or recombinant TonEBP/OREBP or imaging of TonEBP/OREBP tagged with green fluorescent protein in living cells (Miyakawa *et al.*, 1999; Tong *et al.*, 2006; Woo *et al.*, 2000). These methods involve selection of individual cells or fields of cells, followed by various types of measurement of immunostaining or fluorescence. Another method that is both versatile and quantitative involves extraction of the proteins separately from nuclear and cytoplasmic compartments, followed by Western analysis and quantitation of the nuclear:cytoplasmic distribution or n:c ratio (Zhang *et al.*, 2005). The n:c ratio method and its uses are described later. The subcellular localization of TonEBP/OREBP is tonicity dependent. In cultured cells at normotonicity, TonEBP/OREBP is in both cytoplasmic and nuclear compartments, but when tonicity is increased by raising osmolality to 450 to 500 mOsm/kg, most of the TonEBP/OREBP is nuclear (Ko *et al.*, 2000; Lopez-Rodriguez *et al.*, 2001; Miyakawa *et al.*, 1999; Tong *et al.*, 2006). When tonicity is lowered to 135 to 250 mOsm/kg, TonEBP/OREBP is mostly cytoplasmic (Tong *et al.*, 2006; Woo *et al.*, 2000). Ataxia telangiectasia-mutated kinase (ATM) is activated by high NaCl (Irarrazabal *et al.*, 2004). The n:c ratio method was used to show that its activation contributes to nuclear localization of TonEBP/OREBP (Zhang *et al.*, 2005). The n:c method was also used to demonstrate that the N terminus of TonEBP/OREBP contains the domains necessary for tonicity-dependent nuclear localization. Recombinant TonEBP/OREBP, containing only amino acids 1 to 547, moves into the nucleus when NaCl is added, indicating that the C terminus is not required for translocation (Zhang *et al.*, 2005). The N terminus contains an NLS, an NES, and an auxiliary export domain (Tong *et al.*, 2006).

The N terminus of TonEBP/OREBP also contains the TonEBP/OREBP DNA-binding domain (Ko *et al.*, 2000; Lopez–Rodriguez *et al.*, 2001). Electrophoretic mobility shift assays (EMSAs), described later, are used to characterize TonEBP/OREBP binding to OREs. TonEBP/OREBP dimerization is necessary for this DNA binding (Lopez–Rodriguez *et al.*, 2001). Binding of TonEBP/OREBP to OREs does not require its own phosphorylation, as isolated TonEBP/OREBP binds to an ORE regardless of treatment with a general phosphatase (Dahl *et al.*, 2001). Other studies indicate involvement of serine/threonine kinases, but not tyrosine kinases in the signaling pathway that leads to ORE binding of TonEBP/OREBP (Aida *et al.*, 1999). A modification of the EMSA protocol, referred to as a supershift, is also described later. It was used to identify other proteins in the complex with TonEBP/OREBP that binds to OREs (Irarrazabal *et al.*, 2004).

In addition to entering the nucleus and binding to DNA, transcription factors must also transactivate target genes. The transactivation domains of transcription factors recruit and interact with the specific protein components of its enhanceosome, as well as with RNA polymerase and other members of the basal transcriptional complex. Transactivational activity is regulated by posttranslational modifications, by interaction with signaling molecules, and by association with other proteins, including other transcription factors (Groner *et al.*, 2000; Macian *et al.*, 2001). Use of a binary transactivational assay, described later, facilitates identification of members of the enhanceosome, of signaling molecules responsible for posttranslational modifications to the transcription factor, and of interacting proteins. This binary assay was used to demonstrate that TonEBP/OREBP transactivational activity is regulated by high NaCl (Ferraris *et al.*, 2002b) as well as that multiple TonEBP/OREBP TADs and modulation domains within TonEBP/OREBP act in synergy to transactivate target genes (Lee *et al.*, 2003).

The binary transactivation assay has also been used to identify factors that are activated by hypertonicity and contribute to increased transactivational activity of TonEBP/OREBP. These include the mitogen–activated protein kinase, p38 (Han *et al.*, 1994; Ko *et al.*, 2002), a member of the Src family of tyrosine kinases, Fyn (Kapus *et al.*, 1999; Ko *et al.*, 2002), the serine/threonine kinases, PKAc (Ferraris *et al.*, 2002a) and ATM (Irarrazabal *et al.*, 2004), the phosphoinositide 3-kinase, PI3K-IA (Irarrazabal *et al.*, 2006), and reactive oxygen species (Zhang *et al.*, 2004; Zhou *et al.*, 2005). As is the case for other transcription factors, full activation of TonEBP/OREBP requires a combination of signals. Inhibition of each of these signals, alone, only partially reduces the hypertonicity–induced increase of TonEBP/OREBP activity.

1.3. Cell culture

For reproducible results, each of the following protocols requires thorough knowledge of the growth characteristics of the chosen cell type. Careful control of seeding densities and cell confluence levels is essential and is best achieved by careful counting of cells and consistent feeding and treatment. We used HEK293 cells at passages 40 to 46 in the protocols that follow. Experimental hypotonic medium (200 mOsm/kg) is NaCl-free medium (Biofluids) to which NaCl is added. Hypertonic medium (500 mOsm/kg) is normotonic medium (300 mOsm/kg) to which NaCl is added. Media substituted during an experiment are equilibrated in the incubator for at least 2 h to minimize shifts of pH or temperature and plates are handled individually to minimize exposure to room air.

1.4. Transfection

Transfection of adherent cells using various agents such as Lipofectamine (Invitrogen) requires optimization of amounts of plasmid or reagent and cell number and is specific for cell type. Recommendations are provided by suppliers. When using different experimental treatments, it is often desirable to have equal transfection efficiencies. This can be achieved by combining cells after transfection and then dividing them for the experimental treatments. Another consideration is the use of transient versus stable transfection. Each has advantages and disadvantages. A major difference is that only a fraction of the cells express transiently transfected constructs, whereas all of the cells express stably transfected constructs, as long as antibiotic selection is maintained. Cells stably transfected with reporter constructs are useful for high-throughput screening. However, sometimes it is desirable to cotransfect a second construct. If this is done transiently in cells already stably expressing another construct, any effects of the transiently transfected construct will be diluted because of the background of cells that do not contain it. In this case transient cotransfection is preferable.

Constructs that express recombinant proteins should be titrated to determine amounts of DNA necessary to obtain similar levels of protein expression and, even if this is done, it is still desirable to measure the amounts of proteins that are expressed in order to normalize the results. Transfection efficiency is often estimated by cotransfection and subsequent normalization to a second reporter construct. However, this may be misleading. In our experience even reporter constructs that have minimal promoters and are stripped of most consensus-binding sites still have responded to changes in tonicity, inhibitors, and so on, introducing unwanted bias.

Cells that do not express a particular gene (null cells) or express only a nonfunctional protein are particularly useful as they can be reconstituted to express the wild-type protein or one containing mutations. Cells transfected with an empty vector provide a useful control. Wild-type cells, even from matched littermates, can behave very differently from null cells, so strict comparison may be difficult.

2. QUANTITATION OF THE NUCLEAR:CYTOPLASMIC DISTRIBUTION RATIO OF THE TRANSCRIPTION FACTOR TonEBP/OREBP IN HEK293 CELLS

2.1. Overview

This protocol utilizes cells that express native TonEBP/OREBP or cells that additionally express TonEBP/OREBP as a recombinant protein fused to a V5 epitope (Invitrogen). Antibody to the epitope tag allows recognition of the recombinant TonEBP/OREBP. Recombinant TonEBP/OREBP is introduced by transient or stable transfection (protocols given later) and can be wild type or with nucleotide modifications to generate amino acid deletions, truncations, or changes. For example, we cloned N-terminal amino acids 1 to 547 of TonEBP/OREBP into pcDNA6-V5-His (Invitrogen). Using an antibody against V5, we compared the behavior of the truncated protein with the full-length protein (amino acids 1–1531) to demonstrate that amino acids 1 to 547 are sufficient for tonicity-dependent nuclear translocation (Zhang *et al.*, 2005).

The following method is based on preparing nuclear and cytoplasmic protein extracts using a commercially available kit, NePer (Pierce), slightly modifying the supplier's instructions. Proteins are separated by denaturing gel electrophoresis and transferred to a membrane and then TonEBP/OREBP is detected with an antibody. Signals from the TonEBP/OREBP bands are quantitated and the amount of TonEBP/OREBP in each compartment is calculated.

2.2. Nuclear:Cytoplasmic distribution ratio protocol

1. Seed 5×10^6 cells/10-cm dish and allow to grow. Once cells are 80% confluent, change the medium to the same medium or one that is 200 or 500 mOsm/kg.
2. Harvest the cells after 30 to 120 min. Stressing cells longer than 4 h at high NaCl may add an additional complication to a translocation experiment, as TonEBP/OREBP protein abundance increases because of enhanced stability of its mRNA at high NaCl (Cai *et al.*, 2005b).
3. Preweigh tubes. The rest of the protocol is performed on ice, with all reagents and tubes being precooled. Add protease (Complete Mini,

Roche) and phosphatase inhibitors (phosphatase inhibitor cocktails 1 and 2, Sigma) to phosphate-buffered saline (PBS; 200, 300, and 500 mOsm/kg) and the NePER reagents, CERI, CERII, and NER.

4. Treat one dish at a time until all cell pellets are in PBS, leaving the other dishes in the incubator. Aspirate medium and rinse gently with an equal volume of PBS. Aspirate the PBS, and replace with 1 ml PBS. Scrape cells and transfer to a 1.5-ml tube. Centrifuge at 500g for 4 min, decant the supernatant, blot, and use a pipette to remove any residual supernatant.

5. Weigh each tube and leave it on ice until all dishes have been treated. Add 100 μl CERI /20-mg pellet to each tube. Vortex 5 s, place on ice 10 min, and vortex. Add 5.5 μl CERII/20-mg pellet, vortex 5 s, ice 1 min, vortex 5 s, centrifuge at 16,000g for 5 min, and transfer supernatant (cytoplasmic extract) to a 1.5-ml tube. Add 10% of the volume of CERI to the remaining pellet, vortex, centrifuge, and combine this supernatant with the initial cytoplasmic extract.

6. Resuspend the nuclear pellet in 50 μl NER/20 mg original cell pellet. Vortex 5 s and ice 10 min. Repeat for 40 min (total of five vortexes). Centrifuge at 16,000g for 10 min and transfer supernatant (nuclear extract) to a 1.5-ml tube.

7. Weigh all tubes to determine lysate volume (1 mg $\approx$ 1 μl), determine protein concentrations, and store extracts at $-80°$.

8. Separate proteins on 4 to 12% Novex Tris-glycine gels and transfer to nitrocellulose membranes (Invitrogen). After transfer, assess equal loading by Coomassie stain of the gel proteins. Perform Western blot analysis using a quantitative system such as the Odyssey infrared imaging system (Licor). The adequacy of the separation of nuclear from cytoplasmic proteins can be confirmed by immunoblotting for proteins known to be nuclear or cytoplasmic. For example, 99.9% of p38 is present in the cytoplasmic fraction and 90.5% of poly(ADP-ribose) polymerase-1 is in the nuclear fraction at 200, 300, and 500 mOsm/kg (Zhang et al., 2005).

9. From the relative quantities of TonEBP/OREBP and the relative volumes of the cytoplasmic or nuclear extract, calculate the relative total amounts of TonEBP/OREBP in the cytoplasmic and nuclear fractions and the n:c ratio.

   a. lysate volume (μl) · protein concentration (μg/μl) = total protein in a fraction (μg)

   b. protein loaded in gel (μg)/total protein in fraction (μg) = percentage of protein loaded

   c. densitometry of TonEBP/OREBP band/percentage of protein loaded = TonEBP/OREBP in cytoplasm or nuclear fraction

   d. TonEBP/OREBP in nucleus/TonEBP/OREBP in cytoplasm = n:c ratio

3. Nonradioactive EMSA of Nuclear Extracts of HEK293 Cells for Binding of the Complex of TonEBP/OREBP and Associated Proteins to ORE Elements

3.1. Overview

We use a double-stranded DNA probe, biotinylated at each 5′ end. Addition of a molar excess of an identical double-stranded DNA that is not biotinylated competes for binding with the biotinylated probe, demonstrating specificity (Irarrazabal et al., 2004). Supershift of the complex by a specific antibody is used to test whether a given protein is part of the complex bound to the ORE element. Controls include absence of antibody or cellular extract. The protocol is a modified LightShift chemiluminescent electrophoretic mobility-shift assay (LS EMSA) (Pierce).

3.2. Nuclear protein extraction

1. Seed 7.5×10^6 cells/15-cm dish and allow to grow. Once cells are 80% confluent, change the medium to the same medium or one that is 500 mOsm/kg. Follow the n:c distribution ratio protocol given earlier to step 6, doubling the volumes of CERI and CERII.
2. Resuspend the nuclear pellet in 100 μl of lysis buffer (Lee et al., 2002). Lysis buffer is 50 mM Tris-Cl, pH 7.5, 150 mM NaCl, 1 mM EDTA, 1% Triton X-100, and 1 mM dithiothreitol (DTT). Add DTT and phosphatase inhibitor cocktails 1 and 2 (Sigma) and protease inhibitor (Roche) cocktails just before use.
3. Vortex for 15 s, ice 10 min, and repeat for 40 min (total of five vortexes). Centrifuge at 16,000g for 10 min, transfer supernatant to a clean tube, and store at −80°.

3.3. Electromobility shift assay with supershift

3.3.1. Preparation of probe and nonbiotinylated competitor
Anneal equimolar amounts of complementary 5′ biotinylated or nonbiotinylated oligonucleotides (−1238 to −1104 of the human AR gene containing three OREs in native gene context) (Irarrazabal et al., 2004). Prepare annealing reaction in 1× NEB restriction enzyme buffer #2 (New England Biolabs), total volume of 50 μl. Heat to 95° for 4 min, 70° in a beaker containing 500 ml of water for 10 min, and then allow to cool slowly to room temperature in the same beaker. Aliquot and store at −20°. Prepare working dilutions at time of EMSA.

3.3.2. Agarose gel preparation

Prepare 100 ml 0.4% agarose (Seakem Gold, Cambrex) in 0.5× Tris borate EDTA (TBE) buffer (10×; KD Medical). Measure agarose mixture by weight, cover beaker with plastic wrap having several small puncture holes, and microwave at highest setting until agarose is dissolved. Bring back to original weight with distilled water, allow to cool (15 to 20 min), and pipette 55 ml into a 9 × 11-cm gel cast. Set comb and polymerize gel at 4°. Place in apparatus and cover with 0.5× TBE running buffer to a depth of 2 to 3 mm. Flush wells by pipetting running buffer in and out of each well.

3.3.3. Binding reaction and electrophoresis

1. Dilute protein extracts in extraction buffer (see earlier discussion). It is advisable to titrate each extract in advance for the amount of protein needed to shift 100 fmol of probe in the presence of variable amounts of nonspecific competitor DNA [poly(dAdT)]. Do not exceed 3 μl of extract per binding reaction. During titration, each binding reaction should contain the same volume of extract. Select conditions that yield a tight band.
2. Dilute probe to 100 fmol/μl; nonbiotinylated competitor to 1000 fmol/μl. Prepare 2× binding buffer (Lee *et al.*, 2002) (5 mM MgCl$_2$, 50 mM KCl, 20 mM HEPES, pH 7.9, 5% [v/v] glycerol, and 1 mM DTT; add DTT just before use).
3. Prepare binding reactions. Pipette, in order, into a 0.5-ml tube for each reaction: ultrapure H$_2$O (volume calculated to bring final volume to 20 μl), 10 μl 2× binding buffer, poly(dAdT), protein extract, and antibody, as indicated (typically 1 μg anti–TonEBP/OREBP [NFAT5, Affinity Bioreagents]). If EMSA includes supershift, incubate for 1 h at 4°. Add nonbiotinylated competitor DNA (1000 fmol) to appropriate tubes and incubate for 10 min at room temperature. Add biotinylated probe DNA (100 fmol) to every tube. Incubate for 20 min at room temperature.

3.3.4. Electrophoresis, membrane transfer, and cross-linking

1. Add 5 μl of 5× loading buffer (LS EMSA) to each tube. Load gel, run 80 V, 2 h, 4°.
2. Transfer DNA to a nylon membrane (Hybond-N, Amersham Life Science) using 4° prechilled 0.5× TBE. It may be necessary to cut the gel to fit the transfer cassette (Bio-Rad Mini-trans blot cell with ice chamber). For TonEBP/OREBP, a cut may be made 0.25 in. below the bromphenol blue dye. In the apparatus, the nylon membrane must be on the cathode side and the agarose on the anode side and the apparatus three-fourths full with transfer buffer. Use the ice chamber to keep the buffer cold. Transfer at 200 mA for 20 min (Bio-Rad Power Pac 200).

3. Once transfer is complete, place nylon membrane (bromphenol blue side up) on a dry filter paper for 5 to 10 s to absorb excess buffer without drying. Autocross-link (UV Stratalinker 1800, Stratagene).

3.3.5. Chemiluminescence detection

1. Warm LightShift blocking buffer and 4× wash buffer to 37 to 50° in a water bath until dissolved. Use between room temperature and 50°. Warm the substrate equilibration buffer to room temperature.
2. Block membrane in 20 ml blocking buffer, 15 min, with gentle shaking.
3. Prepare conjugate/blocking buffer solution, replace blocking buffer, and incubate 15 min with gentle shaking.
4. Prepare 1× wash buffer, transfer membrane to fresh container, and rinse with 20 ml of this buffer. Repeat four times in the same container, 5 min each, with gentle shaking.
5. Transfer membrane to a fresh container and add 30 ml LightShift substrate equilibration buffer. Incubate 5 min with gentle shaking.
6. Prepare LightShift substrate working solution. Carefully blot an edge of the membrane to remove excess buffer and place in a fresh container, DNA side up. Pour the LightShift substrate working solution onto the membrane to cover the surface completely. Incubate 5 min without shaking.
7. Remove membrane and wick away excess solution but do not allow drying. Wrap membrane in plastic wrap, avoiding bubbles and wrinkles. Expose to X-ray film for 0.5 to 10 s.

4. QUANTITATION OF TRANSCRIPTIONAL AND TRANSACTIVATIONAL ACTIVITIES OF TonEBP/OREBP USING TRANSIENT OR STABLE TRANSFECTION OF HEK293 CELLS

4.1. Reporter constructs

Transcription reporters contain a cognate DNA-binding element upstream of a reporter gene. Our various ORE reporter constructs contain at least one ORE cloned upstream of the *Photinus pyralis* luciferase cDNA in a commercially available vector such as pGL3 (Promega). The constructs are (1) nucleotides −3497 to +27 of the 5′-flanking region of the rabbit aldose reductase gene that contain three OREs and an adjacent AP1 site in native gene context (ARLuc9); the promoter alone (nucleotides −209 to +27, ARLuc6) is the control (Ferraris *et al.*, 1994). (2) Nucleotides −1233 to −1105 of the human *AR* gene, containing three OREs and an adjacent AP1 site in native context (Zhou *et al.*, 2005) upstream of a minimal IL-2 promoter (Trama *et al.*, 2000). An otherwise identical construct with each of the three OREs mutated to prevent TonEBP/OREBP binding is the control (Zhou et al., 2005). (3) Two tandem copies of the human ORE-X sequence (Ferraris et al., 1999) are inserted into the XhoI site of the human IL-2 minimal promoter (Trama et al., 2000) to produce ORE-X IL2min-GL3. For stable transfection, the GL3-based constructs were modified to express the blasticidin resistance gene (Bsd) by insertion of nucleotides 1934 to 3204 of pCMV/Bsd (Invitrogen, Carlsbad, CA) into XmnI/BsaI sites (e.g., IL2min-GL3-Bsd [Zhou et al., 2005]). Insertion into XmnI/BsaI sites eliminates ampicillin resistance, so selection in bacteria is also achieved with blasticidin.

Transactivation reporters are based on the functional independence of domains of transcription factors (Ptashne, 1988). Transactivation domains remain functional when fused to heterologous proteins (Sadowski et al., 1988; Webster et al., 1988). The reporter system consists of two vectors that are cotransfected. The first is an expression vector in which the TonEBP/OREBP TAD is fused to the yeast GAL4 DNA-binding domain (GAL4dbd). It is generated by in-frame insertion of the sequence coding for amino acids 872 to 1271 or amino acids 548 to 1531 of TonEBP/OREBP into pFA-CMV (Stratagene) (Ferraris et al., 2002b). Unmodified pFA-CMV expresses only the GAL4dbd and is used as a control. The second vector is a reporter containing GAL4-binding sites (Kakidani and Ptashne, 1988; Sadowski et al., 1988; Webster et al., 1988). It contains five tandem repeats of the yeast GAL4-binding site (upstream activating sequence, UAS), a minimal promoter (TATATA), and the P. pyralis luciferase gene in pFR-Luc (Stratagene). For stable transfection in mammalian cells, the GAL4UAS/TATATA sequence from pFR-Luc was cloned into IL2min-GL3-Bsd (see earlier discussion), which contains the blasticidin resistance gene to create GAL4UASGL3_Bsd. pFA-CMV constructs already express neomycin resistance and were not modified for selection in mammalian cells.

4.2. Overview

Transcriptional reporter assays can measure activity of the native TonEBP/OREBP or, in addition, of recombinant TonEBP/OREBP that is transfected. In contrast, transactivation activity reporters function independently of the native transcription factor. Otherwise, the assays are similar. In the following, unless specifically limited to one type of assay, constructs are simply referred to as "reporters."

4.3. Transient transfections

1. Seed 5×10^6 cells/10-cm dish and allow to grow until 90 to 95% confluent.

2. For each dish, prepare complexes as follows.

    a. Add DNA to Opti-MEM I (Invitrogen) to a total volume of 1.5 ml. For ORE-dependent transcription reporters, 2 to 5 μg of DNA is sufficient. For binary transactivation reporters, use 5 μg pFR-Luc and 150 ng GAL4dbd-548-1531. If recombinant proteins that might affect reporter activity are to be cotransfected, the DNA amounts should be titrated in advance.

    b. Add Lipofectamine 2000 (0.5 μl/μg DNA) to Opti-MEM I (total volume of 1.5 ml). Incubate 5 min at room temperature and then combine with the DNA/Opti-MEM I solution. Mix gently and incubate at room temperature for 20 min.

3. Replace medium in a 10-cm dish with 7 ml of fresh medium without antibiotics. Add the Lipofectamine:DNA complex dropwise, rocking back and forth to mix. Incubate overnight at 37°, 5% CO_2. (To generate stable transfectants, proceed to selection protocol that follows.)

4. Trypsinize cells and seed 5.5 × 10^5/well into 6-well plates. Let cells grow for 4 to 6 h, change the medium to the same medium or one that is 200 or 500 mOsm/kg, and incubate 16 to 24 h.

5. Rinse cells with PBS (200, 300, or 500 mOsm/kg). Aspirate PBS and add 100 μl passive lysis buffer (Promega). Incubate on ice at least 20 min.

6. Scrape cells, transfer lysate to a 1.5-ml tube, and leave on ice until all dishes are processed. Centrifuge 5 min 20,000g. Transfer supernatant to fresh tubes and keep on ice.

7. Use 5 μl of lysate for bicinchoninic acid (BCA) protein determination and 20 μl for luciferase assay (Luciferase Assay System, Promega).

4.4. Stable transfection

1. Test untransfected cells to determine the concentration of antibiotic(s) that kills 99% of cells within 7 to 10 days. For example, 5 μg/ml blasticidin is sufficient for HEK293. When used in combination, antibiotic concentrations should be reduced. For example, 2.5 μg/ml blasticidin plus 150 μg/ml neomycin is sufficient for HEK293. Proceed with the transient transfection protocol given earlier, using two 10-cm dishes for each reporter. Following overnight incubation, remove transfection complexes and add fresh medium. After 24 h, and from this point on, add medium containing the selected concentration of antibiotic(s). Change the medium every 2 days.

2. After 10 days, trypsinize the cells and, using one of the dishes, test reporter activity. From the remaining dish, count the cells and seed into a 96-well plate at 1 cell per well.

3. Allow cells to grow until single colonies appear ($\approx$7 days). When colonies have grown to 30 to 50% confluence ($\approx$2 weeks), trypsinize and

transfer to 24-well plates. When cells reach 50% confluence ($\approx$4 weeks), transfer to 6-well plates. Sufficient cells are now present to allow identification of which clones have reporter activity and continue expanding them. Promising clones should be stored in liquid nitrogen as the selection process continues.

4.5. Reporter assays using stably transfected reporter cells

1. Grow cells in 10-cm dishes, maintaining antibiotics. Trypsinize and seed 5×10^3 cells/well in a 96-well plate having a clear bottom (Packard Viewplate), omitting antibiotics from this point.
2. After 24 h, change the medium to the same medium or one that is 200 or 500 mOsm/kg and incubate 16 to 18 h. Cells should not be confluent.
3. Rinse cells with PBS (200, 300, or 500 mOsm/kg). Aspirate PBS and add 30 μl of passive lysis buffer (Promega). Incubate on ice at least 20 min.
4. Mix and transfer 5 μl lysate to a 96-well plate for BCA protein determination. Add 80 μl of luciferase assay buffer (Luciferase Assay System, Promega) directly to view plate for analysis. Analyze on a plate reader such as a Victor^3 multilabel counter (Perkin-Elmer).

ACKNOWLEDGMENT

This research was supported by the Intramural Research Program of the NIH, NHLBI.

REFERENCES

Aida, K., Tawata, M., Ikegishi, Y., and Onaya, T. (1999). Induction of rat aldose reductase gene transcription is mediated through the cis-element, osmotic response element (ORE): Increased synthesis and/or activation by phosphorylation of ORE-binding protein is a key step. *Endocrinology* **140,** 609–617.

Bode, J. G., Gatsios, P., Ludwig, S., Rapp, U. R., Haussinger, D., Heinrich, P. C., and Graeve, L. (1999). The mitogen-activated protein (MAP) kinase p38 and its upstream activator MAP kinase kinase 6 are involved in the activation of signal transducer and activator of transcription by hyperosmolarity. *J. Biol. Chem.* **274,** 30222–30227.

Borsook, D., Konradi, C., Falkowski, O., Comb, M., and Hyman, S. E. (1994). Molecular mechanisms of stress-induced proenkephalin gene regulation: CREB interacts with the proenkephalin gene in the mouse hypothalamus and is phosphorylated in response to hyperosmolar stress. *Mol. Endocrinol.* **8,** 240–248.

Burg, M. B., Kwon, E. D., and Kultz, D. (1997). Regulation of gene expression by hypertonicity. *Annu. Rev. Physiol.* **59,** 437–455.

Cai, Q., Dmitrieva, N. I., Ferraris, J. D., Brooks, H. L., van Balkom, B. W., and Burg, M. (2005a). Pax2 expression occurs in renal medullary epithelial cells *in vivo* and in cell culture, is osmoregulated, and promotes osmotic tolerance. *Proc. Natl. Acad. Sci. USA* **102,** 503–508.

Cai, Q., Ferraris, J. D., and Burg, M. B. (2005b). High NaCl increases TonEBP/OREBP mRNA and protein by stabilizing its mRNA. *Am. J. Physiol. Renal Physiol.* **289,** F803–F807.

Caruccio, L., Bae, S., Liu, A. Y., and Chen, K. Y. (1997). The heat-shock transcription factor HSF1 is rapidly activated by either hyper- or hypo-osmotic stress in mammalian cells. *Biochem. J.* **327,** 341–347.

Clerk, A., and Sugden, P. H. (1997). Cell stress-induced phosphorylation of ATF2 and c-Jun transcription factors in rat ventricular myocytes. *Biochem. J.* **325,** 801–810.

Cohen, D. M., Wasserman, J. C., and Gullans, S. R. (1991). Immediate early gene and HSP70 expression in hyperosmotic stress in MDCK cells. *Am. J. Physiol.* **261,** C594–C601.

Cohen, P. (2000). The regulation of protein function by multisite phosphorylation: A 25 year update. *Trends Biochem. Sci.* **25,** 596–601.

Cyert, M. S. (2001). Regulation of nuclear localization during signaling. *J. Biol. Chem.* **276,** 20805–20808.

Dahl, S. C., Handler, J. S., and Kwon, H. M. (2001). Hypertonicity-induced phosphorylation and nuclear localization of the transcription factor TonEBP. *Am. J. Physiol. Cell Physiol.* **280,** C248–C253.

De Cesare, D., Fimia, G. M., and Sassone-Corsi, P. (1999). Signaling routes to CREM and CREB: Plasticity in transcriptional activation. *Trends Biochem. Sci.* **24,** 281–285.

Dmitrieva, N., Kultz, D., Michea, L., Ferraris, J., and Burg, M. (2000). Protection of renal inner medullary epithelial cells from apoptosis by hypertonic stress-induced p53 activation. *J. Biol. Chem.* **275,** 18243–18247.

Ferraris, J. D., Persaud, P., Williams, C. K., Chen, Y., and Burg, M. B. (2002a). cAMP-independent role of PKA in tonicity-induced transactivation of tonicity-responsive enhancer/osmotic response element-binding protein. *Proc. Natl. Acad. Sci. USA* **99,** 16800–16805.

Ferraris, J. D., Williams, C. K., Martin, B. M., Burg, M. B., and Garcia-Perez, A. (1994). Cloning, genomic organization, and osmotic response of the aldose reductase gene. *Proc. Natl. Acad. Sci. USA* **91,** 10742–10746.

Ferraris, J. D., Williams, C. K., Ohtaka, A., and Garcia-Perez, A. (1999). Functional consensus for mammalian osmotic response elements. *Am. J. Physiol.* **276,** C667–C673.

Ferraris, J. D., Williams, C. K., Persaud, P., Zhang, Z., Chen, Y., and Burg, M. B. (2002b). Activity of the TonEBP/OREBP transactivation domain varies directly with extracellular NaCl concentration. *Proc. Natl. Acad. Sci. USA* **99,** 739–744.

Gallazzini, M., Ferraris, J. D., Kunin, M., Morris, R. G., and Burg, M. B. (2006). Neuropathy target esterase catalyzes osmoprotective renal synthesis of glycerophosphocholine in response to high NaCl. *Proc. Natl. Acad. Sci. USA* **103,** 15260–15265.

Garnovskaya, M. N., Mukhin, Y. V., Vlasova, T. M., and Raymond, J. R. (2003). Hypertonicity activates Na+/H+ exchange through Janus kinase 2 and calmodulin. *J. Biol. Chem.* **278,** 16908–16915.

Gatsios, P., Terstegen, L., Schliess, F., Haussinger, D., Kerr, I. M., Heinrich, P. C., and Graeve, L. (1998). Activation of the Janus kinase/signal transducer and activator of transcription pathway by osmotic shock. *J. Biol. Chem.* **273,** 22962–22968.

Groner, B., Fritsche, M., Stocklin, E., Berchtold, S., Merkle, C., Moriggl, R., and Pfitzner, E. (2000). Regulation of the trans-activation potential of STAT5 through its DNA-binding activity and interactions with heterologous transcription factors. *Growth Horm. IGF Res.* **10**(Suppl. B), S15–S20.

Han, J., Lee, J. D., Bibbs, L., and Ulevitch, R. J. (1994). A MAP kinase targeted by endotoxin and hyperosmolarity in mammalian cells. *Science* **265,** 808–811.

Hasler, U., Jeon, U. S., Kim, J. A., Mordasini, D., Kwon, H. M., Feraille, E., and Martin, P. Y. (2006). Tonicity-responsive enhancer binding protein is an essential regulator of aquaporin-2 expression in renal collecting duct principal cells. *J. Am. Soc. Nephrol.* **17,** 1521–1531.

Holmberg, C. I., Tran, S. E., Eriksson, J. E., and Sistonen, L. (2002). Multisite phosphorylation provides sophisticated regulation of transcription factors. *Trends Biochem. Sci.* **27,** 619–627.

Huang, Z., and Tunnacliffe, A. (2005). Gene induction by desiccation stress in human cell cultures. *FEBS Lett.* **579,** 4973–4977.

Irarrazabal, C. E., Burg, M. B., Ward, S. G., and Ferraris, J. D. (2006). Phosphatidylinositol 3-kinase mediates activation of ATM by high NaCl and by ionizing radiation: Role in osmoprotective transcriptional regulation. *Proc. Natl. Acad. Sci. USA* **103,** 8882–8887.

Irarrazabal, C. E., Liu, J. C., Burg, M. B., and Ferraris, J. D. (2004). ATM, a DNA damage-inducible kinase, contributes to activation by high NaCl of the transcription factor TonEBP/OREBP. *Proc. Natl. Acad. Sci. USA* **101,** 8809–8814.

Kakidani, H., and Ptashne, M. (1988). GAL4 activates gene expression in mammalian cells. *Cell* **52,** 161–167.

Kapus, A., Szaszi, K., Sun, J., Rizoli, S., and Rotstein, O. D. (1999). Cell shrinkage regulates Src kinases and induces tyrosine phosphorylation of cortactin, independent of the osmotic regulation of Na^+/H^+ exchangers. *J. Biol. Chem.* **274,** 8093–8102.

Ko, B. C., Lam, A. K., Kapus, A., Fan, L., Chung, S. K., and Chung, S. S. (2002). Fyn and p38 signaling are both required for maximal hypertonic activation of the OREBP/TonEBP. *J. Biol. Chem.* **277,** 46085–46092.

Ko, B. C., Turck, C. W., Lee, K. W., Yang, Y., and Chung, S. S. (2000). Purification, identification, and characterization of an osmotic response element binding protein. *Biochem. Biophys. Res. Commun.* **270,** 52–61.

Ko, B. C. B., Ruepp, B., Bohren, K. M., Gabbay, K. H., and Chung, S. S. (1997). Identification and characterization of multiple osmotic response sequences in the human aldose reductase gene. *J. Biol. Chem.* **272,** 16431–16437.

Lam, A. K., Ko, B. C., Tam, S., Morris, R., Yang, J. Y., Chung, S. K., and Chung, S. S. (2004). Osmotic response element-binding protein (OREBP) is an essential regulator of the urine concentrating mechanism. *J. Biol. Chem.* **279,** 48048–48054.

Lee, S. D., Colla, E., Sheen, M. R., Na, K. Y., and Kwon, H. M. (2003). Multiple domains of TonEBP cooperate to stimulate transcription in response to hypertonicity. *J. Biol. Chem.* **278,** 47571–47577.

Lee, S. D., Woo, S. K., and Kwon, H. M. (2002). Dimerization is required for phosphorylation and DNA binding of TonEBP/NFAT5. *Biochem. Biophys. Res. Commun.* **294,** 968–975.

Lin, K. W., Nam, S. Y., Toh, W. H., Dulloo, I., and Sabapathy, K. (2004). Multiple stress signals induce p73beta accumulation. *Neoplasia* **6,** 546–557.

Lopez-Rodriguez, C., Aramburu, J., Jin, L., Rakeman, A. S., Michino, M., and Rao, A. (2001). Bridging the NFAT and NF-kappaB families: NFAT5 dimerization regulates cytokine gene transcription in response to osmotic stress. *Immunity* **15,** 47–58.

Lopez-Rodriguez, C., Aramburu, J., Rakeman, A. S., and Rao, A. (1999). NFAT5, a constitutively nuclear NFAT protein that does not cooperate with Fos and Jun. *Proc. Natl. Acad. Sci. USA* **96,** 7214–7219.

Macian, F., Lopez-Rodriguez, C., and Rao, A. (2001). Partners in transcription: NFAT and AP-1. *Oncogene* **20,** 2476–2489.

Miyakawa, H., Woo, S. K., Dahl, S. C., Handler, J. S., and Kwon, H. M. (1999). Tonicity-responsive enhancer binding protein, a Rel-like protein that stimulates transcription in response to hypertonicity. *Proc. Natl. Acad. Sci. USA* **96,** 2538–2542.

Nakayama, Y., Peng, T., Sands, J. M., and Bagnasco, S. M. (2000). The TonE/TonEBP pathway mediates tonicity-responsive regulation of UT-A urea transporter expression. *J. Biol. Chem.* **275,** 38275–38280.

Nemeth, Z. H., Deitch, E. A., Szabo, C., and Hasko, G. (2002). Hyperosmotic stress induces nuclear factor–kappaB activation and interleukin-8 production in human intestinal epithelial cells. *Am. J. Pathol.* **161,** 987–996.

Ptashne, M. (1988). How eukaryotic transcriptional activators work. *Nature* **335,** 683–689.

Sadowski, I., Ma, J., Triezenberg, S., and Ptashne, M. (1988). GAL4-VP16 is an unusually potent transcriptional activator. *Nature* **335,** 563–564.

Schmitz, M. L., Bacher, S., and Kracht, M. (2001). I kappa B-independent control of NF-kapa B activity by modulatory phosphorylations. *Trends Biochem. Sci.* **26**, 186–190.

Takenaka, M., Preston, A. S., Kwon, H. M., and Handler, J. S. (1994). The tonicity-sensitive element that mediates increased transcription of the betaine transporter gene in response to hypertonic stress. *J. Biol. Chem.* **269**, 29379–29381.

Thiel, G., and Cibelli, G. (2002). Regulation of life and death by the zinc finger transcription factor Egr-1. *J. Cell Physiol.* **193**, 287–292.

Tong, E. H., Guo, J. J., Huang, A. L., Liu, H., Hu, C. D., Chung, S. S., and Ko, B. C. (2006). Regulation of nucleocytoplasmic trafficking of transcription factor OREBP/TonEBP/NFAT5. *J. Biol. Chem.* **281**, 23870–23879.

Torban, E., and Goodyer, P. (1998). What PAX genes do in the kidney. *Exp. Nephrol.* **6**, 7–11.

Trama, J., Lu, Q., Hawley, R. G., and Ho, S. N. (2000). The NFAT-related protein NFATL1 (TonEBP/NFAT5) is induced upon T cell activation in a calcineurin-dependent manner. *J. Immunol.* **165**, 4884–4894.

Vega, S., Morales, A. V., Ocana, O. H., Valdes, F., Fabregat, I., and Nieto, M. A. (2004). Snail blocks the cell cycle and confers resistance to cell death. *Genes Dev.* **18**, 1131–1143.

Webster, N., Jin, J. R., Green, S., Hollis, M., and Chambon, P. (1988). The yeast UASG is a transcriptional enhancer in human HeLa cells in the presence of the GAL4 trans-activator. *Cell* **52**, 169–178.

Wiese, S., Schliess, F., and Haussinger, D. (1998). Osmotic regulation of MAP-kinase activities and gene expression in H4IIE rat hepatoma cells. *Biol. Chem.* **379**, 667–671.

Wollnik, B., Kubisch, C., Maass, A., Vetter, H., and Neyses, L. (1993). Hyperosmotic stress induces immediate-early gene expression in ventricular adult cardiomyocytes. *Biochem. Biophys. Res. Commun.* **194**, 642–646.

Woo, S. K., Dahl, S. C., Handler, J. S., and Kwon, H. M. (2000). Bidirectional regulation of tonicity-responsive enhancer binding protein in response to changes in tonicity. *Am. J. Physiol. Renal Physiol.* **278**, F1006–F1012.

Woo, S. K., Lee, S. D., Na, K. Y., Park, W. K., and Kwon, H. M. (2002). TonEBP/NFAT5 stimulates transcription of HSP70 in response to hypertonicity. *Mol. Cell Biol.* **22**, 5753–5760.

Xi, H., and Kersh, G. J. (2004). Sustained early growth response gene 3 expression inhibits the survival of CD4/CD8 double-positive thymocytes. *J. Immunol.* **173**, 340–348.

Yancey, P. H., Clark, M. E., Hand, S. C., Bowlus, R. D., and Somero, G. N. (1982). Living with water stress: Evolution of osmolyte systems. *Science* **217**, 1214–1222.

Ying, Z., Reisman, D., and Buggy, J. (1996). AP-1 DNA binding activity induced by hyperosmolality in the rat hypothalamic supraoptic and paraventricular nuclei. *Brain Res. Mol. Brain Res.* **39**, 109–116.

Zhang, Z., Dmitrieva, N. I., Park, J. H., Levine, R. L., and Burg, M. B. (2004). High urea and NaCl carbonylate proteins in renal cells in culture and *in vivo*, and high urea causes 8-oxoguanine lesions in their DNA. *Proc. Natl. Acad. Sci. USA* **101**, 9491–9496.

Zhang, Z., Ferraris, J., Irarrazabal, C. E., Dmitireva, N. I., Park, J. H., and Burg, M. B. (2005). Ataxia-telangiectasia mutated (ATM), a DNA damage-inducible kinase, contributes to high NaCl-induced nuclear localization of the transcription factor TonEBP/OREBP. *Am. J. Physiol. Renal Physiol.* **289**, F506–F511.

Zhou, X., Ferraris, J. D., Cai, Q., Agarwal, A., and Burg, M. B. (2005). Increased reactive oxygen species contribute to high NaCl-induced activation of the osmoregulatory transcription factor TonEBP/OREBP. *Am. J. Physiol. Renal Physiol.* **289**, F377–F385.

CHAPTER SEVENTEEN

HYPEROSMOTIC INDUCTION OF MITOGEN-ACTIVATED PROTEIN KINASE SCAFFOLDING

Thomas L. Hilder, Michael H. Malone, *and* Gary L. Johnson

Contents

Abstract

Eukaryotic cells respond to hyperosmotic conditions by expunging water from the cell, leading to cell shrinkage. This is counteracted by adaptive responses that restore cell volume and strengthen the cytoskeletal architecture. In the budding

Department of Pharmacology and Lineberger Comprehensive Cancer Center, University of North Carolina at Chapel Hill, Chapel Hill, North Carolina

Methods in Enzymology, Volume 428

ISSN 0076-6879, DOI: 10.1016/S0076-6879(07)28017-6

yeast *Saccharomyces cerevisiae*, this response is mediated primarily by the mitogen-activated protein kinase (MAPK) cascade CDC42-STE50-STE11-Pbs2-Hog1. In mammalian cells, MAPK scaffold proteins facilitate the efficiency of signaling within the cascade by placing a kinase near its substrate and also regulate the subcellular localization of the signaling. Our laboratory has discovered a scaffold that coordinates the analogous Hog1 signal in mammalian cells, termed OSM (*o*smosensing *s*caffold for *M*EKK3). OSM organizes a complex consisting of the small GTPase Rac, MEKK3, and MKK3 for the activation of p38 MAPK. Interactions among OSM, Rac, and MEKK3 are augmented in response to sorbitol and are also localized to membrane ruffles, sites of rapid actin turnover. Suppression of the expression of OSM or MEKK3 by RNA interference strongly inhibits the sorbitol-dependent activation of p38. Furthermore, mutations in OSM were concurrently found to cause cerebral cavernous malformations (CCM), a disease of the central nervous system characterized by thin-walled, leaky blood vessels that become hemorrhagic. Our laboratory has also demonstrated that Krit1, another gene harboring mutations that lead to CCM, binds OSM and its interaction is enhanced in response to sorbitol in a similar manner as the MEKK3–OSM interaction. This chapter describes the cell biological and biochemical methods used for assaying protein–protein interactions in live cells using fluorescence resonance energy transfer, *in vitro* kinase assays for MEKK3-MKK3-p38 pathway members, and gene suppression by RNA interference to study hyperosmotic stress-dependent signaling.

1. INTRODUCTION

The mitogen–activated protein kinase (MAPK) family responds to a diverse number of stimuli that lead to changes in cell proliferation, differentiation, survival, and motility, among others. Activation of MAPKs (Erk1/2, p38, JNK, Erk5, Erk7) occurs in a three kinase phosphorelay cascade in which 1 of 20 MAP/Erk kinase kinase kinases (MEKK or MAP3K) phosphorylates 1 or more of 7 MAP2K to activate one of the five MAPK families (Uhlik *et al.*, 2004). One well-characterized MAPK pathway in the yeast *Saccharomyces cerevisiae* is the Hog1 pathway. High osmolarity results in activation of the CDC42-STE50-STE11-Pbs2-Hog1 cascade; this in turn activates a number of transcription factors, kinases, and other enzymes that allow for yeast survival and adaptation (Saito and Tatebayashi, 2004). In mammalian cells, the homologous pathway that is activated in response to hyperosmotic stress includes Rac-MEKK3-MKK3-p38 (Uhlik *et al.*, 2003). Substrates of p38 in mammalian cells include transcription factors, controlling hyperosmotic stress–induced gene transcription (Bell *et al.*, 2000; Garmyn *et al.*, 2001; Sheikh-Hamad *et al.*, 1998), and kinases such as MAPK–activated protein kinase-2 (MAPKAPK2), which phosphorylates HSP27 to stabilize actin filaments under conditions of stress (Guay *et al.*, 1997). In addition to p38, Erk1/2 and Erk5 MAPKs have been shown to be activated in response to hyperosmotic stress (Fusello *et al.*, 2006; Wang *et al.*, 2006). Therefore, because hyperosmotic stress induces a substantial efflux of water from the cell, the activation of multiple MAPKs is critical for survival as the cell adapts and responds to its environment.

In many instances, MAPK signals are mediated by scaffold proteins with no enzymatic activity that bind and organize multiple components of the cascade, thereby regulating the duration of the signal as well as its subcellular localization (Johnson *et al.*, 2005). A scaffold cloned by our laboratory, termed OSM (*o*smosensing *s*caffold for *M*EKK3), was shown to bind the small GTPase Rac, MEKK3, and MKK3 (Fig. 17.1). OSM is required for mediating p38 activation in response to sorbitol, as short interfering RNA (siRNA) knockdown of OSM (or MEKK3) reduces the sorbitol-dependent activation of p38 significantly (Uhlik *et al.*, 2003). Furthermore, using fluorescence resonance energy transfer (FRET) to measure protein–protein interactions in live cells, it was shown that the interaction of OSM and MEKK3 is enhanced at membrane ruffles in response to sorbitol (Uhlik *et al.*, 2003). Another scaffold termed KSR (*k*inase *s*uppressor of *R*as), which binds Raf, Mek, and Erk, was shown to be required for the sorbitol-dependent activation of Erk, as mouse embryonic fibroblasts derived from KSR$^{-/-}$ animals display reduced Erk activation as compared to heterozygous cells (Fusello *et al.*, 2006). Therefore, hyperosmotic stress-induced activation and localization of MAPKs appear to be mediated by multiple scaffold proteins.

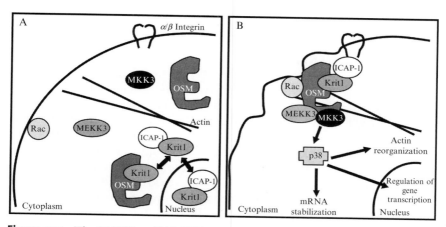

Figure 17.1 The MAPK scaffold OSM organizes a complex at membrane ruffles for p38 activation in response to sorbitol. (A) In resting cells, OSM and its binding partners are diffusely localized in the cytoplasm. Low levels of OSM and Krit1 interact in resting cells, as measured by FRET (Zawistowski *et al.*, 2005). Krit1 is also capable of shuttling to the nucleus and interacting with the integrin-binding protein ICAP-1. (B) Following sorbitol treatment, OSM organizes a complex at membrane ruffles consisting of the small GTPase Rac, Krit1, MEKK3, and MKK3 for the activation of p38 MAPK. OSM also binds F-actin and the two proteins colocalize at membrane ruffles (Uhlik *et al.*, 2003).

Concurrent with our identification of OSM as a scaffold for p38 activation, Liquori and colleagues (2003) mapped mutations in the gene MGC4607 (the human orthologue of OSM) from nine families with cerebral cavernous malformations (CCMs). CCMs are vascular lesions of the central nervous system characterized by dilated, thin-walled, leaky vessels that become hemorrhagic and lead to seizures and stroke and result from sporadic or inherited mutations in one of three genes (protein in parenthesis): *ccm1* (Krit1), *ccm2* (OSM, malcavernin), or *ccm3* (PDCD10, CCM3) (Bergametti *et al.*, 2005; Denier *et al.*, 2004; Laberge-le Couteulx *et al.*, 1999; Liquori *et al.*, 2003; Sahoo *et al.*, 1999). It has since been demonstrated that OSM binds through its phosphotyrosine-binding (PTB) domain to Krit1, as an engineered point mutation or an OSM patient mutation within the PTB domain disrupts this interaction (Zawistowski *et al.*, 2005). Like OSM and MEKK3, the interaction of OSM and Krit1 is enhanced at membrane ruffles in response to sorbitol as measured by FRET (Zawistowski *et al.*, 2005). It is currently unknown what role Krit1 plays in the regulation and/or localization of the MEKK3-MKK3-p38 cascade; however, the scaffolds Krit1 and OSM, along with MEKK3, are coordinately translocated to membrane ruffles in response to hyperosmotic stress.

This chapter presents methods used to assess the redistribution and interaction of proteins in live cells, enzymatic assays for MAPKs, and RNA interference of MAPK scaffolds in the context of hyperosmotic stress–induced MAPK activation.

2. RNA INTERFERENCE

The use of short interfering RNA, a gene-specific, double-stranded RNA (dsRNA) oligonucleotide, allows for degradation of the homologous cellular RNA, thereby suppressing the translation of a protein of interest to nearly undetectable levels (Hannon, 2002; Hannon and Rossi, 2004). The extent of protein knockdown depends on the stability of the oligonucleotide, the efficiency of siRNA uptake into the cell, the potency of the oligonucleotide toward its target, and the incubation time following introduction of the siRNA into the cell. Using the following transfection protocols, near complete knockdown of MAPK scaffolds or MAP3Ks in our laboratory has been observed as soon as 24 h posttransfection of the oligonucleotides and persists to at least 72 h. In some cells that proliferate rapidly and have poor uptake of siRNA, the oligonucleotides can be effectively "diluted out" by 72 h as the cells lacking oligonucleotides overrun the population. However, this is usually the exception.

We have had considerable success with the suppression of MAPK scaffolds using siRNAs purchased from Dharmacon (Lafayette, CO). The role of

OSM in sorbitol-mediated p38 activation was determined using a single dsRNA (Uhlik *et al.*, 2003); however, the use of ON-TARGETplus SMARTpool siRNAs from Dharmacon has given us a much greater knock-down of OSM (data not shown). Oligonucleotides with no target ("scrambled" sequence) are available from Dharmacon as a negative experimental control. The following protocols were used in our laboratory for the successful introduction of siRNAs into a number of cell lines, including mouse embryonic fibroblasts, RAW264.7 macrophages, and HEK293 cells.

2.1. Electroporation of siRNA into adherent cells

Rinse cells growing in log phase in phosphate-buffered saline (PBS) and trypsinize. Stop trypsinization by adding an equal volume of serum-containing medium without antibiotics. Spin 4×10^6 cells per oligonucleotide. Remove the trypsin/medium solution and resuspend the cell pellet in 0.2 ml of the Nucleofector solution (Amaxa Biosystems, Cologne, Germany) specific to the cell line used. Immediately add 100 to 300 pmol of siRNA, mix, and add to the Nucleofector cuvette. Electroporate with the appropriate program for the cell type used. Electroporation conditions for numerous cell lines have been optimized by Amaxa and are available online. Following electroporation, remove the cell suspension and add to 0.8 ml of culture medium supplemented with serum but without antibiotics in a 1.5-ml microfuge tube. Allow the cells to recover at 37° for 5 to 10 min. Plate the cells at a low enough density to allow 48 to 72 h growth without reaching confluence. Harvest the cells in a suitable lysis buffer and determine knockdown of the target protein by immunoblotting.

2.2. Transfection of siRNA into adherent cells

Dilute 50 to 200 pmol of oligonucleotide in 0.5 ml OptiMEM (Invitrogen, Carlsbad, CA) and mix gently. Dilute 0.01 ml of Lipofectamine 2000 or OligofectAMINE (Invitrogen) in 0.5 ml OptiMEM. Incubate both mixtures for 5 min at room temperature, mix both solutions, and incubate at room temperature for 20 min. Rinse cells at 30 to 50% confluency (in 60-mm culture dishes) in antibiotic-free mediumand replace with 4 ml of this medium. Add the siRNA mixture to the cells, incubate 48 to 72 h, and then assess knockdown of the target protein by immunoblotting. For example, to determine the role the MAPK scaffold OSM plays in response to hyperosmotic stress, HEK293 cells are transfected with control or OSM siRNAs. Following a 24-h incubation, cells are serum starved for another 24 h and then are treated with 0.2 M sorbitol for 15 min. Cells are lysed in ice-cold lysis buffer (50 mM Tris-HCl, pH 7.5, 100 mM NaCl, 50 mM NaF, 5 mM Na$_4$P$_2$O$_7$, 1 mM Na$_3$VO$_4$, 0.5% Triton X-100, 1 mM EDTA, 1 mM EGTA, and supplemented with protease inhibitors), and OSM levels are measured by immunoblotting (Uhlik *et al.*, 2003). p38 activity is measured as described later.

3. *In Vitro* Kinase Assays

The ability to measure kinase activity is critical for determining the effect of hyperosmotic stress-induced MAPK scaffolding. A number of commercially available antibodies are directed toward the activating phosphorylation sites within a kinase and can serve as a relative measure of kinase activity. These include p-p38 (*Thr180/Tyr182*) and p-MKK3/6 (*Ser189/207*) from Cell Signaling Technology (Danvers, MA). However, immunoblotting is relatively qualitative and can only show changes in kinase activity over a limited dynamic range. Additionally, phospho-specific antibodies have not been generated for all members of the MAPK family, particularly at the level of the MAP3Ks. Therefore, *in vitro* kinase assays allow for a more quantitative assessment of kinase activity.

3.1. MEKK3 assay

To determine the sorbitol-dependent activation of MEKK3, serum starve the cells of interest for 3 h and then treat with vehicle or 0.2 M sorbitol for 5 to 15 min. Rinse the cells twice in PBS and lyse in ice-cold MEKK3 lysis buffer (20 mM Tris-HCl, pH 7.4, 1% Nonidet P-40 [NP-40], 135 mM NaCl, 10% glycerol, 0.1 mM EDTA, 0.1 mM EGTA, plus protease inhibitors). For MEKK3 assays, Triton X-100 cannot be substituted for NP-40 as MEKK3 is inactive in this detergent (Widmann *et al.*, 2001). Immunoprecipitate MEKK3 with the anti-MEKK3 antibody (BD Biosciences, San Jose, CA) from 1 mg of lysate in 1 ml of MEKK3 lysis buffer for 1 to 2 h and then add 0.03 ml of recombinant protein G–Sepharose 4B (Invitrogen). Alternatively, epitope-tagged MEKK3 subcloned into a mammalian expression vector such as pcDNA3 or pCMV5 can be transfected into cells 24 to 48 h prior to treatment of cells with sorbitol and immunoprecipitated with antibodies against the epitope tag or MEKK3. Collect the immunoprecipitated complex by centrifugation and wash three times with cold buffer B and two times with MEKK3 kinase buffer (20 mM HEPES, pH 7.4, 10 mM MgCl$_2$, 1 mM dithiothreitol [DTT], 0.1 mM Na$_3$VO$_4$, and 10 mM β-glycerophosphate). Remove all liquid from the immunoprecipitates, and begin the assay by adding 0.05 ml of MEKK3 kinase buffer containing 0.5 μg His-MKK6 K/M (kinase inactive MKK6 purified from *Escherichia coli* using standard methods) and 1 μl 10 mCi/ml [γ-^{32}P]ATP (MP Biomedicals, Irvine, CA). Incubate the assay at 30° for 20 min and stop the assay by adding 2× SDS-PAGE buffer. Load the assay on a gel; after running, stain the gel with Coomassie or silver, dry, and expose to X-ray film. Quantitate the amount of phosphate incorporated into the substrate by excising the His-MKK6 band from the dried gel and subjecting it to liquid scintillation counting. This assay as described was used to demonstrate that sorbitol induces a twofold increase in MEKK3 kinase activity following a 15-min sorbitol treatment (Uhlik et al., 2003).

3.2. p38 MAPK assay

The p38 MAPK assay is performed essentially as just described with the following differences. Endogenous p38 MAPK can be immunoprecipitated or FLAG-p38α in pCMV5 can be transfected prior to the experiment. Following sorbitol treatment, rinse the cells in PBS and lyse in p38 lysis buffer (20 mM Tris-HCl, pH 7.5, 150 mM NaCl, 20 mM NaF, 0.2 mM Na_3VO_4, 1% Triton X-100, 0.5% NP-40, 1 mM EDTA, 1 mM EGTA, and supplemented with protease inhibitors). Following immunoprecipitation, wash three times in lysis buffer and twice in p38 kinase buffer (25 mM HEPES, pH 7.4, 25 mM β-glycerophosphate, 25 mM $MgCl_2$, 4 mM DTT, 0.1 mM Na_3VO_4). Remove all liquid from the immunoprecipitates, and begin the assay by adding 38 μl kinase buffer, 1 μl bacterially expressed and purified recombinant, inactive, human His-ATF2 (1 μg/μl), and 1 μl [γ-^{32}P]ATP. Incubate for 20 min at 30° and stop the assay by adding 40 μl 2× SDS-PAGE buffer. Run on a 10% SDS-PAGE gel and analyze as described earlier. This assay was performed following the cotransfection of FLAG-p38 MAPK with wild-type or dominant negative Rac, the results of which demonstrated that inactive Rac reduces p38 activity substantially in both untreated and sorbitol-treated cells (Uhlik et al., 2003).

4. MEASURING DYNAMIC INTERACTIONS BETWEEN MAPK SCAFFOLD MEMBERS USING FRET

By coordinating when and where proteins interact within the cell, scaffold proteins significantly influence biological responses to physiological stimuli. Measuring the Förster fluorescence resonance energy transfer between two chimeric fluorescent proteins is a quantitative approach to determining the spatial and temporal regulation of protein interactions (Jares-Erijman and Jovin, 2003; Pollok and Heim, 1999). FRET occurs when an excited donor fluorophore promotes the excitation of an adjacent acceptor fluorophore without emission of a photon (Forster, 1948). For example, excitation of the donor cyan fluorescent protein (CFP) with 435 nm of light normally causes emission of blue fluorescence around 477 nm. However, in the presence of a nearby acceptor, excited CFP will transfer the excited energy to YFP, which will then emit yellow fluorescence around 527 nm (Fig. 17.2). Because this

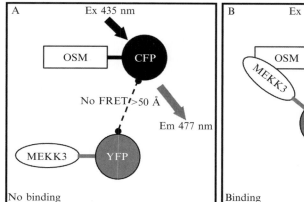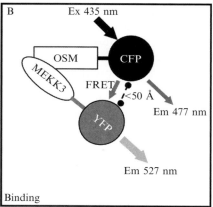

Figure 17.2 Fluorescence resonance energy transfer. When two proteins are >50 Å apart, excitation of CFP at 435 nm leads to normal CFP emission at 477 nm (A). However, when two proteins are bound, bringing the CFP donor in close proximity (less than 50 Å) to the YFP acceptor molecule, excitation of CFP leads to an emission of fluorescence energy capable of exciting YFP, leading to a measurable emission at 527 nm (B). Ex, excitation; Em, emission.

energy transfer does not involve emission of a photon, the acceptor and donor molecules must be very close to each other. The probability of an excited donor molecule transferring its energy to an acceptor molecule is inversely related to the sixth power of the distance between donor and acceptor. Thus FRET only occurs when donor and acceptor are within close proximity of each other (≈50 Å for fluorescent protein pairs). Technological advances in fluorescent proteins, band-pass filters, and epifluorescent and confocal microscopes have made monitoring the spatial and temporal organization of signal transduction complexes by FRET accessible to most investigators. This section discusses aspects of experimental design, instrumentation, data acquisition, and analysis found critical to performing a successful FRET experiment.

4.1. Considerations for constructing fluorescent chimeric proteins

The most common donor:acceptor pair for live-cell FRET analysis is CFP: YFP. Newly engineered variants of CFP and YFP are available that have increased brightness (Cerulean and mCitrine) or better FRET efficiency (cyPet and yPet), and the development of improved fluorescent proteins is the focus of intensive research (Nguyen and Daugherty, 2005; Shaner *et al.*, 2005). The benefits of using a new variant should always be considered before designing a new fluorescent chimera.

Although GFP-based fluorescent proteins fold into a compact β-barrel structure, steric hindrance can prevent protein–protein interactions from occurring, particularly if protein-interaction domains are located near the end that is fused to the fluorescent protein. For this reason it is advisable to make both amino- and carboxy-terminal versions of the fusion proteins and assay each potential combination of donor:acceptor pair.

4.2. Selecting a cell system to answer meaningful biological questions

Before sitting in front of a microscope, one must carefully choose a cell type, a transfection strategy, and a method for controlling the abundance of expressed proteins in order to obtain biologically meaningful FRET data. To be biologically relevant, the cell type used should express all components of the signal transduction pathway of interest and respond in an appropriate manner to physiological stimuli. For example, bone marrow-derived macrophages produce proinflammatory cytokines in response to bacterial lipopolysaccharide, making them suitable cells for studying the organization of signaling complexes that promote cytokine gene transcription. In addition, ideal cells for FRET imaging are flat, tightly adherent, and easily transfected.

Whether to express fluorescent proteins by stable or transient transfection is often determined by the transfection efficiency. Transient transfection is preferred if transfection efficiency is good (>30%) because it allows a number of transfection conditions to be screened quickly in order to obtain cells that express the fluorescent proteins at an appropriate level (discussed later). However, if a chosen cell type transfects poorly, making stable cell lines may ultimately be the least labor-intensive.

The ability to regulate the abundance of expressed protein is also critical for obtaining interpretable results. Overexpression of fluorescently tagged proteins can force interactions with low-affinity interaction partners that do not normally occur at endogenous expression levels. For this reason, fluorescent proteins should be expressed close to endogenous levels, and expressing fluorescent chimeras at endogenous levels in cells derived from knockout mice or utilizing RNA interference or knockouts is best. While ideal, this approach is not always feasible or practical and investigators must be careful to balance overexpression with obtaining a suitable fluorescence signal. Finally, optimal energy transfer between acceptor and donor fluorophore is obtained when the expression ratios are approximately 1:1 (Berney and Danuser, 2003). Therefore, the abundance CFP and YFP fusion proteins should be approximately equal and this ratio should be fairly constant across multiple samples.

4.3. Microscope configuration

Many different wide-field and confocal microscope configurations may be used for FRET measurements in living cells. Each particular system has its own advantages and disadvantages and are reviewed elsewhere (Sekar and Periasamy, 2003). Regardless of the microscope used, we recommend a three-channel FRET analysis paradigm that uses three excitation/emission filter sets, one each for imaging CFP (CFP excitation/CFP emission), YFP (YFP excitation/YFP emission), and FRET (CFP excitation/YFP emission). Our epifluorescence microscope system consists of a Zeiss Axiovert 200M inverted fluorescence microscope, a xenon arc lamp, an encased stage with temperature, CO_2, and humidity controls, the CoolsnapHQ cooled charge-coupled device (CCD) camera (Photometrics), and an optical air table to isolate room vibrations. Images are collected using a 100× 1.4 NA Apochromat objective using three filter sets from Chroma Technology Corporation: YFP (Set #41028), CFP (Set #31044v2), and FRET (Set #86002v1). These filter sets are rotated into the optical path by means of a computer-controlled motorized turret. Microscope control, image capture, and processing are performed with Slidebook v 4.0 (Intelligent Imaging Innovations, Inc.).

4.4. Cell culture

Transfection of fluorescent chimeras should be performed as appropriate for the chosen cell type. Typically, cells are plated on 25-mm circular #1 thickness cover glasses, placed within a well of a six-well plate, transfected with a cationic transfection reagent, or electroporated and then plated. Cells are typically imaged 24 h after transfection. The following samples should be prepared.

Sample	*Purpose*
YFP-fusion alone	Control for spectral cross talk
CFP-fusion alone	Control for spectral cross talk
CFP-fusion + YFP (free)	Negative control for FRET
YFP-fusion + CFP (free)	Negative control for FRET
CFP-fusion + YFP-fusion	Experimental sample

Spectral cross talk is unique to each microscope configuration and should be measured for each fluorophore. It should be monitored frequently and recalculated as described later whenever changes to the microscope optics are made. In addition to the aforementioned samples, it is also extremely beneficial to express a set of CFP/YFP fusion proteins known to interact and provide ample energy transfer to use as a positive control.

4.5. Setting image acquisition parameters

Determining the optimal acquisition settings for FRET imaging can be tricky. The exposure settings for a CCD camera (or gain settings for a photomultiplier tube) must be *identical* for each channel (CFP, YFP, and FRET). This can be problematic, as YFP is often brighter than CFP under the same exposure settings. Thus, it is difficult to optimize the exposure of CFP without saturating the YFP signal. Therefore, we typically set the exposure settings using the YFP channel such that both YFP and CFP images have a signal-to-noise ratio that is at least 2:1. Because oxygen radicals are produced with high-intensity light, phototoxicity can be minimized by choosing longer exposures with dim illumination (provided by neutral density filters). However, exposures for live-cell FRET imaging should be as short as possible to minimize artifacts introduced by the movement of protein complexes during image acquisition. In cases of dim fluorescence, the recorded signal can often be increased by increasing the detector gain, binning data (for CCD detectors), or increasing the scan time (for confocal microscopes). Once acquisition parameters have been determined, they should not be changed during the course of the experiment.

4.6. Image acquisition

If prolonged time-lapse images are to be acquired, use of an environmentally controlled stage chamber or stage is recommended to control temperature, carbon dioxide, and humidity. Round cover glasses are removed from wells of a six-well plate using angled forceps and placed cell side up, on a round live cell imaging chamber, and then bathed in 3 ml complete medium without phenol red. If a means of controlling carbon dioxide is not available, medium should contain 25 mM HEPES, pH 7.4. For each sample, images for each of three channels (CFP, YFP, and FRET) are acquired. Choose only fields of cells that do not saturate the detector and have a fluorescence signal that is greater than 10% of the maximum bit level detected by your detector (above 25 for a 255-bit level 8-bit detector, 410 for a 4095-level 12-bit detector, or 655 for a 65,535-level 16-bit detector). To monitor the redistribution of protein–protein interactions in response to hyperosmotic stress, begin collecting a series of time-lapse three-channel images. After three or more time points have been collected, pause the acquisition and carefully add 1 ml sorbitol-containing medium near the edge of the dish. Great care must be taken to avoid touching any component of the microscope so that neither the field of view nor the focus is changed. Resume capture immediately after sorbitol addition and continue capture until no further changes in morphology or protein localization are observed.

4.7. Image processing

All captured images should be corrected for uneven illumination (flat–field correction) and then a local background should be subtracted. The registration of each channel should be verified and corrected, if necessary, using an automatic image alignment feature found in most digital image analysis software.

4.8. Determining spectral cross talk coefficients

Many image processing packages have plug–ins that will calculate FRET efficiencies, including cross talk coefficients. These coefficients can also be calculated manually using the following procedure.

Starting with an image acquired from CFP-only expressing cells, select the CFP-containing pixels by creating a selection mask covering the pixels whose CFP intensity is greater than 10% of the brightest pixel. For example, if the brightest pixel has a value of 2354, the mask should select pixels whose intensity is between 235 and 2354. For pixels defined by the mask, plot the intensity of FRET versus CFP intensity, fit this curve to a line, and record the slope. If your software package does not perform linear regression analysis easily, you may divide the average FRET intensity of the masked pixels by the average CFP intensity. Repeat for the other two fields and average the results. The average slope (or FRET:CFP ratio) is the CFP spectral cross talk coefficient. Repeat this process with cells expressing only YFP, measuring the slope of a plot of FRET versus YFP intensity or a ratio of the average FRET to average YFP intensities to obtain the YFP spectral cross talk coefficient.

4.9. Calculating FRET indices

While a multitude of approaches to calculating FRET efficiencies and indices have been described (Berney and Danuser, 2003), we have had good success using the method of Xia and Liu (2001). This method calculates a corrected FRET efficiency value ($FRET^C$) using Eq. (17.1),

$$FRET^C = I_{FRET} - (I_{YFP} \times a) - (I_{CFP} \times b) \qquad (17.1)$$

where I represents the fluorescence intensity of the indicated channel, and a and b are the empirically derived YFP and CFP spectral cross talk coefficients, respectively. To enable comparisons between different samples, $FRET^C$ is then normalized by the intensity of CFP and YFP using Eq. (17.2):

$$FRET^{NC} = \frac{I_{FRET} - (I_{YFP} \times a) - (I_{CFP} \times b)}{\sqrt{I_{YFP} \times I_{CFP}}} \qquad (17.2)$$

Corrected and normalized FRET efficiencies ($FRET^{NC}$) are called FRET indices and can be compared across multiple samples and experiments. If time-lapse imaging was performed, the entire image stack is processed in the same manner as a single image.

To create FRET images, $FRET^{NC}$ must be measured pixel by pixel. Many commercial image processing packages will allow pixel-by-pixel FRET measurements; however, plug-ins for ImageJ are also available for free download at the ImageJ Web site (http://rsb.info.nih.gov/ij/) (Feige et al., 2005). While some software will calculate both $FRET^C$ and $FRET^{NC}$ at the pixel level, other packages only calculate $FRET^C$. In this case, one can display the $FRET^C$ image and calculate average $FRET^{NC}$ for individual cells or by fields of cells.

To calculate $FRET^{NC}$ averaged over an entire field, mask the image set to select pixels containing both CFP and YFP at intensities greater than 10% of the maximum intensity for each channel. Calculate the average intensity of the pixels within this mask for the CFP, YFP, and $FRET^C$ channels and export these data to Excel. For each field, $FRET^{NC}$ can be calculated using Eq. (17.3):

$$FRET^{NC} = \frac{FRET^C_{avg}}{\sqrt{I_{YFPavg} \times I_{CFPavg}}} \qquad (17.3)$$

For each sample, the $FRET^{NC}$ values for each image set are averaged and the standard error of the mean is calculated for comparison between other samples.

4.10. Data interpretation

Like many other quantitative measurements, most FRET experiments will give some numerical result whether the experiment was successfully performed or not. Knowing the magnitude and precision of $FRET^{NC}$ in both experimental samples and negative controls is critical to determining whether a measurable interaction is occurring between the tagged proteins. To determine whether a measurable interaction is occurring between two proteins, the $FRET^{NC}$ for the experimental sample should be statistically higher than the $FRET^{NC}$ of the negative control samples. It is important to note that while $FRET^{NC}$ values are normalized to correct for differences in fluorescent protein expression between samples, they are not normalized to correct for differences in microscope configuration. Therefore, $FRET^{NC}$

values can only be compared when obtained from the same microscope system.

As is often the case with scaffold proteins, the interaction between two signaling molecules may not be seen when FRET^{NC} of whole cells or fields of cells are averaged. Interactions often occur only at a specific region of the cell, and averaging this spatially restricted FRET signal over the entire cell may preclude its detection. As with any experimental approach, confirmation of a positive FRET interaction is best provided by a complementary technique, such as immunoprecipitation of endogenous proteins.

Fluorescence resonance energy transfer was used by our laboratory to demonstrate a number of interactions that are enhanced in response to hyperosmotic stress. CFP-OSM and YFP-MEKK3 were cotransfected in COS7 cells and then were left untreated or treated with 0.2 M sorbitol for 10 min. Both fluorescent proteins relocalized to ruffle-like structures at the cell periphery in response to sorbitol, and their interaction was confirmed by the FRET^C images (Uhlik *et al.*, 2003). In another study, the interaction between OSM and Krit1 was determined in COS7 cells. The FRET^{NC} value for CFP-OSM cotransfected with YFP alone was 0.35×10^{-5}; however, for CFP-OSM and YFP-Krit1 the value was 7.43×10^{-5}, and following sorbitol treatment for 5 min the value for the CFP-OSM and YFP-Krit1 interaction increased to 15.77×10^{-5} (Zawistowski *et al.*, 2005). Not only did the FRET^{NC} calculation reveal a twofold increase in the interaction between these proteins in response to sorbitol, the FRET^C image showed a clear redistribution of these two proteins from a diffusely cytosolic localization to the cell periphery in response to hyperosmotic stress. Therefore, FRET provides both a visual representation of protein–protein interactions and their stimulus-dependent redistribution and a quantitative value representing the amount of protein–protein interaction within live cells.

REFERENCES

Bell, L. M., Leong, M. L., Kim, B., Wang, E., Park, J., Hemmings, B. A., and Firestone, G. L. (2000). Hyperosmotic stress stimulates promoter activity and regulates cellular utilization of the serum- and glucocorticoid-inducible protein kinase (Sgk) by a p38 MAPK-dependent pathway. *J. Biol. Chem.* **275**, 25262–25272.

Bergametti, F., Denier, C., Labauge, P., Arnoult, M., Boetto, S., Clanet, M., Coubes, P., Echenne, B., Ibrahim, R., Irthum, B., Jacquet, G., Lonjon, M., *et al.* (2005). Mutations within the programmed cell death 10 gene cause cerebral cavernous malformations. *Am. J. Hum. Genet.* **76**, 42–51.

Berney, C., and Danuser, G. (2003). FRET or no FRET: A quantitative comparison. *Biophys. J.* **84**, 3992–4010.

Denier, C., Goutagny, S., Labauge, P., Krivosic, V., Arnoult, M., Cousin, A., Benabid, A. L., Comoy, J., Frerebeau, P., Gilbert, B., Houtteville, J. P., Jan, M., *et al.* (2004). Mutations within the MGC4607 gene cause cerebral cavernous malformations. *Am. J. Hum. Genet.* **74,** 326–337.

Feige, J. N., Sage, D., Wahli, W., Desvergne, B., and Gelman, L. (2005). PixFRET, an ImageJ plug-in for FRET calculation that can accommodate variations in spectral bleed-throughs. *Microsc. Res. Tech.* **68,** 51–58.

Forster, T. (1948). Intermolecular energy migration and fluorescence. *Ann. Phys. (Leipzig)* **2,** 55–75.

Fusello, A. M., Mandik-Nayak, L., Shih, F., Lewis, R. E., Allen, P. M., and Shaw, A. S. (2006). The MAPK scaffold kinase suppressor of Ras is involved in ERK activation by stress and proinflammatory cytokines and induction of arthritis. *J. Immunol.* **177,** 6152–6158.

Garmyn, M., Mammone, T., Pupe, A., Gan, D., Declercq, L., and Maes, D. (2001). Human keratinocytes respond to osmotic stress by p38 map kinase regulated induction of HSP70 and HSP27. *J. Invest. Dermatol.* **117,** 1290–1295.

Guay, J., Lambert, H., Gingras-Breton, G., Lavoie, J. N., Huot, J., and Landry, J. (1997). Regulation of actin filament dynamics by p38 map kinase-mediated phosphorylation of heat shock protein 27. *J. Cell Sci.* **110**(Pt. 3), 357–368.

Hannon, G. J. (2002). RNA interference. *Nature* **418,** 244–251.

Hannon, G. J., and Rossi, J. J. (2004). Unlocking the potential of the human genome with RNA interference. *Nature* **431,** 371–378.

Jares-Erijman, E. A., and Jovin, T. M. (2003). FRET imaging. *Nat. Biotechnol.* **21,** 1387–1395.

Johnson, G. L., Dohlman, H. G., and Graves, L. M. (2005). MAPK kinase kinases (MKKKs) as a target class for small-molecule inhibition to modulate signaling networks and gene expression. *Curr. Opin. Chem. Biol.* **9,** 325–331.

Laberge-le Couteulx, S., Jung, H. H., Labauge, P., Houtteville, J. P., Lescoat, C., Cecillon, M., Marechal, E., Joutel, A., Bach, J. F., and Tournier-Lasserve, E. (1999). Truncating mutations in CCM1, encoding KRIT1, cause hereditary cavernous angiomas. *Nat. Genet.* **23,** 189–193.

Liquori, C. L., Berg, M. J., Siegel, A. M., Huang, E., Zawistowski, J. S., Stoffer, T., Verlaan, D., Balogun, F., Hughes, L., Leedom, T. P., Plummer, N. W., Cannella, M., *et al.* (2003). Mutations in a gene encoding a novel protein containing a phosphotyrosine-binding domain cause type 2 cerebral cavernous malformations. *Am. J. Hum. Genet.* **73,** 1459–1464.

Nguyen, A. W., and Daugherty, P. S. (2005). Evolutionary optimization of fluorescent proteins for intracellular FRET. *Nat. Biotechnol.* **23,** 355–360.

Pollok, B. A., and Heim, R. (1999). Using GFP in FRET-based applications. *Trends Cell Biol.* **9,** 57–60.

Sahoo, T., Johnson, E. W., Thomas, J. W., Kuehl, P. M., Jones, T. L., Dokken, C. G., Touchman, J. W., Gallione, C. J., Lee-Lin, S. Q., Kosofsky, B., Kurth, J. H., Louis, D. N., *et al.* (1999). Mutations in the gene encoding KRIT1, a Krev-1/rap1a binding protein, cause cerebral cavernous malformations (CCM1). *Hum. Mol. Genet.* **8,** 2325–2333.

Saito, H., and Tatebayashi, K. (2004). Regulation of the osmoregulatory HOG MAPK cascade in yeast. *J. Biochem. (Tokyo)* **136,** 267–272.

Sekar, R. B., and Periasamy, A. (2003). Fluorescence resonance energy transfer (FRET) microscopy imaging of live cell protein localizations. *J. Cell Biol.* **160,** 629–633.

Shaner, N. C., Steinbach, P. A., and Tsien, R. Y. (2005). A guide to choosing fluorescent proteins. *Nat. Methods* **2,** 905–909.

Sheikh-Hamad, D., Di Mari, J., Suki, W. N., Safirstein, R., Watts, B. A., 3rd, and Rouse, D. (1998). p38 kinase activity is essential for osmotic induction of mRNAs for

HSP70 and transporter for organic solute betaine in Madin-Darby canine kidney cells. *J. Biol. Chem.* **273,** 1832–1837.

Uhlik, M. T., Abell, A. N., Cuevas, B. D., Nakamura, K., and Johnson, G. L. (2004). Wiring diagrams of MAPK regulation by MEKK1, 2, and 3. *Biochem. Cell Biol.* **82,** 658–663.

Uhlik, M. T., Abell, A. N., Johnson, N. L., Sun, W., Cuevas, B. D., Lobel-Rice, K. E., Horne, E. A., Dell'Acqua, M. L., and Johnson, G. L. (2003). Rac-MEKK3-MKK3 scaffolding for p38 MAPK activation during hyperosmotic shock. *Nat. Cell Biol.* **5,** 1104–1110.

Wang, X., Finegan, K. G., Robinson, A. C., Knowles, L., Khosravi-Far, R., Hinchliffe, K. A., Boot-Handford, R. P., and Tournier, C. (2006). Activation of extracellular signal-regulated protein kinase 5 downregulates FasL upon osmotic stress. *Cell Death Differ.* **13,** 2099–2108.

Widmann, C., Sather, S., Oyer, R., Johnson, G. L., and Dreskin, S. C. (2001). *In vitro* activity of MEKK2 and MEKK3 in detergents is a function of a valine to serine difference in the catalytic domain. *Biochim. Biophys. Acta* **1547,** 167–173.

Xia, Z., and Liu, Y. (2001). Reliable and global measurement of fluorescence resonance energy transfer using fluorescence microscopes. *Biophys. J.* **81,** 2395–2402.

Zawistowski, J. S., Stalheim, L., Uhlik, M. T., Abell, A. N., Ancrile, B. B., Johnson, G. L., and Marchuk, D. A. (2005). CCM1 and CCM2 protein interactions in cell signaling: Implications for cerebral cavernous malformations pathogenesis. *Hum. Mol. Genet.* **14,** 2521–2531.

CHAPTER EIGHTEEN

OSMOREGULATION OF BILE FORMATION

Ralf Kubitz *and* Dieter Häussinger

Contents

Abstract

Bile secretion by liver parenchymal cells is the result of vectorial transcellular transport of solutes and involves the coordinated action of transport proteins at the basolateral (sinusoidal) and apical (canalicular) membranes of the hepatocyte. A complex network of signals controls uptake and efflux transporters on a long- and a short-term timescale, including regulation at the level of gene transcription, protein translation and maturation, covalent modification, and dynamic localization of transporter proteins, as well as substrate availability. Evidence has shown that the hepatocellular hydration state exerts powerful control on the transcellular transport of solutes, such as conjugated bile acids and glucuronide and glutathione conjugates. This is of physiological significance because liver cell hydration is a dynamic parameter, which changes within minutes under the influence of hormones, nutrients, and oxidative stress. Thus, osmoregulation of bile formation is of physiological and pathophysiological interest.

Clinic of Gastroenterology, Hepatology and Infectiology, Heinrich-Heine-University Düsseldorf, Düsseldorf, Germany

Methods in Enzymology, Volume 428
ISSN 0076-6879, DOI: 10.1016/S0076-6879(07)28018-8

1. Networks of Transport Systems

Bile formation is the result of transport activity of many transport systems. According to their substrate specificities, groups of transporters form functional networks. For example, bile acid transport is mediated by the Na^+-coupled uptake transporter Na^+-taurocholate cotransporting polypeptide (Ntcp/Slc10a1) (Hagenbuch and Meier, 1994; Stieger et al., 1994). Ntcp/NTCP (for the protein in animals and humans, respectively) represents the major uptake system for conjugated bile salts and is functionally complemented by members of the organic anion transporting polypeptide (Oatp/OATP) family (Hagenbuch and Meier, 2003), such as Oatp1/Oatp1a1, Oatp2/Oatp1a4, and Oatp4/Oatp1b2 in rats or OATP2/OATP1B1 and OATP8/OATPB3 in humans (Fig. 18.1). Net bile salt uptake is further determined by efflux transporters at the sinusoidal membrane of hepatocytes, such as the ATP-binding cassette (ABC) transporter multidrug resistance associated protein 4 (Mrp4/MRP4) and Mrp3/MRP3, which transport bile salts back into the blood at the expense of ATP (Rius et al., 2003, 2006). Bile acid secretion into the canaliculus is driven by the bile salt export pump Bsep, which belongs

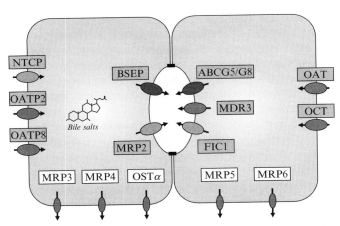

Figure 18.1 Hepatobiliary transporter of liver parenchymal cells. Uptake and efflux transporters are localized in the sinusoidal membrane of liver parenchymal, while the canalicular membrane harbors efflux transporters, which transport their substrates at the expense of ATP. (Left) Transporters involved in bile salt transport (among other substrates). (Right) Transporters of other substrates than bile salts. NTCP, sodium taurocholate cotransporting polypeptide; OATP, organic anion transporting polypeptide; OAT, organic anion transporter; OCT, organic cation transporter; BSEP, bile salt export pump; ABCG5/G8, ABC transporter subfamily G members 5 and 8; MDR, multidrug resistance 3; FIC1, familial intrahepatic cholestasis 1; MRP, multidrug resistance associated protein; OST, organic solute transporter.

to the ABC transporter family. Likewise, the conjugate export pump Mrp2 brings about canalicular secretion of glucuronide and glutathione conjugates. Thus, vectorial bile acid transport is accomplished by the concerted action of Ntcp/Bsep, whereas other organic anions are transported by the Oatp/Mrp2 system. In general, the transport capacity of sinusoidal transporters exceeds that of canalicular transport systems. This led to the widely accepted view that the canalicular excretion step is rate limiting for overall transcellular transport of most cholephilic compounds under physiological conditions. However, the control strength theory (Groen *et al.*, 1982) has not yet been applied to transcellular transport. Thus, the possibility is not ruled out that significant control on bile formation is also exerted at the step of uptake across the sinusoidal membrane. This may be relevant, especially at physiologically low bile acid concentrations or under pathophysiologic conditions, which are accompanied by a downregulation of Ntcp expression (Green *et al.*, 1996; Simon *et al.*, 1996).

2. OSMOREGULATION OF CANALICULAR SECRETION

Transporter insertion and retrieval into and from the canalicular membrane is a major mechanism of short-term regulation of bile formation (Häussinger *et al.*, 2000; Kubitz *et al.*, 1997). By this mechanism, the total number of transporter molecules in the canalicular and sinusoidal membrane and therefore transport capacity (V_{max}) can change within minutes. One major regulator of this process are changes in hepatocellular hydration, which occur in response to hormones, nutrients, oxidative stress, or changes in ambient osmolarity (for review, see Häussinger, 1996; Lang *et al.*, 1998). In perfused rat liver, cell shrinkage inhibits, whereas cell swelling stimulates taurocholate (TC) excretion into bile, regardless of whether cell volume is modified by anisotonic exposure, insulin, cumulative amino acid uptake, or ethanol (Hallbrucker *et al.*, 1992; Häussinger *et al.*, 1992, 2000). Regulation of TC excretion into bile by the hepatocellular hydration state is due to rapid changes of transport capacity: a 10% increase of hepatocyte water content doubles the V_{max} of TC excretion into bile within minutes, whereas hyperosmotic cell shrinkage decreases V_{max}. The swelling–induced increase in transport capacity is abolished in the presence of colchicine, indicating the requirement of intact microtubules (Häussinger *et al.*, 1993). As shown by immunohistochemistry (Schmitt *et al.*, 2001), hyperosmotic hepatocyte shrinkage triggers a rapid retrieval of Bsep from the canalicular membrane and its transfer into an intracellular vesicular compartment, whereas cell swelling triggers the insertion of intracellularly stored Bsep into the canalicular membrane. Likewise, hyperosmotic hepatocyte shrinkage triggers retrieval of the canalicular conjugate pump Mrp2 from the canalicular membrane, whereas hypoosmotic hepatocyte swelling stimulates a rapid insertion of Mrp2 into the canalicular membrane (Dombrowski *et al.*, 2000; Kubitz *et al.*, 1997), which is also reflected at the functional level, that is, corresponding changes in the canalicular secretion of the glutathione conjugates (Dombrowski *et al.*, 2000; Wettstein *et al.*, 1995). Interestingly, following hyperosmotic hepatocyte shrinkage, Bsep and Mrp2 are retrieved into different intracellular vesicular compartments. As shown by immunohistochemistry, only 15% of the retrieved vesicles contained both Bsep and Mrp2, whereas the remainder of vesicles contained either Mrp2 or Bsep (Schmitt *et al.*, 2001); accordingly, BSEP and MRP2 can be found in different vesicles in human livers (Fig. 18.2). This may either indicate different retrieval mechanisms for the two transporters in response to hyperosmolarity or suggest the existence of canalicular membrane domains being enriched in either Bsep or Mrp2. In support of the latter suggestion, immunocytochemistry of pseudocanaliculi formed between two adjacent HepG2 cells indeed shows canalicular membrane patches with enrichment of either Bsep or Mrp2 and only a few patches with transporter colocalization (Kubitz and Häussinger, unpublished result). Also, the expression of Bsep and Mrp2 was found to be osmoregulated. Hypoosmotic hepatocyte swelling increases Bsep and Mrp2 expression, whereas hyperosmotic cell shrinkage decreases (Kubitz *et al.*, 1999a; Warskulat *et al.*, 1999).

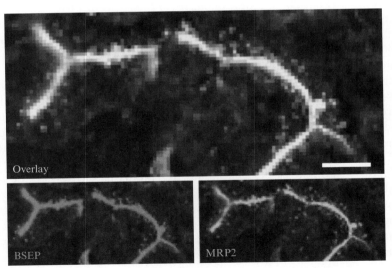

Figure 18.2   Transporter distribution in human liver. Cryofixed human liver (from surgically resected liver tissue) was stained for the bile salt export pump BSEP (red) and the bilirubin transporter MRP2 (green). Most immunoreactivity is found in a tubular configuration of the canaliculi, and some is found in vesicular structures within the cells. The two transporter proteins are distributed unequally within these vesicles. Bar: 10 μm. (See color insert.)

3. Osmosensing and Osmosignaling Pathways Toward Canalicular Secretion

Osmoregulation of canalicular secretion requires structures that pick up changes in hepatocyte hydration (osmosensing) and transmit this information toward effector sites (osmosignaling). Studies have identified the integrin system as one major osmosensor in hepatocytes (Häussinger et al., 2003; Schliess et al., 2004; vom Dahl et al., 2003). Integrins are a family of extracellular matrix (ECM) adhesion molecules involved in "mechano-transduction" and growth factor signaling (Aplin et al., 1998; Hynes, 2002; Ingber, 1997; Miranti and Brugge, 2002). In liver the most important integrins are $\alpha_1\beta_1$, $\alpha_5\beta_1$, and $\alpha_9\beta_1$ (Carloni et al., 2001; Hsu et al., 2001; Torimura et al., 2001). As shown by immunohistochemistry, hypoosmotic exposure of rat hepatocytes leads to the rapid appearance of the active conformation of the $\beta_1$subunit in the plasma membrane, indicating integrin activation in response to hepatocyte swelling (Häussinger et al., 2003; Schliess et al., 2004; vom Dahl et al., 2003). Downstream consequences of hypoosmotic integrin activation are the activation of Src kinases and of mitogen-activated protein kinases Erks and p38^{MAPK} (Häussinger et al., 2003; vom Dahl et al., 2003). Dual activation of both Erks and p38^{MAPK} is required for the choleretic effect of cell swelling, and choleresis induced by hypoosmotic cell swelling is abolished in the presence of inhibitors of either the integrin system or Src or one of the two MAP kinases (Häussinger et al., 2003; Kurz et al., 2001; Noe et al., 1996; Schmitt et al., 2001). Interestingly, TC also induces swelling and it was suggested that an increased load of this physiological bile acid to the liver may stimulate canalicular bile acid excretion via a feedforward regulation as a consequence of a swelling-induced recruitment of Bsep to the canalicular membrane (Häussinger et al., 1992; Noe et al., 1996). Such a response would accelerate enterohepatic circulation of bile acids after ingestion of a meal, which by itself also triggers nutrient-driven hepatocyte swelling, for example, by the concentrative uptake of amino acids into hepatocytes. The role of integrins as osmosensors is underlined by the fact that integrin-inhibitory peptides exhibiting a RGD motif fully abolish osmosignaling toward MAP kinases and the stimulation of bile formation, which is otherwise triggered by hypoosmotic swelling. Such peptides prevent integrin binding to the RGD attachment sites of ECM proteins such as fibronectin, thereby impairing the dynamics of integrin/matrix interactions, which are essential for effective mechanotransduction (Ruoslahti, 1996). Interestingly, tauroursodeoxycholate also activates nonosmotically integrin-dependent osmosensing and osmosignaling pathways, which may explain the choleretic action of this bile acid (Häussinger et al., 2003; Kurz et al., 2001).

Endosomes have been identified as an osmosensing compartment, which is activated in response to hyperosmotic hepatocyte shrinkage (Reinehr *et al.*, 2006). Here, a hyperosmolarity-induced endosomal acidification was shown to trigger ceramide formation within seconds, which in turn activates protein kinase Cζ, which results in an activation of NADPH oxidase isoforms because of an activating phosphorylation of p47^phox. As a consequence, hyperosmotic hepatocyte shrinkage produces oxidative stress. It is not yet clear to what extent this oxidative stress response is involved in the hyperosmotic retrieval of canalicular transport systems; however, exogenously added hydroperoxides were shown to induce the retrieval of Mrp2 from the canalicular membrane (Schmitt *et al.*, 2000). In line with the suggestion that oxidative stress may contribute to the cholestatic action of hyperosmotic hepatocyte shrinkage is also the finding that toxic, hydrophobic bile acids, which are known to be cholestatic, also induce oxidative stress via NADPH oxidase activation (Becker *et al.*, 2007; Reinehr *et al.*, 2005).

4. OSMOREGULATION OF SINUSOIDAL BILE ACID UPTAKE

Less is known about the regulation of sinusoidal transporters by cell volume changes. Under hypoosmotic conditions, Ntcp, the main bile salt uptake transporter, is stimulated by insertion into the sinusoidal membrane in a phosphoinositide 3-kinase and protein kinase B–dependent manner (Webster *et al.*, 2000), whereas MAP kinases, which are part of the signaling pathway in canalicular transporter insertion (Noe *et al.*, 1996), are not involved. Whether other sinusoidal uptake transporter such as OATPs or efflux transporters such as MRP4 or OSTα/β are regulated at the level of localization in response to cell volume changes has to be addressed in the future.

5. METHODS USED TO STUDY OSMOREGULATED TRANSPORTER INSERTION/RETRIEVAL

Various techniques were employed in order to study the subcellular localization of transporter proteins in hepatocytes. These are reviewed elsewhere (Kubitz *et al.*, 2005) and include immunofluorescent studies on isolated cells or cryosectioned tissues using confocal laser-scanning microscopy (Kubitz *et al.*, 1997, 1999b; Mühlfeld *et al.*, 2003; Schmitt *et al.*, 2000, 2001), studies at the level of electron microscopy (Beuers *et al.*, 2001; Dombrowski *et al.*, 2000), the use of differential centrifugation (Kipp and Arias, 2000) or sucrose gradient centrifugation (Mukhopadhyay *et al.*, 1998), and measurement of transport capacity as a surrogate marker of

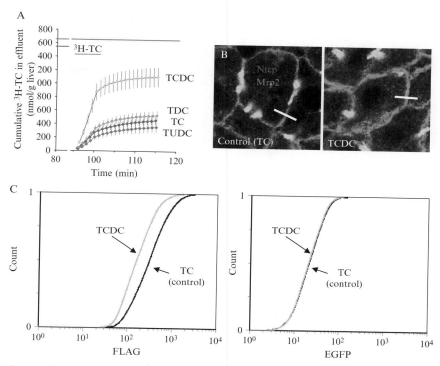

Figure 18.3 Bile salt-dependent modulation of bile salt uptake. (A) Rat livers were perfused with 25 μmol/liter of bile salts [taurocholate (TC), taurochenodeoxycholate (TCDC), taurodeoxycholate (TDC), or tauroursodeoxycholate (TUDC)] for 30 min. After a washout period of 20 min, a 5-min pulse of radiolabeled taurocholate ([^{3}H]TC) was administered. Recovery of radioactivity in the effluent revealed that TCDC but not the other bile salts reduced net uptake of [^{3}H]TC. (B) TCDC induced a retrieval of Ntcp, as shown by immunofluorescence of rat liver slices after perfusion with 25 μmol/liter TCDC + 100 μmol/liter TC compared to 25 μmol/liter TC + 100 μmol/liter TC (control). (C) HepG2 cells were transfected with double-tagged Ntcp. An extracellular FLAG tag was used to quantify Ntcp at the cell membrane by an anti–FLAG antibody, while intracellular EGFP was used to measure total Ntcp. As compared to TC (control, black line), incubation with TCDC (100 μmol/liter; 1 h; gray line) induced a shift of FLAG fluorescence toward lower values, indicating a reduction of membrane-bound Ntcp. Total Ntcp was not changed, as shown by constant EGFP fluorescence under both conditions. (See color insert.)

transporter availability (Häussinger *et al.*, 1992, 1993; Kubitz *et al.*, 2004), as well as biotinylation of membrane proteins (Webster *et al.*, 2000). Furthermore, a method that uses a FLAG-tagged Ntcp in combination with flow cytometry in order to determine the distribution of Ntcp between plasma membrane and intracellular sites has been developed (Kubitz *et al.*, 2005) (Fig. 18.3). An automated, self-learning system has been introduced that picks up fluorescence profiles across the plasma and canalicular membranes in

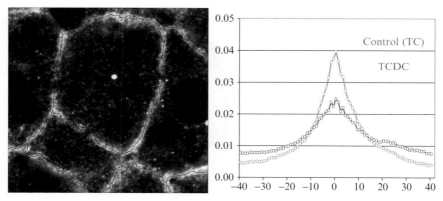

Figure 18.4 Quantification of transporter distribution by fluorescence densitometry. (A) Rat livers were perfused with 25 μmol/liter of TC (control) or TCDC. After immunostaining with an anti-Ntcp antibody, fluorescence pictures were recorded by confocal scanning microscopy (as shown in Fig. 18.3B). Pictures were analyzed by a semiautomatic software: single cells were identified by their border strips. Thereafter, distribution of fluorescence intensity perpendicular to all straight stretches of single cells was measured and averaged as shown on the right side. In TCDC-treated livers, fluorescence profiles of Ntcp were flattened as compared to TC (control)-treated livers. (See color insert.)

immunostained liver sections. Flattening or narrowing of these fluorescence profiles is detected automatically in a multitude of hepatocytes and can be taken as a measure for transporter internalization or membrane insertion, respectively (Fig. 18.4). Here, addition of the hydrophobic bile acid taurochenodeoxycholate, which induces cell shrinkage (Becker *et al.*, 2007), reduces the net uptake of taurocholate (Fig. 18.3A), alters immunoreactivity of Ntcp in liver slices (Fig. 18.3B), and results in a flattening of the Ntcp fluorescence profile (Fig. 18.4) across the plasma membrane, suggestive of an internalization of Ntcp in response to this bile acid. Accordingly, in Ntcp-transfected HepG2 cells, reduction of membrane-bound Ntcp can be measured by flow cytometry (Fig. 18.3C). Such a regulation may protect hepatocytes from bile acid overload, which would otherwise result in apoptotic hepatocyte injury.

6. Concluding Remarks

Several independent techniques can be used in order to demonstrate transporter insertion or retrieval in hepatocytes in response to osmotic challenges, all of them supporting the concept that bile formation is controlled at the short-term timescale by a regulated insertion/retrieval of transporter molecules. Future developments may aim toward the visualization of such transporter movements in the living cell and the disclosure of protein–protein interactions involved in this dynamic process; fluorescence resonance energy

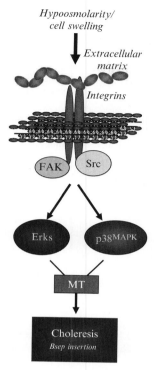

Figure 18.5 Cell swelling–induced signaling pathway. Cell swelling triggers integrin attachment to the extracellular matrix (ECM) and integrin activation. Focal adhesion kinases (FAK) and src kinases are thereby activated. Further signal transduction involves the dual activation of Erk-type and p38-type MAP kinases. Eventually, transporter bearing vesicles are inserted into the target membrane in a microtubule (MT)-dependent manner, along with an increase of transporter activity. A similar mechanism is activated nonosmotically by tauroursodeoxycholate and triggers choleresis. (See color insert.)

transfer techniques may have here some potential. Furthermore, much has to be learned about the osmosensing and osmosignaling pathways and their nonosmotic activation, which will give new insights into the development of choleresis and cholestasis in health and disease (Fig. 18.5).

ACKNOWLEDGMENTS

Our own studies reported herein were supported by Deutsche Forschungsgemeinschaft through Sonderforschungsbereich SFB 575 "Experimental Hepatology" (Düsseldorf). Contributions of Dr. V. Keitel, A. Helmer, Professor T. Berlage, and Matthias Jungmann are acknowledged.

REFERENCES

Aplin, A. E., Howe, A., Alahari, S. K., and Juliano, R. L. (1998). Signal transduction and signal modulation by cell adhesion receptors: The role of integrins, cadherins, immunoglobulin-cell adhesion molecules, and selectins. *Pharmacol. Rev.* **50,** 197–263.

Becker, S., Reinehr, R., Graf, D., vom Dahl, S., and Häussinger, D. (2007). Hydrophobic bile salts induce hepatocyte shrinkage via NADPH oxidase activation. *Cell. Physiol. Biochem.* **19,** 89–98.

Beuers, U., Bilzer, M., Chittattu, A., Kullak-Ublick, G. A., Keppler, D., Paumgartner, G., and Dombrowski, F. (2001). Tauroursodeoxycholic acid inserts the apical conjugate export pump, Mrp2, into canalicular membranes and stimulates organic anion secretion by protein kinase C-dependent mechanisms in cholestatic rat liver. *Hepatology* **33,** 1206–1216.

Carloni, V., Mazzocca, A., Pantaleo, P., Cordella, C., Laffi, G., and Gentilini, P. (2001). The integrin, alpha6beta1, is necessary for the matrix-dependent activation of FAK and MAP kinase and the migration of human hepatocarcinoma cells. *Hepatology* **34,** 42–49.

Dombrowski, F., Kubitz, R., Chittattu, A., Wettstein, M., Saha, N., and Häussinger, D. (2000). Electron-microscopic demonstration of multidrug resistance protein 2 (Mrp2) retrieval from the canalicular membrane in response to hyperosmolarity and lipopolysaccharide. *Biochem. J.* **348**(Pt. 1), 183–188.

Green, R. M., Beier, D., and Gollan, J. L. (1996). Regulation of hepatocyte bile salt transporters by endotoxin and inflammatory cytokines in rodents. *Gastroenterology* **111,** 193–198.

Groen, A. K., van der Meer, R., and Westerhoff, H. V. (1982). "Control of Metabolic Fluxes" (H. Sies, ed.). Academic Press, London.

Hagenbuch, B., and Meier, P. J. (1994). Molecular cloning, chromosomal localization, and functional characterization of a human liver Na+/bile acid cotransporter. *J. Clin. Invest.* **93,** 1326–1331.

Hagenbuch, B., and Meier, P. J. (2003). The superfamily of organic anion transporting polypeptides. *Biochim. Biophys. Acta* **1609,** 1–18.

Hallbrucker, C., Lang, F., Gerok, W., and Häussinger, D. (1992). Cell swelling increases bile flow and taurocholate excretion into bile in isolated perfused rat liver. *Biochem. J.* **281** (Pt. 3), 593–595.

Häussinger, D. (1996). The role of cellular hydration in the regulation of cell function. *Biochem. J.* **313**(Pt. 3), 697–710.

Häussinger, D., Hallbrucker, C., Saha, N., Lang, F., and Gerok, W. (1992). Cell volume and bile acid excretion. *Biochem. J.* **288**(Pt. 2), 681–689.

Häussinger, D., Kurz, A. K., Wettstein, M., Graf, D., vom, D. S., and Schliess, F. (2003). Involvement of integrins and Src in tauroursodeoxycholate-induced and swelling-induced choleresis. *Gastroenterology* **124,** 1476–1487.

Häussinger, D., Saha, N., Hallbrucker, C., Lang, F., and Gerok, W. (1993). Involvement of microtubules in the swelling-induced stimulation of transcellular taurocholate transport in perfused rat liver. *Biochem. J.* **291**(Pt. 2), 355–360.

Häussinger, D., Schmitt, M., Weiergräber, O., and Kubitz, R. (2000). Short-term regulation of canalicular transport. *Semin. Liver Dis.* **20,** 307–321.

Hsu, S. L., Cheng, C. C., Shi, Y. R., and Chiang, C. W. (2001). Proteolysis of integrin alpha5 and beta1 subunits involved in retinoic acid-induced apoptosis in human hepatoma Hep3B cells. *Cancer Lett.* **167,** 193–204.

Hynes, R. O. (2002). Integrins: Bidirectional, allosteric signaling machines. *Cell* **110,** 673–687.

Ingber, D. E. (1997). Integrins, tensegrity, and mechanotransduction. *Gravit. Space Biol. Bull.* **10,** 49–55.

Kipp, H., and Arias, I. M. (2000). Newly synthesized canalicular ABC transporters are directly targeted from the Golgi to the hepatocyte apical domain in rat liver. *J. Biol. Chem.* **275**, 15917–15925.

Kubitz, R., D' Urso, D., Keppler, D., and Häussinger, D. (1997). Osmodependent dynamic localization of the multidrug resistance protein 2 in the rat hepatocyte canalicular membrane. *Gastroenterology* **113**, 1438–1442.

Kubitz, R., Helmer, A., and Häussinger, D. (2005). Biliary transport systems: Short-term regulation. *Methods Enzymol.* **400**, 542–557.

Kubitz, R., Saha, N., Kühlkamp, T., Dutta, S., vom, D. S., Wettstein, M., and Häussinger, D. (2004). Ca^{2+}-dependent protein kinase C isoforms induce cholestasis in rat liver. *J. Biol. Chem.* **279**, 10323–10330.

Kubitz, R., Warskulat, U., Schmitt, M., and Häussinger, D. (1999a). Dexamethasone- and osmolarity-dependent expression of the multidrug-resistance protein 2 in cultured rat hepatocytes. *Biochem. J.* **340**(Pt. 3), 585–591.

Kubitz, R., Wettstein, M., Warskulat, U., and Häussinger, D. (1999b). Regulation of the multidrug resistance protein 2 in the rat liver by lipopolysaccharide and dexamethasone. *Gastroenterology* **116**, 401–410.

Kurz, A. K., Graf, D., Schmitt, M., vom Dahl, S., and Häussinger, D. (2001). Tauroursodesoxycholate-induced choleresis involves p38(MAPK) activation and translocation of the bile salt export pump in rats. *Gastroenterology* **121**, 407–419.

Lang, F., Busch, G. L., Ritter, M., Volkl, H., Waldegger, S., Gulbins, E., and Häussinger, D. (1998). Functional significance of cell volume regulatory mechanisms. *Physiol. Rev.* **78**, 247–306.

Miranti, C. K., and Brugge, J. S. (2002). Sensing the environment: A historical perspective on integrin signal transduction. *Nat. Cell Biol.* **4**, E83–E90.

Mühlfeld, A., Kubitz, R., Dransfeld, O., Häussinger, D., and Wettstein, M. (2003). Taurine supplementation induces Mrp2 and Bsep expression in rats and prevents endotoxin-induced cholestasis. *Arch. Biochem. Biophys.* **413**, 32–40.

Mukhopadhyay, S., Ananthanarayanan, M., Stieger, B., Meier, P. J., Suchy, F. J., and Anwer, M. S. (1998). Sodium taurocholate cotransporting polypeptide is a serine, threonine phosphoprotein and is dephosphorylated by cyclic adenosine monophosphate. *Hepatology* **28**, 1629–1636.

Noe, B., Schliess, F., Wettstein, M., Heinrich, S., and Häussinger, D. (1996). Regulation of taurocholate excretion by a hypo-osmolarity-activated signal transduction pathway in rat liver. *Gastroenterology* **110**, 858–865.

Reinehr, R., Becker, S., Braun, J., Eberle, A., Grether-Beck, S., and Häussinger, D. (2006). Endosomal acidification and activation of NADPH oxidase isoforms are upstream events in hyperosmolarity-induced hepatocyte apoptosis. *J. Biol. Chem.* **281**, 23150–23166.

Reinehr, R., Becker, S., Keitel, V., Eberle, A., Grether-Beck, S., and Häussinger, D. (2005). Bile salt-induced apoptosis involves NADPH oxidase isoform activation. *Gastroenterology* **129**, 2009–2031.

Rius, M., Hummel-Eisenbeiss, J., Hofmann, A. F., and Keppler, D. (2006). Substrate specificity of human ABCC4 (MRP4)-mediated cotransport of bile acids and reduced glutathione. *Am. J. Physiol. Gastrointest. Liver Physiol.* **290**, G640–G649.

Rius, M., Nies, A. T., Hummel-Eisenbeiss, J., Jedlitschky, G., and Keppler, D. (2003). Cotransport of reduced glutathione with bile salts by MRP4 (ABCC4) localized to the basolateral hepatocyte membrane. *Hepatology* **38**, 374–384.

Ruoslahti, E. (1996). RGD and other recognition sequences for integrins. *Annu. Rev. Cell Dev. Biol.* **12**, 697–715.

Schliess, F., Reissmann, R., Reinehr, R., vom Dahl, S., and Häussinger, D. (2004). Involvement of integrins and Src in insulin signaling toward autophagic proteolysis in rat liver. *J. Biol. Chem.* **279**, 21294–21301.

Schmitt, M., Kubitz, R., Lizun, S., Wettstein, M., and Häussinger, D. (2001). Regulation of the dynamic localization of the rat Bsep gene-encoded bile salt export pump by anisoosmolarity. *Hepatology* **33**, 509–518.

Schmitt, M., Kubitz, R., Wettstein, M., vom Dahl, S., and Häussinger, D. (2000). Retrieval of the mrp2 gene encoded conjugate export pump from the canalicular membrane contributes to cholestasis induced by tert-butyl hydroperoxide and chlorodinitrobenzene. *Biol. Chem.* **381**, 487–495.

Simon, F. R., Fortune, J., Iwahashi, M., Gartung, C., Wolkoff, A., and Sutherland, E. (1996). Ethinyl estradiol cholestasis involves alterations in expression of liver sinusoidal transporters. *Am. J. Physiol.* **271**, G1043–G1052.

Stieger, B., Hagenbuch, B., Landmann, L., Hoechli, M., Schröder, A., and Meier, P. J. (1994). *In situ* localisation of the hepatocytic Na$^+$/taurocholate cotransporting polypeptide in rat liver. *Gastroenterology* **107**, 1781–1787.

Torimura, T., Ueno, T., Kin, M., Harad, R., Nakamura, T., Sakamoto, M., Kumashiro, R., Yano, H., Kojiro, M., and Sata, M. (2001). Laminin deposition to type IV collagen enhances haptotaxis, chemokinesis, and adhesion of hepatoma cells through beta1-integrins. *J. Hepatol.* **35**, 245–253.

vom Dahl, S., Schliess, F., Reissmann, R., Gorg, B., Weiergraber, O., Kocalkova, M., Dombrowski, F., and Häussinger, D. (2003). Involvement of integrins in osmosensing and signaling toward autophagic proteolysis in rat liver. *J. Biol. Chem.* **278**, 27088–27095.

Warskulat, U., Kubitz, R., Wettstein, M., Stieger, B., Meier, P. J., and Häussinger, D. (1999). Regulation of bile salt export pump mRNA levels by dexamethasone and osmolarity in cultured rat hepatocytes. *Biol. Chem.* **380**, 1273–1279.

Webster, C. R., Blanch, C. J., Phillips, J., and Anwer, M. S. (2000). Cell swelling-induced translocation of rat liver Na(+)/taurocholate cotransport polypeptide is mediated via the phosphoinositide 3-kinase signaling pathway. *J. Biol. Chem.* **275**, 29754–29760.

Wettstein, M., Noe, B., and Häussinger, D. (1995). Metabolism of cysteinyl leukotrienes in the perfused rat liver: The influence of endotoxin pretreatment and the cellular hydration state. *Hepatology* **22**, 235–240.

CHAPTER NINETEEN

OSMOSIGNALING AND VOLUME REGULATION IN INTESTINAL EPITHELIAL CELLS

Christina H. Lim, Alice G. M. Bot, Hugo R. de Jonge, *and* Ben C. Tilly

Contents

Abstract

Most cells have to perform their physiological functions under a variable osmotic stress, which, because of the relatively high permeability of the plasma membrane for water, may result in frequent alterations in cell size. Intestinal epithelial cells are especially prone to changes in cell volume because of their high capacity of salt and water transport and the high membrane expression of various nutrient transporters. Therefore, to avoid excessive shrinkage or swelling, enterocytes, like most cell types, have developed efficient mechanisms to maintain osmotic balance. This chapter reviews selected model systems that can be used to investigate cell volume regulation in intestinal epithelial cells, with emphasis on the regulatory volume decrease, and the methods available to study the

Department of Biochemistry, Erasmus University Medical Center, Rotterdam, The Netherlands

Methods in Enzymology, Volume 428
ISSN 0076-6879, DOI: 10.1016/S0076-6879(07)28019-X

compensatory redistribution of (organic) osmolytes. In addition, a brief summary is presented of the pathways involved in osmosensing and osmosignaling
in the intestine.

1. INTRODUCTION

Intestinal epithelial cells, like most cell types, are constantly exposed to
a variable osmotic stress caused by the uptake or release of osmotically active
substances (amino acids, sugars, etc.), the formation or degradation of
macromolecules (proteins, glycogen), hormone-induced alterations in cellular metabolism, or changes in the osmolarity of the surrounding fluid.
Because of the relatively high water permeability of the plasma membrane,
osmotic imbalances will immediately trigger a redistribution of cellular
water and, consequently, induce a rapid change in cell volume. To protect
the cells against the potentially deleterious effects of alterations in size,
almost all cell types have developed compensatory mechanisms to prevent
excessive shrinkage or swelling and to restore their original volume
(reviewed by Lang *et al.*, 1998; Mongin and Orlov, 2001; O'Neill, 1999;
Pasantes-Morales *et al.*, 2006b; Strange, 2004; Wehner *et al.*, 2003). In
general, adjustment of intracellular osmolarity is achieved by the activation
of transport pathways in the plasma membrane, leading to a net accumulation (regulatory volume increase [RVI]) or loss (regulatory volume decrease
[RVD]) of osmotically active substances. Whereas the RVI involves the net
uptake of NaCl through stimulation of Na^+/H^+ and Cl^-/HCO_3^- exchangers or the activation of Na^+-K^+-$2Cl^-$ and Na^+-Cl^- symporters, the RVD
largely depends on the release of KCl through K^+- and Cl^--selective ion
channels or by the activation of K^+-Cl^- symporters. In addition, an organic
osmolyte release pathway was found to be activated in response to osmotic
swelling in a number of cell models, resulting in the release of small organic
molecules such as taurine and betaine (for reviews, see Junankar and Kirk,
2000; Lambert, 2004; Pasantes-Morales *et al.*, 2002).

2. PROPERTIES OF ION CHANNELS
AND TRANSPORTERS ACTIVATED DURING
RVD IN INTESTINAL EPITHELIAL CELLS

Hypoosmotic stimulation of intestinal epithelial cells will result in a
rapid increase in cell volume and a subsequent activation of specific K^+ and
Cl^- channels, leading to a net efflux of KCl, which promotes the loss of
cellular water and restoration of the original cell volume. In cultured

Intestine 407 cells, recovery of cell size is achieved within a 1- to 2-min period after applying a hypoosmotic stimulus (Hazama et al., 1988; Tilly et al., 1993). For isolated intact intestinal crypts, however, considerably longer periods are needed to obtain full volume recovery (Mignen et al., 1999; O'Brien et al., 1991). Whereas the K^+ conductance involved has been identified as a Ca^{2+}-dependent K^+ channel, sensitive to Ba^{2+}, quinidine, tetraethylammonium, and charybdotoxin (Grunnet et al., 2002; Hazama et al., 1988; Niemeyer et al., 2001; Wang et al., 2003), the molecular identity of the anion channel (volume-regulated anion channel [VRAC]), despite its ubiquitous expression throughout the animal kingdom and its marked biophysical profile, has not yet been elucidated. Several potential candidates have been proposed, which include MDR-1/P-glycoprotein, ClC-2, ClC-3, and ICln; however, none of them meet all the electrical and pharmacological criteria of VRAC (Jentsch et al., 2002; Nilius and Droogmans, 2003; Okada, 1997, 2006; Sardini et al., 2003).

The cell swelling-activated chloride conductance has been studied in numerous different cell types and its electrical characteristics were found to be very similar in all models investigated. Key features of VRAC are (1) strong outward rectification and prominent inactivation of the conductance at depolarizing potentials, (2) a permeability sequence that corresponds to Eisenman's sequence I ($SCN^- > I^- > NO_3^- > Br^- > Cl^- > F^- >$ gluconate), and (3) inhibition by common Cl^- channel blockers such as 4-acetamido-4'-isothiocyanostilbene (SITS), 4,4'-diisothiocyanatostilbene-2,2'-disulfonic acid (DIDS), 5-nitro-2-(3-phenyl-propylamino)-benzoate (NPPB), diphenylamine-2-carboxylate (DPC), and niflumic acid (reviewed by Jentsch et al., 2002; Nilius and Droogmans, 2003; Okada, 1997, 2006; Pasantes-Morales et al., 2006b). Finally, the hypotonicity-provoked VRAC activation in Intestine 407 cells was found to be potentiated, but not elicited, by Ca^{2+}-mobilizing hormones, as well as by micromolar concentrations of ATP (Tilly et al., 1994; Van der Wijk et al., 1999). In contrast, millimolar concentrations of ATP, as well as the purinoceptor antagonists suramin and reactive blue, were found to inhibit VRAC (Van der Wijk et al., 1999). This inhibition is most prominent at depolarizing membrane potentials and does not involve purinoceptor activation and Ca^{2+} signaling.

In addition to the activation of K^+ and Cl^- channels, the release of small organic osmolytes contributes significantly to the RVD response in many cell models. In a number of tissues and cell types, the organic osmolyte release pathway was found to be the major mechanism involved in cell volume correction (Junankar and Kirk, 2000; Lambert, 2004; Pasantes-Morales et al., 2002, 2006b). In Intestine 407 cells, osmotic cell swelling was found to promote the release of taurine independently of the activation of VRAC. Unlike activation of the cell swelling-induced anion conductance, activation of the organic anion release pathway occurred only after a distinct lag time of approximately 30 to 60 s (Tomassen et al., 2004).

In addition, the threshold for activation of taurine release was reached only at a relatively strong hypotonic stimulation. This suggests that, in Intestine 407 cells, the release of organic osmolytes acts a second line of defense.

3. Osmosensing and Signaling

To date, very little is known about the molecular identity of the receptor involved in osmosensing in vertebrates. A specific class of osmo-receptors belonging to the family of histidine kinases has been identified in fungi, yeast, and plants cells (Grefen and Harter, 2004; Hohmann, 2002; Urao *et al.*, 2000). Histidine kinases, however, as a part of the so-called two-component systems, are rarely expressed in mammalian cells. Several alternative mechanisms for osmosensing have been proposed, including macromolecular crowding and signaling via the extracellular matrix and integrin receptors, as well as activation of mechanosensitive ion channels (Häussinger *et al.*, 2006; Mongin and Orlov, 2001; Okada, 1997; Pasantes-Morales *et al.*, 2006b). For intestinal epithelial cells, the role of integrins and their effects on cytoskeletal rearrangements are of a particular interest because of the sensitivity of VRAC in these cells for agents that affect the integrity of the cytoskeleton and by the observation of several groups that osmotic cell swelling is accompanied by a cytoskeletal reorganization (Carton *et al.*, 2003; Cornet *et al.*, 1993; Foskett and Spring, 1985; Pedersen *et al.*, 1999; Tilly *et al.*, 1996).

Although differences may exist between cell types, several general features of cell signaling in response to hypoosmotic stress have emerged. Especially the requirement of protein tyrosine phosphorylation for the activation of VRAC has been well documented. In the presence of tyrosine kinase inhibitors such as herbimycin A, genistein, or tyrphostins, the cell swelling-activated anion conductance was found to be largely reduced, whereas (per) vanadate inhibition of phosphotyrosine phosphatases, to reduce phosphotyrosine phosphatase activity, potentiated the anion efflux triggered by nonsaturating hypoosmotic stimulation (Shi *et al.*, 2002; Sorota, 1995; Tilly *et al.*, 1993; Voets *et al.*, 1998). Many hypotonicity-activated (targets of) tyrosine kinases have been reported, including the focal adhesion kinase, receptor tyrosine kinases such as the epidermal growth factor receptor, phosphatidylinositol 3-kinase, and members of the families of src and MAP kinases (for review, see Pasantes-Morales *et al.*, 2006a). A direct activation of VRAC, however, has only been established for the src-like p56lck tyrosine kinase in Jurkat T lymphocytes. For these cells, strong evidence exists that p56lck is both essential and adequate for channel activation (Lepple-Wienhues *et al.*, 1998). This notion is supported by our observation that the hypotonicity-provoked anion efflux in

Intestine 407 cells is largely reduced after treating the cells with damnacanthal, an inhibitor of p56lck (B. C. Tilly, unpublished results). In Caco-2 colonocytes, however, targeting c-src to caveolae was found to inhibit VRAC (Trouet et al., 2001).

In addition to tyrosine kinases and/or phosphatases, the involvement of G proteins in the activation of the RVD has been reported. Indeed, activation of G proteins by the intracellular administration of GTPγS rapidly induced an anion-selective current in several cell types, including human HT29cl19A colonocytes (Tilly et al., 1991; Voets et al, 1998). Furthermore, a regulatory role for the Ras-related G-protein p21Rho in the activation of VRAC has been reported for Intestine 407 cells and various other cell types (Carton et al., 2002; Estevez et al., 2001; Pedersen et al., 2002; Tilly et al., 1996; Voets et al., 1998). Most plausibly, p21Rho exerts its function through the induction of cytoskeletal remodeling. In Intestine 407 cells, the related p12Ras and its downstream target Raf-1, which are also activated during osmotic cell swelling, did not affect VRAC regulation but were found to be essential for activation of the Erk-1/2 map kinases (Van der Wijk et al., 1998). Notably, although p21Rho is essential for VRAC regulation, studies in bovine endothelial cells have shown that Rho activation alone is not sufficient to induce opening of the channel (Carton et al., 2002). The hypotonicity-induced release of organic osmolytes was not sensitive to tyrosine kinase or phosphatase inhibition and did not require p21Rho or PtdIns-3-kinase activity, indicating that the efflux is regulated independently of VRAC (Tomassen et al., 2004).

Osmotic cell swelling is often accompanied by an increase in vesicle (re-)cycling and exocytosis (Bruck et al., 1992; Okada et al., 1992; Štrbák and Greer, 2002; Van der Wijk et al., 2003). In Intestine 407 cells and several other cell types, an increase in volume leads to the extracellular release of ATP, an almost universal cellular response to mechanical stress, through a mechanism involving exocytosis (Hazama et al., 1999; Van der Wijk et al., 1998, 2003). Although extracellular ATP is not directly involved in the regulation of VRAC, purinergic receptor activation was found to trigger the hypotonicity-provoked stimulation of Erk-1/2 in Intestine 407 cells (Van der Wijk, 1998, 1999). It has been reported that exocytosis also contributes to the development of the compensatory anion conductance by the recruitment of additional anion channels to the plasma membrane (Lim et al., 2006).

Osmotic cell swelling activates several distinct signaling pathways that are often also associated with (neuro-)hormone and/or growth factor-associated signal transduction. Whereas some of these cascades are coupled to the regulation of VRAC, others, such as the ATP-provoked activation of Erk-1/2, are apparently not involved in channel regulation and, although speculative, may have a function in restoring cellular homeostasis and in maintaining cell viability. Importantly, with the notable exception of p56lck in Jurkat T cells, activation of these signaling molecules by itself is not sufficient to activate VRAC. Linking these widely utilized signaling modules to an independent, but yet unidentified, "volume sensor" not only preserves the specificity of the response, but also couples the RVD to pathways activated by hormonal stimulation, thereby facilitating the correction of small changes in cell volume that may occur during hormone-induced changes in cellular metabolism.

4. MODEL SYSTEMS

To investigate cellular responses to osmotic challenges in intestinal epithelial cells, a variety of different model systems can be used that vary in their complexity and in their resemblance to native intestinal epithelium. The human Intestine 407 (Henle and Deinhardt, 1957) and rat IEC-6 (Quaroni *et al.*, 1979) cells are examples of relatively fast propagating epithelial cells, which can be easily maintained in culture using standard conditions. Although they grow as monolayers, the cells are not interconnected by junctional complexes and do not develop into a semitight or "leaky" epithelium, making them unsuitable for transepithelial potential or current measurements in Ussing chambers. They are, however, good model cells for patch clamp and radioisotope efflux studies, as well as for biochemical analysis of signal transduction cascades and optical imaging. In addition, as compared to many other enterocyte or colonocyte cell types, the expression of proteins can be modified relatively easily using modern molecular genetic tools. Intestine 407 cells are very suitable in investigating cell swelling-activated anion currents because these cells lack expression of any other Cl^- conductances aside VRAC, such as Ca^{2+}- and voltage-sensitive Cl^- channels or cAMP/protein kinase A-sensitive CFTR Cl^- channels (Hazama and Okada, 1988; Kubo and Okada, 1992; Tilly *et al.*, 1993). In addition, P-glycoprotein expression was not detected in the subclone used in our laboratory.

The colonocyte cell lines Caco-2 (Fogh *et al.*, 1977), HT29-cl.19A (Augeron *et al.*, 1986), and T84 (Dharmsathaphorn *et al.*, 1984; Murakami and Masui, 1980) are all derived from lung metastasis of human colon rectal cancers. As compared to Intestine 407 and IEC-6 cells, they more closely resemble native enterocytes in that they express the cystic fibrosis transmembrane conductance regulator (CFTR, the chloride channel defective in cystic fibrosis patients), which is present ubiquitously in the apical membrane of intestinal epithelial cells, and, when cultured on a permeable support for at least 2 weeks, form semitight epithelia and develop junctional complexes. When grown on Costar Transwell polycarbonate cell culture inserts, these cell lines are not only suitable for Ussing chamber experiments, but can also be used to determine independently the isotope efflux from the apical and basolateral parts of the cells.

Enterocytes, villi, or crypts, isolated from freshly excised intestine, are preparations of the native epithelial cells lining the gut. Although villi and crypt preparations can be used to study changes in volume (Mignen *et al.*, 1999; O'Brien *et al.*, 1991), for instance, using time-lapse video imaging, and to unravel the signaling pathways triggered by such a change biochemically, these models are less suitable to investigate ionic responses using physiological techniques such as radioisotope efflux assays or patch clamping. This is partly because of impaired cell viability, a short life span after isolation, and difficulties in attaching to a substrate. Suspensions of villus epithelial cells, however, have been used successfully to investigate changes in cell volume by electronic cell sizing using a Coulter counter (MacLeod and Hamilton, 1990). In addition, this preparation was found suitable to determine alterations in intracellular pH and $[Ca^{2+}]$, as well as to quantitate $^{22}Na^+$ uptake and $^{86}Rb^+$ efflux in hypotonicity or Na^+ nutrient-stimulated cells (MacLeod and Hamilton, 1996, 1999; MacLeod *et al.*, 1992).

Perhaps the most physiological model for intact intestinal epithelium is isolated muscle-stripped mucosa, which can be obtained from small laboratory animals (Andres *et al.*, 1985). This model can also be used to study and compare the different regions of the intestine (Diener *et al.*, 1996; Hörger *et al.*, 1998). In our laboratory, the preparation of muscle-stripped mucosa for short circuit measurements in Ussing chambers is performed as described by Andres *et al.* (1985). Briefly, under full anesthesia, the intestine is removed and subsequently flushed extensively with ice-cold modified Meyler solution (108 mM NaCl, 4.7 mM KCl, 1.3 mM CaCl$_2$, 1 mM MgCl$_2$, 20 mM NaHCO$_3$, 0.8 mM Na$_2$HPO$_4$, 0.4 mM NaH$_2$PO$_4$, 20 mM HEPES, and 10 mM glucose, pH 7.4). Sections of approximately 3 to 4 cm are cut of the area of interest and placed on a glass rod. Using a blunt scalpel, a superficial longitudinal incision is made and the serosa and muscle layers are removed mechanically using tweezers. Thereafter, the tissue is sliced open, mounted on a circular tissue holder, and placed between the two Ussing chamber half cells. Both half chambers are perfused continuously with modified Meyler solution at 37° and gassed with 95% O$_2$/5% CO$_2$.

5. Measuring Ionic Responses

5.1. Whole cell patch clamp

The single electrode patch clamp technique allows direct evaluation of the magnitude and the electrical properties of currents under voltage clamp conditions (Hamill *et al.*, 1981; Sakmann and Neher, 1984). In its whole cell configuration, this method has been used to study volume-sensitive channels under well-defined (ionic) conditions in numerous cell models, leading to the identification of K^+ channels involved and to a detailed electrical characterization of VRAC (see earlier discussion). Although whole cell patch clamping is a well-established technique, widely used by many different research groups around the world, and almost all of the equipment and software needed is available commercially, investigators who are planning to introduce this technique in their own laboratory are strongly advised to get into contact with a laboratory specialized in membrane electrophysiology. This is because of the special requirements needed to isolate the setup from environmental electrical noise and mechanical vibrations. Because a high-resistance seal between the pipette and the cell membrane can be obtained easily when the cell has a smooth surface (Hamill *et al.*, 1981; Neher, 1981), it is much easier to study cultured cells at low densities than using fully differentiated enterocytes containing numerous microvilli and an apical mucous layer. As a consequence, isolated villi and crypts, as well as stripped intact intestinal epithelium, are less suitable for patch clamp analysis. In combination with fluorescence microscopy, patch clamping is very useful to study channel activation in cells transfected with fluorescently tagged vectors, making it possible to identify cells with high expression levels. Notably, in contrast to the isotope efflux assays and short circuit measurements (described later), the RVD is not functional under whole cell patch clamp conditions and the cell swelling-induced currents last as long as the bathing solution remains hypotonic with respect to the pipette solution (see Nilius and Droogmans, 2003).

Whole cell recordings can be obtained after disruption of the membrane patch directly under the pipette by mild negative pressure, creating a direct access between the cell interior and the microelectrode. Although this allows us to control the intracellular composition, it could also result in a dilution of essential cellular components and lead to a run down of the currents. To avoid diffusion of small molecules, but not ions, the perforated patch configuration (Levitan and Kramer, 1990) can be used, in which the membrane patch under the pipette is not disrupted but is instead permeabilized using pore-forming antibiotics (nystatin, amphothericin).

To investigate VRAC activation in Intestine 407 cells, we use a bathing isotonic solution composed of 110 mM CsCl, 5 mM MgSO$_4$, 3.5 mM sodium gluconate, 12 mM HEPES, 8 mM Tris-HCl, and 100 mM mannitol at pH 7.4 (Van der Wijk *et al.*, 1999). The intracellular pipette solution contains 110 mM CsCl, 2 mM MgSO$_4$, 25 mM HEPES, 1 mM EGTA, 1 mM Na$_2$ATP, and 50 mM mannitol, pH 7.4. Patch pipettes are pulled from borosilicate glass (Clark Electromedical Instruments, Pangbourne, Berks, UK) and heat polished and have a resistance of 2 to 3 MΩ. Whole cell current recordings are made using a RK-300 amplifier (Bio-Logic, Claix, France) and digitized using a Digidata 1200 AD converter (Axon Instruments Inc., Foster City, CA).

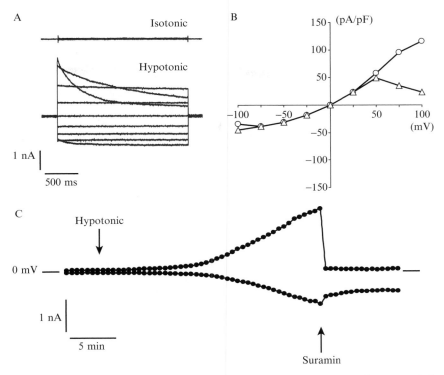

Figure 19.1 Volume-sensitive chloride currents from Intestine 407 cells. (A) Whole cell Cl⁻ currents in control (isotonic) and hypoosmotically stimulated cells in response to step pulses (2-s duration) from −100 to 100 mV (25-mV increments). (B) Current-to-voltage relationship of hypotonicity-provoked Cl⁻ currents immediately after the onset of the pulse (circles) and after reaching their plateau values (triangles). (C) Time course of hypotonicity-provoked anion currents at −100 and +100 mV, as well as the voltage-dependent inhibition of the currents by suramin (100 μM).

To monitor the development of the current, alternating step pulses (100-ms duration) from 0 to 100 mV are applied every 30 s. Voltage dependence of whole cell current is monitored by applying step pulses (2-s duration, 7-s interval) from −100 to +100 mV with 25-mV increments (Fig. 19.1). For command pulse control, data acquisition, and analysis, pCLAMP 9 software (Axon Instruments, Union City, CA) was used. All data are sampled at 5 kHz after being low pass filtered at 500 Hz.

5.2. Isotope efflux assay

The radioisotope efflux assay is a rapid and easy method to quantitate alterations in ionic conductances, as well as to determine the release of small organic osmolytes from cultures of cells (Tomassen *et al*, 2004; Vaandrager

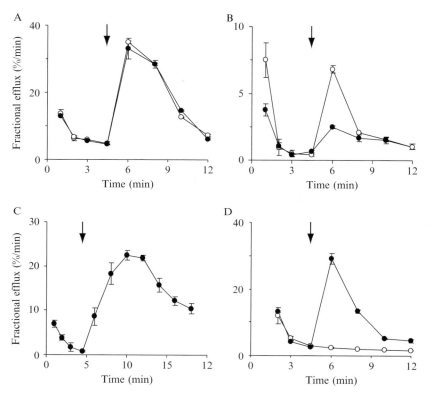

Figure 19.2 Radioisotope efflux from Intestine 407 cells. $^{125}I^-$ (A) and $^{86}Rb^+$ (B) efflux from isotope-loaded hypoosmotically stimulated control (open symbols) and phorbol 12-myristate 12-acetate-treated (PMA; 200 nM for 5 min; closed symbols) cultures. A marked inhibition of the $^{86}Rb^+$, but not the $^{125}I^-$, efflux was observed in PMA-treated cells. (C) Hypotonicity-provoked efflux of [^{3}H]taurine. (D) Efflux of $^{45}Ca^{2+}$ from isotope-loaded cells stimulated with bradykinin (closed symbols) or a hypoosmotic (open symbols) medium. All data are expressed as mean ± SEM for $n = 3$. Arrows indicate time points of hypotonic stimulation (70% tonicity, A–D) or addition of bradykinin (1 μM, D).

et al., 1991; Fig. 19.2). As a tracer for chloride conductance, $^{36}Cl^-$, $^{125}I^-$, or both can be used. Cation conductances can be evaluated by using $^{86}Rb^+$, $^{22}Na^+$, or $^{45}Ca^{2+}$ for, respectively, K^+, Na^+, and Ca^{2+} channel activity. An increase in the efflux of a specific isotope indicates the activation and opening of that particular channel, but does not give accurate information about the magnitude or the direction of the current. For instance, stimulation of $^{45}Ca^{2+}$-loaded Intestine 407 cells with bradykinin will result in a rapid and transient radioisotope efflux, indicative for a transient opening of Ca^{2+} channels, which is against the electrochemical gradient for Ca^{2+} (Tilly *et al*, 1994; Fig. 19.2C). In addition, the efflux of tracer can occur through exchange with ions in the bath and does not require the presence of a shunt conductance, for example, an almost normal hypotonicity-provoked efflux of $^{125}I^-$ can be observed from Intestine 407 cells treated with phorbol esters to block the potassium conductance (Fig. 19.2A and B).

To perform isotope efflux assays, cells can be grown in tissue culture dishes, 12- or 24-multiwell plates, or tissue culture-treated polycarbonate filters (Costar Transwell). Although optimal labeling conditions may vary per cell type, we routinely load our cells with 20 to 50 $\mu Ci/ml$ $^{36}Cl^-$, 0.5 to 1 $\mu Ci/ml$ $^{45}Ca^{2+}$, 5 to 10 $\mu Ci/ml$ $^{125}I^-$, and/or 0.5 to 1 $\mu Ci/ml$ $^{86}Rb^+$ in modified Meyler solution for 2 h at 37° under a humidified atmosphere of 95% air/5% CO_2. Prior to the experiment, cells are washed three times with a low sodium isotonic medium containing mannitol (80 mM NaCl, 5 mM KCl, 1.3 mM $CaCl_2$, 1 mM $MgCl_2$, 10 mM glucose, 95 mM mannitol, and 20 mM HEPES, pH 7.4). The radioisotope efflux from cells grown in wells is determined by consecutive replacement (1- to 2-min intervals) of the medium (1 ml/well for 24-well plates) with either isotonic or hypotonic medium at 37°. Hypotonic buffers are prepared by adjusting the mannitol concentration. The residual isotope is determined at the end of the experiment by dissolving the cells in 0.5 M NaOH. Alternatively, the cells can be incubated for an additional 60 min in 1 ml of buffer to determine the amount of free diffusible tracer still present after full equilibration. This procedure is especially useful if a considerable amount of the tracer is immobilized intracellularly. When cells are grown on filter insets, the efflux both from the mucosa (apical efflux, upper compartment) and from the serosa (basolateral efflux, lower compartment) can be determined by replacing the medium in the upper compartment as described earlier and by simultaneously transferring the filter inset to a consecutive well containing fresh medium. Isotope efflux can be quantitated by β-scintillation ($^{36}Cl^-$ or $^{45}Ca^{2+}$) or γ-scintillation ($^{125}I^-$ with or without $^{86}Rb^+$) counting. To compare the magnitude of the efflux during the experiment, data are expressed as "fractional efflux per minute" (Vaandrager et al., 1991).

A similar approach can be used to study the release of organic osmolytes. For Intestine 407 cells, loading for 2 h with 0.1 $\mu Ci/ml$ [^3H]taurine or [^3H] choline was found sufficient to get a reasonable intracellular accumulation (Tomassen et al., 2004). To quantitate the actual amount of taurine released, medium fractions can be collected at the various time points and analyzed by reverse-phase HPLC after precolumn derivatization with o-phthaldialdehydeas, as described previously.

5.3. Ussing chamber experiments

The "Ussing chamber" technique is particularly convenient for studying ionic responses in intact muscle-stripped intestinal epithelium or in filter-grown monolayers of cultured enterocytes that have developed a junctional complex. This method not only allows a direct evaluation of the transepithelial potential or current but can also be used to measure changes in the epithelial resistance (reviewed by de Jonge *et al.*, 2004; Hug, 2002). In comparison with patch clamp studies and radioisotope efflux assays, Ussing chamber experiments have the advantage that intact cells and epithelia, not impaled with microelectrodes, are studied and that, unlike tracer studies, the actual transepithelial current, as well as its direction, can be determined. A disadvantage of this method, however, is that the observed current reflects the sum of the magnitudes and directions of all individual anion and cation currents from both mucosal (apical) or serosal (basolateral) sides. Therefore, to evaluate a specific current, inhibitors should be used or conditions should be chosen that nullify other contributing conductances.

Since the first description of the Ussing chamber by the Danish physiologist Hans Ussing (Ussing and Zerahn, 1951), a number of different designs have been reported, some available commercially, that fit pieces of isolated intestinal epithelium, filter inserts with cultured cells, or even small pieces of tissue obtained from biopsies. In general, the chamber consists of two separate compartments connected by a small hole that can contain the tissue or the filter insert. Each compartment has two electrodes: one (platinum) for current injections and one for transepithelial voltage recording (calomel or Ag/AgCl electrode). Using an amplifier suitable for Ussing chamber studies, the transepithelial potential difference (PD) or the short circuit current (Isc, under voltage clamp conditions), as well as by injecting a current pulse of known magnitude the transepithelial electrical resistance (TEER) can be determined. The most suitable chamber designs have half cells, which can be perfused continuously and independently of each other, making changes in the composition of the serosal and mucosal medium possible, and are temperature controlled.

To study transepithelial potentials and short circuit currents in murine or rat intestinal epithelium, small sections of stripped mucosa are mounted in Ussing chambers (0.3-cm^2 exposed area) perfused with modified Meyler solution gassed continuously with 95% O_2/5% CO_2 at 37°. The potential difference is measured using calomel electrodes (Radiometer, Copenhagen, Denmark) connected to each of the half chambers by KCl-agar bridges; current (1–10 μA) is injected through platinum electrodes. The electrodes are connected to a DVC–1000 dual channel voltage/current clamp amplifier (WPI, Berlin, Germany). Data are digitized using a Digidata 1322A AD converter (Axon Instruments) and analyzed using AxoScope software (Axon Instruments). Two Ussing chambers are used simultaneously to record the PD or Isc. The PD is expressed relative to the potential of the serosal side of the tissue; an Isc current from mucosa to serosa is designated positive. The Isc is measured continuously with the transepithelial voltage set at the PD value. To evoke hyperosmotically induced changes in Isc, mannitol could be added to either the serosal and/or the mucosal side of the tissue. To avoid junction potentials, however, it is better to study hypotonicity-provoked currents in

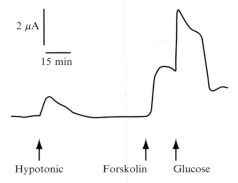

Figure 19.3 Short circuit current recording of stimulated murine muscle-stripped ileum mounted in an Ussing chamber. Arrows indicate time points of the shift to a hypotonic medium (70% tonicity, serosal and mucosal) or the addition of forskolin (10 μM, serosal), an activator of CFTR-mediated Cl^- secretion, or glucose (10 mM, mucosal), initiating glucose-coupled Na^+ uptake. Upward deflections represent a net current from the mucosal to the serosal side of the chamber.

chambers perfused with a low sodium and mannitol-containing modified Meyler solution. A hypoosmotic stimulus can then be applied by switching the perfusion medium to one containing less mannitol. The magnitude of the Isc current in response to mucosal glucose, a consequence of electrogenic Na^+-driven glucose uptake, will help the experimenter evaluate tissue viability. Figure 19.3 shows an example of osmotically induced changes in Isc in murine ileum.

5.4. Fluorometric quantification of intracellular ion concentrations

To date, fluorescent probes are used frequently as easy and noninvasive tools to study changes in the intracellular concentrations of ions or to determine alterations in membrane potential and intracellular pH in many different model systems. In addition, several of these indicators have been proven to be suitable for high-throughput quantitative studies. The use of fluorescent anion indicators has been boosted by a search for pharmacological potentiators and correctors of F508delCFTR, the most common mutant CFTR chloride channel identified in cystic fibrosis patients. Most fluorescent anion indicators, such as SPQ and MQAE (Molecular Probes, Eugene, OR), are methoxyquinolinium derivatives and have a rather poor plasma membrane permeability (Verkman, 1990). As a consequence, loading of the cells with these probes may take many hours. A halide-sensitive derivative of green fluorescent protein (YFP-H148Q/I152L) has been generated that could be expressed in target cells, thereby avoiding long periods of loading (Galietta *et al.*, 2001). Although these indicators are very useful in studying channel activation under isotonic conditions, the fluorescence signal is sensitive to changes in probe concentration, limiting their suitability for studies of ion concentration changes in response to changes in cell volume.

6. Concluding Remarks

To date, a wide variety of model systems and techniques are available to study osmosensing and osmosignaling in intestinal epithelial cells. However, despite many years of intensive research by multiple groups, important questions, such as the identity of the volume sensor, VRAC, and the transporter/channel involved in organic osmolyte release, remain to be resolved. In addition, our present knowledge about the processes linking the putative osmosensor to its effectors is rather limited. Perhaps newly emerging technologies, such as gene knockdown with small interfering RNAs (see Eckstein, 2005) in either cells or intact animals, may help in elucidating these questions.

REFERENCES

Andres, H., Bock, R., Bridges, R. J., Rummel, W., and Schreiner, J. (1985). Submucosal plexus and electrolyte transport across rat colonic mucosa. *J. Physiol.* **364,** 301–312.
Augeron, C., Maoret, J. J., Laboisse, C. L., and Grasset, E. (1986). Permanently differentiated cell clones isolated from the human colonic adenocarcinoma cell line HT-29: Possible models for the study of ion transport and mucuc production. *In* "Ion Gradient Coupled Transport" (F. Alvarado and C. H. Van Os, eds.), pp. 363–366. Elsevier, New York.
Bruck, R., Haddad, P., Graf, J., and Boyer, J. L. (1992). Regulatory volume decrease stimulates bile flow, bile acid excretion, and exocytosis in isolated perfused rat liver. *Am. J. Physiol.* **262,** G806–G812.
Carton, I., Hermans, D., and Eggermont, J. (2003). Hypotonicity induces membrane protrusions and actin remodeling via activation of small GTPases Rac and Cdc42 in Rat-1 fibroblasts. *Am. J. Physiol.* **285,** C935–C944.
Carton, I., Trouet, D., Hermans, D., Barth, H., Aktories, K., Droogmans, G., Jorgensen, N. K., Hoffmann, E. K., Nilius, B., and Eggermont, J. (2002). RhoA exerts a permissive effect on volume-regulated anion channels in vascular endothelial cells. *Am. J. Physiol.* **283,** C115–C125.
Cornet, M., Lambert, I. H., and Hoffmann, E. K. (1993). Relation between cytoskeleton, hypo-osmotic treatment, and volume regulation in Ehrlich ascites tumor cells. *J. Membr. Biol.* **131,** 55–66.
de Jonge, H. R., Ballmann, M., Veeze, H., Bronsveld, I., Stanke, F., Tümmler, B., and Sinaasappel, M. (2004). *Ex vivo* CF diagnosis by intestinal current measurements (ICM) in small aperture, circulating Ussing chambers. *J. Cyst. Fibros.* **3,** 159–163.

Dharmsathaphorn, K., McRoberts, J. A., Mandel, K. G., Tisdale, L. D., and Masui, H. (1984). A human colonic tumor cell line that maintains vectorial electrolyte transport. *Am. J. Physiol.* **246**, G204–G208.

Diener, M., Bertog, M., Fromm, M., and Scharrer, E. (1996). Segmental heterogeneity of swelling-induced Cl⁻ transport in rat small intestine. *Pflüg. Arch.* **432**, 293–300.

Eckstein, F. (2005). Small non-coding RNAs as magic bullets. *Trends Biochem. Sci.* **8**, 445–452.

Estevez, A. Y., Bond, T., and Strange, K. (2001). Regulation of $I_{(Cl,swell)}$ in neuroblastoma cells by G protein signaling pathways. *Am. J. Physiol.* **281**, C89–C98.

Fogh, J., Fogh, J. M., and Orfeo, T. (1977). One hundred and twenty-seven cultured human tumor cell lines producing tumors in nude mice. *J. Natl. Cancer Inst.* **59**, 221–226.

Foskett, J. K., and Spring, K. R. (1985). Involvement of calcium and cytoskeleton in gallbladder epithelial cell volume regulation. *Am. J. Physiol.* **248**, C27–C36.

Galietta, L. J. V., Haggie, P. M., and Verkman, A. S. (2001). Green fluorescent protein-based halide indicators with improved chloride and iodide affinities. *FEBS Lett.* **499**, 220–224.

Grefen, C., and Harter, K. (2004). Plant two-component systems: Principles, functions, complexity and cross talk. *Planta* **219**, 733–742.

Grunnet, M., MacAulay, N., Jorgensen, N. K., Olesen, S. P., and Klaerke, D. A. (2002). Regulation of cloned, Ca^{2+}-activated K^+ channels by cell volume changes. *Pflüg. Arch.* **444**, 167–177.

Hamill, O. P., Marty, A., Neher, E., Sakmann, B., and Sigworth, F. J. (1981). Improved patch-clamp techniques for high-resolution current recording from cells and cell-free membrane patches. *Pflüg. Arch.* **391**, 85–100.

Häussinger, D., Reinehr, R., and Schliess, F. (2006). The hepatocyte integrin system and cell volume sensing. *Acta Physiol. (Oxf.)* **187**, 249–255.

Hazama, A., and Okada, Y. (1988). Ca^{2+} sensitivity of volume-regulatory K^+ and Cl^- channels in cultured human epithelial cells. *J. Physiol.* **402**, 687–702.

Hazama, A., Shimizu, T., Ando-Akatsuka, Y., Hayashi, S., Tanaka, S., Maeno, E., and Okada, Y. (1999). Swelling-induced, CFTR-independent ATP release from a human epithelial cell line: Lack of correlation with volume-sensitive Cl^- channels. *J. Gen. Physiol.* **114**, 525–533.

Henle, G., and Deinhardt, F. (1957). The establishment of strains of human cells in tissue culture. *J. Immunol.* **79**, 54–59.

Hohmann, S. (2002). Osmotic stress signaling and osmoadaptation in yeasts. *Microbiol. Mol. Biol. Rev.* **66**, 300–372.

Hörger, S., Schultheiß, G., and Diener, M. (1998). Segment-specific effects of epinephrine on ion transport in the colon of the rat. *Am. J. Physiol.* **275**, 1367–1376.

Hug, M. J. (2002). Transepithelial measurements using the Ussing chamber: The online Virtual Repository of Cystic Fibrosis European Network 2002. http://central.igc.gulbenkian.pt/cftr/vr/d/hug_transepithelial_measurements_using_the_ussing_chamber.pdf.

Jentsch, T. J., Stein, V., Weinreich, F., and Zdebik, A. A. (2002). Molecular structure and physiological function of chloride channels. *Physiol. Rev.* **82**, 503–568.

Junankar, P. R., and Kirk, K. (2000). Organic osmolyte channels: A comparative view. *Cell. Physiol. Biochem.* **10**, 355–360.

Kubo, M., and Okada, Y. (1992). Volume-regulatory Cl^- channel currents in cultured human epithelial cells. *J. Physiol.* **456**, 351–371.

Lambert, I. H. (2004). Regulation of the cellular content of the organic osmolyte taurine in mammalian cells. *Neurochem. Res.* **29**, 27–63.

Lang, F., Busch, G. L., Ritter, M., Völkl, H., Waldegger, S., Gulbins, E., and Häussinger, D. (1998). Functional significance of cell volume regulatory mechanisms. *Physiol. Rev.* **78**, 247–306.

Lepple-Wienhues, A., Szabo, I., Laun, T., Kaba, N. K., Gulbins, E., and Lang, F. (1998). The tyrosine kinase p56lck mediates activation of swelling-induced chloride channels in lymphocytes. *J. Cell Biol.* **141,** 281–286.

Levitan, E. S., and Kramer, R. H. (1990). Neuropeptide modulation of single calcium and potassium channels detected with a new patch clamp configuration. *Nature* **348,** 545–547.

Lim, C. H., Schoonderwoerd, K., Kleijer, W. J., De Jonge, H. R., and Tilly, B. C. (2006). Regulation of the cell swelling-activated chloride conductance by cholesterol-rich membrane domains. *Acta Physiol. (Oxf.)* **187,** 295–303.

MacLeod, R. J., and Hamilton, J. R. (1990). Regulatory volume increase in mammalian jejunal villus cells is due to bumetanide-sensitive NaKCl2 cotransport. *Am. J. Physiol.* **258,** G665–G674.

MacLeod, R. J., and Hamilton, J. R. (1996). Activation of Na^+/H^+ exchange is required for regulatory volume decrease after modest "physiological" volume increases in jejunal villus epithelial cells. *J. Biol. Chem.* **271,** 23138–23145.

MacLeod, R. J., and Hamilton, J. R. (1999). Increases in intracellular pH and Ca(2+) are essential for K(+) channel activation after modest 'physiological' swelling in villus epithelial cells. *J. Membr. Biol.* **172,** 47–58.

MacLeod, R. J., Lembessis, P., and Hamilton, J. R. (1992). Differences in Ca^{2+}-mediation of hypotonic and Na^+-nutrient regulatory volume decrease in suspensions of jejunal enterocytes. *J. Membr. Biol.* **130,** 23–31.

Mignen, O., Le Gall, C., Harvey, B. J., and Thomas, S. (1999). Volume regulation following hypotonic shock in isolated crypts of mouse distal colon. *J. Physiol.* **515,** 501–510.

Mongin, A. A., and Orlov, S. N. (2001). Mechanisms of cell volume regulation and possible nature of the cell volume sensor. *Pathophysiology* **8,** 77–88.

Murakami, H., and Masui, H. (1980). Hormonal control of human colon carcinoma cell growth in serum-free medium. *Proc. Natl. Acad. Sci. USA* **77,** 3464–3468.

Neher, E. (1981). Unit conductance studies in biological membranes. *In* "Techniques in Cellular Physiology" (P. F. Baket ed.). Elsevier/North Holland Biomedical Press, Amsterdam.

Niemeyer, M. I., Cid, L. P., and Sepúlveda, F. V. (2001). K^+ conductance activated during regulatory volume decrease. *Comp. Biochem. Physiol. A Mol. Integr. Physiol.* **130,** 565–575.

Nilius, B., and Droogmans, G. (2003). Amazing chloride channels: An overview. *Acta Physiol. Scand.* **177,** 119–147.

O'Brien, J. A., Waiters, R. J., and Sepúlveda, F. V. (1991). Regulatory volume decrease in small intestinal crypts is inhibited by K^+ and Cl^- channel blockers. *Biochim. Biophys. Acta* **1070,** 501–504.

Okada, Y. (1997). Volume expansion-sensing outward rectifier Cl^- channel: Fresh start to the molecular identity and volume sensor. *Am. J. Physiol.* **273,** C755–C789.

Okada, Y. (2006). Cell volume-sensitive chloride channels: Phenotypic properties and molecular identity. *Contrib. Nephrol.* **152,** 9–24.

Okada, Y., Hazama, A., Hashimoto, A., Maruyama, Y., and Kubo, M. (1992). Exocytosis upon osmotic swelling in human epithelial cells. *Biochim. Biophys. Acta* **1107,** 201–205.

O'Neill, W. C. (1999). Physiological significance of volume-regulatory transporters. *Am. J. Physiol.* **276,** C995–C1011.

Pasantes-Morales, H., Franco, R., Ochoa, L., and Ordaz, B. (2002). Osmosensitive release of neurotransmitter amino acids: Relevance and mechanisms. *Neurochem. Res.* **27,** 59–65.

Pasantes-Morales, H., Lezama, A. P., and Ramos-Mandujano, G. (2006a). Tyrosine kinases and osmolyte fluxes during hyposmotic swelling. *Acta Physiol. (Oxf.)* **187,** 93–102.

Pasantes-Morales, H., Lezama, A. P., Ramos-Mandujano, G., and Tuz, K. L. (2006b). Mechanisms of cell volume regulation in hypo-osmolality. *Am. J. Med.* **119,** S4–S11.

Pedersen, S. F., Beisner, K. H., Hougaard, C., Willumsen, B. M., Lambert, I. H., and Hoffmann, E. K. (2002). Rho family GTP binding proteins are involved in the regulatory volume decrease process in NIH3T3 mouse fibroblasts. *J. Physiol.* **541,** 779–796.

Pedersen, S. F., Mills, J. W., and Hoffmann, E. K. (1999). Role of the F-actin cytoskeleton in the RVD and RVI processes in Ehrlich ascites tumor cells. *Exp. Cell Res.* **252,** 63–74.

Quaroni, A., Wands, J., Trelstad, R. L., and Isselbacher, K. J. (1979). Epithelioid cell cultures from rat small intestine: Characterization by morphologic and immunologic criteria. *J. Cell Biol.* **80,** 248–265.

Sakmann, B., and Neher, E. (1984). Patch clamp techniques for studying ionic channels in excitable membranes. *Annu. Rev. Physiol.* **46,** 455–472.

Sardini, A., Amey, J. S., Weylandt, K. H., Nobles, M., Valverde, M. A., and Higgins, C. F. (2003). Cell volume regulation and swelling-activated chloride channels. *Biochim. Biophys. Acta* **1618,** 153–162.

Shi, C., Barnes, S., Coca-Prados, M., and Kelly, M. E. (2002). Protein tyrosine kinase and protein phosphatase signaling pathways regulate volume-sensitive chloride currents in a nonpigmented ciliary epithelial cell line. *Invest. Ophthalmol. Vis. Sci.* **43,** 1525–1532.

Sorota, S. (1995). Tyrosine protein kinase inhibitors prevent activation of cardiac swelling-induced chloride current. *Pflüg. Arch.* **431,** 178–185.

Strange, K. (2004). Cellular volume homeostasis. *Adv. Physiol. Educ.* **28,** 155–159.

Štrbák, V., and Greer, M. A. (2002). Regulation of hormone secretion by acute cell volume changes: Ca^{2+}-independent hormone secretion. *Cell. Physiol. Biochem.* **10,** 393–402.

Tilly, B. C., Edixhoven, M. J., Tertoolen, L. G. J., Morii, N., Saitoh, Y., Narumiya, S., and de Jonge, H. R. (1996). Activation of the osmo-sensitive chloride conductance involves $p21^{rho}$ and is accompanied by a transient reorganization of the F-actin cytoskeleton. *Mol. Biol. Cell.* **7,** 1419–1427.

Tilly, B. C., Edixhoven, M. J., van den Berghe, N., Bot, A. G., and de Jonge, H. R. (1994). Ca^{2+}-mobilizing hormones potentiate hypotonicity-induced activation of ionic conductances in Intestine 407 cells. *Am. J. Physiol.* **267,** C1271–C1278.

Tilly, B. C., Kansen, M., van Gageldonk, P. G. M., van den Berghe, N., Bijman, J., Galjaard, H., and de Jonge, H. R. (1991). G-proteins mediate intestinal chloride channel activation. *J. Biol. Chem.* **266,** 2036–2040.

Tilly, B. C., van den Berghe, N., Tertoolen, L. G. J., Edixhoven, M. J., and de Jonge, H. R. (1993). Protein tyrosine phosphorylation is involved in osmoregulation of ionic conductances. *J. Biol. Chem.* **268,** 19919–19922.

Tomassen, S. F. B., Fekkes, D., de Jonge, H. R., and Tilly, B. C. (2004). Osmotic swelling provoked release of organic osmolytes in human intestinal epithelial cells: Relation to chloride channel activation and regulation by PKC. *Am. J. Physiol.* **286,** C1417–C1422.

Trouet, D., Carton, I., Hermans, D., Droogmans, G., Nilius, B., and Eggermont, J. (2001). Inhibition of VRAC by c-Src tyrosine kinase targeted to caveolae is mediated by the Src homology domains. *Am. J. Physiol.* **281,** C248–C256.

Urao, T., Yamaguchi-Shinozaki, K., and Shinozaki, K. (2000). Two-component systems in plant signal transduction. *Trends Plant Sci.* **5,** 67–74.

Ussing, H. H., and Zerahn, K. (1951). Active transport of sodium as the source of electric current in the short-circuited isolated frog skin. *Acta Physiol. Scand.* **23,** 110–127.

Vaandrager, A. B., Bajnath, R., Groot, J. A., Bot, A. G. M., and De Jonge, H. R. (1991). Ca^{2+} and cAMP activate different chloride efflux pathways in HT-29.cl19A colonic epithelial cell line. *Am. J. Physiol.* **261,** G958–G965.

Van der Wijk, T., de Jonge, H. R., and Tilly, B. C. (1999). Osmotic cell swelling-induced ATP release mediates the activation of Erk-1/2 but not the activation of osmo-sensitive anion channels. *Biochem. J.* **343,** 579–586.

Van der Wijk, T., Dorrestijn, J., Narumiya, S., Maassen, J. A., de Jonge, H. R., and Tilly, B. C. (1998). Osmotic swelling-induced activation of the extracellular signal-regulated protein kinases Erk-1 and Erk-2 involves the Ras/Raf signalling pathway. *Biochem. J.* **331,** 863–869.

Van der Wijk, T., Tomassen, S. F. B., Houtsmuller, A. B., de Jonge, H. R., and Tilly, B. C. (2003). Increased vesicle recycling in response to osmotic cell-swelling; cause and consequence of hypotonicity-provoked ATP release. *J. Biol. Chem.* **278,** 40020–40025.

Verkman, A. S. (1990). Development and biological applications of chloride-sensitive fluorescent indicators. *Am. J. Physiol.* **259,** C375–C388.

Voets, T., Manolopoulos, V., Eggermont, J., Ellory, C., Droogmans, G., and Nilius, B. (1998). Regulation of a swelling-activated chloride current in bovine endothelium by protein tyrosine phosphorylation and G proteins. *J. Physiol.* **506,** 341–352.

Wang, J., Morishima, S., and Okada, Y. (2003). IK channels are involved in the regulatory volume decrease in human epithelial cells. *Am. J. Physiol.* **284,** C77–C84.

Wehner, F., Olsen, H., Tinel, H., Kinne-Saffran, E., and Kinne, R. K. H. (2003). Cell volume regulation: Osmolytes, osmolyte transport, and signal transduction. *Rev. Physiol. Biochem. Pharmacol.* **148,** 1–80.

CHAPTER TWENTY

OSMOTIC REGULATION OF CELLULAR GLUCOSE UPTAKE

Philippe Gual, Teresa Gonzalez, Thierry Gremeaux,
Yannick Le Marchand-Brustel, *and* Jean-François Tanti

Contents

Abstract

This chapter describes various approaches allowing the study of hyperosmolarity in the functions of 3T3-L1 adipocytes. Hyperosmolarity mimics insulin responses, such as glucose uptake and membrane ruffling, but also antagonizes these insulin effects, which can be evaluated in 3T3-L1 adipocytes. The molecular mechanisms of these effects can be also investigated by measuring the

INSERM U 568; University of Nice Sophia-Antipolis, Nice, France

Methods in Enzymology, Volume 428
ISSN 0076-6879, DOI: 10.1016/S0076-6879(07)28020-6

activation of different signaling pathways: (i) the phosphorylation of docking proteins on tyrosine and serine residues (serines 307 and 632), (ii) the phosphorylation of serine/threonine kinases, and (iii) the activation of phosphatidylinositol 3-kinase.

1. INTRODUCTION

Exposure of many mammalian cell types to metabolic or osmotic stress results in an acute increase in the rate of glucose uptake. This adaptive response allows the cells to maintain or regain their ATP levels by increasing flux through the glycolytic pathway. However, like several other insulinomimetic agents, hyperosmolarity not only partly activates several insulin-specific biological responses, but also induces a state of insulin resistance. Conditions dehydrating insulin target tissues such as hyperosmolarity or amino acid deprivation are frequently associated with insulin resistance. Sepsis and burn injury are associated with dehydration–induced insulin resistance. Furthermore, in rat epididymal fat cells, hyperosmotic stress markedly reduces *in vitro* insulin–induced glucose transport (Komjati *et al.*, 1988). In perfused rat liver, hyperosmolarity impairs insulin–mediated cell swelling and reverses the proteolysis inhibition induced by insulin (Vom Dahl *et al.*, 1991). In 3T3-L1 adipocytes, pretreatment with sorbitol strongly decreases the ability of insulin to stimulate glucose uptake, lipogenesis, glycogen synthesis, and membrane ruffling (Chen *et al.*, 1999; Gual *et al.*, 2003a).

Glucose uptake induced by insulin in muscle and adipose tissues is due to the translocation of the glucose transporter Glut 4 from an intracellular pool to the plasma membrane (Bryant *et al.*, 2002; Saltiel and Kahn, 2001). These biological responses require tyrosine phosphorylation of insulin receptor substrate-1 (IRS1), which leads to the binding and activation of phosphatidylinositol 3-kinase (PI 3-kinase). Downstream effectors of PI 3-kinase such as protein kinase B (PKB) or atypical PKC are involved in Glut 4 translocation. Furthermore, it has been shown that insulin-induced Glut 4 translocation also requires activation of a second pathway, which is completely independent of PI 3-kinase activity. In adipocytes, the Cbl protooncogene, associated with Cbl-associated protein and adapter protein containing PH and SH2 domain, is phosphorylated in response to insulin and regulates glucose uptake. Once phosphorylated, Cbl recruits the adapter protein Crk-II in a complex with C3G, a GDP to GTP exchange factor for TC10, a Rho family GTPase, allowing for its activation, which regulates the traffic of Glut 4–containing vesicles. GTP-bound TC10 could participate in Glut 4 translocation through a modification of cortical actin or a stimulation of actin polymerization at the level of Glut 4 compartments

(Scheme 20.1) (Dugani and Klip, 2005; Gual *et al.*, 2003b; Saltiel and Pessin, 2002; Watson *et al.*, 2004).

In both adipocyte (Scheme 20.1) and muscle cells, hyperosmolarity promotes glucose uptake by multiple mechanisms that do not require the PI 3-kinase/PKB pathway (Gual *et al.*, 2003b) but are dependent on the cell type. In muscle, osmotic stress induces glucose uptake by stimulation of AMP-kinase and/or inhibition of Glut 4 endocytosis. In adipocytes, activation of the Grb2-associated binder-1 (Gab1)-dependent signaling pathway plays an important role in osmotic stress-mediated glucose uptake (Gual *et al.*, 2003a; Janez *et al.*, 2000). Upon sorbitol stimulation, the phosphorylated Gab1 recruits Crk-II via its SH2 domain. The Crk-II SH3 domains are constitutively associated with C3G, a GDP-to-GTP exchange factor for several small GTP-binding proteins, including TC10 (Gual *et al.*, 2002). The activation of TC10 and remodeling of cortical actin are required for osmotic shock-mediated Glut 4 translocation and glucose uptake (Gual *et al.*, 2002).

Apart from its insulin-like effects, hyperosmolarity leads to cellular insulin resistance (Fig. 20.1) mediated by both prevention of PKB activation

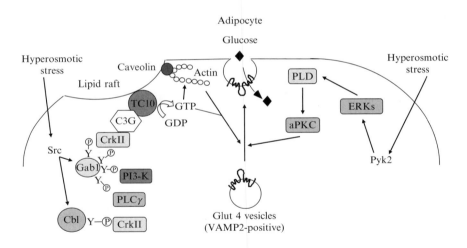

Scheme 20.1 Signaling pathways activated by hyperosmotic stress to induce Glut 4 translocation in 3T3-L1 adipocytes and adipose cells. Osmotic stress induces glucose uptake by activation of Gab1-dependent signaling pathways. Sorbitol promotes the activation of cytosolic Src kinase, which phosphorylates Gab1 on the tyrosine residue. Phosphorylated Gab1 recruits the Crk-II/C3G complex. C3G, a guanine nucleotide exchange factor, exchanges GDP for GTP on TC10. Activated GTP-bound TC10 could then modify the cortical actin structure or stimulate the actin polymerization on Glut 4 compartments. Osmotic stress recruits vesicles containing both Glut 4 and VAMP2 in adipocytes. Glut 4 translocation stimulated by hyperosmotic stress could also depend on PYK2 activity, which leads to the activation of ERK, PLD, and finally atypical PKC. Adapted from Gual *et al.* (2003b).

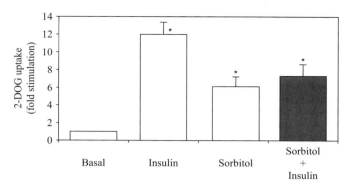

Figure 20.1 Hyperosmolarity and glucose transport. After serum starvation, 3T3-L1 adipocytes were treated without or with insulin (100 n*M*) or sorbitol (600 m*M*) for 20 min (empty bars) or pretreated with sorbitol (600 m*M*) for 40 min before a 20-min insulin stimulation (100 n*M*) (gray bar). Uptake of [2-³H]deoxyglucose was measured during a 3-min period. Means ± SEM of three independent experiments are shown.

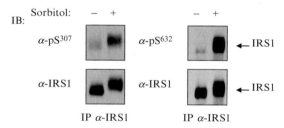

Figure 20.2 Hyperosmotic stress triggers the phosphorylation of IRS1 on serines 307 and 632. 3T3-L1 adipocytes were stimulated without or with 600 m*M* sorbitol for 20 min at 37°. Cell lysates were then immunoprecipitated (IP) using anti-IRS1 antibodies. Immunoprecipitated proteins were resolved by SDS-PAGE and blotted (IB) using anti-pSer³⁰⁷ IRS1 or anti-pSer⁶³² IRS1 antibodies, as indicated. The membrane was then stripped and probed with anti-IRS1 antibodies. Representative autoradiographs are shown.

(Chen *et al.*, 1999) and inhibition of the IRS1 function (Gual *et al.*, 2003a). Chen *et al.* (1999) have reported that hyperosmolarity prevents insulin-induced PKB activation. They suggest that a calyculin A- or okadaic acid-sensitive protein phosphatase leads to the deactivation of PKB (Chen *et al.*, 1999). Furthermore, a short-term osmotic stress induces the phosphorylation of IRS1 on serine 307 by an mTOR-dependent pathway (Gual *et al.*, 2003a) but also on serine 632 (Fig. 20.2). This, in turn, leads to a decrease in early proximal signaling events induced by physiological insulin concentrations (Gual *et al.*, 2003a). However, prolonged osmotic stress alters IRS1 and IRS2 functions by inducing their degradation (Gual *et al.*,

2003a,b; Rui *et al.*, 2001), which could contribute to the downregulation of insulin action.

2. Hyperosmolarity and Glucose Transport

In muscle and adipose cells, hyperosmolarity triggers the cell surface accumulation of Glut 4, leading to an increase in glucose transport (Dugani and Klip, 2005; Gual *et al.*, 2003b; Saltiel and Pessin, 2002; Watson *et al.*, 2004). While the hyperosmotic stress mimics insulin responses, it also antagonizes insulin effects. Both effects can be evaluated in 3T3–L1 adipocytes (see Scheme 20.1).

2.1. Differentiation of 3T3-L1 adipocytes

3T3–L1 fibroblasts are a continuous substrain of 3T3 (Swiss albino) developed through clonal isolation (CL-173, ATCC). The cells undergo a preadipose to adipose-like conversion as they progress from a rapidly dividing to a confluent and contact inhibited state. After differentiation, Glut 4 translocation and glucose uptake can be induced in response to insulin. The differentiation protocol is as follows.

3T3–L1 fibroblasts are grown in six-well plates or 100-mm dishes in Dulbecco's modified Eagle's medium (DMEM) containing 25 mM glucose and 10% calf serum and induced to differentiate in adipocytes. Two days after confluence, medium is changed for a differentiation buffer, which includes DMEM, 25 mM glucose, 10% fetal calf serum supplemented with isobutylmethylxanthine (0.5 mM), dexamethasone (0.25 μM), thiazolidinediones (10 μM), and insulin (5 μg/ml). The medium is removed after 2 days and replaced with DMEM, 25 mM glucose, 10% fetal calf serum supplemented with insulin, and thiazolidinediones (10 μM) for 2 more days. Then, the cells are fed every 2 days with DMEM, 25 mM glucose, and 10% fetal calf serum. 3T3-L1 adipocytes are used 8 to 15 days after the beginning of the differentiation protocol. Sixteen hours before each experiment, the medium is changed to serum-free DMEM supplemented with 0.5% bovine serum albumin (BSA).

2.2. Glucose uptake induced in response to hyperosmolarity

The technique allows for the measurement of glucose uptake using a glucose analog (2-deoxyglucose, DOG). The labeled 2-DOG is transported into the cells with high affinity, phosphorylated but not further metabolized. The labeled 2-DOG phosphate is trapped in the cell, and the rate of uptake can be taken as a measure of unidirectional transport.

1. After serum starvation, cells (six-well plates) are washed three times with 1 ml of Krebs–Ringer phosphate (KRP) buffer (10 mM phosphate buffer, pH 7.4, 1.25 mM MgSO$_4$, 1.25 mM CaCl$_2$, 136 mM NaCl, 4.7 mM KCl) at 37°.
2. Cells are incubated in 900 μl of KRP buffer supplemented with 0.2% BSA without or with sorbitol (600 mM) for 20 min.
3. Measurement of glucose transport is initiated by the addition of 100 μl of KRP buffer supplemented with 0.2% BSA and containing 1 mM deoxy-D-glucose and 0.5 μCi of 2-[^{3}H]deoxy-D-glucose.
4. The reaction is stopped after 3 min by rapid aspiration, and cells are washed four times with ice-cold phosphate-buffered saline (PBS).
5. Cells are lysed with 700 μl of KRP and 1% Triton X-100. After brief sonication, 300 μl of lysates is mixed with 3 ml of liquid scintillation, and the radioactivity associated with the cells is counted in a β-counter using the tritium program.
6. Radioactivity is normalized by measuring the protein concentration in each sample by the bicinchoninic acid (BCA) assay (Pierce).

Do not forget to count an aliquot of the radioactive deoxyglucose solution (using a commercial scintillation fluid), which will allow expressing the results in absolute amounts of glucose taken up by the cells.

2.3. Study of pathways involved in the hyperosmotic effect on glucose uptake

To determine the pathways required for hyperosmotic stress–induced glucose transport or the study agents capable of preventing this response, 3T3-L1 adipocytes are pretreated with a specific inhibitor or agents (at step 2) before stimulation with sorbitol (600 mM). Deoxyglucose uptake is then measured as described earlier. The TC10 activity per se can be inhibited by *Clostridium difficile* toxin B (1 μg/ml), an inhibitor of the Rho family proteins. Latrunculin B (10 μM) (Calbiochem), an agent that sequesters actin monomers, or jasplakinolide (10 μM) (Molecular Probes, Eugene, OR), a stabilizer of filamentous actin, also inhibits sorbitol-induced glucose uptake. It is important to note that latrunculin B does not alter the early proximal osmotic shock signaling events such as Gab1 phosphorylation and its association with Crk-II. After serum starvation, 3T3-L1 adipocytes are treated without or with *C. difficile* toxin B (1 μg/ml), latrunculin B (10 μM), or jasplakinolide (10 μM) for 120 min at 37°. Cells are then either left untreated or stimulated with 600 mM sorbitol for 20 min at 37°.

2.4. Inhibition of insulin-stimulated glucose uptake by hyperosmotic stress

Like several other insulinomimetic agents, hyperosmolarity not only partially activates several insulin-specific biological responses, but also induces a state of insulin resistance. To evaluate this, the following protocol is used.

1. After serum starvation, cells (six-well plates) are then washed three times with 1 ml of KRB buffer at 37°.
2. Cells are incubated in 800 μl of KRP buffer supplemented with 0.2% BSA without or with sorbitol (600 mM) for 40 min.
3. One hundred microliters of insulin (5 nM in KRP buffer supplemented with 0.2% BSA) is added per well of six-well plates for 15 min.
4. Deoxyglucose uptake is then determined as described previously.

3. HYPEROSMOLARITY AND MEMBRANE RUFFLING

Hyperosmotic stress induced an increase of nearly twofold in the number of cells with membrane ruffles (Gual *et al.*, 2003a; Janez *et al.*, 2000). Insulin stimulation promoted a fourfold increase in the number of cells harboring membrane ruffles, an effect that was totally abolished by sorbitol treatment. This indicates that sorbitol pretreatment inhibits insulin-induced membrane ruffling markedly, suggesting that inhibition of IRS1-associated PI 3-kinase activity by hyperosmotic stress appears sufficient to alter this insulin effect (Gual *et al.*, 2003a,b).

3.1. Membrane ruffling assay

As described previously (Barres *et al.*, 2006; Gual *et al.*, 2003a), 3T3-L1 adipocytes are grown and differentiated on glass coverslips. After overnight serum starvation, the medium is replaced by serum-free medium supplemented without or with 600 mM sorbitol before 20 min of insulin stimulation (0.5 nM). Cells are washed twice with ice-cold PBS and fixed with 4% paraformaldehyde for 20 min on ice. After two washes with ice-cold PBS, cells are permeabilized with PBS containing 0.1% Triton X-100 and 1% bovine serum albumin for 30 min at room temperature. After three washes with ice-cold PBS, cells are incubated with Texas red-phalloidin (Molecular Probes, Inc.) in PBS/0.1% Triton X-100/1% BSA for 30 min at room temperature. Cells are then washed twice with ice-cold PBS, and coverslips are mounted in 20 μl of Mowiol onto glass slides. Cells are examined using a Leica confocal microscope equipped with a Leica confocal laser-scanning imaging. Cells are studied at a magnification of 40× using a 1.0 to 0.50 oil immersion objective. Series of images are collected along the z axis and examined. In each condition, 200 cells in random fields should be examined by two different persons blind to the origin of the images. Cells that show clustering of actin staining at the periphery are scored as positive for membrane ruffles.

4. Hyperosmolarity and Signaling Pathways

In 3T3-L1 adipocytes, osmotic stress promotes the activation of cytosolic Src kinase, which phosphorylates Gab1 on the tyrosine residue. Phosphorylated Gab1 recruits the Crk-II/C3G, leading to the activation of TC10, which could then modify the cortical actin structure or stimulate the actin polymerization on Glut 4 compartments. Glut 4 translocation stimulated by hyperosmotic stress could also depend on PYK2 activity, which leads to the activation of ERK, PLD, and finally atypical PKC (Gual *et al.*, 2003b).

Acute osmotic stress (from 10 to 30 min of sorbitol cell treatment) induces the phosphorylation of IRS1 on Ser^{307} by a mTOR-dependent pathway (Gual *et al.*, 2003a), which in turn leads to an impairment of IRS1 functions induced by physiological insulin concentrations (Gual *et al.*, 2003a, 2005). Hyperosmotic stress also promotes activation of a PKB phosphatase that maintains PKB in an inactive state in response to insulin. However, prolonged osmotic stress (4 h of sorbitol cell treatment) alters IRS function by inducing their degradation, thus contributing to the downregulation of insulin action (Gual *et al.*, 2003a, 2005).

4.1. Preparation of total cell lysates

The activation of SRC, ERK, JNK, p38 mitogen-activated protein kinase, PKC, and PKB can be evaluated directly by determining their phosphorylation levels by Western blotting from total cell lysates using phospho-specific antibodies. In parallel, the total amount of IRS proteins can be evaluated directly by Western blotting from total cell lysates.

3T3-L1 adipocytes are serum starved overnight in DMEM/0.5% BSA. 3T3-L1 adipocytes are incubated in serum-free medium supplemented or not with 600 mM sorbitol and subsequently treated with or without a low concentration of insulin (0.2 nM). To study the effect of pharmacological inhibitors, cells are pretreated for 30 min with various inhibitors in serum-free medium followed by incubation in serum-free medium without or with 600 mM sorbitol and pharmacological inhibitors. Cells are subsequently washed with ice-cold buffer (20 mM Tris, pH 7.4, 150 mM NaCl, 10 mM EDTA, 100 mM NaF, 10 mM Na$_4$P$_2$O$_7$, and 2 mM sodium orthovanadate) before solubilization for 30 min at 4° in lysis buffer containing phosphatase and protease inhibitors to keep on the level of phosphorylation of proteins

(20 mM Tris, pH 7.4, 150 mM NaCl, 5 mM EDTA, 150 mM NaF, 2 mM sodium orthovanadate, 0.5 mM phenylmethylsulfonyl fluoride [PMSF], protease inhibitors cocktail, 100 nM okadaic acid, and 1% Triton X-100). Clarified lysates are obtained after centrifugation (15 min at 15,000g at 4°), and protein concentration is measured by the BCA assay.

The addition of okadaic acid, a potent inhibitor of serine/threonine protein phosphatases PP1 and PP2A, in the lysis buffer preserves the phosphorylation of proteins on serine/threonine residues.

4.2. Immunoprecipitation of docking proteins (Gab1 or IRS1)

The phosphorylation levels of Gab1 and IRS1 and their association with signaling molecules can be evaluated after specific immunoprecipitation followed by immunoblotting with appropriate antibodies (Fig. 20.2).

Clarified lysates (0.5–1 mg of proteins) prepared as described earlier are incubated for 3 h at 4° with appropriate antibodies preadsorbed on protein G–Sepharose (4 μg of antibodies/sample). After washes with lysis buffer, immune pellets are resuspended in Laemmli buffer and proteins are separated by SDS–PAGE using a 7.5 or 10% resolving gel and transferred to a polyvinylidene difluoride membrane.

4.3. Western blotting assays

The membrane is blocked with saline buffer (10 mM Tris, pH 7.4, and 140 mM NaCl) containing 5% (w/v) bovine serum albumin for 2 h at room temperature and blotted overnight at 4° with commercial antibodies at the dilution indicated in the manufacturer's instructions and at 1 μg/ml for the anti-pS[307] or anti-pS[632]-IRS1 antibodies (home made). After incubation with horseradish peroxidase-conjugated secondary antibodies, proteins are detected by enhanced chemiluminescence. In some cases, the membrane is stripped for 30 min at 50° in 62 mM Tris, pH 6.8, 100 mM 2-mercaptoethanol, and 2% SDS and is reprobed with the appropriate antibodies.

5. HYPEROSMOLARITY AND PHOSPHATIDYLINOSITOL 3-KINASE ACTIVITY

The insulin effect on glucose uptake requires the tyrosine phosphorylation of IRS1 and the recruitment and activation of the PI 3-kinase. Hyperosmotic stress antagonizes insulin-mediated IRS1 phosphorylation, IRS1-associated PI 3-kinase activity, and subsequently Glut 4 translocation and glucose uptake (Gual $et\ al.$, 2003a,b). The following assay evaluates the PI 3-kinase activity associated with IRS1 or p85 $in\ vitro$.

5.1. Preparation of phosphatidylinositol

L-α-Phosphatidylinositol (bovine liver in ammonium salt in chloroform) can be purchased from Sigma. The required amount of PI (10-μg/sample, prepare amount for $n + 4$ samples) is added to a 1.5-ml microcentrifuge tube, and the organic solvent is removed by evaporation using a Speed–Vac apparatus. Five microliters per sample of sonication buffer (10 mM HEPES, pH 7.4, 1 mM EGTA) is added, and the PI is suspended by sonication on ice for 10 min (20 × 30-s pulses, the samples are placed on ice to avoid overheating); alternatively, a sonication water bath can be used.

5.2. Immunoprecipitation of phosphatidylinositol 3-kinase

3T3–L1 adipocytes are starved overnight in 2 ml of serum-free medium containing 0.5% BSA. Cells are then stimulated with 600 mM sorbitol for 40 min and then stimulated without or with 0.2 nM insulin for 5 min at 37°. Each condition is realized in triplicate. Plates are put on ice and the medium is removed quickly. Cells are washed twice with 2 ml of ice-cold freshly prepared buffer A (20 mM Tris-HCl, pH 7.4, 150 mM NaCl, 10 mM EDTA, 10 mM Na$_4$P$_2$O$_7$, 100 mM NaF, and 2 mM sodium orthovanadate). Buffer A is removed, and cells are scraped in 1 ml of ice-cold lysis buffer (buffer A containing 1% Nonidet P-40, 10 μg/ml aprotinin, 1 mM PMSF) and are transferred to 1.5-ml microcentrifuge tubes. The tubes are rocked for 1h at 4° and are then centrifuged at 15,000g for 15 min at 4° to sediment insoluble material. The supernatant fraction is transferred to new 1.5-ml microcentrifuge tubes, and the protein concentration is measured in each sample. Three hundred to 500 μg of proteins is then immunoprecipitated with 5 μg of rabbit antibodies against the p85 subunit of the PI 3-kinase or against IRS1 (Upstate Biotechnology, Lake Placid, NY) preadsorbed on protein A–Sepharose beads (wet volume of protein A–Sepharose beads: 20 μl). Tubes are shaken for 3 h at 4° and then the immunoprecipitates are collected by centrifugation for 2 min at 15,000g. Tubes are placed on ice, and the immunoprecipitates are washed twice with each of the following ice-cold buffers, which have to be freshly prepared:

i. Phosphate-buffered saline containing 200 $\mu$$M$ sodium orthovanadate and 1% Nonidet P-40
ii. 100 mM Tris-HCl, pH 7.4, 500 mM LiCl, 200 $\mu$$M$ sodium orthovanadate
iii. 10 mM Tris-HCl, pH 7.4, 1 mM EDTA, 200 $\mu$$M$ sodium orthovanadate

Then, the last wash is removed completely using a Hamilton syringe and 5 μl (10 μg) of sonicated L-α-phosphatidylinositol in 10 mM HEPES, pH 7.5, 1 mM EGTA is added directly on dried protein A–Sepharose beads.

5.3. Measurement of PI 3-kinase activity in immunoprecipitation

All the following steps should be performed in agreement with the guidelines for handling radioactive samples.

PI 3-kinase activity is measured at room temperature by the addition of 25 μl of a reaction mixture consisting of 40 mM HEPES, pH 7.4, 20 mM MgCl$_2$, 80 μM ATP, 5 μCi [γ-^{32}P]ATP (Amersham BioSciences, PB10168, 10 mCi/ml, 13,000 Ci/mmol). The reaction is stopped after 20 min by the addition of 40 μl 4 N HCl, and phospholipids are extracted into chloroform by the addition of 160 μl of chloroform:methanol (1:1, vol/vol) with shaking for 10 min. The organic and aqueous phases are separated by centrifugation at 15,000g for 2 min, and organic phase–containing phospholipids (70–80 μl) are collected in microcentrifuge tubes using a Hamilton syringe or an automatic pipette. The organic phase could be stored at $-20°$ if necessary on a practical point of view. All the preceding steps have to be performed on fresh cells. For each sample, the chloroformic phase is evaporated using a Speed–Vac apparatus, and the phospholipids are resuspended in 10 μl of chloroform and are spotted onto silica gel 60 TLC plates at 3 cm of the bottom of the plate (a maximum of 15 samples are spotted on a plate). TLC plates are developed by chromatography in chloroform/methanol/ammoniac/water (60 ml/ 47 ml/4.4 ml /8.8 ml) until the migration front reaches 1 cm of the top of the plate (this step should be performed under a chemical hood). The plate is dried and radiolabeled lipids are visualized by autoradiography. Quantification is performed by densitometry scanning of the autoradiogram.

6. Conclusion

This chapter summarizes the current approaches permitting the study of hyperosmolarity on membrane ruffling, Glut 4 translocation, glucose transport, and signaling pathways in 3T3-L1 adipocytes. It should be noted that hyperosmotic stress (usually induced by high extracellular concentrations of sorbitol) is not a physiological stimulus, but is used as a tool that could lead to the discovery of novel molecular mechanisms of glucose transport and also for a better understanding of the molecular mechanisms of cellular insulin resistance.

ACKNOWLEDGMENTS

This work was supported by grants from the Institut National de la Santé et de la Recherche Médicale (INSERM, France), the University of Nice, the Fondation Bettencourt- Schueller, the Région Provence Alpes Côte d'Azur, the Conseil Général des Alpes-Maritimes, and the Comité Doyen Jean Lépine (Nice, France). Part of the work was supported by a grant from ALFEDIAM-Takeda Laboratories (Puteaux, France) to J.-F. Tanti. P. Gual was successively supported by a Fellowship from La Ligue Contre le Cancer and from ALFEDIAM.

REFERENCES

Barres, R., Grémeaux, T., Gual, P., Gonzalez, T., Gugenheim, J., Tran, A., Le Marchand-Brustel, Y., and Tanti, J. F. (2006). Enigma interacts with adaptor protein with PH and SH2 domains to control insulin-induced actin cytoskeleton remodeling and glucose transporter 4 translocation. *Mol. Endocrinol.* **20,** 2864–2875.

Bryant, N. J., Govers, R., and James, D. E. (2002). Regulated transport of the glucose transporter GLUT4. *Nat. Rev. Mol. Cell. Biol.* **3,** 267–277.

Chen, D., Fucini, R. V., Olson, A. L., Hemmings, B. A., and Pessin, J. E. (1999). Osmotic shock inhibits insulin signaling by maintaining Akt/protein kinase B in an inactive dephosphorylated state. *Mol. Cell. Biol.* **19,** 4684–4694.

Dugani, C. B., and Klip, A. (2005). Glucose transporter 4: Cycling, compartments and controversies. *EMBO Rep.* **6,** 1137–1142.

Gual, P., Gonzalez, T., Grémeaux, T., Barres, R., Le Marchand-Brustel, Y., and Tanti, J. F. (2003a). Hyperosmotic stress inhibits insulin receptor substrate-1 function by distinct mechanisms in 3T3-L1 adipocytes. *J. Biol. Chem.* **278,** 26550–26557.

Gual, P., Le Marchand-Brustel, Y., and Tanti, J. F. (2003b). Positive and negative regulation of glucose uptake by hyperosmotic stress. *Diabetes Metab.* **29,** 566–575.

Gual, P., Le Marchand-Brustel, Y., and Tanti, J. F. (2005). Positive and negative regulation of insulin signaling through IRS-1 phosphorylation. *Biochimie* **87,** 99–109.

Gual, P., Shigematsu, S., Kanzaki, M., Grémeaux, T., Gonzalez, T., Pessin, J. E., Le Marchand-Brustel, Y., and Tanti, J. F. (2002). A Crk-II/TC10 signaling pathway is required for osmotic shock-stimulated glucose transport. *J. Biol. Chem.* **277,** 43980–43986.

Heydrick, S. J., Gautier, N., Olichon-Berthe, C., Van Obberghen, E., and Le Marchand-Brustel, Y. (1995). Early alteration of insulin stimulation of PI 3-kinase in muscle and adipocyte from gold thioglucose obese mice. *Am. J. Physiol.* **268,** E604–E612.

Janez, A., Worrall, D. S., Imamura, T., Sharma, P. M., and Olefsky, J. M. (2000). The osmotic shock-induced glucose transport pathway in 3T3-L1 adipocytes is mediated by Gab-1 and requires Gab-1-associated phosphatidylinositol 3-kinase activity for full activation. *J. Biol. Chem.* **275,** 26870–26876.

Komjati, M., Kastner, G., Waldhausl, W., and Bratusch-Marrain, P. (1988). Detrimental effect of hyperosmolality on insulin-stimulated glucose metabolism in adipose and muscle tissue *in vitro*. *Biochem. Med. Metab. Biol.* **39,** 312–318.

Rui, L., Fisher, T. L., Thomas, J., and White, M. F. (2001). Regulation of insulin/insulin-like growth factor-1 signaling by proteasome-mediated degradation of insulin receptor substrate-2. *J. Biol. Chem.* **276,** 40362–40367.

Saltiel, A. R., and Kahn, C. R. (2001). Insulin signalling and the regulation of glucose and lipid metabolism. *Nature* **414,** 799–806.

Saltiel, A. R., and Pessin, J. E. (2002). Insulin signaling pathways in time and space. *Trends Cell Biol.* **12,** 65–71.

Vom Dahl, S., Hallbrucker, C., Lang, F., Gerok, W., and Häussinger, D. (1991). Regulation of liver cell volume and proteolysis by glucagon and insulin. *Biochem. J.* **278,** 771–777.

Watson, R. T., Kanzaki, M., and Pessin, J. E. (2004). Regulated membrane trafficking of the insulin-responsive glucose transporter 4 in adipocytes. *Endocr. Rev.* **25,** 177–204.

CHAPTER TWENTY-ONE

EFFECTS OF OSMOLYTES ON PROTEIN FOLDING AND AGGREGATION IN CELLS

Zoya Ignatova* *and* Lila M. Gierasch[†]

Contents

Abstract

Nature has developed many strategies to ensure that the complex and challenging protein folding reaction occurs *in vivo* with adequate efficiency and fidelity for the success of the organism. Among the strategies widely employed in a huge range of species and cell types is the elaboration of small organic molecules called osmolytes that offset the potentially damaging effects of osmotic stress. While considerable knowledge has been gained *in vitro* regarding the influence of osmolytes on protein structure and folding, it is of great interest to probe the effects of osmolytes in cells. We have developed an in-cell fluorescent-labeling method that enables the study of protein stability and also protein aggregation *in vivo*. We utilize a genetically encoded tag called a tetra-Cys motif that binds specifically to a bis-arsenical fluorescein-based dye "FlAsH"; we inserted the tetra-Cys motif into a protein of interest in such a

* Max Planck Institute for Biochemistry, Martinsried, Germany
† Departments of Biochemistry & Molecular Biology and Chemistry, University of Massachusetts, Amherst, Massachusetts

Methods in Enzymology, Volume 428
ISSN 0076-6879, DOI: 10.1016/S0076-6879(07)28021-8

way that the FlAsH signal reported on the state of folding or aggregation of the protein. Then, we designed protocols to assess how various osmolytes influence the stability and propensity to aggregate of our protein of interest. These are described here. Not only are there potential biotechnological applications of osmolytes in the quest to produce greater quantities of well-folded proteins, but also osmolytes may serve as tools and points of departure for therapeutic intervention in protein folding and aggregation diseases. Having *in vivo* methods to analyze how osmolytes affect folding and aggregation enhances our ability to further these goals greatly.

1. INTRODUCTION

Protein folding is a remarkably complex physicochemical process via which a polymer of amino acids that samples many, many conformations in its unfolded state adopts a well-packed and essentially unique native fold. Realization of high yields of properly folded proteins *in vitro* requires optimization of conditions, for example, using solutions that are highly diluted, in order to facilitate folding and minimize inappropriate interactions between partially folded chains. Folding *in vivo* must occur effectively in the face of the complexity of the heterogeneous cellular environment. Among the issues that confront a growing or newly synthesized polypeptide chain in the cell are the high concentrations of macromolecules and macromolecular assemblies, and the resulting confinement of folding molecules to highly restricted volumes. Additionally, many cells must contend with stresses that alter their environments drastically and yet not lose the capacity to function, which requires folding to continue to occur with fidelity. Among the major natural challenges faced are the variations in water content and perturbed ionic homeostasis caused by osmotic stress. Given the complexity of the folding reaction, the challenges of the cellular environment, and the vulnerability of partially folded proteins to aggregation, it is not surprising that a great deal of "wasteful" *de novo* synthesis occurs in the cell: about 30% of newly synthesized proteins are misfolded conformers degraded by the proteasome shortly after their release from ribosomes (Schubert *et al.*, 2000). In addition, some fraction of misfolded proteins escapes the cellular quality control machinery, succumbs to inappropriate intermolecular interactions, and forms thermodynamically and metabolically stable aggregates. Stress conditions, for example, thermal, oxidative or osmotic stress, or alterations in primary structure by mutation, can facilitate the aggregation process.

Protein misfolding and deposition of extracellular or intracellular aggregates constitute a hallmark of the neurodegenerative pathologies, including Alzheimer's, Huntington's, and Parkinson's diseases. Despite differences in the proteins involved and variations in the genetic background and inheritance, all of these diseases are associated with aging-dependent disruption of the cellular folding homeostasis (Cohen *et al.*, 2006; Gidalevitz *et al.*, 2006).

Failure in protein folding also has implications in the research laboratory and in the biotechnology industry: off-pathway aggregation limits yields of recombinant proteins. Considerable evidence indicates that aggregates formed upon expression of proteins in bacteria (termed "inclusion bodies") have the same fundamental cross-β structure characteristic of disease-associated aggregates in humans (Chiti and Dobson, 2006; Fink, 1998; Oberg *et al.*, 1994; Speed *et al.*, 1996). Together with the understanding that amyloid fibril formation is a generic property of the polypeptide backbone (Dobson, 2001), this suggests a common pathway of aggregate formation. In light of this premise, common conditions or strategies should be effective to suppress aggregation in either bacterial or mammalian cells.

As described elsewhere in this volume, a set of cellular "coping mechanisms" is triggered by osmotic stress. A key characteristic of these responses is the cellular import or synthesis of small organic molecules called osmoprotectants or osmolytes, which function to offset the deleterious effects of dehydration (Wood, 1999). Osmolytes have a direct impact on protein stability and solubility (Bolen, 2004; Yancey, 2005) and may modulate both the tendency of proteins to aggregate and the nature of the aggregates formed (Ignatova and Gierasch, 2006; Kanapathipillai *et al.*, 2005; Liu *et al.*, 2005). This chapter describes experimental approaches to understanding how osmolytes influence the process of protein folding, the stability of proteins, and the propensity of proteins to aggregate in cells.

After sensing osmotic up- or downshift, cells readjust their volumes and restore osmotic homeostasis by biosynthesis or uptake of protective osmolytes (Wood, 1999). According to the compatibility hypothesis (Brown and Simpson, 1972; Yancey *et al.*, 1982), most osmolytes are interchangeable, that is, protective effects can be achieved with a variety of osmolytes. It is not known why organisms employ such a broad spectrum of osmolytes; the particular osmolyte deployed may be a function of nutrient conditions and dietary limitations, or alternatively, the properties of each osmolyte might be helpful with a certain type of stress (Yancey, 2005). Preferential exclusion of osmolytes from the surfaces of native proteins and the consequent preferential hydration of the protein surface comprise the accepted universal molecular basis for the stabilization of proteins (Cayley and Record, 2003; Timasheff, 1998); together these phenomena have been dubbed the "solvophobic" effect (Bolen, 2004). Studies of the physical chemical origins of osmolyte action provide a formalism to account for variations in the mode of action of different osmolytes: the combination of favorable or unfavorable backbone and side chain interactions varies from osmolyte to osmolyte (Auton and Bolen, 2005; Bolen, 2004; Cayley and Record, 2003; Courtenay *et al.*, 2000). For instance, the dominance of unfavorable interactions with the peptide backbone over favorable interactions with side chains is the origin of the high effectiveness of trimethylamine-N-oxide to induce structure in proteins and to force proteins to fold (Bolen, 2004; Pradeep and Udgaonkar, 2004). In contrast, proline has significant favorable interactions with the side chain and less unfavorable interactions with the backbone (Auton and Bolen, 2005), which explains why it is the least effective osmolyte in inducing structure, but the most effective in the solubilization of native proteins.

The relative contributions of peptide backbone solvation and osmolyte–side chain interactions will also determine the preferential stabilization or destabilization of aggregation-prone states of proteins, as well as different aggregated species along the aggregation pathway. It is not surprising, therefore, that quite contrasting effects of natural osmolytes on aggregation have been reported: from promoting aggregation by induction of structure (Scaramozzino et al., 2006; Uversky et al., 2001; Yang et al., 1999) to inhibition of aggregation (Ignatova and Gierasch, 2006; Kanapathipillai et al., 2005; Liu et al., 2005; Singer and Lindquist, 1998). Inhibition of aggregation by osmolytes can arise because of solubilization of the native state (Ignatova and Gierasch, 2006) or because of changes in the stability of aggregation-prone species or alterations of the aggregation pathway. Ameliorative effects have been associated with osmolyte-induced formation of nontoxic aggregated species (Davies et al., 2006; Furusho et al., 2005; Kanapathipillai et al., 2005; Tanaka et al., 2004). The impact of osmolytes on aggregation can be extremely complex and subtle: trehalose counteracts the toxicity of Aβ40 while it increases the susceptibility of Aβ42 to form toxic oligomers (Liu et al., 2005). The emerging view that intermediate oligomeric species, rather than mature aggregates, trigger neurotoxicity in the devastating neurodegenerative diseases (Ross and Poirier, 2004; Walsh and Selkoe, 2004) underlines the biomedical urgency to systematically address the fundamental effects of different osmoprotectants on cellular protein folding and misfolding.

2. ANALYSIS OF EFFECTS OF OSMOLYTES *IN VIVO*

The direct monitoring of folding/misfolding processes *in vivo* is challenged by the high background of other cellular constituents, as well as by multifaceted interactions that can be established intermolecularly among the same or different protein species. Ideally, to characterize folding/misfolding *in vivo* the signal for the protein of interest needs to be observable over the cellular background and sensitive to conformational states.

This chapter describes an approach developed to determine *in vivo* and *in vitro* protein stability (Ignatova and Gierasch, 2004, 2005) and how this approach may be used to monitor the effects of various osmolytes on folding and aggregation propensities both *in vivo* and *in vitro* using a well-behaved model system. By incorporating a specific sequence called a tetra-cysteine motif (Cys-Cys-Xxx-Yyy-Cys-Cys generally, here Cys-Cys-Gly-Pro-Cys-Cys) into the internal Ω loop of the 136 amino acid cellular retinoic acid-binding protein (CRABP), we have engineered a binding site for the fluorescence dye "FlAsH." We found that the FlAsH fluorescence signal reports on the state of folding of this protein both *in vivo* and *in vitro*: the denatured ensemble and aggregated species are hyperfluorescent compared to the native state. The tetra-Cys sequence slightly diverges from the optimal Cys-Cys-Pro-Gly-Cys-Cys motif described based on peptide studies (Griffin *et al.*, 2000); we elected to retain the Gly-Pro that is present in the Ω loop of CRABP to minimize sequence alterations. The CRABP variant carrying the tetra-Cys motif was indistinguishable in structure and function from the native counterpart whether FlAsH labeled or unlabeled (Ignatova and Gierasch, 2004). Additionally, FlAsH fluorescence, as reported earlier, can be used as a real-time readout of the solubility and aggregation propensity of CRABP *in vivo* (Ignatova and Gierasch, 2004). Mutation of the helix-terminating residue Pro39 to Ala slows down the folding and unfolding of CRABP (Eyles and Gierasch, 2000), and P39A tetra-Cys CRABP shows a high tendency to form inclusion bodies *in vivo* and aggregates *in vitro* (Ignatova and Gierasch, 2004). Upon induction of protein synthesis, we observed differences in the expression pattern for the soluble tetra-Cys CRABP and the aggregation-prone P39A tetra-Cys CRABP (Ignatova and Gierasch, 2004). For the soluble protein, the time course of FlAsH fluorescence after induction of synthesis reports on its expression, and a plateau in fluorescence signal is observed at about 180 min, corresponding to the termination of *de novo* biosynthesis when the cells enter stationary phase. The fluorescence time course of P39A tetra-Cys CRABP reported on both synthesis and aggregate formation: an abrupt rise in FlAsH fluorescence was seen at 120 min. This rise is a consequence of the accumulation of misfolded and aggregated species, as indicated as well by cell fractionation studies, which showed that unlike tetra-Cys CRABP, P39A tetra-Cys CRABP partitions between soluble and insoluble fractions (Ignatova and Gierasch, 2004).

The approach described here is not necessarily general and directly applicable to a wide array of proteins. Each protein would present its own design constraints so that (1) the tetra-Cys motif is tolerated without structural perturbation, and (2) the FlAsH quantum yield is sensitive to the folding of the protein host. Nonetheless, careful design and willingness to carry out multiple trials are likely to yield suitable FlAsH-binding sites in proteins of interest. We have explored a wide array of sites for tetra-Cys motifs and learned that the geometric properties of the binding sites are crucial to the FlAsH fluorescence characteristics (B. Krishnan and L. Gierasch, manuscript in preparation). A clear advantage provided by this system is the direct readout of aggregation in intact cells. This allowed us to directly follow the influence of different types of osmolytes on folding and aggregation *in vivo* (see later).

2.1. Culture growth and induction of osmolyte uptake *in vivo*

1. Transform pET16b plasmid bearing tetra-Cys CRABP or P39A tetra-Cys CRABP under the T7 promoter into *Escherichia coli* BL21(DE3) cells, carrying the DE3 lysogen for high-level expression of T7 polymerase. For expression, use only freshly transformed cells. A single colony is inoculated in LB medium and grown overnight at 30°. This and all subsequent growths are in the presence of 100 μg/ml ampicillin. The cells are sterile harvested (2060g, 15 min, 4°), resuspended, and brought up to 25× the original volume in fresh LB medium.

2. Grow cultures in LB medium at 37° until $OD_{600} = 0.5$. After gentle lysozyme pretreatment (50 ng/ml for 10 min on ice), switch to an equal volume of fresh, modified M9 minimal medium (4 g/liter glucose, 20 mM KH$_2$PO$_4$, 40 mM K$_2$HPO$_4$, 10 mM NH$_4$Cl, 1 mM MgSO$_4$, 5 mM NaCl, 100 μM CaCl$_2$, pH 7.4) in order to control exogenous osmolyte concentrations and then preload with FlAsH dye along with EDT to suppress the labeling of endogenous cysteine pairs. In the usual protocol, cell aliquots of 1 ml are labeled with 0.2 μM FlAsH-EDT$_2$ and 1 μM EDT. The volume of the aliquots can be adjusted depending on the need for subsequent experiments.

3. After one generation, at $OD_{600} = 1.0$, induce protein synthesis with 0.4 mM isopropyl-β-D-thiogalactoside (IPTG). Simultaneous with the induction of either tetra-Cys CRABP protein, osmolyte uptake can be stimulated by increasing the osmolality of the nutrient medium through the addition of sterile NaCl to a concentration of 300 mM in order to activate the ProP transporter. To stimulate the uptake of exogenous osmolytes over the synthesis of endogenous osmolytes, supply the medium with 20 mM sterile osmolyte (here proline or glycine betaine).

4. To monitor thermodynamic stability *in vivo*, 2 h after induction with IPTG add urea from a sterile stock (9 M urea in 10 mM Tris-HCl buffer, pH 7.5) to the medium at various final concentrations (not higher than 3 M end concentration). Adjust the volumes of all samples to equal amounts with a sterile-modified M9 medium. To maintain constant osmotic stress during stability measurements, adjust the final sodium chloride concentration of each sample to 300 mM with a sterile 5 M NaCl stock solution. Incubate at 37° for at least 75 min to ensure equilibration (but no longer than 120 min) and subject the bulk cell suspensions to fluorescence measurements at 530 nm (excitation 500 nm). From each point subtract a blank of FlAsH-labeled CRABP-expressing cells (with no tetra-Cys motif) and treat with the corresponding urea concentration in modified M9 medium. Compare the thermodynamic

stability in the presence of osmolytes to the stability in cells cultured in a balanced medium (modified M9 medium) without exposing them to osmotic stress.

5. Monitor the aggregation time course of an aggregation-prone protein by FlAsH fluorescence of the bulk suspension at 530 nm (excitation 500 nm) in a preequilibrated cuvette holder at 37°. Fluorescence of cells treated identically but with a plasmid bearing a wild-type CRABP without the tetra-Cys motif is used as a blank, and the measured value is subtracted from each point in the time course. Osmolyte uptake, for example, activation of the ProP transporter with NaCl and addition of exogenous osmolyte in the medium, can also be induced at any point during protein expression (no longer than 5 h).

2.2. Partitioning into insoluble and soluble cell fractions

Ten-milliliter aliquots of induced protein-expressing cells are harvested (2060g, 10 min at 4°) and resuspended in 1.5 ml 50 mM phosphate buffer, pH 8.0, containing 300 mM NaCl. When samples are analyzed over the time course of expression, the cell concentration should be adjusted to the same value by dilution with fresh medium. We usually use OD values at 600 nm as arbitrary units for the cell number. The cells are lysed by sonication (3 min, 30% duty cycle) after treatment with lysozyme (500 μg/ml, 30 min on ice) and DNase (50 μg/ml, 15 min on ice). The cell homogenate is separated into soluble and insoluble fractions by centrifugation at 27,000g for 30 min. The insoluble pellet is solubilized in 1.5 ml of the same buffer containing 8 M urea. Both fractions are analyzed on SDS–PAGE, and bands are quantified for the protein of interest in each fraction by optical densitometry (Bio-Rad).

For tighter control of the uptake of osmolytes (proline, glycine betaine), we used a ProP deletion strain (*E. coli* WG710) with an exogenous copy of the ProP autotransporter on the pDC80 (Amp^R) plasmid under the control of the arabinose promoter (Racher *et al.*, 2001). Both tetra-Cys CRABP and P39A tetra-Cys CRABP were subcloned in a pHSG398 vector (Takara Bio, Tokyo, Japan) plasmid bearing chloramphenicol resistance and under the control of the P_{tac} promotor. The ProP transporter can be induced with 0.2% arabinose and activated with 300 mM NaCl concomitantly with induction of the biosynthesis of the tetra-Cys CRABP proteins at OD$_{600}$ = 1.0. Pulse induction (0.2% arabinose) and activation (300 mM NaCl) of the ProP transporter 10 min before the induction of the protein with 0.4 mM IPTG and subsequent change to a fresh MOPS medium (4 g/liter glycerol, 40 mM MOPS, 4 mM tricine, 15 mM KH$_2$PO$_4$, 1.5 mM K$_2$HPO$_4$, 15 mM NaCl, 9.5 mM NH$_4$Cl, 0.5 mM MgSO$_4$, 0.2 mM FeSO$_4$, 50 μM CaCl$_2$, pH 7.4; additionally supplied with 300 mM NaCl) yielded the same results.

Exogenous osmoprotectants, readily available in the nutrient medium, have priority over the metabolically costly biosynthesis of endogenous osmolytes, which ensures more rapid osmoadaptation of the cells (Wood, 1999). Although the accumulation of exogenous osmolytes is faster (the transporters are activated within a minute), one might wish to test the contribution of endogenous osmolytes to the stability and solubility of proteins *in vivo*. Exposing the cells to osmotic upshift without providing any exogenous osmolyte will force them to counteract the osmotic stress by the synthesis of endogenous osmoprotectants. We followed this procedure: *E. coli* BL21(DE3) cells transformed with pET16b plasmid bearing the tetra-Cys CRABP or P39A tetra-Cys CRABP under the T7 promoter are cultured at 37° in modified M9 minimal medium and preloaded with FlAsH as described earlier. At the time of induction (0.4 m*M* IPTG), the osmolality of the medium is increased by the addition of 300 m*M* sterile NaCl. The influence of the endogenously synthesized osmolytes (here trehalose) on the aggregation propensity is monitored by following FlAsH fluorescence of the bulk suspension at 530 nm (excitation 500 nm) at 37°.

2.3. Variations in the effect of different osmolytes *in vivo*

We have tested the effect of three osmoprotectants that belong to different chemical categories: proline–α-amino acids and derivatives; glycine betaine–methylammonium and methylsulfonium solutes; and trehalose–carbohydrates (Yancey, 2005). Despite a similar physiological function, namely to counteract osmotic stress, they show differences in their effects on the folding/aggregation pattern of CRABP, our model protein. Both proline and glycine betaine enhanced the stability of the soluble tetra-Cys CRABP (Fig. 21.1A) and aggregation-prone P39A tetra-Cys CRABP (Fig. 21.1B). Unlike proline, glycine betaine decreased somewhat the slope of the melting curve near the C_m (related to the *m* value). This change might arise from a greater collapse of the denatured state in the presence of glycine betaine and is consistent with the preferential exclusion of glycine betaine from the protein surface (Cayley and Record, 2003; Hong *et al.*, 2005).

In a comparative sense, proline is reported to be a highly effective osmolyte in solubilization (Bolen, 2004). Strikingly, we find that it can completely combat the aggregation of P39A tetra-Cys CRABP when present early in cells (Fig. 21.2A and C) (Ignatova and Gierasch, 2006). The rise in FlAsH fluorescence (as seen for P39A tetra-Cys CRABP grown in low salt medium), indicating the formation of a highly fluorescent-aggregated species, disappears when the cells experience osmotic stress in the presence of proline; the resulting FlAsH fluorescence curve superimposes on the profile of the soluble tetra-Cys CRABP (Fig. 21.2A) (Ignatova and Gierasch, 2006). The osmotic stress influences neither the cell density nor the expression level of the protein (see tetra-Cys CRABP in the presence and absence of proline,

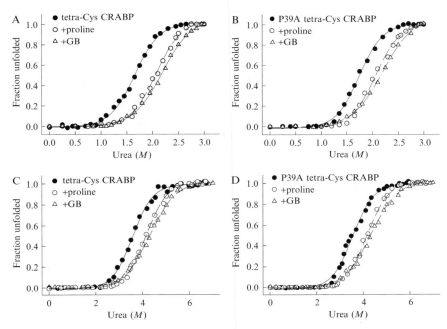

Figure 21.1 Thermodynamic stability. *E. coli* BL21(DE) cells expressing tetra-Cys CRABP (A) and P39A tetra-Cys CRABP (B), labeled with FlAsH-EDT$_2$, and exposed to osmotic stress (with 300 mM NaCl) at the time of induction were subjected to urea stability measurements as described in the text. The high osmolality medium was additionally supplied with 20 mM proline or glycine betaine. Cells cultured in a salt-balanced medium served as control. The *in vitro* thermodynamic stability was measured with 7 μM purified tetra-Cys CRABP (C) or P39A tetra-Cys CRABP (D), prelabeled with FlAsH-EDT$_2$. Samples were equilibrated at different urea concentrations containing either 500 mM proline or 500 mM glycine betaine (GB) at 37°, and the unfolding transitions were monitored by Trp fluorescence. Urea unfolding of both proteins without any osmolyte is used for comparison. The actual urea concentration was determined by measuring the index of refraction, and curves were analyzed by a two-state model using the linear extrapolation method (Ignatova *et al.*, 2007). (A and C contain data published previously [Ignatova and Gierasch, 2006]; all other results are unpublished.)

Fig. 21.2C), and the observed effect is solely because of the soluble expression of P39A tetra-Cys CRABP, as evidenced from the fractionation analysis (Fig. 21.2B). When *E. coli* cells are subjected to osmotic stress, the secondary membrane transporter ProP is turned on; it then mediates the active uptake of compatible solutes from the external medium (e.g., proline, glycine betaine, ectoine) (Wood *et al.*, 2001). To test the effect of osmolytes in a more controllable way, we expressed the tetra-Cys CRABP and P39A tetra-Cys CRABP in an *E. coli* ProP deletion mutant (Racher *et al.*, 2001), in which the function of the ProP transporter is reconstituted on a plasmid behind the arabinose-inducible promoter. Simultaneous induction (with

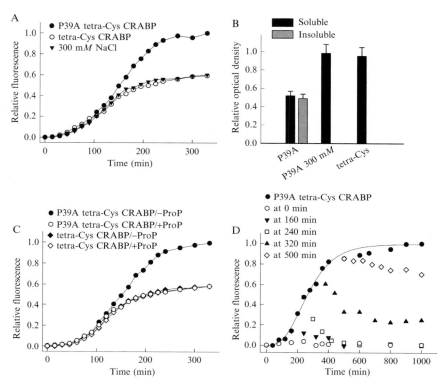

Figure 21.2 Proline suppresses aggregation of P39A tetra-Cys CRABP *in vivo* and *in vitro*. (A) Time evolution of the bulk FlAsH fluorescence at 530 nm (excitation 500 nm) of *E. coli* BL21 (DE3) cells expressing P39A tetra-Cys CRABP and tetra-Cys CRABP. At the time of induction with 0.4 mM IPTG, one aliquot of the P39A tetra-Cys CRABP-expressing cells was transferred to high osmolality medium (containing 300 mM NaCl) additionally supplied with 20 mM proline. (B) Proline accumulated in response to osmotic upshift changes the partitioning pattern of P39A tetra-Cys CRABP (P39A 300 mM): the protein is present only in the soluble fraction. P39A tetra-Cys CRABP (P39A) cultured in a low-salinity medium is distributed almost equally between soluble and insoluble fractions. Note the contrasting behavior of the completely soluble tetra-Cys CRABP. Samples are from 240 min postinduction. (C) Time evolution of the aggregation of P39A tetra-Cys CRABP monitored by the bulk FlAsH fluorescence signal of labeled cells, under depleted (-ProP) and upregulated (+ProP) ProP transporter conditions in the *E. coli* strain WG710. Note that the expression of the soluble tetra-Cys CRABP is not influenced by the amount of proline influx; it remains the same in the absence (-ProP) and upon overexpression of ProP transporter (+ProP). (D) Proline (500 mM) added at different times during the *in vitro* aggregation of 15 μM labeled P39A tetra-Cys CRABP inhibits its aggregation. The time evolution of *in vitro* P39A tetra-Cys CRABP aggregation was monitored by FlAsH fluorescence (modified from Ignatova and Gierasch, 2006).

arabinose) and activation of the ProP transporter (with 300 m*M* NaCl) and then IPTG induction of P39A tetra-Cys CRABP expression led to the same effect: the characteristic fluorescence enhancement associated with aggregation is lost as seen for *E. coli* BL21(DE3) cells in a high-salt medium, and the fluorescence curve overlaps that of the soluble tetra-Cys CRABP (Fig. 21.2C). In contrast, providing glycine betaine, which is also a substrate of the ProP transporter, instead of proline in the medium decreases the aggregation of P39A tetra-Cys CRABP in *E. coli* BL21(DE3), but cannot abolish it completely (Fig. 21.3A). The fluorescence rise compared to the

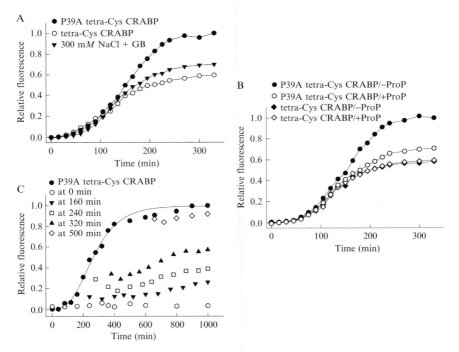

Figure 21.3 Glycine betaine is not able to completely abolish the aggregation of P39A tetra-Cys CRABP *in vivo* and *in vitro*. (A) Time evolution of the bulk FlAsH fluorescence at 530 nm (excitation 500 nm) of *E. coli* BL21(DE3) cells expressing P39A tetra-Cys CRABP, cultured in a low salinity medium and in high salinity (300 m*M*) supplied with 20 m*M* glycine betaine (GB). For comparison, signal from fully soluble tetra-Cys CRABP is also shown. (B) Time evolution of the *in vivo* aggregation of P39A tetra-Cys CRABP monitored by the bulk FlAsH fluorescence signal of labeled cells, under depleted (-ProP) and upregulated (+ProP) ProP transporter conditions in the *E. coli* strain WG710. The expression of soluble tetra-Cys CRABP is not influenced by the amount of the glycine betaine influx; it remains the same in the absence (-ProP) and upon accumulation of glycine betaine through the ProP transporter (+ProP). (C) Glycine betaine (500 m*M*) added at different times during the *in vitro* aggregation of 15 μ*M* labeled P39A tetra-Cys CRABP. The *in vitro* P39A tetra-Cys CRABP aggregation was monitored by FlAsH fluorescence (unpublished results of Ignatova and Gierasch).

tetra-Cys CRABP fluorescence reveals the presence of a significant fraction of highly fluorescent aggregate species. Consistent with this interpretation are results in the ProP deletion strain: ProP-induced uptake of glycine betaine after the cells sense osmotic stress cannot completely prevent aggregation (Fig. 21.3B).

The endogenously synthesized osmolyte trehalose is ineffective in preventing the aggregation of P39A tetra-Cys CRABP in cells (Fig. 21.4A): the fluorescence curve is almost indistinguishable from that of cells grown in low-salinity medium. The inability of trehalose to combat aggregation cannot be attributed to a time delay in trehalose synthesis; parallel studies *in vitro* show that trehalose added at early times of aggregation increases the time for nucleus formation, but cannot inhibit aggregation (Fig. 21.4B).

The different mechanisms of action of various osmolytes, driven by variations in the contributions of the osmolyte-backbone preferential solvent exclusion and the favorable/unfavorable interactions with protein side chains, can be used to explain the observed differences on P39A tetra-Cys CRABP aggregation *in vivo*. While glycine betaine is preferentially excluded from anionic surfaces (Felitsky *et al.*, 2004), the solubilization effect of proline is a consequence of the preferential destabilization of species with a greater backbone area (Auton and Bolen, 2005; Ignatova and Gierasch, 2006).

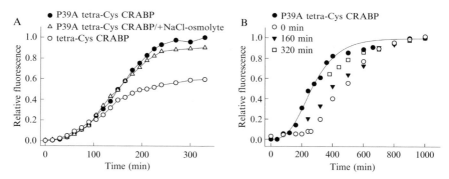

Figure 21.4 Endogenously synthesized trehalose cannot prevent P39A tetra-Cys CRABP aggregation *in vivo* and *in vitro*. (A) Time evolution of the bulk FlAsH fluorescence at 530 nm (excitation 500 nm) of *E. coli* BL21(DE3) cells expressing P39A tetra-Cys CRABP and tetra-Cys CRABP. P39A tetra-Cys CRABP aggregation was monitored in medium with low salinity, high salinity (300 mM NaCl) without any exogenous osmolyte (P39A tetra-Cys CRABP/+NaCl-osmolyte). Modified from Ignatova and Gierasch (2006). (B) Trehalose (500 mM) was added at different times during the *in vitro* aggregation of 15 μM P39A FlAsH-labeled tetra-Cys CRABP, and aggregation was monitored by FlAsH fluorescence (unpublished result).

3. Comparison of the Influence of Osmolytes *In Vivo* with That *In Vitro*

Because the unfolded and folded states display different quantum yields of FlAsH dye fluorescence, we can use FlAsH fluorescence to follow folding and aggregation *in vitro* (Fig. 21.2D) (Ignatova and Gierasch, 2005). In order to do so, we first express and purify the tetra–Cys containing proteins, label them with FlAsH, and carry out the desired experiment, be it urea titration or aggregation kinetics. As background, CRABP and its tetra–Cys variants show two–state unfolding equilibria upon urea titration (Ignatova and Gierasch, 2004). Thus, the perturbation to their stability can be assessed using the FlAsH dye signal, or alternatively in a more traditional way, the signal from intrinsic tryptophans, and the impact of osmolytes explored.

The *in vitro* aggregation of P39A tetra–Cys CRABP shows kinetics consistent with a nucleated polymerization mechanism initiated from a monomeric nucleus (Ignatova and Gierasch, 2005). Although the Pro39 to Ala mutation retards both folding and unfolding, it does not destabilize the native state, arguing that a stabilized intermediate is the aggregation-prone species. Once folded properly, P39A tetra–Cys CRABP does not aggregate spontaneously, suggesting that partial unfolding is required to initiate its aggregation. Usually, we destabilize the protein with mild urea concentrations (1.8 M), thus accelerating the kinetics of unfolding and shifting the equilibrium between unfolded and native states toward the unfolded (and partially folded) state(s). In addition, the partial destabilization enables us to study *in vitro* aggregation kinetics at reasonable timescales.

3.1. Expression and purification

Each tetra–Cys construct, cloned into the pET16b vector and transformed into *E. coli* BL21(DE3) cells, is cultured in 1 liter LB medium at 37° with constant shaking (200 rpm). At $OD_{600} = 1.0$, IPTG is added to 0.4 mM, and the culture is incubated further under the same conditions for 4 h. Bacteria are collected by centrifugation (2060g, 15 min, 4°), and tetra–Cys CRABP (or other variant) is purified from the soluble fraction of the cell extract on a Ni^{2+}-nitrilotriacetic acid agarose column (Qiagen, Valencia, CA) (Clark *et al.*, 1998). A typical purification from the soluble fraction yields 7 to 10 mg of protein from a 1-liter culture. While P39A tetra–Cys CRABP partitions nearly equally between soluble and insoluble cell fractions, we isolated it only from the soluble fraction to ensure the homogeneity of the samples used for *in vitro* studies. Concentrations of the pure protein solutions are determined spectrophotometrically using the ε_{280} of 21,750 $M^{-1}cm^{-1}$ for

tetra-Cys CRABP. Purified protein is stored in aliquots up to 200 μM at 4°
and used within 2 weeks.

3.2. *In vitro* stability measurements

Purified tetra-Cys CRABP and P39A tetra-Cys CRABP are labeled with
FlAsH (Ignatova and Gierasch, 2004, 2005) and ethanedithiol (FlAsH:EDT
ratio 1:5) at room temperature in 10 mM HEPES, pH 7.5, containing 1 mM
tris(carboxyethyl)phosphine for 1 to 2 h in the dark. When labeling purified
protein, we usually use either an equimolar ratio of FlAsH:protein, or 2:1.
Prelabeled proteins (7 μM) are equilibrated for a minimum of 6 h in solutions of various urea concentrations containing 500 mM proline or glycine
betaine, and the unfolding transitions are monitored by Trp fluorescence
(excitation at 280 nm; emission at 350 nm).

3.3. *In vitro* aggregation kinetics

To initiate aggregation, the labeled P39A tetra-Cys CRABP is destabilized
by the addition of 10 mM HEPES containing 8.5 M urea (pH 7.8) preincubated at 37° to a final urea concentration of 1.8 M. Samples, usually 200
to 250 μl total volume, are incubated without stirring and subjected peri-
odically to fluorescence measurement at 530 nm upon excitation at 500 nm
(bandwidth 2 nm). Prior to withdrawal of an aliquot for measurement
(120 μl), the reaction mixture is vortexed gently to ensure a homogeneous
distribution of the aggregates over the entire volume of the sample. The
temperature of the cuvette holder is maintained at 37° with a water bath.
Blank samples treated in the same way but lacking protein serve as negative
controls, and their values are subtracted from that of the sample. At different
time points after initiation of the *in vitro* aggregation of P39A tetra-Cys
CRABP, osmolytes (proline, glycine betaine, or trehalose) are added to a
final concentration of 500 mM, and the aggregation is monitored by FlAsH
fluorescence. The rationale for using a 500 mM concentration of the osmolytes in the *in vitro* experiments is based on literature data showing that
osmotic stress induced by the osmolality of the nutrient medium used in our
in vivo experiments causes an accumulation of approximately 500 mM
osmolyte (Cayley and Record, 2003).

3.4. Different osmolytes have different effects on folding and aggregation *in vitro*

The *in vitro* experiments recapitulated the observed effect of osmolytes
in vivo, and the conclusions about the effects of osmolytes are remarkably
parallel to the observations made *in vivo*. The incubation time required for

equilibrium to be reached *in vitro* urea titrations of CRABP and variants is significantly longer than *in vivo* (Ignatova *et al.*, 2007). Nonetheless, the influence of osmolytes on the stability of CRABP and derivatives is very similar *in vitro* to that observed *in vivo*: both proline and glycine betaine stabilized tetra-Cys CRABP (Fig. 21.1C) and P39A tetra-Cys CRABP (Fig. 21.1D). Whereas proline appears to have no effect on the cooperativity of unfolding (*m* value was not changed), glycine betaine modestly decreased the urea dependency of the transition expressed.

Among the three osmolytes tested on P39A tetra-Cys CRABP aggregation, proline has the most pronounced solubilization effect: when added at an early time point, either during the lag phase or in early exponential aggregate growth, aggregation was fully inhibited (Fig. 21.2D). Glycine betaine and trehalose are less effective in preventing the aggregation of P39A tetra-Cys CRABP: glycine betaine increased the lag phase and the effect was marginal in the case of trehalose (Figs. 21.3C and 21.4B). When added at later times, in the late exponential phase of already formed aggregates, proline inhibited further aggregate growth, but could not solubilize the existing aggregate as indicated by a relatively stable fluorescent signal (Fig. 21.2D). In contrast, trehalose and glycine betaine stimulated aggregate formation when added in the late aggregation phase (Figs. 21.3C and 21.4B).

On the basis of these experiments the effect of different osmolytes on aggregation can be rationalized as follows: (1) proline disfavors the aggregation of P39A tetra-Cys CRABP *in vivo* and *in vitro* by stabilizing and solubilizing the native state while destabilizing aggregation-prone and small soluble-aggregated species; (2) glycine betaine and, to some extent trehalose, retard aggregation, presumably by destabilizing and thus reducing the concentration of the aggregation-prone monomeric intermediate; and (3) in the later phases of aggregation, all the osmoprotectants diminish in their influence, most likely because they have little impact on larger aggregates. These larger species sequester substantial backbone surface area, which effectively decreases the solvophobic osmolyte backbone contribution. The reported efficacy of trehalose on alleviating the aggregation of trinucleotide repeat disease proteins (Davies *et al.*, 2006; Tanaka *et al.*, 2004) can be reconciled with our observations. Trehalose, in general, is thought to be an aggregation-protective osmolyte, as it stabilizes stress-denatured proteins in a partially folded state, which can be reactivated further by molecular chaperones (Singer and Lindquist, 1998). Taken together, these results demonstrate that various osmolytes can affect species in the aggregation pathway differently, and these differential effects are most probably based on the relative tendency of a given osmolyte to be preferentially excluded from the protein surface, the magnitudes of osmolyte–side chain interactions, and the properties of the particular aggregated (or aggregation-prone) species.

4. Conclusions

It is crucial to understand the stability and aggregation propensity of proteins in the context of a cell in order to explore their physiologically relevant behaviors. The naturally occurring small organic molecules that are produced in cells to modulate protein properties in response to stresses such as osmotic shock have their action in this same *in vivo* environment. We have developed methods that enable the study of osmolyte effects on protein folding and aggregation *in vivo*. We observed stabilizing effects of proline and glycine betaine that are comparable to those *in vitro*. Additionally, we have found that proline in particular has a strongly inhibitory effect on aggregation both *in vivo* and *in vitro*. These findings serve as a point of departure for further studies that may help optimize protocols for protein expression in biotechnological applications and enable further dissection of the mechanism of aggregation and its modulation under physiological conditions, which may help therapeutic design for protein misfolding and aggregation diseases.

ACKNOWLEDGMENTS

The authors acknowledge support from the National Institutes of Health, Grant GM027616 and a 2006 NIH Director's Pioneer Award (5 DPI OD00945) to LMG, and DFG-project IG73/4–1 and the Heisenberg award IG73/1-1 to ZI.

REFERENCES

Auton, M., and Bolen, D. W. (2005). Predicting the energetics of osmolyte-induced protein folding/unfolding. *Proc. Natl. Acad. Sci. USA* **102**, 15065–15068.

Bolen, D. W. (2004). Effects of naturally occurring osmolytes on protein stability and solubility: Issues important in protein crystallization. *Methods* **34**, 312–322.

Brown, A. D., and Simpson, J. R. (1972). Water relations of sugar-tolerant yeasts: The role of intracellular polyols. *J. Gen. Microbiol.* **72**, 589–591.

Cayley, S., and Record, M. T., Jr. (2003). Roles of cytoplasmic osmolytes, water, and crowding in the response of *Escherichia coli* to osmotic stress: Biophysical basis of osmo-protection by glycine betaine. *Biochemistry* **42**, 12596–12609.

Chiti, F., and Dobson, C. M. (2006). Protein misfolding, functional amyloid, and human disease. *Annu. Rev. Biochem.* **75**, 333–366.

Clark, P. L., Weston, B. F., and Gierasch, L. M. (1998). Probing the folding pathway of a β-clam protein with single-tryptophan constructs. *Fold. Des.* **3**, 401–412.

Cohen, E., Bieschke, J., Perciavalle, R. M., Kelly, J. W., and Dillin, A. (2006). Opposing activities protect against age-onset proteotoxicity. *Science* **313**, 1604–1610.

Courtenay, E. S., Capp, M. W., Anderson, C. F., and Record, M. T., Jr. (2000). Vapor pressure osmometry studies of osmolyte-protein interactions: Implications for the action of osmoprotectants *in vivo* and for the interpretation of "osmotic stress" experiments *in vitro*. *Biochemistry* **39**, 4455–4471.

Davies, J. E., Sarkar, S., and Rubinsztein, D. C. (2006). Trehalose reduces aggregate formation and delays pathology in a transgenic mouse model of oculopharyngeal muscular dystrophy. *Hum. Mol. Genet.* **15**, 23–31.

Dobson, C. M. (2001). Protein folding and its links with human disease. *Biochem. Soc. Symp.* 1–26.

Eyles, S. J., and Gierasch, L. M. (2000). Multiple roles of prolyl residues in structure and folding. *J. Mol. Biol.* **301**, 737–747.

Felitsky, D. J., Cannon, J. G., Capp, M. W., Hong, J., Van Wynsberghe, A. W., Anderson, C. F., and Record, M. T., Jr. (2004). The exclusion of glycine betaine from anionic biopolymer surface: Why glycine betaine is an effective osmoprotectant but also a compatible solute. *Biochemistry* **43**, 14732–14743.

Fink, A. L. (1998). Protein aggregation: Folding aggregates, inclusion bodies and amyloid. *Fold. Des.* **3**, R9–R23.

Furusho, K., Yoshizawa, T., and Shoji, S. (2005). Ectoine alters subcellular localization of inclusions and reduces apoptotic cell death induced by the truncated Machado-Joseph disease gene product with an expanded polyglutamine stretch. *Neurobiol. Dis.* **20**, 170–178.

Gidalevitz, T., Ben-Zvi, A., Ho, K. H., Brignull, H. R., and Morimoto, R. I. (2006). Progressive disruption of cellular protein folding in models of polyglutamine diseases. *Science* **311**, 1471–1474.

Griffin, B. A., Adams, S. R., Jones, J., and Tsien, R. Y. (2000). Fluorescent labeling of recombinant proteins in living cells with FlAsH. *Methods Enzymol.* **327**, 565–578.

Hong, J., Capp, M. W., Saecker, R. M., and Record, M. T., Jr. (2005). Use of urea and glycine betaine to quantify coupled folding and probe the burial of DNA phosphates in lac repressor-lac operator binding. *Biochemistry* **44**, 16896–16911.

Ignatova, Z., and Gierasch, L. M. (2004). Monitoring protein stability and aggregation *in vivo* by real-time fluorescent labeling. *Proc. Natl. Acad. Sci. USA* **101**, 523–528.

Ignatova, Z., and Gierasch, L. M. (2005). Aggregation of a slow-folding mutant of a β-clam protein proceeds through a monomeric nucleus. *Biochemistry* **44**, 7266–7274.

Ignatova, Z., and Gierasch, L. M. (2006). Inhibition of protein aggregation *in vitro* and *in vivo* by a natural osmoprotectant. *Proc. Natl. Acad. Sci. USA* **103**, 13357–13361.

Ignatova, Z., Krishnan, B., Bombardier, J. P., Marcelino, A. M. C., Hong, J., and Gierasch, L. M. (2007). From the test tube to the cell: Exploring the folding and aggregation of a β-clam protein. *Biopolymers* **88**, 157–163.

Kanapathipillai, M., Lentzen, G., Sierks, M., and Park, C. B. (2005). Ectoine and hydro-xyectoine inhibit aggregation and neurotoxicity of Alzheimer's β-amyloid. *FEBS Lett.* **579**, 4775–4780.

Liu, R., Barkhordarian, H., Emadi, S., Park, C. B., and Sierks, M. R. (2005). Trehalose differentially inhibits aggregation and neurotoxicity of β-amyloid 40 and 42. *Neurobiol. Dis.* **20**, 74–81.

Oberg, K., Chrunyk, B. A., Wetzel, R., and Fink, A. L. (1994). Nativelike secondary structure in interleukin-1β inclusion bodies by attenuated total reflectance FTIR. *Biochemistry* **33**, 2628–2634.

Pradeep, L., and Udgaonkar, J. B. (2004). Osmolytes induce structure in an early intermediate on the folding pathway of barstar. *J. Biol. Chem.* **279**, 40303–40313.

Racher, K. I., Culham, D. E., and Wood, J. M. (2001). Requirements for osmosensing and osmotic activation of transporter ProP from *Escherichia coli*. *Biochemistry* **40**, 7324–7333.

Ross, C. A., and Poirier, M. A. (2004). Protein aggregation and neurodegenerative disease. *Nat. Med.* **10**(Suppl.), S10–S17.

Scaramozzino, F., Peterson, D. W., Farmer, P., Gerig, J. T., Graves, D. J., and Lew, J. (2006). TMAO promotes fibrillization and microtubule assembly activity in the C-terminal repeat region of tau. *Biochemistry* **45**, 3684–3691.

Schubert, U., Anton, L. C., Gibbs, J., Norbury, C. C., Yewdell, J. W., and Bennink, J. R. (2000). Rapid degradation of a large fraction of newly synthesized proteins by proteasomes. *Nature* **404**, 770–774.

Singer, M. A., and Lindquist, S. (1998). Multiple effects of trehalose on protein folding *in vitro* and *in vivo*. *Mol. Cell* **1**, 639–648.

Speed, M. A., Wang, D. I., and King, J. (1996). Specific aggregation of partially folded polypeptide chains: The molecular basis of inclusion body composition. *Nat. Biotechnol.* **14**, 1283–1287.

Tanaka, M., Machida, Y., Niu, S., Ikeda, T., Jana, N. R., Doi, H., Kurosawa, M., Nekooki, M., and Nukina, N. (2004). Trehalose alleviates polyglutamine-mediated pathology in a mouse model of Huntington disease. *Nat. Med.* **10**, 148–154.

Timasheff, S. N. (1998). Control of protein stability and reactions by weakly interacting cosolvents: The simplicity of the complicated. *Adv. Protein Chem.* **51**, 355–432.

Uversky, V. N., Li, J., and Fink, A. L. (2001). Trimethylamine-N-oxide-induced folding of α-synuclein. *FEBS Lett.* **509**, 31–35.

Walsh, D. M., and Selkoe, D. J. (2004). Oligomers on the brain: The emerging role of soluble protein aggregates in neurodegeneration. *Protein Pept. Lett.* **11**, 213–228.

Wood, J. M. (1999). Osmosensing by bacteria: Signals and membrane-based sensors. *Microbiol. Mol. Biol. Rev.* **63**, 230–262.

Wood, J. M., Bremer, E., Csonka, L. N., Kraemer, R., Poolman, B., van der Heide, T., and Smith, L. T. (2001). Osmosensing and osmoregulatory compatible solute accumulation by bacteria. *Comp. Biochem. Physiol. A Mol. Integr. Physiol.* **130**, 437–460.

Yancey, P. H. (2005). Organic osmolytes as compatible, metabolic and counteracting cytoprotectants in high osmolarity and other stresses. *J. Exp. Biol.* **208**, 2819–2830.

Yancey, P. H., Clark, M. E., Hand, S. C., Bowlus, R. D., and Somero, G. N. (1982). Living with water stress: Evolution of osmolyte systems. *Science* **217**, 1214–1222.

Yang, D. S., Yip, C. M., Huang, T. H., Chakrabartty, A., and Fraser, P. E. (1999). Manipulating the amyloid-β aggregation pathway with chemical chaperones. *J. Biol. Chem.* **274**, 32970–32974.

CHAPTER TWENTY-TWO

SIMULATIONS OF MACROMOLECULES IN PROTECTIVE AND DENATURING OSMOLYTES: PROPERTIES OF MIXED SOLVENT SYSTEMS AND THEIR EFFECTS ON WATER AND PROTEIN STRUCTURE AND DYNAMICS

David A. C. Beck, Brian J. Bennion, Darwin O. V. Alonso, *and* Valerie Daggett

Contents

Abstract

Rarely is any solution simply solute and water. *In vivo*, solutes, such as proteins and nucleic acids, swim in a sea of water, salts, ions, small molecules, and lipids, not to mention other macromolecules. *In vitro*, virtually all solutions

Department of Medicinal Chemistry, University of Washington, Seattle, Washington

Methods in Enzymology, Volume 428

ISSN 0076-6879, DOI: 10.1016/S0076-6879(07)28022-X

contain a mixture of aqueous solvents, or "cosolvents" [i.e., solvent(s) in addition to water], that can alter the dynamics, behavior, solubility, and stability of proteins and nucleic acids. We have developed models for a number of cosolvents, including the denaturant urea and the small chemical chaperone trimethylamine *N*-oxide (TMAO). This chapter examines the models for these two cosolvents in the context of experimental data. The direct and indirect effects of these molecules on water and protein are studied with molecular dynamics simulations. These observations and conclusions are drawn from simulations of these molecules in pure water and as a cosolvent for the protein chymotrypsin inhibitor 2. Urea-induced denaturation occurs initially through attack of the protein by water and hydration of hydrophobic protein moieties as a result of disruption of the hydrogen bonding network of water by urea. This indirect denaturing effect of urea is followed by more direct action as urea replaces some waters involved in the initial hydration of the hydrophobic core and subsequently binds to polar residues and the protein main chain to compete with the intraprotein hydrogen bonds. In the case of TMAO, we find that it encourages water–water interactions, thereby stabilizing the protein as a result of the increased penalty for the hydration of hydrophobic residues.

1. INTRODUCTION

Experimental studies of protein and nucleic acid dynamics and folding typically necessitate the use of solutions with denaturants or stabilizing osmolytes (Matouschek *et al.*, 1990, 1989; Pace and Tanford, 1968; Tanford, 1968; Tanford *et al.*, 1966). Despite the prolific use of these molecules in experimental studies, the number of molecular dynamics (MD) simulations of proteins employing such molecules is relatively small (Alonso and Daggett, 1995, 1998; Bennion and Daggett, 2003, 2004; Bennion *et al.*, 2004; Brooks and Nilsson, 1993; Caballero-Herrera *et al.*, 2005; Caflisch and Karplus, 1999; Kang *et al.*, 1995; Patel *et al.*, 1998; Schiffer *et al.*, 1995; Smith *et al.*, 2005; Tirado-Rives *et al.*, 1997; Vishnyakov *et al.*, 1999; Zhang *et al.*, 2001), particularly when compared with the large number of MD studies using pure water.

The mechanisms by which denaturants such as urea promote protein denaturation have been conceptually reduced to two categories: indirect and direct effects. In the case of indirect action, the denaturant alters the hydrophobicity of the solvent environment (Collins and Washabaugh, 1985; Zou *et al.*, 1998). With respect to direct action, the denaturant binds to the protein, often with preference for particular amino acids (Duffy *et al.*, 1993; Makhatadze and Privalov, 1992; Nozaki and Tanford, 1970; Robinson and Jencks, 1965; Whitney and Tanford, 1965). External to that debate is the concern that not all denaturants share the same mechanism of action (i.e., direct/indirect) in destabilizing proteins (Shimizu *et al.*, 2000). Two previous theoretical studies have cited direct interactions as the primary mechanism by which urea destabilizes proteins (Caflisch and Karplus, 1999; Tirado-Rives *et al.*, 1997). Taken together, however, these two studies are contradictory and do not specifically exclude indirect effects of urea. In fact, the simulations were too short to actually observe denaturation. More recently, simulations from our group observed urea-induced denaturation and have shed light on both direct and indirect effects of urea (Bennion and Daggett, 2003, 2004).

Also of great interest are the mechanisms by which small molecules such as trimethylamine *N*-oxide (TMAO) act as protein stabilization/renaturation agents, also known as protective osmoloytes. TMAO counteracts the effect of urea, typically in a molar ratio of 1:2 TMAO to urea. Combined theoretical and experimental techniques have provided evidence to highlight the fundamentals of the indirect mechanism of action of TMAO at low concentrations (Bennion and Daggett, 2004; Bennion *et al.*, 2004; Zou *et al.*, 2002). This work is expanded upon here.

Our continued interest in creating realistic representations of solvent environments for use in simulations of protein folding/unfolding, function, and dynamics has led us to develop models for several of the cosolvents used most commonly in experimental studies: guanidinium hydrochloride, trifluroethanol, urea, TMAO, and methanol. Our force field models not only reproduce structural characteristics of solutions, such as experimental radial distribution functions (RDF), but also the dynamic behavior of the individual solvent components, such as self-diffusion rates. These models have been constructed in keeping with the guiding developmental philosophy of our force field (Levitt *et al.*, 1995, 1997). Namely, all molecular models have full internal flexibility, all atoms are represented explicitly, and interactions are atom centered. This chapter presents models for urea and TMAO, compares our results with experiment, and catalogs the direct and indirect effects of these cosolvents (approximately 8 *M* urea and 4 *M* TMAO) on water and the protein chymotrypsin inhibitor 2 (CI2). These data are presented and explored in the context of highlighting direct and indirect effects of urea denaturation and how TMAO modulates this effect.

2. METHODS

Force field models of cosolvents have been developed with the goal of providing a description of solute and solvent that is in keeping with the tradition of the Lifson consistent force field (Lifson and Warshel, 1968). The form of the nonbonded interaction potential from the force field of Levitt *et al.* (1995) was used, as were charges when possible. This was done to avoid problems associated with mismatched nonbonded potential parameters that may arise when dissimilar potentials are employed. To demonstrate that our models are sufficiently robust to describe dynamics and structure of solvent systems across a range of cosolvent concentrations, we have utilized common concentrations found in protein/solvent experimental work for the validation simulations.

2.1. Urea

Urea potential function parameters have been presented and described previously (Zou *et al.*, 2002). An initial parameter set, including the Lennard–Jones radii and well depth, bond distances, angles, and torsion parameters, is taken directly from our standard Levitt *et al.* (1995) force field. A previous study of urea had determined that the default atom-centered charges did not reproduce the experimental nitrogen-centered RDF well (Laidig and Daggett, 1996). A new charge set is derived from *ab initio* molecular orbital calculations. The resulting atomic charges produce an average dipole moment of 5.1 ± 0.2 D at integer concentrations from 1 to 8 M. This is in comparison with a gas–phase value of 4.6 D (McClellan, 1963) and a solution value of 5.7 D (Gaumann, 1958). The urea starting structure for simulation is created by *in vacuo* steepest descent (SD) minimization of distance-geometry generated coordinates using our modeling and simulation program: *in lucem* Molecular Mechanics (*il*mm) (Beck *et al.*, 2006).

2.2. Trimethylamine *N*-oxide

Our model of TMAO is based on *ab initio* studies reported by Noto *et al.* (1995) and subsequently modified and described elsewhere (Zou *et al.*, 2002). In gas phase, our TMAO model exhibits a molecular dipole moment of 5.16 D—in good agreement with the experimental value of 5.0 D (McClellan, 1963) and theoretical values of 4.96 D (Haaland *et al.*, 1991) and 5.2 D (Radom *et al.*, 1977). Our partial charge assignments yield a net-neutral molecule. Simulation starting configurations of TMAO are obtained from *in vacuo* SD minimization of a distance geometry-generated configuration in *il*mm.

2.3. Chymotrypsin inhibitor 2

The starting structure for CI2 simulations is the crystal structure of the protein with PDB code 1ypc (Harpaz *et al.*, 1994). Hydrogen atoms are added to the crystal structure as appropriate for a neutral pH environment.

2.4. Molecular dynamics simulations

The complete list of molecular dynamics trajectories performed can be found in Table 22.1. All simulations in this study are performed using *il*mm. The potential functions (Levitt *et al.*, 1995, 1997) and fundamental protocols have been described in detail elsewhere (Beck and Daggett, 2004). Simulations are performed in the NVE*p* ensemble with a 2-fs time step under periodic boundary conditions. During the simulation, frames are saved at 1-ps intervals for later analysis. The nonbonded interaction list is updated every 6 fs (three steps). The first nanosecond of every trajectory is allocated to equilibration and discarded from most analyses, except where noted.

For simulations of urea and TMAO cosolvent–water mixtures, 108 cosolvent molecules are used and cubic periodic box dimensions with the appropriate number of waters are selected to achieve the desired molarity. The cosolvent molecules are placed on a face–centered cubic lattice after each is rotated randomly about its geometric center. Waters are placed randomly in the box to minimize the difference between the MD system density and the experimental density, after random rotation about their geometric centers, such that no two water oxygens are closer than 1.6 Å and no water oxygen is within 1.9 Å of any cosolvent heavy atom. These simulations are performed at 298 K.

In the case of cosolvent, water, and protein systems, a solvent box is prepared without protein that contains cosolvent molecules in water at the experimental measured density for the desired concentration and temperature (333 K). In the case of 8 M urea, the density of 1.103 g/ml is used, and for 4 M TMAO, the density of 1.015 g/ml is used (Bennion and Daggett, 2004). The protein is then placed in the center of the solvent box, and waters and urea that overlap with the protein are removed. This process results in the removal of a number of cosolvent molecules. For the CI2 in the urea system, the final urea concentration is 7.15 M, and for CI2 in TMAO, the final TMAO concentration is 3.87 M. We follow the preparation procedure, including the minimization and dynamics protocols, as described previously (Bennion and Daggett, 2003). The protein in water simulation is prepared from the same CI2 starting structure with our standard protocols (Beck and Daggett, 2004). Simulations of CI2 in pure water and 7.15 M urea are conducted at 333 K using *il*mm. The simulation of CI2 in 3.87 M TMAO is performed in ENCAD for the first 10 ns and the final 21 ns are calculated in *il*mm. The temperature of the CI2 simulations is selected based on experimental conditions used to denature the protein (Jackson and Fersht, 1991). We note that 333 K is well below the melting temperature of CI2.

2.5. Analyses

Radial distribution function, diffusion and water population, and orientation graphics are computed by using methods described elsewhere (Beck *et al.*, 2003). Water relaxation times are derived by fitting a single exponential

Table 22.1 Properties of molecular dynamics simulation systems

System[a]	No. of cosolvent molecules	No. of waters	Box volume (Å^3)	Solution density (g/ml)		Sim. length (ns)[c]
				Expt.[b]	MD	
Pure water	0	279	8373	0.997	0.997	10
1 M urea	108	5725	179342	1.015	1.015	10
2 M urea	108	2730	89671	1.031	1.031	10
3 M urea	108	1730	59781	1.046	1.046	10
4 M urea	108	1230	44836	1.061	1.061	10
6 M urea	108	731	29890	1.092	1.092	10
8 M urea	108	482	22418	1.123	1.124	10
1 M TMAO	108	5544	179342	1.000	1.000	10
2 M TMAO	108	2559	89671	1.004	1.004	10
3 M TMAO	108	1566	59781	1.009	1.009	10
4 M TMAO	108	1073	44836	1.016	1.016	10
CI2 in water	0	3199	109514	0.983	0.983	50
CI2 in 7.15 M urea	493	2103	114457	1.103	1.084	50
CI2 in 3.87 M TMAO	271	2446	145759	1.015	1.023	30

[a] Molarities reported are accurate to within 1×10^{-4}.
[b] Experimental densities for urea and TMAO (Zou et al., 2002).
[c] Length of simulation used for analysis. All simulations were 1 ns longer than the reported period with the first nanosecond designated for equilibration and ignored during subsequent analysis. Structures were saved at 1-ps intervals, yielding 1000 structures per nanosecond.

decay function to vector autocorrelation traces. The vector from oxygen to the midpoint of the hydrogens is used for computation of the autocorrelation trace. Residence times are derived by fitting a single exponential decay function to residence time distribution functions where incidences of single-frame occupancies are neglected (Impey *et al.*, 1983). Hydrogen bonds are considered to be intact if the donor-to-acceptor distance is less than 2.6 Å and the donor–proton–acceptor angle is within 35° of linearity. For the purposes of Table 22.3, we utilize the hydrogen bond cutoff model of Wernet and colleagues (2004). This model is derived from spectroscopic studies of water and results in a smooth hydrogen bonding cutoff definition as a function of donor and acceptor distance (r) and donor–proton–acceptor angle (θ). Specifically, the cutoff is defined by any combination of r and θ that is less than the bounding values from the function $r(\theta) = -0.00044\theta^2 + 3.3$ Å, which has the form of an inverted parabola with a maximum at $r = 3.3$ Å and a linear θ angle (i.e., $\theta = 0$) (Wernet *et al.*, 2004). For contact analyses, a contact is defined as two heavy atoms within 4.6 Å or two nonpolar carbon atoms within 5.4 Å of each other. All analyses are performed in *il*mm.

3. RESULTS

3.1. Pure water

The F3C water model provides an accurate depiction of both the structure and the dynamics of water as a function of temperature (Beck *et al.*, 2003; Day *et al.*, 2002; Levitt *et al.*, 1997). The pure water diffusion of the F3C model is 0.23 Å²/ps at 298 K (Beck *et al.*, 2003), which is in very good agreement with experiment 0.23 to 0.25 Å²/ps (Krynicki *et al.*, 1978). The mean water–water residence time was 1 ± 1 ps, and the mean water angular relaxation time was 2 ± 1 ps. Two-thirds of all waters in a pure-water simulation had at least one well-structured hydrogen bonding partner.

3.2. Urea in water

Comparison with experimentally measured diffusion rates provides a stringent test of the dynamic characteristics of the *in silico* models by testing both the behavior of the urea and its interaction with water to yield the bulk properties of the combined solution. Over the range of concentrations for which experimental results are available, the observed diffusion of urea decreases nearly linearly with increasing concentration (Fig. 22.1A). Our model correctly reproduces the drop in diffusion with increasing concentration, but predicts slightly low urea diffusion coefficients for the 1, 2, 3, and 4 *M* urea solutions. However, the diffusion values calculated from simulations

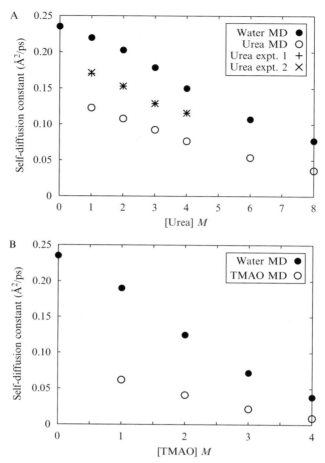

Figure 22.1   Molecular self-diffusion for cosolvents and water as a function of increasing cosolvent concentration. (A) Urea at concentrations of 1, 2, 3, 4, 6, and 8 M and (B) TMAO at concentrations of 1, 2, 3, and 4 M. Experimental data for urea are from Gostling and Akely (1952).

are all within 0.03 Å^2/ps of the experimental values (Gosting and Akely, 1952). The general downward sloping trend of experimental data is reproduced by the urea model. The diffusion of water in these solutions also decreases linearly throughout the concentration range.

The structure of aqueous solutions of urea has been the subject of much discussion but little consensus (Turner *et al.*, 1991). The interpretation of diffraction data from solution usually requires either a good starting model or an approximate spherical symmetry in the local environment. Aqueous solutions of urea have neither. Some early simulations suggested that urea–urea multimers are important in describing the solution structure (Astrand *et al.*, 1994; Cristinziano *et al.*, 1989; Duffy *et al.*, 1993; Tanaka *et al.*, 1985). Other studies found little or no urea dimerization (Boek and Briels, 1993; Chitra and Smith, 2000; Shimizu *et al.*, 2000; Tanaka *et al.*, 1985; Tsai *et al.*, 1996).

Figure 22.2A presents the total nitrogen-centered radial distribution function $G_{N-N}(R)$ from our simulations, along with one derived from experimental structure factors (Turner *et al.*, 1991). The first narrow peak is shifted to a shorter distance than in experiment, as seen in previous models (Boek and Briels, 1993; Laidig and Daggett, 1996; Tsai *et al.*, 1996), but the peak has a similar height, whereas the second, broad peak with the correct shape and height is shifted away from the origin. The other urea-based RDFs (data not shown) are in general agreement with the available experimental information and resemble models presented previously by us and many others (Astrand *et al.*, 1994; Bennion and Daggett, 2003; Boek and Briels, 1993; Chitra and Smith, 2000; Cristinziano *et al.*, 1989; Duffy *et al.*, 1993; Shimizu *et al.*, 2000; Tanaka *et al.*, 1984, 1985; Tsai *et al.*, 1996). No sustained urea dimerization was observed during the course of our urea/water simulations.

The water–water distribution functions are largely unaffected by the presence of urea in our simulations, although the conformation of water in the immediate vicinity of urea is perturbed, as has been seen previously (Boek and Briels, 1993; Chitra and Smith, 2000; Laidig and Daggett, 1996; Tirado-Rives *et al.*, 1997; Tsai *et al.*, 1996). The peaks in all three water–water distributions, $G_{O-O}(R)$ presented in Fig. 22.2B, $G_{O-H}(R)$ (data not shown), and $G_{H-H}(R)$ (data not shown), are located at the same distances as in pure water, with each peak height accentuated in the urea solution.

The water–water residence time was also largely unaffected by the concentration of urea (Table 22.2). At 8 M urea, this residence time was 1 ps, which is identical to that of pure water. However, at 8 M urea, the water angular relaxation time was double that of pure water to 4 ps (vs 2 ps). There was an approximately linear upward trend in this quantity as a function of urea concentration. Most notably, as the urea concentration increased, there was a marked disruption in the structure of the first coordination shell of water, as reflected by the number of broken hydrogen bonds (Table 22.3). There was little change in the water–water donor–proton–acceptor angular distributions (Fig. 22.3C).

Not surprisingly, there were two dominant hydration sites per urea molecule (Fig. 22.4A), one with the water protons oriented toward the carbonyl oxygen and the other with the water oxygen oriented toward the amide protons. Two subdominant hydration sites that coordinate with the second hydration shell can be found on the amide protons. No significant hydration sites were found out of the plane of the urea.

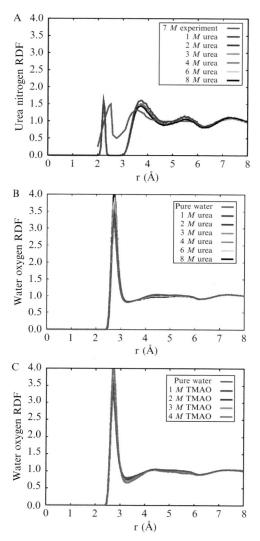

Figure 22.2   Radial distribution functions as a function of increasing cosolvent concentration. (A) Nitrogen-centered radial distribution function for urea from simulations at 1, 2, 3, 4, 6, and 8 *M* urea and experiment (Turner *et al.*, 1991); (B) water oxygen radial distribution function for urea at 1, 2, 3, 4, 6, and 8 *M*; and (C) water oxygen radial distribution function for TMAO at 1, 2, 3, and 4 *M*. (See color insert.)

3.3. Trimethylamine *N*-oxide in water

Water diffusion decreases significantly in the presence of TMAO in the simulations (Fig. 22.1B). A reduction in water diffusion of 30% by 1 *M* TMAO has been observed experimentally (Clark *et al.*, 1982). In

Table 22.2 Water–water interaction properties in various solvent systems

System	Water–water residence time (ps)	Water–water angular relaxation time (ps)
Pure water	1	2
4 M TMAO	1	19
8 M urea	1	4

Table 22.3 Hydrogen bonding structure of the first water coordination shell

System	% of waters with double donor	% of waters with single donor	% of waters with no donor
Pure water	16.0	51.0	32.9
1 M urea	14.9	49.9	35.2
2 M urea	13.5	48.6	37.9
3 M urea	12.2	47.0	40.8
4 M urea	10.9	45.2	43.9
6 M urea	8.4	40.7	50.9
8 M urea	6.0	34.9	59.1
1 M TMAO	15.7	50.4	33.9
2 M TMAO	15.5	49.4	35.1
3 M TMAO	15.6	48.0	36.4
4 M TMAO	15.7	46.6	37.7
CI2 in water	15.1	50.7	34.2
CI2 in 7.15 M urea	5.1	32.7	62.2
CI2 in 3.87 M TMAO	15.9	51.9	32.2

Fig. 22.1C, our simulations of 1 M TMAO solutions are in agreement with the experimental value of 0.17 $Å^2$/ps (Gostling and Akely, 1952). Simulations of 1, 2, 3, and 4 M TMAO demonstrate a nearly linear decrease in water self-diffusion with respect to concentration (Fig. 22.1B).

Radial distribution functions of water oxygen atoms were calculated for each TMAO solution simulation. The positions of the first and second peaks in the RDFs of each simulation are equivalent to those of pure water (Fig. 22.2C). This indicates that the general tetrahedral structure of water remains intact. The height of the first peak in the TMAO simulations is larger ($\approx$7%) than in pure water.

The reduction in water diffusion is a result of more persistent intermolecular interactions. Analyses show that average lifetimes of hydrogen bonds between water and TMAO are longer (7 ps) than water–water lifetimes

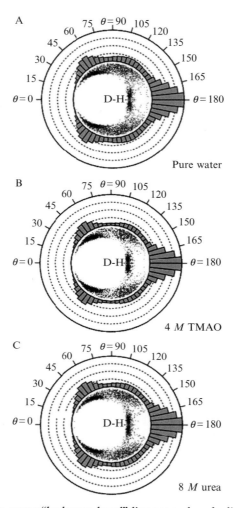

Figure 22.3 Water–water "hydrogen bond" distance and angle distributions for various cosolvent simulation systems. The D-H in the center of each plot represents the donor and proton; in this case the water oxygen and hydrogen. From the D in the inner center to the edge of the circle represents 3.6 Å. As a representative subset of data, 2000 points are plotted in the center circle, showing the position and angle of an acceptor (note the upper/lower half symmetry). Angular distributions are represented by the histograms surrounding the circle. Each histogram bin is weighted by the sine of the solid angle to account for conical volume variation. (A) Data from a pure water simulation at 298 K; (B) and (C) show data from a 4 *M* TMAO simulation and an 8 *M* urea simulation, respectively.

(1 ps) (Zou *et al.*, 2002). Water orientation relaxation times increase with increasing concentrations of TMAO to a value of 19 ps at 4 *M* TMAO (Table 22.2).

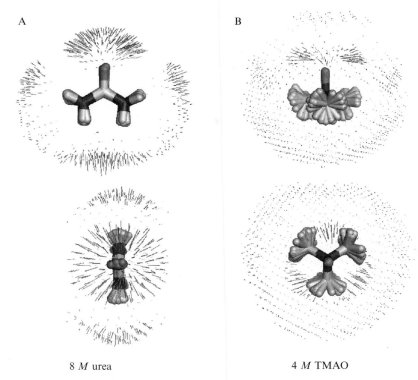

8 *M* urea   4 *M* TMAO

Figure 22.4 Hydration site population and orientation from cosolvent simulations. (a) Urea and (b) TMAO. The population color scale increases from lowest (red) to highest (blue). Only the top 15% populated hydration sites are displayed; i.e., only sites with a $G_{XO}(r)$ above 1.8 pairs are shown. (See color insert.)

Unlike urea, TMAO does not substantially disturb the hydrogen bonding of the hydration shell (Table 22.3; see Zou *et al.*, 2002). There also is no effect on water–water residence times (Table 22.2). The three-dimensional structure of TMAO that is presented to hydration water, which has been discussed previously, nicely supports the tetrahedral arrangement of waters (Zou *et al.*, 2002). Only minor evidence of a shift toward linear hydrogen bonds in the water–water hydrogen bonding structure of the first coordination shell is observed in donor–proton–donor angular distributions shown in Fig. 22.3B. However, Table 22.3 shows that as the concentration of TMAO is increased, 3 to 5% of single donor hydrogen bonds are "broken" or "melted" such that no hydrogen bond donors are present.

The time-averaged hydration site population and water orientation depiction shown in Fig. 22.4B reveals the expected hydration site off the oxygen. Because of the molecular symmetry of TMAO, what appears as three hydration sites is actually one with threefold radial symmetry.

This depiction is rather interesting, as the entire hydration shell can be seen (short red and white vectors). As seen in Fig. 22.4B, an upturned semisphere encapsulates the hydrophobic methyl groups, forcing water to optimize interactions with itself rather than with the molecule. Also visible is a "belt" that adjoins the highly populated hydration site to the lower "bowl." Finally, a "cap" forming the second hydration shell can be seen. The lack of persistent orientation or high population sites in this second shell is a result of the symmetry of the highly populated hydration site.

3.4. Chymotrypsin inhibitor 2 in water, urea, and TMAO

The native state simulation of CI2 in pure water at 333 K was stable with a mean Cα RMSD of 2.3 ± 0.4 Å over the final 50 ns (Fig. 22.5). The simulation reveals a relatively stable hydrophobic cluster (i.e., all residues in contact 87% of the time) consisting of residues Trp 5, Leu 8, Ala 16, Val 19, Ile 20, Ile 29, Val 31, Val 47, Leu 49, Val 51, Ile 57, and Pro 61. This cluster contains residues from every secondary structure component of CI2 with the exception of the active site loop and the turn connecting the second and third β strands (β2 and β3). This cluster is marginally less stable at 333 K than in simulations at lower temperatures (Day *et al.*, 2002). When these residues were not in contact, it was not that water had breached the protein core. Instead dynamic noncore residues had inserted themselves into the core, disrupting packing without unfolding the protein. There is, however, a subset of this cluster that is present 100% of the time at 333 K consisting of residues Ala 16 and Ile 20 in the α helix, Val 47 and Leu 49 in β2, and Pro 61 in β3. These residues describe a concise nucleus expected to be present in native structures of CI2 at 333 K.

In the simulation of CI2 in 7.15 M urea, this reduced hydrophobic core remained intact until 5.429 ns when water began to attack the contact between residues Ile 20 and Val 47 (Fig. 22.5). Before this time, the mean Cα RMSD was 2.5 ± 0.4 Å, after which the mean Cα RMSD was 6.3 ± 1.6 Å. The large change in Cα RMSD was obvious in the cluster analysis involving multidimensional scaling of the pairwise Cα RMSD matrix (data not shown), which is indicative of an unfolding transition state (TS) (Day and Daggett, 2005; Li and Daggett, 1994, 1996). The mean Ca RMSD of the TS (3.1 Å) is somewhat lower than what has been described previously (Day and Daggett, 2005, 2007; Day *et al.*, 2002).

In the starting structure for this simulation, there is one water that is hydrogen bonded to the main chain of residue Leu 49. At 5.429 ns, the TS, water attacks the core and distorts the α helix. Ten picoseconds later (5.439 ns), two more waters have begun to open the gap between the α helix and the β sheet. Additionally, these waters recruit a urea that remains on the periphery of the core until approximately 10.5 ns, when, along with two other urea molecules, it completely disrupts the hydrophobic cluster.

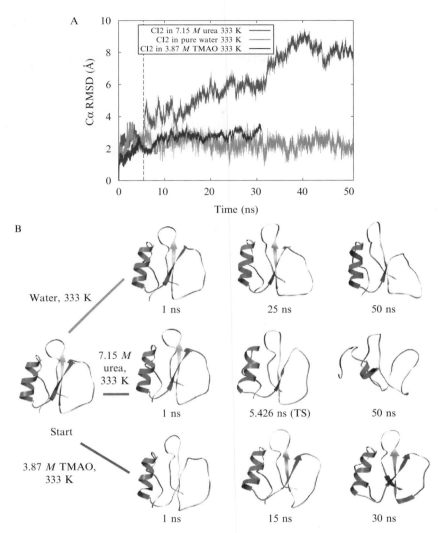

Figure 22.5 Cα RMSD as a function of time and snapshots from simulations of CI2 in pure water, 7.15 *M* urea, and 3.87 *M* TMAO. (a) The Cα RMSD of simulation structures to the crystal structure are plotted in green for CI2 in pure water control, red for CI2 in 7.15 *M* urea, and blue for CI2 in 3.87 *M* TMAO. Simulations of CI2 in pure water and 3.87 *M* TMAO at 333 K are stable and well folded; however, in the presence of 7.15 *M* urea, the protein begins to unfold, resulting in the large Cα RMSD. (b) Snapshots of CI2 from the pure water, 7.15 *M* urea, and 3.87 *M* TMAO simulations. The protein is shown as ribbons. In pure water and TMAO, CI2 maintains its native hydrophobic core and overall topology over the course of the simulations. In urea, at the beginning of the simulation (1 ns) CI2 is well structured and waters are excluded from the core. However, at the unfolding TS, which occurs at 5.429 ns, water begins to enter the core of the protein and the active site loop has changed conformation. After this time, urea molecules begin to join and replace waters that have penetrated and disrupted the core (50 ns). (See color insert.)

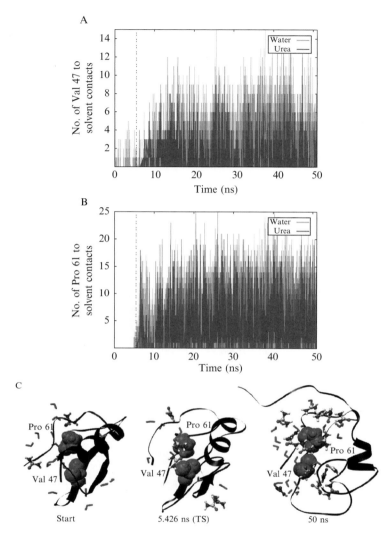

Figure 22.6 Water and urea contacts to CI2 core residues as a function of simulation time accompanied by snapshots of unfolding. Simulation of CI2 in 7.15 *M* urea: (a) The number of heavy atom contacts between CI2's Val 47 and water (cyan) and urea (red) as the MD simulation progresses. (b) The number of heavy atom contacts between CI2's Pro 61 and water (cyan) and urea (red) vs simulation time. (c) Snapshots of CI2 showing hydration waters and urea within 5.0 Å of Val 47 and Pro 61. The protein is shown as ribbons with two of the hydrophobic residues of the core shown in space filling. At the beginning of the simulation, there are no urea to Val 47 or Pro 61 contacts, although several water contacts can be observed (a and c). After water attacks the core at the TS (vertical bar in a and b at 5.426 ns), urea molecules are observed entering into the hydrophobic nucleus. By the end of the simulation (50 ns) the protein is denatured and the core is solvated by urea. (See color insert.)

Water and urea exchange throughout the rest of the simulation (51 ns) until these core residues are entirely solvated.

The number of residue-to-solvent contacts as a function of simulation time was calculated for all of the specified core residues. The plots for two of these residues, Val 47 and Pro 61, are particularly illustrative of the urea-induced unfolding process (Fig. 22.6). For both residues, water is the first to solvate these residues. In the case of Val 47, the mobilization of water into the core begins almost immediately (Fig. 22.6A). However, for Pro 61 (Fig. 22.6B), hydration occurs shortly (tens of picoseconds) before the unfolding TS at 5.429 ns. After the initial disruption of the core and active site loop, urea molecules also begin to enter the core and eventually replace some water molecules (Fig. 22.6C).

In our simulation of CI2 in 3.87 M TMAO at 333 K, CI2 was stable with a mean Cα RMSD of 2.2 $\pm$ 0.6 Å over the final 30 ns (Fig. 22.5). The hydrophobic cluster identified from the 333 K pure water simulation of CI2 (Trp 5, Leu 8, Ala 16, Val 19, Ile 20, Ile 29, Val 31, Val 47, Leu 49, Val 51, Ile 57, and Pro 61) was intact 85% of the time. This simulation absolutely conserved the very stable concise hydrophobic nucleus formed by Ala 16, Ile 20, Val 47, Leu 49, and Pro 61. By this metric and others, including the previously mentioned Cα RMSD, CI2 is well folded in 3.87 M TMAO.

The average water residence time in the proximity of polar and charged protein residues was affected by the addition of urea, but not TMAO. CI2 in water and 3.87 M TMAO, waters that were within 3.5 Å to polar or charged groups, had an average residence time of 19.2 $\pm$ 8.1 and 21.0 $\pm$ 6.7 ps. This is in sharp contrast with the lower value from CI2 in 7.15 M urea of 9.6 $\pm$ 4.2 ps.

A comparison of the percentage of double, single, and no hydrogen donor waters in pure water and CI2 simulations also reveals differences (Table 22.3). CI2, in water alone, has the slight effect of shifting the balance of single and double hydrogen bond donors away from that of pure water toward config-urations with less energetically favorable interactions. While this effect is minor with CI2 and water alone (1 to 2% overall), when 7.15 M urea is present the percentage of broken hydrogen bond networks in water increases more dramatically than 8 M urea or CI2 alone. Also of note, the addition of 3.87 M TMAO not only prevents degradation of the hydrogen bonding network by CI2, but yields networks consistent with pure water (sans protein).

4. DISCUSSION

Studies support the idea that denaturing cosolvents alter protein sta-bility by direct interaction of the denaturant with the solute (Makhatadze and Privalov, 1992). With respect to urea, previous studies have shown that urea binds to the polar residues and exposed main-chain groups of solutes (Bennion and Daggett, 2003; Caflisch and Karplus, 1999; Smith *et al.*, 2005; Tirado-Rives *et al.*, 1997). Polar interactions of urea with regions of the main chain of CI2 are long lived, ranging from 20 to 30 ps. They often exceed fast-phase water residence times with the same amino group (5 to 10 ps, results not shown). These interactions result in a drop in diffusion of the cosolvent in the hydration shell (3.5 Å) with respect to bulk.

In 7.15 M urea, urea–protein residence times to hydrophobic groups are 2 to 8 ps, depending on which protein group is considered and the form of the decay fit function: single exponential decay leans toward 8 ps, whereas a double exponential puts a fast phase at 2 ps. In the same simulation system, no significant attenuation in urea diffusion is seen in the hydration shell relative to bulk. These numbers are in reasonable agreement with the same values for water. In the case of CI2, sustained urea–protein contacts were not observed until after the TS (5.429 ns) at which time the protein reached its transition state of unfolding. After that time, there was an average of 14 urea molecules in contact with CI2; 21% made hydrogen bonds as donor or acceptor to the protein with the rest making other nonspecific contacts.

It is possible that urea dimerization, proposed at higher concentrations (Astrand *et al.*, 1994) and subsequent surface binding of the dimer, could act as a screen for hydrophobic interactions. This would result in a decrease in the hydrophobic effect in the shielded region. However, in this study, as with previous computational studies (Bennion and Daggett, 2003; Sokolic *et al.*, 2002), no significant evidence for this phenomenon exists.

Based on our simulations, direct effects of urea become pronounced only after the protein has begun to deform in response to indirect solvent effects and direct attack by water. At this point, the protein has yielded interactions that are normally intraprotein to the cosolvent. This view is consistent with the observations of others (Bennion and Daggett, 2003; Pace, 1986; Smith *et al.*, 2005).

With indirect effects, cosolvents alter the structure and dynamics of water, such as changing the number of hydrogen bonds per water and changing the water–water hydrogen bond residence time. We look to changes in the statistics of water–water interactions to shed light on these effects. Figure 22.3 demonstrates several points about the effects of cosolvent on the water–water matrix. First, there is no change in the water–water hydrogen bond distance for any of the cosolvents. This is confirmed by the radial distribution functions for the water oxygen (Fig. 22.2 for urea and TMAO), which are in good agreement with experiments (Idrissi, 2005; Soper, 2000; Soper and Finney, 1993; Soper and Luzar, 1996; Turner *et al.*, 1991; Yamaguchi *et al.*, 2000).

In the presence of the highest concentrations of TMAO, the angular distribution of hydrogen bonds is subtly shifted toward linear arrangements, especially for the more structured interactions with a donor–acceptor distance less than 2.6 Å (Fig. 22.3B). This is especially interesting, as many observed that the methyl groups in methanol and ethanol (Dixit *et al.*, 2002; Raschke and Levitt, 2005) and butane/octane (Laidig *et al.*, 1998) create clathrate-like (also known as ice-like) structures in the water surrounding them. Simulations of ice Ih (and its melting) performed by our group also exhibited a shift toward linear orientations of hydrogen bonds (unpublished results) when compared to pure liquid water at 298 K. As the TMAO behavior with respect to methyl clustering is similar to that of methanol, which is well documented experimentally (Dixit *et al.*, 2001, 2002), it is very likely that these water "clathrates" are present in solutions of TMAO, even at concentrations as low as 1 *M* where TMAO dimerization is observed in occasional snapshots. Urea, however, does not cause any noticeable redistribution of angular distributions.

The RDFs for these systems exhibit exaggerated first peak heights for all of the cosolvents at all concentrations (Fig. 22.2). This effect is concentration dependent, that is, the peak height (the number of first shell partners) increases with concentration. It would be a mistake, however, to assume that these partners are participating in true hydrogen bonds. With the angular shifts depicted in Fig. 22.3 and the direct calculation of hydrogen bonding partners (either by 2.6 Å and <35° from linearity or by the definition of Wernet and colleagues [2004]), it is clear that only in TMAO do *some* of the new partners actually make hydrogen bonds. In contrast, in urea, as the concentration increases, there is a dramatic decrease in unambiguous hydrogen bonds in the first shell, such that at 8 *M* urea, only about 1 in 20 waters makes two hydrogen bonds to neighboring water, 1 in 3 making one, and the rest none. Yet the first shell may have up to double the number of partners compared with pure water.

The slow diffusion of water in TMAO, the 14-ps relaxation times of water, and the relatively long-lived TMAO–water hydrogen bonds (~10 ps) point to favorable TMAO–water and water–water enthalpies accompanied by a loss of water entropy. Thus, the optimization of water structure discourages the hydration of water of nonpolar groups. Contrast this with urea, where the diffusion is slow, as is relaxation, but these are not accompanied by a favorable enthalpic contribution (i.e., urea exhibits a large number of broken and short-lived hydrogen bonds). As with direct effects, we must consider the role of cosolvent concentration. For virtually all water-related properties there is a concentration gradient that shifts the statistics away from those of pure water as the concentration of cosolvent increases.

The unfolding of CI2 in 7.15 *M* urea illustrates how the hydrophobic core of a protein is disrupted by a denaturing osmolyte. In accordance with the aforementioned perturbation of water structure by indirect action of urea, the hydrophobic effect is reduced. This enables the waters in the TS depiction of Fig. 22.5 to break down the cluster of aliphatic side chains formed by Ala 16, Ile 20, Val 47, Leu 49, and Pro 61. There was a distinct lack of urea molecules present within the core. However, very shortly thereafter, the urea molecules began to displace waters that had cracked open the core. It is only after this time, when the hydrophobic core is already disrupted, that direct action of urea is observed in the form of nonspecific urea to side-chain interactions and urea to main-chain hydrogen bonds, which continue to unfold the protein and stabilize the denatured state.

Finally, it is interesting that CI2 alone has the effect of disrupting the optimization of the water hydrogen bonding network (Table 22.3). While this effect is relatively minor, it is exaggerated by the addition of high concentrations of urea. In the presence of 8 M urea only, the percentage of no hydrogen bond donor waters is 59.1, but in a slightly lower concentration (7.15 M) with CI2, this number increases to 62.2%. This perturbation effect of CI2 on water is counteracted, however, by the addition of 3.87 M TMAO. In such conditions, the water hydrogen bonding network is more like that of pure water than water and CI2. The directionality of these changes relative to pure water and the size of the changes indicate that the effect of urea on water is compounded by CI2, whereas the effect of TMAO on water structure is enhanced by the presence of the protein.

5. Conclusions

Our results provide a molecular basis for the ability of urea to solubilize hydrophobic solutes and therefore denature proteins. The disordering of water around denaturing osmolytes has the effect of "preloading" the aqueous environment with water molecules that are akin to those found in the hydrophobic hydration shell. With a decrease in the number of hydrogen bonds per water molecule, the water in the urea solutions is in effect "primed" to accept nonpolar moieties with less energetic penalty. That is, the hydrogen bonding of the water is compromised by the denaturants in much the same way as by a hydrophobic moiety, such that there is little to no additional penalty to introduce nonpolar groups into the aqueous denaturant. These urea-influenced waters have fewer water–water hydrogen bonds and are orientationally constrained in comparison with bulk water. Water molecules in the solvation layer of a nonpolar molecule face entropic and energetic penalties; water orientation is restricted and the number of possible hydrogen bonds is reduced when compared with bulk water. However, in the presence of denaturant, water has enthalpic compensation for the loss of hydrogen bonding partners that occurs upon solvation of hydrophobic molecules. Also, direct interactions of denaturant with the solute occur at higher concentrations. These interactions exclude solvation water, thereby increasing the entropy of first shell water. Consequently, the hydrophobic effect is reduced within the sphere of influence of urea molecules. Counteraction of the effect of denaturants by small molecule osmolytes such as TMAO leads to improved water–water hydrogen bonding and in this way discourages the introduction of nonpolar groups to prevent exposure of core hydrophobics and the initiation of the unfolding of proteins.

ACKNOWLEDGMENTS

We are grateful for financial support provided by the NIH (GM50789). UCSF chimera was used to prepare protein images (Pettersen et al., 2004).

REFERENCES

Alonso, D. O. V., and Daggett, V. (1995). Molecular dynamics simulations of protein unfolding and limited refolding: Characterization of partially unfolded states of ubiquitin in 60% methanol and in water. *J. Mol. Biol.* **247,** 501–520.

Alonso, D. O. V., and Daggett, V. (1998). Molecular dynamics simulations of hydrophobic collapse of ubiquitin. *Protein Sci.* **7,** 860–874.

Astrand, P. O., Wallqvist, A., and Karlstrom, G. (1994). Molecular dynamics simulation of 2-M aqueous urea solutions. *J. Phys. Chem. B* **98,** 8224–8233.

Beck, D. A. C., Alonso, D. O. V., and Daggett, V. (2003). A microscopic view of peptide and protein solvation. *Biophys. Chem.* **100,** 221–237.

Beck, D. A. C., Alonso, D. O. V., and Daggett, V. (2006). *"in lucem* Molecular Mechanics." University of Washington, Seattle, WA.

Beck, D. A. C., and Daggett, V. (2004). Methods for molecular dynamics simulations of protein folding/unfolding in solution. *Methods* **34,** 112–120.

Bennion, B. J., and Daggett, V. (2003). The molecular basis for the chemical denaturation of proteins by urea. *Proc. Natl. Acad. Sci. USA* **100,** 5142–5147.

Bennion, B. J., and Daggett, V. (2004). Counteraction of urea-induced protein denaturation by trimethylamine N-oxide: A chemical chaperone at atomic resolution. *Proc. Natl. Acad. Sci. USA* **101,** 6433–6438.

Bennion, B. J., DeMarco, M. L., and Daggett, V. (2004). Preventing misfolding of the prion protein by trimethylamine N-oxide. *Biochemistry* **43,** 12955–12963.

Boek, E. S., and Briels, W. J. (1993). Molecular-dynamics simulations of aqueous urea solutions: Study of dimer stability and solution structure, and calculation of the total nitrogen radial-distribution function Gn(R). *J. Chem. Phys.* **98,** 1422–1427.

Brooks, C. L., and Nilsson, L. (1993). Promotion of helix formation in peptides dissolved in alcohol and water-alcohol mixtures. *J. Am. Chem. Soc.* **115,** 11034–11035.

Caballero-Herrera, A., Nordstrand, K., Berndt, K. D., and Nilsson, L. (2005). Effect of urea on peptide conformation in water: Molecular dynamics and experimental characterization. *Biophys. J.* **89,** 842–857.

Caflisch, A., and Karplus, M. (1999). Structural details of urea binding to barnase: A molecular dynamics analysis. *Struct. Fold. Design* **7,** 477–488.

Chitra, R., and Smith, P. E. (2000). Molecular dynamics simulations of the properties of cosolvent solutions. *J. Phys. Chem. B* **104,** 5854–5864.

Clark, M. E., Burnell, E. E., Chapman, N. R., and Hinke, J. A. M. (1982). Water in barnacle muscle. 4. Factors contributing to reduced self-diffusion. *Biophys. J.* **39**, 289–299.

Collins, K. D., and Washabaugh, M. W. (1985). The Hofmeister effect and the behavior of water at interfaces. *Q. Rev. Biophys.* **18**, 323–422.

Cristinziano, P., Lelj, F., Amodeo, P., Barone, G., and Barone, V. (1989). Stability and structure of formamide and urea dimers in aqueous solution: A theoretical study. *J. Chem. Soc.-Faraday Trans. I* **85**, 621–632.

Day, R., Bennion, B. J., Ham, S., and Daggett, V. (2002). Increasing temperature accelerates protein unfolding without changing the pathway of unfolding. *J. Mol. Biol.* **322**, 189–203.

Day, R., and Daggett, V. (2005). Sensitivity of the folding/unfolding transition state ensemble of chymotrypsin inhibitor 2 to changes in temperature and solvent. *Protein Sci.* **14**, 1242–1252.

Day, R., and Daggett, V. (2007). Direct observation of microscopic reversibility in single-molecule protein folding. *J. Mol. Biol.* **366**, 677–686.

Dixit, S., Crain, J., Poon, W. C. K., Finney, J. L., and Soper, A. K. (2002). Molecular segregation observed in a concentrated alcohol-water solution. *Nature* **416**, 829–832.

Dixit, S. S., Poon, W. C. K., Crain, J., Finney, J. L., and Soper, A. K. (2001). Hydrophobicity and liquid state structure in aqueous solutions of methanol. *Am. Chem. Soc.* **221**, U250–U250. [Abstract]

Duffy, E. M., Severance, D. L., and Jorgensen, W. L. (1993). Urea: Potential functions, log-P, and free-energy of hydration. *Israel J. Chem.* **33**, 323–330.

Gaumann, T. (1958). Dielektrische Messungen an polaren Gemischen. 6. Mitteilung Harnstoffe. *Helvetica Chimica Acta* **41**(7), 1956–1970.

Gosting, L. J., and Akely, D. F. (1952). A study of the diffusion of urea in water at 25 with the Gouy Interference Method. *J. Am. Chem. Soc.* **74**(8), 2058–2060.

Haaland, A., Thomassen, H., and Stenstrom, Y. (1991). Trimethylamine N-oxide, structure and bonding investigated by gas electron-diffraction and abinitio Mo calculations. *J. Mol. Struct.* **263**, 299–310.

Harpaz, Y., Elmasry, N., Fersht, A. R., and Henrick, K. (1994). Direct observation of better hydration at the N-terminus of an alpha-helix with glycine rather than alanine as the N-cap residue. *Proc. Natl. Acad. Sci. USA* **91**, 311–315.

Idrissi, A. (2005). Molecular structure and dynamics of liquids: Aqueous urea solutions. *Spectrochim. Acta Part A* **61**, 1–17.

Impey, R. W., Madden, P. A., and Mcdonald, I. R. (1983). Hydration and mobility of ions in solution. *J. Phys. Chem.* **87**, 5071–5083.

Jackson, S. E., and Fersht, A. R. (1991). Folding of chymotrypsin inhibitor 2. 1. Evidence for a two-state transition. *Biochemistry* **30**, 10428–10435.

Kang, N. S., Yu, J. Y., and Jhon, M. S. (1995). Effect of cosolvent on a tripeptide solution using molecular-dynamics simulations. *J. Mol. Struct.* **355**, 201–209.

Krynicki, K., Green, C. D., and Sawyer, D. W. (1978). Pressure and temperature dependence of self-diffusion in water. *Faraday Discussions* 199–208.

Laidig, K. E., and Daggett, V. (1996). Testing the modified hydration-shell hydrogen-bond model of hydrophobic effects using molecular dynamics simulation. *J. Phys. Chem.* **100**, 5616–5619.

Laidig, K. E., Gainer, J. L., and Daggett, V. (1998). Altering diffusivity in biological solutions through modification of solution structure and dynamics. *J. Am. Chem. Soc.* **120**, 9394–9395.

Levitt, M. (1990). "ENCAD (Energy Calculations and Dynamics)." Molecular Applications Group, Palo Alto, CA.

Levitt, M., Hirshberg, M., Sharon, R., and Daggett, V. (1995). Potential energy function and parameters for simulations of the molecular dynamics of proteins and nucleic acids in solution. *Comput. Phys. Commun.* **91**, 215–231.

Levitt, M., Hirshberg, M., Sharon, R., Laidig, K. E., and Daggett, V. (1997). Calibration and testing of a water model for simulation of the molecular dynamics of proteins and nucleic acids in solution. *J. Phys. Chem. B* **101,** 5051–5061.

Li, A., and Daggett, V. (1994). Characterization of the transition state of protein unfolding by use of molecular dynamics: Chymotrypsin inhibitor 2. *Proc. Natl. Acad. Sci. USA* **91,** 10430–10434.

Li, A., and Daggett, V. (1996). Identification and characterization of the unfolding transition state of chymotrypsin inhibitor 2 by molecular dynamics simulations. *J. Mol. Biol.* **257,** 412–429.

Lifson, S., and Warshel, A. (1968). Consistent force field for calculations of conformations vibrational spectra and enthalpies of cycloalkane and N-alkane molecules. *J. Chem. Phys.* **49,** 5116.

Makhatadze, G. I., and Privalov, P. L. (1992). Protein interactions with urea and guanidinium chloride: A calorimetric study. *J. Mol. Biol.* **226,** 491–505.

Matouschek, A., Kellis, J. T., Jr., Serrano, L., Bycroft, M., and Fersht, A. R. (1990). Transient folding intermediates characterized by protein engineering. *Nature* **346,** 440–445.

Matouschek, A., Kellis, J. T., Serrano, L., and Fersht, A. R. (1989). Mapping the transition-state and pathway of protein folding by protein engineering. *Nature* **340,** 122–126.

McClellan, A. L. (1963). "Tables of Experimental Dipole Moments." Freeman and Company, San Francisco.

Noto, R., Martorana, V., Emanuele, A., and Fornili, S. L. (1995). Comparison of the water perturbations induced by 2 small organic solutes: Ab-initio calculations and molecular-dynamics simulation. *J. Chem. Soc.-Faraday Trans.* **91,** 3803–3808.

Nozaki, Y., and Tanford, C. (1970). Solubility of amino acids, diglycine, and triglycine in aqueous guanidine hydrochloride solutions. *J. Biol. Chem.* **245,** 1648.

Pace, C. N. (1986). Determination and analysis of urea and guanidine hydrochloride denaturation curves. *Methods Enzymol.* **131,** 266–280.

Pace, N. C., and Tanford, C. (1968). Thermodynamics of unfolding of beta-lactoglobulin A in aqueous urea solutions between 5 and 55 degrees. *Biochemistry* **7,** 198–208.

Patel, A. B., Srivastava, S., Phadke, R. S., Coutinho, E., and Kamath, S. (1998). Conformation of renin substrate (angiotensinogen) in water is different from DMSO: A H-1 NMR and molecular dynamics study. *Magn. Reson. Chem.* **36,** 285–295.

Pettersen, E. F., Goddard, T. D., Huang, C. C., Couch, G. S., Greenblatt, D. M., Meng, E. C., and Ferrin, T. E. (2004). UCSF chimera: A visualization system for exploratory research and analysis. *J. Comput. Chem.* **25,** 1605–1612.

Radom, L., Binkley, J. S., and Pople, J. A. (1977). Molecular structure of ammonia oxide (Nh3O): Abinitio study. *Austr. J. Chem.* **30,** 699–703.

Raschke, T. M., and Levitt, M. (2005). Nonpolar solutes enhance water structure within hydration shells while reducing interactions between them. *Proc. Natl. Acad. Sci. USA* **102,** 6777–6782.

Robinson, D. R., and Jencks, W. P. (1965). The effect of compounds of urea–guanidinium class on activity coefficient of acetyltetraglycine ethyl ester and related compounds. *J. Am. Chem. Soc.* **87,** 2462–2470.

Schiffer, C. A., Dotsch, V., Wuthrich, K., and Vangunsteren, W. F. (1995). Exploring the role of the solvent in the denaturation of a protein: A molecular-dynamics study of the DNA-binding domain of the 434-repressor. *Biochemistry* **34,** 15057–15067.

Shimizu, A., Fumino, K., Yukiyasu, K., and Taniguchi, Y. (2000). NMR studies on dynamic behavior of water molecule in aqueous denaturant solutions at 25 degrees C: Effects of guanidine hydrochloride, urea and alkylated ureas. *J. Mol. Liquids* **85,** 269–278.

Smith, L. J., Jones, R. M., and van Gunsteren, W. F. (2005). Characterization of the denaturation of human alpha-lactalbumin in urea by molecular dynamics simulations. *Proteins Struct. Funct. Bioinform.* **58,** 439–449.

Sokolic, F., Idrissi, A., and Perera, A. (2002). Concentrated aqueous urea solutions: A molecular dynamics study of different models. *J. Chem. Phys.* **116**, 1636–1646.

Soper, A. K. (2000). The radial distribution functions of water and ice from 220 to 673 K and at pressures up to 400 MPa. *Chem. Phys.* **258**, 121–137.

Soper, A. K., and Finney, J. L. (1993). Hydration of methanol in aqueous solution. *Phys. Rev. Lett.* **71**, 4346–4349.

Soper, A. K., and Luzar, A. (1996). Orientation of water molecules around small polar and nonpolar groups in solution: A neutron diffraction and computer simulation study. *J. Phys. Chem.* **100**, 1357–1367.

Tanaka, H., Nakanishi, K., and Touhara, H. (1985). Computer experiments on aqueous solutions. 7. Potential energy function for urea dimer and molecular-dynamics calculation of 8 mol-percent aqueous solution of urea. *J. Chem. Phys.* **82**, 5184–5191.

Tanaka, H., Touhara, H., Nakanishi, K., and Watanabe, N. (1984). Computer experiment on aqueous solution. 4. Molecular-dynamics calculation on the hydration of urea in an infinitely dilute aqueous solution with a new urea-water pair potential. *J. Chem. Phys.* **80**, 5170–5186.

Tanford, C. (1968). Protein denaturation. *Adv. Protein Chem.* **23**, 121–282.

Tanford, C., Pain, R. H., and Otchin, N. S. (1966). Equilibrium and kinetics of the unfolding of lysozyme (muramidase) by guanidine hydrochloride. *J. Mol. Biol.* **15**, 489–504.

Tirado-Rives, J., Orozco, M., and Jorgensen, W. L. (1997). Molecular dynamics simulations of the unfolding of barnase in water and 8 M aqueous urea. *Biochemistry* **36**, 7313–7329.

Tsai, J., Gerstein, M., and Levitt, M. (1996). Keeping the shape but changing the charges: A simulation study of urea and its iso-steric analogs. *J. Chem. Phys.* **104**, 9417–9430.

Turner, J., Finney, J. L., and Soper, A. K. (1991). Neutron diffraction studies of structure in aqueous solutions of urea and tetramethylammonium chloride and in methanol. *Zeitschrift Naturforschung Sect. a J. Phys. Sci.* **46**, 73–83.

Vishnyakov, A., Widmalm, G., Kowalewski, J., and Laaksonen, A. (1999). Molecular dynamics simulation of the alpha-D-Manp-(1 → 3)-beta-D-Glcp-OMe disaccharide in water and water DMSO solution. *J. Am. Chem. Soc.* **121**, 5403–5412.

Wernet, P., Nordlund, D., Bergmann, U., Cavalleri, M., Odelius, M., Ogasawara, H., Naslund, L. A., Hirsch, T. K., Ojamae, L., Glatzel, P., Pettersson, L. G. M., and Nilsson, A. (2004). The structure of the first coordination shell in liquid water. *Science* **304**, 995–999.

Whitney, P. L., and Tanford, C. (1965). Recovery of specific activity after complete unfolding and reduction of an antibody fragment. *Proc. Natl. Acad. Sci. USA* **53**, 524–532.

Yamaguchi, T., Benmore, C. J., and Soper, A. K. (2000). The structure of subcritical and supercritical methanol by neutron diffraction, empirical potential structure refinement, and spherical harmonic analysis. *J. Chem. Phys.* **112**, 8976–8987.

Zhang, Z. Y., Zhu, Y. J., and Shi, Y. Y. (2001). Molecular dynamics simulations of urea and thermal-induced denaturation of S-peptide analogue. *Biophys. Chem.* **89**, 145–162.

Zou, Q., Bennion, B. J., Daggett, V., and Murphy, K. P. (2002). The molecular mechanism of stabilization of proteins by TMAO and its ability to counteract the effects of urea. *J. Am. Chem. Soc.* **124**, 1192–1202.

Zou, Q., Habermann-Rottinghaus, S. M., and Murphy, K. P. (1998). Urea effects on protein stability: Hydrogen bonding and the hydrophobic effect. *Proteins Struct. Funct. Genet.* **31**, 107–115.

CHAPTER TWENTY-THREE

Application of the Transfer Model to Understand How Naturally Occurring Osmolytes Affect Protein Stability

Matthew Auton *and* D. Wayne Bolen

Contents

Abstract

A primary thermodynamic goal in protein biochemistry is to attain a predictive understanding of the energetic changes responsible for solvent-induced folding and unfolding. This chapter demonstrates the use of Tanford's transfer model to predict solvent-dependent cooperative protein folding/unfolding free energy changes (m values). This approach provides a thermodynamic description of these free energy changes in terms of individual contributions from the peptide backbone and residue side chains. The quantitative success of the transfer

Department of Biochemistry and Molecular Biology, University of Texas Medical Branch, Galveston, Texas

Methods in Enzymology, Volume 428
ISSN 0076-6879, DOI: 10.1016/S0076-6879(07)28023-1

model has been hindered for many years because of unresolved issues involving proper measurement of the group transfer-free energies of amino acid side chains and the peptide backbone unit. This chapter demonstrates what is necessary to design experiments properly so that reliable values of group transfer-free energies are obtainable. It then demonstrates how to derive a prediction of the *m* value for the description of protein folding/unfolding cooperativity and that the calculated values using the transfer model agree quite well with experimentally measured values.

1. INTRODUCTION

How cosolvents bring about the effects they have on protein stability is one of the oldest issues in biophysical chemistry. Hofmeister ions, organic solvents, denaturants, and stabilizers all have been investigated, but with varying degrees of success in understanding their actions on proteins. In the latter half of the twentieth century, research in the biology of adaptation uncovered a new class of solutes known as organic osmolytes that appeared to protect the cells of organisms against environmentally induced water stress conditions. Because the water stress conditions are often destabilizing to proteins, osmolytes having the ability to stabilize proteins against denaturing conditions were selected by nature to adapt the organism to the stress. The discovery of protecting osmolytes gave impetus to understanding how organic osmolytes stabilize proteins. Often in evolution a mechanism is found to solve a biological problem and is used frequently to solve similar problems. With the expectation that the principles underlying the mechanism are likely to be the same for all organic osmolytes, we hypothesized that by restricting our studies to naturally occurring osmolytes we would have a better opportunity at uncovering underlying principles than through the study of solutes not passed through the filter of natural selection. This chapter presents the transfer–free energy approach used in investigating how these agents affect protein stability and structure and gives the methods we and others developed to implement discovery of, and insight into, the principles of osmolyte action on proteins.

2. INITIAL OBSERVATIONS OF OSMOLYTE EFFECTS ON PROTEINS

Depending on the osmolyte in question, proteins have been observed to have a preference for solvation by either water or the osmolyte. Preferential interaction is a thermodynamic measure of the redistribution of water and osmolyte molecules around the protein surface relative to their distribution in bulk solution. In practice, the measurement of preferential interaction is accomplished by means of a dialysis experiment, under thermodynamic equilibrium, in which the concentration of the osmolyte is measured in the dialyzate and in the compartment containing the protein. In the presence of protein, water and osmolyte distribute around the protein according to their relative preferences for protein groups exposed on the surface. Thus, their concentrations in the vicinity of the protein will be different from their concentrations in bulk solution. The osmolyte concentration differences in the presence and absence of protein in the dialysis experiment can be quantified by densimetric (Lee et al., 1979) or osmotic pressure (Courtenay et al., 2000) measurements, giving a quantity called the preferential interaction parameter. If water is enriched near the protein surface relative to its composition in the bulk solution, a preferential interaction exists between water and the protein surface, resulting in what can be called *preferential hydration* from the perspective of water or *preferential exclusion* from the osmolyte perspective. If the osmolyte is found to be enriched near the protein surface, it is said to preferentially interact with or bind to the protein surface and does so by excluding or replacing water. Thus, if there is preferential binding of one component (water or osmolyte) with the protein, there will necessarily be a preferential exclusion of the other component. In short, the measured preferential interaction parameter can have either a positive or a negative sign representing, respectively, osmolyte preferential binding or preferential exclusion.

The wealth of protein–osmolyte preferential interaction results has been reported by Timasheff and colleagues (Arakawa and Timasheff, 1982, 1983, 1985; Gekko and Timasheff, 1981; Lee and Timasheff, 1981; Timasheff and Xie, 2003; Xie and Timasheff, 1997a,b). More recently, Record and colleagues have reported such measurements (Courtenay et al., 2000; Felitsky et al., 2004; Hong et al., 2004; Zhang et al., 1996). For many proteins these measurements show that protecting osmolytes are excluded preferentially from both native and unfolded states of proteins, whereas the nonprotecting osmolyte, urea, binds these protein species preferentially (Felitsky and Record, 2004; Felitsky et al., 2004; Timasheff, 1992a,b). The preferential interactions appear to be linked to the stabilizing and destabilizing effects that protecting and nonprotecting osmolytes have on proteins. While urea is a well-known protein denaturant, we and others demonstrated that protecting osmolytes have the ability to force intrinsically unstructured proteins to cooperatively fold to native-like protein species that have functional activity (Baskakov and Bolen, 1998a; Gursky, 1999; Henkels and Oas, 2005; Kumar et al., 2001; Mello and Barrick, 2003; Rajagopalan et al., 2005; Wu and Bolen, 2006). The implication is that preferential interactions provide a driving force for folding, with preferential exclusion of protecting osmolytes promoting folding and the preferential binding of urea promoting denaturation. These observations, however, do not provide

mechanistic detail of osmolyte effects on protein stability, they only show
that stability and protein solvation are linked phenomena.

Using preferential interaction parameters determined for native and
denatured protein exposure to the nonprotecting osmolyte urea and to a
protecting osmolyte such as trimethylamine N-oxide (TMAO), Timasheff
(1992a,b) compared how the two types of osmolytes shift the native (N) to
denatured (D) equilibrium ($N \rightleftharpoons D$) in thermodynamic terms (see
Fig. 23.1). Transfer of the native state in water to urea solution lowers the
free energy of the native state because of a net favorable interaction of the
urea with native state surface-exposed groups. The free energy change is
even more favorable for transfer of the denatured state from water to urea
solution because of the larger number of denatured state surface-exposed
groups binding preferentially to the urea. These changes result in a smaller
free energy gap (ΔG) between N and D states in urea compared to that in
water (see Fig. 23.1); the origin of the smaller ΔG illustrates how the
destabilizing effect of urea on proteins arises. In contrast to urea, transfer
of the native and denatured states to protecting osmolyte solution is unfavorable because of net preferential exclusion of the osmolyte at the protein

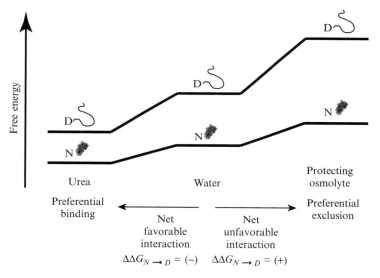

Figure 23.1 The change in chemical potential (free energy) of native and denatured
states when a protecting osmolyte (e.g., sarcosine) or a nonprotecting osmolyte (urea) is
added. The resulting preferential exclusion of a protecting osmolyte from the protein
surface and stabilization of the $N \rightarrow D$ transition is driven by a net unfavorable interaction between the osmolyte and the native and denatured states (right), resulting in a ΔG
increase. The preferential binding of urea to N and D destabilizes $N \rightarrow D$ the transition
(ΔG decreases) and results from a net favorable interaction between urea and the native
and denatured states (left).

surface. Because of greater surface group exposure in the denatured state, the denatured state is more destabilized (i.e., of higher free energy) in the presence of the protecting osmolyte than the native state. This results in greater protein stabilization in the protecting osmolyte than occurs in water (i.e., ΔG is increased in osmolyte solution, see Fig. 23.1).

To explore the relationships between preferential interaction and the effects of osmolytes on protein stability, it is useful to consider which groups exposed on the surface of native and denatured proteins either bind or exclude the osmolyte of interest. Quantitatively, groups that bind osmolyte preferentially (*solvophilic* groups) are characterized by a group transfer-free energy (GTFE) change of negative sign, whereas GTFEs are positive for groups that exclude osmolyte preferentially (*solvophobic* groups). GTFEs are discussed in detail later.

Preferential interaction parameters can be used to experimentally obtain the free energy changes for transfer of the entire native or denatured protein from water to osmolyte solution. An alternative means of obtaining the transfer-free energy of the entire native state from water to osmolyte is to identify the solvent-exposed side chain and backbone groups on the surface of the native state and sum their individual GTFE contributions. The same process can be used to evaluate the water-to-osmolyte transfer-free energy of the entire denatured state. This approach, the transfer-free energy model, was proposed by Tanford in the mid-1960s for use in investigating denaturation by urea, but for several reasons was never fully implemented (Auton and Bolen, 2004). This chapter expands on the model, demonstrating what it takes to make it quantitative and predictive of the effects protecting and nonprotecting osmolytes have on protein stability.

 ## 3. THE TRANSFER-FREE ENERGY MODEL

The premise of the transfer model is that the energetics of osmolyte effects on protein stability can be determined from summing the free energy contributions of the solvophobic and solvophilic groups exposed on the native and denatured states of the protein. A major assumption is that the energetic contributions of the groups are additive. The validity of this assumption is assessed by how well the model predicts thermodynamic properties of the transition. Knowing the GTFEs of all groups on the protein is essential for testing the model and developing an approach that, in principle, offers a residue-level accounting of the effects of osmolytes on protein structure and stability.

As early as the 1930s, Edwin Cohn developed a way to evaluate interactions of protein groups and solute molecules (Cohn and Edsall, 1943). His work framed protein folding in the presence of solutes in terms of the free energy changes that accompany the transfer of amino acid side chains and peptide backbone units from water to various solute-containing solutions. These early ideas were ultimately consolidated into Tanford's transfer model proposing quantification of the thermodynamics of protein–solvent interactions by summing the GTFEs of the constituent solvent-exposed backbone and side chain parts of N and D (Tanford, 1964, 1970)[Eq. (23.1)]:

$$\sum_i n_i \cdot \Delta\alpha_i \cdot \Delta g^o_{tr,i} \tag{23.1}$$

In Eq. (23.1), $\Delta\alpha_i$ is the average fractional change in solvent-accessible surface area of groups of type i upon converting N to D, $\Delta g^o_{tr,i}$ is the free energy of transfer of a backbone unit or side chain group of type i from water to the concentration of denaturant specified, and n_i is the total number of groups of type i present in the protein. Figure 23.2 illustrates the transfer model in terms of a thermodynamic cycle that provides a strategy for evaluating the overall unfolding transition, $N \rightleftharpoons D$, in terms of the residue-specific side chain and backbone groups that become exposed on unfolding. The free energy change of the $N \rightleftharpoons D$ equilibrium at the top of Fig. 23.2, occurring in water alone, is given by $\Delta G^o_{N \rightarrow D}$, while the corresponding equilibrium below it occurs in the presence of 1 M osmolyte. The free energy difference, $\Delta\Delta G = \Delta G^{1M}_{N \rightarrow D} - \Delta G^o_{N \rightarrow D}$ is obtained readily from the experiment by analyzing data from either denaturation of proteins by urea or forced folding of intrinsically unstructured proteins by protecting osmolytes, using the linear extrapolation method (Bolen and Santoro, 1988; Greene and Pace, 1974; Santoro and Bolen, 1988). The aforementioned $\Delta\Delta G$ is equal to the slope (m-value) of linear extrapolation, which is a measure of the osmolyte-dependent cooperativity in either forcing a protein to fold using a protecting osmolyte or forcing it to unfold using the

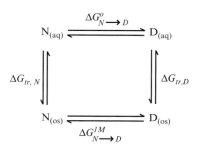

Figure 23.2 Thermodynamic cycle of the transfer model. Horizontal equilibria represent denaturations in the absence (top) and presence (bottom) of 1 M osmolyte. Vertical equilibria represent the transfer from water to 1 M osmolyte of native (left) and denatured (right) states. The relationship $\Delta G^{1M}_{N \rightarrow D} - \Delta G^o_{N \rightarrow D} = \Delta G_{tr,D} - \Delta G_{tr,N}$ is a thermodynamic consequence of the cycle.

nonprotecting osmolyte, urea. The vertical reactions in Fig. 23.2 represent the free energy change of transferring the native and denatured states from water to 1 M osmolyte. The transfer-free energies of N and D complete the thermodynamic cycle, resulting in the relationship $\Delta G_{N \to D}^{1M} - \Delta G_{N \to D}^{o} = \Delta G_{tr,D} - \Delta G_{tr,N}$. Thus, the m value obtained from experiment must also equal $\Delta G_{tr,D} - \Delta G_{tr,N}$, obtained from calculation; this equality provides a means to test the transfer model, including the degree of GTFE additivity [Eq. (23.2)]:

$$m_{\exp} = \Delta G_{tr,D} - \Delta G_{tr,N} = \sum_{i} n_i \cdot \Delta \alpha_i \cdot \Delta g_{tr,i}^{o} \qquad (23.2)$$

Calculating the transfer-free energy difference between D and N (the m value) requires knowledge of the number of groups of type i (n_i), the GTFEs for individual groups ($\Delta g_{tr,i}^{o}$), and the difference between denatured and native state solvent accessibility of each group i in the protein ($\Delta \alpha_i$). What follows are the methods developed to determine these quantities, along with application of the transfer model to predict the energetics of osmolyte-induced protein folding and unfolding.

4. HOW TRANSFER-FREE ENERGIES ARE MEASURED

Following the work of Brønsted and Noyes on solubility, and thermodynamic considerations involving the solubility limits of a solute in two solvents, Lewis and Randall noted the principle that at the solubility limit, the chemical potentials of a solute in the crystal and in the saturated solution are equal (Brønsted, 1920; Lewis and Randall, 1923; Noyes and Bray, 1911a,b). This principle, illustrated in Fig. 23.3, resulted in the ability to evaluate the transfer-free energy of a solute from one solvent to another. At and above the solubility limit, the solute in the crystalline phase is in equilibrium with the solute in the solution phase, and the chemical potentials of the solute in the solution crystal phases are equal. The relationship is expressed in Eq. (23.3a), where $a_{i,w}$ and $a_{i,os}$ represent the chemical activities of the compound in water and osmolyte solution. Taking the traditional approach, where $a_i = s_i \gamma_i$ (explained later), the equation is further expanded as in Eq. (23.3b):

$$\mu_{i,w}^{o} + RT \ln a_{i,w} = \mu_{i,os}^{o} + RT \ln a_{i,os} \qquad (23.3a)$$

$$\mu_{i,w}^{o} + RT \ln s_{i,w} + RT \ln \gamma_{i,w} = \mu_{i,os}^{o} + RT \ln s_{i,os} + RT \ln \gamma_{i,os}$$

$$(23.3b)$$

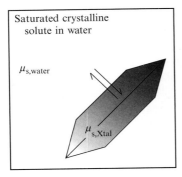 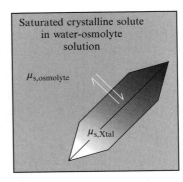

$$\mu_{s,water} = \mu_{s,Xtal} = \mu_{s,osmolyte}$$

Figure 23.3 Chemical potential of the solute at the solubility limit in water (left) and in 1 M osmolyte (right). At saturation, the crystalline solute is in equilibrium with the dissolved solute and the chemical potential of the solute in the crystalline phase is equal to its chemical potential in solution. Because the chemical potential of the crystalline solute is independent of the solution with which it is in contact, the chemical potentials of the solute dissolved in water and in 1 M osmolyte must also be equal. This equality provides the basis for determining transfer-free energies of a solute from one aqueous solution to another.

For the transfer of compound i from water to 1 M osmolyte, $\mu^o_{i,w}$ and $\mu^o_{i,os}$ are the standard chemical potentials assigned to the solute in water and in 1 M osmolyte, $s_{i,w}$ and $s_{i,os}$ are the solubilities of the compound in water and in 1 M osmolyte expressed in molar, molal, or mole fraction units, and $\gamma_{i,w}$ and $\gamma_{i,os}$ represent the activity coefficients of the corresponding concentration scales. The transfer-free energy of the solute from water to 1 M osmolyte is equal to the difference between the standard state chemical potentials of the solute in each solution and gives the preference of the compound for one solution over the other [Eq. (23.4)]:

$$\Delta\mu^o = \mu^o_{i,os} - \mu^o_{i,w} = RT\ln\left(s_{i,w}/s_{i,os}\right) + RT\ln\left(\gamma_{i,w}/\gamma_{i,os}\right) \quad (23.4)$$

Traditionally, the term containing activity coefficients has been ignored because of the difficulty in obtaining these values for three component solutions. While it is possible for this term to make a significant contribution to the total transfer chemical potential, there are conditions (discussed later) in which its contribution becomes negligible. Ignoring activity coefficient contributions leaves the term containing the solubilities, thus defining an apparent transfer-free energy [Eq. (23.5)]:

$$\Delta G^o_{app} = RT\ln\left(s_{i,w}/s_{i,os}\right) \quad (23.5)$$

When the solubility of the compound in 1 M osmolyte is greater than in water, the solubility ratio is <1, ΔG^o_{app} is negative, and the interaction between compound and osmolyte solutions is said to be favorable (solvophilic). Conversely, when the solubility of the compound in 1 M osmolyte is less than in water, the resulting positive defines an unfavorable (solvophobic) interaction between compound and osmolyte solutions. These solvation preferences are illustrated in Fig. 23.4, which shows three experimental data sets for measurement of the solubility of cyclic glycylglycine (CGG, a peptide backbone unit model compound) in water, 1 M urea, and 1 M TMAO. The densities of supernatant solutions are plotted as a function of the total composition of CGG, and the intersection of the fitted lines for the unsaturated solutions and the saturated solutions defines the composition of CGG at its solubility limit in each solvent system. Knowledge of the density at the solubility limit allows the composition at the solubility limit to

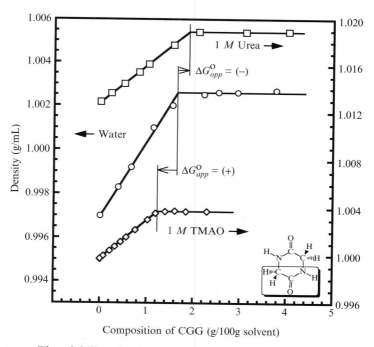

Figure 23.4 The solubility of cyclic glycylglycine in water (○), in 1 M TMAO (◇), and in 1 M urea (□) determined from the density of supernatant solutions as a function of the composition of CGG in water, 1 M TMAO, and 1 M urea. Solubility limits of CGG (vertical lines) are determined from the intersection of the linear fits through unsaturated and saturated data points. An increase in solubility relative to water corresponds to a favorable solvophilic interaction, $\Delta G^0_{app} = (-)$, and a decrease corresponds to an unfavorable solvophobic interaction, $\Delta G^0_{app} = (+)$, between CGG and the osmolyte.

be converted to the conventional units of molarity, molality, and mole fraction commonly used in expressing transfer free energies.

5. The Issue of Activity Coefficients: Solution Nonideality at the Solubility Limit

The primary use of the concept of activity and activity coefficients is to provide a means to account for the dependence of free energy on solution composition (Moelwyn–Hughes, 1961). The thermodynamic activity of a compound in solution is given by $a_i = s_i \gamma_i$, where s_i is the concentration of the compound evaluated in units of molar, molal, or mole fraction and γ_i is the activity coefficient of the solute in the corresponding concentration scale. In dilute real solutions, Henry's law applies, γ_i is unity, and the solution is said to behave in an ideal manner. As the concentration of the compound increases, activity begins to deviate from Henry's law and the activity coefficient represents the magnitude of the deviation (Silbey and Alberty, 2001). At high concentrations of the compound, the activity coefficient can deviate significantly from unity, with the nonideality caused by compound–compound and/or compound–solvent interactions. For the three-component system of a solute in an aqueous osmolyte solution, there are six possible pairwise interactions. These types of component interactions in solution can give rise to thermodynamic nonideality in which the activity of the compound accounts for its concentration effectively being either lower or greater than the concentration expected from Henry's law.

When the compound is fairly soluble its activity coefficient at the solubility limit may make a substantial contribution to the chemical potential of the compound. It is important to note, however, that with respect to transfer-free energies, it is the *ratio* of activity coefficients of the compound in water and 1 M osmolyte at the solubility limit that is important, not the individual activity coefficients themselves [see Eq. (23.4)]. Also, because the solubility of the compound is not very different in water and osmolyte solution, its activity coefficients in water and osmolyte solutions are quite similar. Thus, the ratios of the activity coefficients are found to be close to unity, even when the individual activity coefficients, $\gamma_{i,w}$ and $\gamma_{i,os}$, deviate from unity. By way of example, at the molal solubility limits of glycine in water and 1 M urea, the ratio $(\gamma_{i,w}/\gamma_{i,urea})$ has a value of 0.974, with $\gamma_{i,w} = 0.731$ and $\gamma_{i,urea} = 0.751$. It is reasonable then to expect that for amino acids less soluble than glycine, the ratios of their activity coefficients approach unity and the free energy contributions of this ratio to the transfer chemical potential, $\Delta\mu^o$, approach zero, [Eq. (23.6)]:

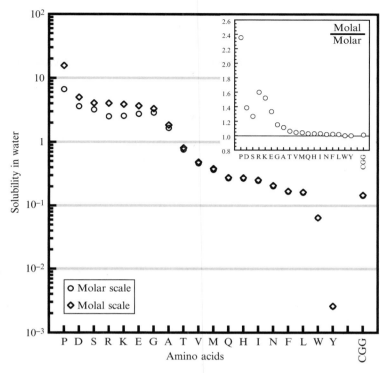

Figure 23.5 Water solubilities of amino acids in water in molar (○) and molal (◊) concentration scales are ranked from greatest to least according to solubility. The solubility of the peptide backbone model compound, cyclic glycylclycine, is included on the far right. (Inset) The ratio of molal solubility over the molar solubility as an index for solution nonideality at the solubility limit.

$$\lim_{s_i \to 0} \left(RT \ln \left(\gamma_{i,w} / \gamma_{i,os} \right) \right) = 0 \qquad (23.6)$$

To illustrate these effects more explicitly, Fig. 23.5 shows the water solubilities of all amino acids and the peptide backbone model compound, cyclic glycylglycine, evaluated on molar and molal scales ranked (left to right) from the greatest to the least soluble. This plot shows that as the solubility of compounds decrease, the molar and molal solubilities become approximately equal and the ratio of the molal to molar solubility approaches unity (inset of Fig. 23.5). The ratio of the molal to molar solubilities of the amino acids provides an indication of amino acid solubilities that are low enough to be representative of an ideal solution such that the limit in Eq. (23.6) is attained. When this criterion is used, we find in Fig. 23.5 that amino acids less soluble than valine are within this solubility regime and that the activity coefficient ratio approaches unity. The peptide backbone model compound, CGG, and more than half of the amino acids, including all the nonpolar residue side chains, fall within this regime. Amino acids with solubilities greater than valine, consisting mostly of polar and charged residues, are expected to have increasingly significant activity coefficient contributions due to nonideality at a high solubility limit. While using solubility as a criterion for deviation from ideality is approximate, it does offer the ability to identify which amino acids will have apparent transfer–free energies that may differ significantly from the transfer chemical potential defined by Eq. (23.3a) and (23.3b). Referring to the inset of Fig. 23.5 and the question of the contribution of the activity coefficient ratio in Eq. (23.3b), an indication of how rapidly the activity coefficient ratio diverges from unity is given by glycine, an amino acid that is significantly more soluble than valine, but with an activity coefficient ratio of 0.974.

6. ADDITIVE, CONCENTRATION SCALE-INDEPENDENT AND MODEL COMPOUND-INDEPENDENT TRANSFER-FREE ENERGIES OF THE PEPTIDE BACKBONE UNIT

Several issues have confounded the application of the transfer model over the years, and only recently have these issues been resolved (Auton and Bolen, 2004). Two of the major issues involve how to eliminate the consequences of choosing a concentration scale to conduct transfer experiments and how to obviate the need for determining activity coefficient ratios called for in the transfer-free energy equations. This section first deals with solving the concentration scale problem and then turns to the issue of activity coefficient ratios.

In determining a transfer-free energy, it is necessary to choose a concentration scale on which to base the evaluation of the quantity. The problem is that for the same compound, different concentration scales can give transfer-free energies that differ in sign (Arnett and McKelvey, 1969). Thermodynamically, this situation is untenable for it means that the question of whether a compound interacts favorably or unfavorably with the osmolyte is dependent on the choice of the concentration scale! To resolve this situation it is first recognized that the molar (c), molal (m), and mole fraction (x) scales can each be expressed as a mole ratio of the compound's solubilities in water and osmolyte [first term on right of Eq. (23.7a), (23.7b), and (23.7c)], plus a term resulting from the definition of the scale (second terms on right in these equations). Here, $\Delta G^{o}_{app,c}$, $\Delta G^{o}_{app,m}$, and $\Delta G^{o}_{app,x}$ are the apparent transfer-free energies using the molar, molal, and mole fraction scales; $n_{i,w}$ and $n_{i,os}$ are the numbers of moles of compound i soluble in water and osmolyte solution; $V_{S,os}$ and $V_{S,w}$ are the total volumes of aqueous solution and osmolyte solution containing the saturating compound i on the molar scale; wt_{os} and wt_w are the total mass of osmolyte and water on the molal scale; and for the mole fraction scale n_w and $n_{w,os}$ are, respectively, the moles of water in the aqueous and osmolyte solutions each containing the saturating compound, while n_{os} is the moles of osmolyte in the osmolyte solution.

$$\Delta G^o_{app,c} = RT \ln \left(n_{i,w}/n_{i,os} \right) + RT \ln \left(V_{S,os}/V_{S,w} \right) \qquad (23.7a)$$
$$\Delta G^o_{app,m} = RT \ln \left(n_{i,w}/n_{i,os} \right) + RT \ln \left(wt_{os}/wt_w \right) \qquad (23.7b)$$
$$\Delta G^o_{app,x} = RT \ln \left(n_{i,w}/n_{i,os} \right) + RT \ln \left(\left(n_{w,os} + n_{os} + n_{i,os} \right) / \left(n_w + n_{i,w} \right) \right)$$
$$(23.7c)$$

Equation (23.7a), (23.7b), and (23.7c) are general and applicable regardless of the solubility of compound i, but these equations become most useful when compound i is relatively insoluble and limiting forms of the expressions apply. By relatively insoluble, we mean that compound i is so dilute in water and osmolyte solution that molality becomes equal to molarity. Examples of compounds with solubilities in which the molality is equal to the molarity are shown in Fig. 23.5. For convenience, in our experimental approach we chose to conduct the solubility experiments on the basis of 100 g of solvent, regardless of whether the solvent is water or osmolyte solution. Thus, in Eq. (23.7b), $wt_{os}/wt_w = 1$. For compounds of such low solubility, Eq. (23.7a), (23.7b), and (23.7c) approach limits as given in Eq. (23.8a), (23.8b), and (23.8c), where the first term on the right corresponds to the intrinsic transfer-free energy of the compound, $\Delta G^o_{int} = RT \ln(n_{i,w}/n_{i,os})$. Also because $n_{i,w}$ and $n_{i,os}$ are very small relative to n_w or n_{os}, the second terms on the right reduce to the ratio of the densities of pure water, (d_w) and osmolyte solution (d_{os}) on the molar scale, and the ratio of the total number of moles of water and osmolyte in 1 M osmolyte and pure water on the mole fraction scale as shown in Eq. (23.8a), (23.8b), and (23.8c):

$$\Delta G^o_{app,c} = RT \ln \left(n_{i,w}/n_{i,os} \right) + RT \ln \left(d_w/d_{os} \right) \qquad (23.8a)$$
$$\Delta G^o_{app,m} = RT \ln \left(n_{i,w}/n_{i,os} \right) + RT \ln (1) \qquad (23.8b)$$
$$\Delta G^o_{app,x} = RT \ln \left(n_{i,w}/n_{i,os} \right) + RT \ln \left(\left(n_{w,os} + n_{os} \right)/n_w \right) \qquad (23.8c)$$

Equation (23.8a), (23.8b), and (23.8c) show that scale-dependent terms are devoid of contributions from compound i and that the second terms on the right are constants for all compounds i of low solubility. Because these terms are constants for all compounds of this solubility class, it is easy to convert from one concentration scale to the other. Accordingly, one can readily evaluate an intrinsic concentration scale-independent transfer-free energy change from the first term on the right side of the equations. The aforementioned treatment shows that by restricting transfer-free energy studies to compounds with solubilities that are sufficiently low, Eq. (23.8a), (23.8b), and (23.8c) apply and the mole ratio, and thus the transfer-free energy, becomes independent of the choice of concentration scale.

It is clear that in selecting models of the peptide backbone for transfer-free energy experiments low solubility is essential. In addition, to evaluate the GTFE for the peptide backbone unit, we select N-acetyl glycine amide peptides of increasing chain length on which to perform the transfer-free energy measurements. With such model compounds we need to ascertain that the solubility is low enough so that concentration scales are not a problem, and also to be assured that the activity coefficient ratios for the compounds of interest are unity.

To be sure, the transfer-free energy quantity of interest, ΔG_{int}^{o}, has associated with it an activity coefficient ratio term [not shown in Eqs. (23.7) and (23.8)]. With the model N-acetyl glycine$_z$ amide peptides, solubility will decrease with increasing chain length (z) and ultimately achieve a solubility range where concentration scale problems are not an issue and the activity coefficient ratios are unity. That is, we expect ΔG_{int}^{o} of the model peptides to reach a point in which ΔG_{int}^{o} changes are due only to the increase in peptide backbone unit length.

Figure 23.6 illustrates this principle for the N-acetyl glycine$_z$ amide linear chain peptides upon transfer from water to 1 M TMAO, 1 M sarcosine, and 1 M urea. Also shown for comparison are results for the peptide model compound, cyclic glycylglycine (CGG). In Fig 23.6A, the solubility of the linear glycine-based peptides in water evaluated on each concentration scale (open symbols) decreases logarithmically as a function of chain length. As seen, the solubility of the other peptide backbone model, CGG (filled symbols), is low and less than that for a linear glycine peptide of 3 internal units in length. In Fig. 23.6B, the intrinsic transfer-free energy is observed to be quite variable among concentration scales for short linear chain peptides, but at chain lengths ≥ 2 internal units, ΔG_{int}^{o}, regardless of concentration scale, changes by a constant factor with each glycine addition in chain length. The slope of the intrinsic transfer-free energy as a function of chain length for 2, 3, and 4 internal units defines a constant increment that represents the transfer-free energy per peptide backbone unit. In Fig. 23.6C, the solubility of CGG is also low enough that the intrinsic transfer-free energy evaluated on each concentration scale is constant.

Table 23.1 gives the peptide backbone transfer-free energy determined by these methods for transfer from water to 1 M concentration of TMAO, sarcosine, and urea (Auton and Bolen, 2004). Comparison of the constant

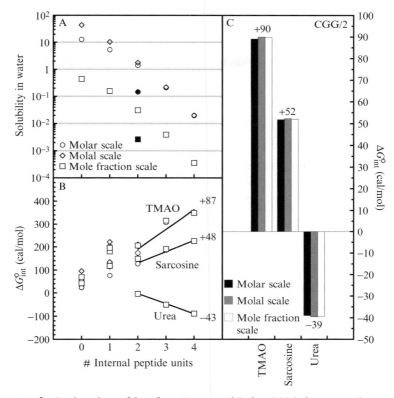

Figure 23.6 Reduced set of data from Auton and Bolen (2004) demonstrating peptide backbone additivity and concentration scale independence of the peptide unit transfer-free energy from water to $1\,M$ TMAO, $1\,M$ sarcosine, and $1\,M$ urea. (A) Solubility of the linear N-acetyl glycine$_z$ amide peptides (open symbols) as a function of the number of internal peptide units ($z = 0$ to 4) and the cyclic glycylglycine (CGG) peptide (filled symbols) at 2 peptide units in molar ($\bigcirc$), molal ($\Diamond$), and mole fraction ($\square$) concentration scales. (B) The constant increment method for determining the peptide unit transfer-free energy from water to $1\,M$ TMAO, $1\,M$ sarcosine, and $1\,M$ urea in which the intrinsic transfer-free energy on molar ($\bigcirc$), molal ($\Diamond$), and mole fraction ($\square$) concentration scales is plotted as a function of the number of internal peptide units in the linear N-acetyl glycine$_z$ amide series of peptides. For $z = 2$, all scales coincide, and the slope of the constant increment represents the transfer-free energy of a single peptide unit (see right ordinate). (C) The intrinsic transfer-free energy of the peptide unit on molar (black), molal (gray), and mole fraction (white) concentration scales defined by dividing the intrinsic transfer-free energy of CGG by two.

increment of the linear glycine–based peptides with the transfer–free energy of the peptide unit defined by CGG shows excellent agreement within experimental errors and demonstrates that not only is the peptide unit transfer–free energy additive (as required by the transfer model), it is

Table 23.1 The molar intrinsic transfer free energy, (cal/mol), of the peptide unit defined by cyclic glycylglycine and by the constant increment method

1M Osmolyte	CGG/2	Constant Increment
TMAO	90	87
Sarcosine	52	48
Urea	–39	–43

Data from Auton and Bolen, 2004.

independent of both the concentration scale and the model compound used to represent the peptide unit (Auton and Bolen, 2004). Success of these methods is a direct consequence of the low solubility of the model compounds.

7. THE TRANSFER-FREE ENERGY OF AMINO ACID SIDE CHAINS

Transfer-free energy values for the 20 common amino acids from water to 1 M osmolyte can be evaluated from solubility measurements such as those shown in Fig. 23.4. From the tabulated values of $\Delta G^o_{tr,i}$, the transfer-free energy of the side chains $(\Delta G^o_{tr,i,SC})$ has traditionally been estimated (Tanford) by subtracting the transfer-free energy of glycine from all other amino acids [Eq. (23.9)]:

$$\Delta G^o_{tr,i,SC} = \Delta G^o_{tr,i} - \Delta G^o_{tr,i,Gly} = RT \ln \left(s_{i,w}/s_{i,os} \right) \\ - RT \ln \left(s_{Gly,w}/s_{Gly,os} \right) \qquad (23.9)$$

It is clear from the previous discussion of the need to use model compounds of low solubility, that many values of $\Delta G^o_{tr,i,SC}$ from tables of side chain transfer-free energies from water to 1 M osmolyte solutions will not meet criteria permitting activity coefficient ratios and concentration scale dependencies to be ignored. Figure 23.5 suggests that amino acids with the solubility of valine or lower likely meet these criteria. Side chains that fall within the acceptable criteria are hydrophobic and comprise the class of side chains that are normally buried in the native state. Those residues that are more soluble than valine, polar and charged side chains, are generally solvent exposed on the native state. It should be noted that the m value for denaturation arises from groups (side chains and backbone units) that become newly exposed on denaturation or newly buried in forcing a protein to fold—those amino acids with solubilities $\leq$ valine. Thus, use of the tables of side chain and backbone transfer-free energies in calculating m values should be reasonably successful.

8. SOLVENT-ACCESSIBLE SURFACE AREAS OF NATIVE AND DENATURED STATES: CALCULATING $\Delta\alpha_i$

To evaluate the contributions of all amino acid side chains and the peptide backbone to the stability of a protein, the transfer-free energies determined from solubility measurements must be scaled by the amount each residue type becomes exposed on unfolding. $\Delta\alpha_i$ is the average fractional change in solvent accessibility of each type of side chain ($^{\Delta}\alpha_i^{SC}$) and the peptide backbone ($^{\Delta}\alpha_i^{BB}$) in going from $N \to D$ [Eq. (23.10)]:

$$\Delta\alpha_i^{\text{BB or SC}} = \frac{\sum_{j=1}^{n_i} \left(ASA_{i,j,D} - ASA_{i,j,N} \right)}{n_i ASA_{i,Gly \cdot X \cdot Gly}} \qquad (23.10)$$

The side chain and backbone solvent-accessible surface area of residues in the native state, $ASA_{i,j,N}$, are calculated from crystal coordinate data obtained from the Protein Data Bank (Berman $et\ al.$, 2000) using the static-accessible surface area algorithm of Lee and Richards as modified by Lesser and Rose with a probe size equal to the hard sphere radius of water, 1.4 Å (Lee and Richards, 1971; Lesser and Rose, 1990; Richards, 1977). Denatured state model-dependent average solvent-accessible surface areas, $ASA_{i,j,D}$, were obtained for individual side chains and backbone from Table 1 of Creamer and colleagues (1997). The table gives ASA values for two denatured state models, one which is equivalent to a random coil in a good solvent (upper bound) and another that models a compact denatured state (lower bound). The denatured state model used here was suggested by Schellman (2003) and takes the mean of ASA for the side chains and backbone units between the upper and the lower bound models. It turns out that Schellman's model is identical in ASA to Goldenberg's model of a self-avoiding random coil (Goldenberg, 2003). Because the side chains and backbone units of the model compounds used in determining transfer-free energies are highly solvent exposed, the ASA of side chain or backbone of residue X in the Gly-X-Gly tripeptide, $ASA_{i,Gly \cdot X \cdot Gly}$, is used in the denominator of Eq. (23.10) as the standard side chain or backbone-solvent accessibility.

9. PUTTING IT ALL TOGETHER: PREDICTING m VALUES FOR THE ENERGETICS OF OSMOLYTE-INDUCED PROTEIN FOLDING/UNFOLDING

From tabulated values of transfer-free energies of the peptide backbone unit and amino acid side chains and calculated from structural coordinates of a protein it is possible to calculate an m value for either urea denaturation or osmolyte-induced forced folding of proteins. Equation (23.11) is the final functional form, combined from Eqs. (23.2) and (23.10) and split into side chain and backbone contributions:

$$m_{calc} = \Delta G_{tr,D}^{1M} - \Delta G_{tr,N}^{1M} = \sum_{i=AA\,type} n_i \Delta g_{tr,i,SC}^{o} \Delta \alpha_i^{SC}$$
$$+ \Delta g_{tr,BB}^{o} \sum_{i=AAtype} n_i \Delta \alpha_i^{BB} \qquad (23.11)$$

Figure 23.7 shows calculated m values versus corresponding m values determined experimentally using the linear extrapolation method for three

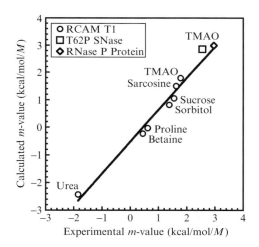

Figure 23.7 Calculated m values versus m values determined experimentally. Reproduced with permission from Auton and Bolen (2005) ©2005 National Academy of Sciences, U.S.A. Calculated m values for folding/unfolding of RCAM T1 (○) as induced by osmolytes (Bolen and Baskakov, 2001) indicated are shown versus m values determined experimentally by means of the linear extrapolation method. Also included are calculated m values versus experimental TMAO-induced m values determined by Henkels and Oas (2005) for P-protein (◇) and by Baskakov and Bolen (1998b) for T62P SNase (□). The slope of the line shown is 1.15 ± 0.07.

proteins that are intrinsically unstructured in aqueous buffer at 25°: reduced and carboxyamidated RNase T1 (RCAM-T1) (Baskakov and Bolen, 1998a; Bolen and Baskakov, 2001), the T62P mutant of staphylococcal nuclease (SNase) (Baskakov and Bolen, 1998b), and the protein component of RNase P (P-protein) (Henkels and Oas, 2005). In the absence of salt and/ or divalent anions, the P-protein is by nature intrinsically unstructured. In contrast, RCAM-T1 and T62P SNase are derived from thermodynamically stable proteins and have been made to be intrinsically unstructured by reducing and blocking the two disulfide bonds in RNase T1 and by replacing threonine with proline in the α helix of SNase. As the temperature decreases below 25°, RCAM-T1 becomes increasingly stable and urea-induced unfolding m values can be determined. The m value for urea represents the quantity determined from urea-induced unfolding at low temperature (5–15°) and extended to 25° (Baskakov and Bolen, 1999).

Figure 23.7 shows a strong correlation between calculated and observed m values over a range of ≈ 5 kcal/mol M^{-1} (Auton and Bolen, 2005). The ability to predict the sign and magnitude of the m value in the presence of seven different osmolytes for three proteins indicates that the assumption of group additivity is valid. The slope of the fit through data is 1.15 ± 0.07, which is not much different from unity. Any deviation of data from unity slope could be attributed to contributions from undetermined activity coefficients, using an ASA model of the denatured state inappropriate for that protein, or other effects. Despite unknown activity coefficients, the apparent transfer-free energies obtained from solubility measurements are shown to successfully predict the m value, a key molecular quantity measuring the cooperativity of the transition from $N \rightarrow D$ or from $D \rightarrow N$.

10. SUMMARY

Since its inception, the transfer model, as proposed by Tanford, has provided a unique way of describing the effects of denaturants and protecting osmolytes on protein stability in terms of the residue-specific contributions of peptide backbone units and amino acid side chains to the cooperativity of the protein folding/unfolding transition. However, the quantitative power of this method was never fully realized because of issues involving solubility, model compound and concentration scale dependence of transfer-free energies, unknown contributions of activity coefficients at the solubility limit, and lack of additivity. As discussed earlier, we have learned how to overcome these unresolved issues and have refined the transfer model so that it is quantitatively predictive of the m values that describe the cooperativity of osmolyte-induced protein folding/unfolding. Nonadditivity and the model compound and concentration scale dependence of GTFEs are all manifestations of high solubility and solution nonideality at the solubility limit. With only limited data available on activity coefficients of three component solutions, it is imperative to use model compounds of low solubility to represent the GTFEs so that the free energy contributions of nonideality are minimal. As it stands, the transfer model is predictive of the osmolyte-induced protein folding/unfolding energetics, but it still has room for improvement. Methods of determining activity coefficients of amino acids in water and osmolyte solutions and of developing improved models of the denatured state solvent accessibility are currently underway to further refine the transfer model in describing the major structural transitions of proteins.

ACKNOWLEDGMENTS

We thank Drs. Jörg Rösgen and Luis Holthauzen for their comments on the manuscript. This work was supported by NIH Grant GM49760 and Welch Foundation Grant H-144.

REFERENCES

Arakawa, T., and Timasheff, S. (1982). Stabilization of protein structure by sugars. *Biochemistry* **21**, 6536–6544.

Arakawa, T., and Timasheff, S. (1983). Preferential interactions of proteins with solvent components in aqueous amino acid solutions. *Arch. Biochem. Biophys.* **224**, 169–177.

Arakawa, T., and Timasheff, S. (1985). The stabilization of proteins by osmolytes. *Biophys. J.* **47**, 411–414.

Arnett, E., and McKelvey, D. R. (1969). Chapter 6. *In* "Solute-Solvent Interactions" (J. F. Coetzee and C. D. Ritchey, eds.). Dekker, New York.

Auton, M., and Bolen, D. (2004). Additive transfer free energies of the peptide backbone unit that are independent of the model compound and the choice of concentration scale. *Biochemistry* **43**, 1329–1342.

Auton, M., and Bolen, D. (2005). Predicting the energetics of osmolyte-induced protein folding/unfolding. *Proc. Natl. Acad. Sci. USA* **102**, 15065–15068.

Baskakov, I., and Bolen, D. (1998a). Forcing thermodynamically unfolded proteins to fold. *J. Biol. Chem.* **273**, 4831–4834.

Baskakov, I., and Bolen, D. (1998b). Monitoring the sizes of denatured ensembles of staphylococcal nuclease proteins: Implications regarding m values, intermediates, and thermodynamics. *Biochemistry* **37**, 18010–18017.

Baskakov, I., and Bolen, D. (1999). The paradox between m values and deltaCp's for denaturation of ribonuclease T1 with disulfide bonds intact and broken. *Protein Sci.* **8**, 1314–1319.

Berman, H., Westbrook, J., Feng, Z., Gilliland, G., Bhat, T., Weissig, H., Shindyalov, I., and Bourne, P. (2000). The Protein Data Bank. *Nucleic Acids Res.* **28**, 235–242.

Bolen, D., and Baskakov, I. (2001). The osmophobic effect: Natural selection of a thermodynamic force in protein folding. *J. Mol. Biol.* **310**, 955–963.

Bolen, D., and Santoro, M. (1988). Unfolding free energy changes determined by the linear extrapolation method. 2. Incorporation of delta G degrees N-U values in a thermodynamic cycle. *Biochemistry* **27**, 8069–8074.

Brønsted, J. (1920). Studies on solubility. 1. The solubility of salts in salt solutions. *J. Am. Chem. Soc.* **42**, 761–786.

Cohn, E., and Edsall, J. (1943). "Proteins, Amino Acids, and Peptides as Ions and Dipolar Ions." ACS Monograph Series No. 90.

Courtenay, E., Capp, M., Anderson, C., and Record, M. J. (2000). Vapor pressure osmometry studies of osmolyte-protein interactions: Implications for the action of osmoprotectants in vivo and for the interpretation of "osmotic stress" experiments *in vitro*. *Biochemistry* **39**, 4455–4471.

Creamer, T., Srinivasan, R., and Rose, G. (1997). Modeling unfolded states of proteins and peptides. II. Backbone solvent accessibility. *Biochemistry* **36**, 2832–2835.

Felitsky, D., Cannon, J., Capp, M., Hong, J., Van Wynsberghe, A. W., Anderson, C., and Record, M. J. (2004). The exclusion of glycine betaine from anionic biopolymer surface: Why glycine betaine is an effective osmoprotectant but also a compatible solute. *Biochemistry* **43**, 14732–14743.

Felitsky, D., and Record, M. J. (2004). Application of the local-bulk partitioning and competitive binding models to interpret preferential interactions of glycine betaine and urea with protein surface. *Biochemistry* **43**, 9276–9288.

Gekko, K., and Timasheff, S. (1981). Mechanism of protein stabilization by glycerol: Preferential hydration in glycerol-water mixtures. *Biochemistry* **20**, 4667–4676.

Goldenberg, D. (2003). Computational simulation of the statistical properties of unfolded proteins. *J. Mol. Biol.* **326**, 1615–1633.

Greene, R. J., and Pace, C. (1974). Urea and guanidine hydrochloride denaturation of ribonuclease, lysozyme, alpha–chymotrypsin, and beta-lactoglobulin. *J. Biol. Chem.* **249**, 5388–5393.

Gursky, O. (1999). Probing the conformation of a human apolipoprotein C-1 by amino acid substitutions and trimethylamine-N-oxide. *Protein Sci.* **8**, 2055–2064.

Henkels, C., and Oas, T. (2005). Thermodynamic characterization of the osmolyte- and ligand-folded states of *Bacillus subtilis* ribonuclease P protein. *Biochemistry* **44**, 13014–13026.

Hong, J., Capp, M., Anderson, C., Saecker, R., Felitsky, D., Anderson, M., and Record, M. J. (2004). Preferential interactions of glycine betaine and of urea with DNA: Implications for DNA hydration and for effects of these solutes on DNA stability. *Biochemistry* **43**, 14744–14758.

Kumar, R., Lee, J., Bolen, D., and Thompson, E. (2001). The conformation of the glucocorticoid receptor af1/tau1 domain induced by osmolyte binds co-regulatory proteins. *J. Biol. Chem.* **276**, 18146–18152.

Lee, B., and Richards, F. (1971). The interpretation of protein structures: Estimation of static accessibility. *J. Mol. Biol.* **55**, 379–400.

Lee, J., Gekko, K., and Timasheff, S. (1979). Measurements of preferential solvent interactions by densimetric techniques. *Methods Enzymol.* **61**, 26–49.

Lee, J., and Timasheff, S. (1981). The stabilization of proteins by sucrose. *J. Biol. Chem.* **256**, 7193–7201.

Lesser, G., and Rose, G. (1990). Hydrophobicity of amino acid subgroups in proteins. *Proteins* **8**, 6–13.

Lewis, G. N., and Randall, M. (1923). "Thermodynamics and the Free Energy of Chemical Substances." McGraw-Hill Book Company, Inc., New York and London.

Mello, C., and Barrick, D. (2003). Measuring the stability of partly folded proteins using TMAO. *Protein Sci.* **12**, 1522–1529.

Moelwyn-Hughes, E. A. (1961). The dissolved state. *In* "Physical Chemistry," p. 833. Pergamon Press, Oxford.

Noyes, A., and Bray, W. (1911a). The effect of salts on the solubility of other salts. I. *J. Am. Chem. Soc.* **33**, 1643–1649.

Noyes, A., and Bray, W. (1911b). The effect of salts on the solubility of other salts. II. *J. Am. Chem. Soc.* **33,** 1650–1663.

Rajagopalan, L., Rösgen, J., Bolen, D., and Rajarathnam, K. (2005). Novel use of an osmolyte to dissect multiple thermodynamic linkages in a chemokine ligand-receptor system. *Biochemistry* **44,** 12932–12939.

Richards, F. (1977). Areas, volumes, packing and protein structure. *Annu. Rev. Biophys. Bioeng.* **6,** 151–176.

Santoro, M. M., and Bolen, D. (1988). Unfolding free energy changes determined by the linear extrapolation method. 1. Unfolding of phenylmethanesulfonyl alpha-chymotrypsin using different denaturants. *Biochemistry* **27,** 8063–8068.

Schellman, J. (2003). Protein stability in mixed solvents: A balance of contact interaction and excluded volume. *Biophys. J.* **85,** 108–125.

Silbey, R. J., and Alberty, R. A. (2001). Phase equilibrium. *In* "Physical Chemistry," p. 204. Wiley, New York.

Tanford, C. (1964). Isothermal unfolding of globular proteins in aqueous urea solutions. *J. Am. Chem. Soc.* **86,** 2050–2059.

Tanford, C. (1970). Protein denaturation. C. Theoretical models for the mechanism of denaturation. *Adv. Protein Chem.* **24,** 1–95.

Timasheff, S., and Xie, G. (2003). Preferential interactions of urea with lysozyme and their linkage to protein denaturation. *Biophys. Chem.* **105,** 421–448.

Timasheff, S. N. (1992a). Stabilization of protein structure by solvent additives. *In* "Stability of Protein Pharmaceuticals, Part B: *In Vivo* Pathways of Degradation and Strategies for Protein Stabilization" (T. J. Ahern and M. C. Manning, eds.), pp. 265–285. Plenum Press, New York.

Timasheff, S. N. (1992b). A physiochemical basis for the selection of osmolytes by nature. *In* "Water and Life: Comparative Analysis of Water Relationships at the Organismic, Cellular, and Molecular Levels" (G. N. Somero, C. B. Osmond, and C. L. Bolis, eds.), pp. 70–84. Springer-Verlag, Berlin.

Wu, P., and Bolen, D. (2006). Osmolyte-induced protein folding free energy changes. *Proteins* **63,** 290–296.

Xie, G., and Timasheff, S. (1997a). The thermodynamic mechanism of protein stabilization by trehalose. *Biophys. Chem.* **64,** 25–43.

Xie, G., and Timasheff, S. (1997b). Mechanism of the stabilization of ribonuclease A by sorbitol: Preferential hydration is greater for the denatured then for the native protein. *Protein Sci.* **6,** 211–221.

Zhang, W., Capp, M., Bond, J., Anderson, C., and Record, M. J. (1996). Thermodynamic characterization of interactions of native bovine serum albumin with highly excluded (glycine betaine) and moderately accumulated (urea) solutes by a novel application of vapor pressure osmometry. *Biochemistry* **35,** 10506–10516.

CHAPTER TWENTY-FOUR

MECHANISMS OF HIGH SALINITY TOLERANCE IN PLANTS

Narendra Tuteja

Contents

Abstract

Among abiotic stresses, high salinity stress is the most severe environmental stress, which impairs crop production on at least 20% of irrigated land worldwide. In response to high salinity stress, various genes get upregulated, the products of which are involved either directly or indirectly in plant protection. Some of the genes encoding osmolytes, ion channels, receptors, components of calcium signaling, and some other regulatory signaling factors or enzymes are able to confer salinity-tolerant phenotypes when transferred to sensitive plants. Overall, the susceptibility or tolerance to high salinity stress in plants is a coordinated action of multiple stress responsive genes, which also cross talk with other components of stress signal transduction pathways. High salinity exerts its negative impact mainly by disrupting the ionic and osmotic

Plant Molecular Biology, International Centre for Genetic Engineering and Biotechnology, Aruna Asaf Ali Marg, New Delhi, India

Methods in Enzymology, Volume 428
ISSN 0076-6879, DOI: 10.1016/S0076-6879(07)28024-3

equilibrium of the cell. In saline soils, high levels of sodium ions lead to plant growth inhibition and even death; therefore, mechanisms of salinity tolerance involve sequestration of Na^+ and Cl^- in vacuoles of the cells, blocking of Na^+ entry into the cell, Na^+ exclusion from the transpiration stream, and some other mechanisms that help in salinity tolerance. Understanding these mechanisms of stress tolerance, along with a plethora of genes involved in the stress signaling network, is important to improve high salinity stress tolerance in crops plants. This chapter first describes the adverse effect of salinity stress and general pathway for the plant stress response, followed by roles of various ion pumps, calcium, SOS pathways, ABA, transcription factors, mitogen-activated protein kinases, glycine betaine, proline, reactive oxygen species, and DEAD-box helicases in salinity stress tolerance. The cross-tolerance between stresses is also mentioned.

1. INTRODUCTION

The world population is increasing rapidly and may reach 6 to 9.3 billion by the year 2050 (http://www.unfpa.org/swp/200/), whereas the crop production is decreasing rapidly because of the negative impact of various environmental stresses; therefore, it is now very important to develop stress-tolerant varieties to cope with this upcoming problem of food security. Among stresses, abiotic stress is the principal cause of decreasing the average yield of major crops by more than 50%, which causes losses worth hundreds of million dollars each year (see Mahajan and Tuteja, 2005). Susceptibility or tolerance to abiotic stresses in plants is a coordinated action of various genes, which are switched on and may cross talk with other components of stress signal transduction pathways. Overall, stress tolerance is a complex phenomenon because plants may go through multiple stresses at the same time during their development. Among abiotic stresses, high salinity stress is the most severe environmental stress, impairing crop production on at least 20% of irrigated land worldwide. In addition, the increased salinity of arable land is expected to have devastating global effects, resulting in up to 50% land loss by the middle of the twenty-first century (see Mahajan and Tuteja, 2005). The amount of salt-affected land is already more than 9000×10^6 ha, which is a threat to agriculture (Flowers, 2004). Furthermore, there is a deterioration of about 2 million ha ($\sim$1%) of world agricultural lands because of salinity each year. Soil salinity may be a result of poor water management, high evaporation, heavy irrigation, and previous exposure to seawater. Seawater contains $\sim$3% NaCl, where Na^+ is $\sim$460 mM, Mg^{2+} is 50 mM, and Cl^- is $\sim$540 mM, and many other ions in small quantities (Mahajan and Tuteja, 2005).

Many crop species are very sensitive to soil salinity and are known as glycophytes, whereas salt-tolerant plants are known as halophytes. In general, glycophytes cannot grow at 100 mM NaCl, whereas halophytes can grow at salinities over 250 mM NaCl. Salinity-sensitive plants restrict the uptake of salt and strive to maintain an osmotic equilibrium by the synthesis of compatible solutes such as prolines, glycine betaine (GB), and sugars. Salinity-tolerant plants have the capacity to sequester and accumulate salt into the cell vacuoles, thus preventing the buildup of salt in the cytosol and maintaining a high cytosolic K^+/Na^+ ratio in their cells. Because of the competition of Na^+ and Cl^- with other nutrients such as K^+, Ca^{2+}, and $NO3^-$, which reduce plant growth by affecting the availability, transport, and partitioning of nutrients, high salinity stress causes nutrient imbalances (reviewed by Hu and Schmidhalter, 2005). Generally, the salinity tolerance is related inversely to the extent of Na^+ accumulation in the shoot. Understanding the molecular mechanism of high salinity stress and subsequently developing salinity-tolerant crops are now essential for solving the current problem of crop yield reduction. Several approaches used to enhance salinity tolerance in cereals and the potential of wild relatives to improve the salinity tolerance of wheat have been reported (reviewed in Colmer et al., 2006; Munns et al., 2006). This chapter describes the adverse effect of salinity stress and the general pathway for the plant stress response. It also describes the roles of various ion pumps, calcium, SOS pathways, abscisic (ABA), transcription factors, mitogen-activated protein kinases, glycine betaine, proline, reactive oxygen species (ROS), and DEAD-box helicases in salinity stress tolerance. The cross-tolerance between stresses is also mentioned.

2. ADVERSE EFFECT OF SALINITY STRESS

The following adverse effects are observed in response to high salinity stress.

1. High salinity interferes with plant growth and development and can also lead to physiological drought conditions and ion toxicity (Zhu, 2002). Therefore, high salinity and drought stresses affect mostly all aspects of plant physiology and metabolism and cause both hyperionic and hyperosmotic stresses, which lead to plant demise.
2. The basic physiology of high salinity stress and drought stress overlaps with each other as a high salt deposition in soil leads to a deposition of a low water potential zone in the soil. This makes it increasingly difficult for the plant to acquire water as well as nutrients.
3. Salinity causes ion-specific stresses resulting in an altered K^+/Na^+ ratio. The external Na^+ can negatively impact intracellular K^+ influx.
4. Salinity leads to a buildup of Na^+ and Cl^- concentrations in the cytosol, which can be ultimately detrimental to the cell. The Na^+ can dissipate the membrane potential and therefore facilitates the uptake of Cl^- down the gradient.

5. Higher concentrations of sodium ions (above 100 mM) are toxic to cell metabolism and can inhibit the activity of many essential enzymes, cell division and expansion, membrane disorganization, and osmotic imbalance, which finally can lead to growth inhibition.

6. Higher concentrations of sodium ions can also lead to a reduction in photosynthesis and the production of reactive oxygen species.

7. Potassium ions are one of the essential elements required for growth. Alterations in K^+ ions (because of the impact of high salinity stress) can disturb the osmotic balance, the function of stomata, and the function of some enzymes.

8. High salinity can also injure cells in transpiring leaves, which leads to growth inhibition. This salt-specific or ion-excess effect of salinity causes a toxic effect of salt inside the plant. The salt can concentrate in the old leaves and the leaves die, which is crucial for the survival of a plant (see Munns et al., 2006).

9. Shoji et al. (2006) demonstrated that high salinity affects cortical microtubule organization and helical growth in Arabidopsis.

3. GENERIC PATHWAY FOR PLANT RESPONSE TO STRESS

Plants can respond to stresses as individual cells and synergistically as a whole organism.

A generic stress signal transduction pathway for the plant stress response is depicted in Fig. 24.1. The stress signal is first perceived at the membrane level by the receptors (G-protein-coupled receptors, ion channel, receptor-like kinase, or histidine kinase), which results in the generation of many secondary signal molecules, such as Ca^{2+}, inositol phosphates, ROS, and abscisic acid ABA. The stress signal then transduces inside the nucleus to induce many stress responsive genes, the products of which ultimately lead to plant adaptation to stress tolerance. The stress responsive genes could be either early or delayed induced genes. Early genes are induced within minutes of stress perception, often express transiently, and their products (e.g., various transcription factors) can activate the expression of delayed genes (e.g., RD [responsive to dehydration], KIN [cold induced], COR [cold responsive]). Overall, gene products are either involved directly in cellular protection against the stress (e.g., late embryogenesis abundant proteins, antifreeze proteins, antioxidant, chaperons, and detoxification enzymes) or involved indirectly in protection (e.g., transcription factor, enzyme of PI metabolism) (Fig. 24.1). Stress-induced gene products are also involved in the generation of regulatory molecules such as ABA, salicylic acid, and ethylene, which can initiate the second round of signaling.

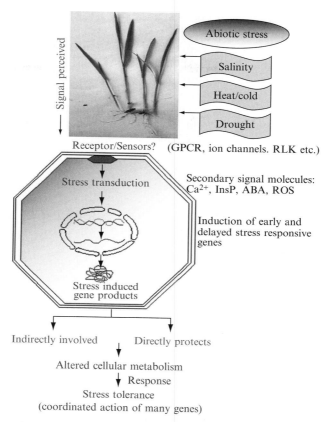

Figure 24.1 Generic pathway for plant response to stress. The extracellular stress signal is first perceived by the membrane receptors and then activates a large and complex signaling cascade intracellularly, including the generation of secondary signal molecules. The signal cascade results in the expression of multiple stress responsive genes, the products of which can provide the stress tolerance directly or indirectly. Overall, the stress response could be a coordinated action of many genes, which may cross talk with each others. GPCR, G-protein-coupled receptor; RLK, receptor-like kinase; InsP, inositol phosphate; ABA, abscisic acid; ROS, reactive oxygen species. (See color insert.)

Overall, the stress signal transduction requires exact coordination of all the signaling molecules, including protein modifiers (methylation, ubiquitination, glycosylation, etc.), adaptors, and scaffolds (see Xiong *et al.*, 2002). In addition to stress tolerance, the response can be either growth inhibition or cell death, which will depend on what kind of genes are up- or down-regulated in response to the stress(es).

The mechanism of salinity tolerance is a very complex phenomenon. Studies have shown that components of various pathways are involved in imparting the salinity tolerance to the plants. An interesting study involving cDNA microarray analysis of 7000 Arabidopsis genes has shown that 194 genes are upregulated under high salinity stress (Seki *et al.*, 2002), suggesting that many transcriptional regulatory mechanisms function in stress signal transduction pathways.

Small molecules such as calcium, glycine betaine, proline, ROS, ABA, and various ion pumps also play important roles in this process. However, studies have shown that some calcium-binding proteins, transcription factors, and enzymes such as protein kinases and helicases also have roles in salinity stress tolerance. These studies are described in the following sections.

4. Ion Pumps, Calcium, and SOS Pathways in Relation to Salinity Stress

High salinity stress causes an imbalance in sodium ions (Na^+) homeostasis, which is maintained by the coordinated action of various pumps, ions, Ca^{2+} sensors, and its downstream interacting partners, which ultimately results in the efflux of excess Na^+ ions. The roles of various ion pumps/channels are depicted in Fig. 24.2. Certain channels show more selectivity to K^+ over Na^+. These include the K inward–rectifying channel, which mediates the influx of K^+ upon plasma membrane hyperpolarization and selectively accumulates K^+ over Na^+ ions. The histidine kinase transporter (HKT) is a low-affinity Na^+ ion transporter, which blocks the entry of Na^+ ions into the cytosol. A nomenclature for HKT, which is the key determinant of plant salinity tolerance, has been set up (Platten *et al.*, 2006). The nonspecific cation channel is a voltage-independent channel, which acts as a gate for the entry of Na^+ into plant cells. Moreover, there is the K^+ outward-rectifying channel, which opens during the depolarization of the plasma membrane and mediates the efflux of K^+ and the influx of Na^+ ions, leading to Na^+ accumulation in the cytosol. The vacuolar Na^+/H^+ exchanger (NHX) helps push excess Na^+ ions into vacuoles. Na^+ extrusion from plant cells is powered by the electrochemical gradient generated by H^+-ATPases, which permit the NHX to couple the passive movement of H^+ inside along the electrochemical gradient and extrusion of Na^+ out of the cytosol. Another pump, the H^+/Ca^{2+} antiporter (CAX1), helps in Ca^{2+} homeostasis (Figs. 24.2 and 24.3) (see Mahajan *et al.*, 2006a; Zhang *et al.*, 2004; Zhu, 2002). A voltage-dependent anion channel has been isolated from *Pennisetum glaucum* and shown to be upregulated in response to multiple stresses, including salinity stress (Desai *et al.*, 2006).

Calcium is one of the principal candidates for functioning as a central node in the overall "signaling web" and plays an important role in providing salinity tolerance to plants. High salinity leads to increased

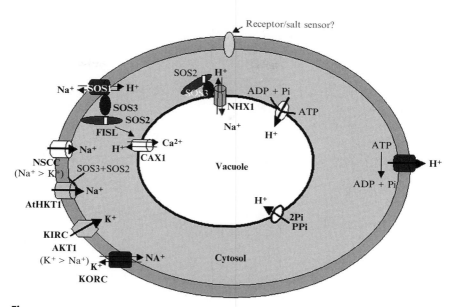

Figure 24.2 Regulation of ion homeostasis by various ion pumps. The salinity stress signal is perceived by a receptor or salt sensor present at the plasma membrane of the cell. This signal is responsible for activating various ion pumps present at plasma and vacuolar membranes. This signal also activates the SOS pathway, the components of which help in regulating some of these pumps. The various pumps/channels are the K^+ inward-rectifying channel (KIRC), histidine kinase transporter (HKT), nonspecific cation channels (NSCC), K^+ outward-rectifying channel (KORC), Na^+/H^+ antiporters (SOS1), vacuolar Na^+/H^+ exchanger (NHX), and H^+/Ca^+ antiporter (CAX1). Na^+ extrusion from plant cells is powered by the electrochemical gradient generated by H^+-ATPases, which permits the Na^+/H^+ antiporters to couple the passive movement of H^+ inside along the electrochemical gradient and extrusion of Na^+ out of the cytosol. The stress signal sensed by SOS3 activates SOS2, which activates SOS1 (for details, see Fig. 24.5 and text). (See color insert.)

cytosolic Ca^{2+}, which initiates the stress signal transduction pathways for stress tolerance. Ca^{2+} release can be primarily from an extracellular source (apoplastic space) as the addition of EGTA or BAPTA blocks calcineurin-mediated activity (Fig. 24.3). Ca^{2+} release may also result from the activation of phospholipase C, leading to the hydrolysis of phosphatidylinositol bisphosphate to inositol trisphosphate and the subsequent release of Ca^{2+} from intracellular Ca^{2+} stores. Furthermore, calcium-binding proteins (calcium sensors) can provide an additional level of regulation in calcium signaling. These sensor proteins recognize and decode the information provided in the calcium signatures and relay the information downstream to initiate a phosphorylation cascade, leading to regulation of gene expression. Wu *et al.* (1996) commenced a mutant screen for *Arabidopsis* plants,

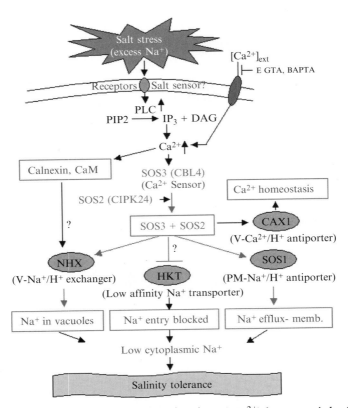

Figure 24.3 Regulation of ion (e.g., Na$^+$, K$^+$, and Ca^{2+}) homeostasis by SOS and related pathways in relation to salinity stress tolerance. High salinity (Na$^+$) stress initiates a calcium signal that activates the SOS pathway. The signal first activates phospholipase C (PLC), which hydrolyses phosphatidylinositol bisphosphate (PIP2) to generate inositol trisphosphate (IP$_3$), and diacylglycerol (DAG) resulting in an increased level of Ca^{2+} ions. This change in cytosolic Ca^{2+} ions is sensed by a calcium sensor such as SOS3, which interacts with the SOS2 protein kinase. This SOS3–SOS2 protein kinase complex phosphorylates SOS1, a Na$^+$/H$^+$ antiporter, resulting in an efflux of excess Na$^+$ ions. The SOS3–SOS2 complex interacts with and influences other salt-mediated pathways, resulting in ionic homeostasis. This complex inhibits HKT1 activity (a low-affinity Na$^+$ transporter), thus restricting Na$^+$ entry into the cytosol. SOS2 also interacts and activates the vacuolar Na$^+$/H$^+$ exchanger (NHX), resulting in the sequestration of excess Na$^+$ ions, further contributing to Na$^+$ ion homeostasis. Calnexin and calmodulin (CaM) or other calcium-binding proteins can also interact and activate the NHX or other transporters. The H$^+$/Ca^{2+} antiporter (CAX1) has been identified as an additional target for SOS2 activity reinstating cytosolic Ca^{2+} homeostasis. (See color insert.)

which were oversensitive to salt stress. As a result of this screen, three genes, *SOS1*, *SOS2*, and *SOS3* (salt overlay sensitive), were identified. The *SOS3* gene (also known as *AtCBL4*) encodes a calcineurin B–like protein (CBL, calcium sensors), which is a Ca^{2+}-binding protein and senses the change in cytosolic Ca^{2+} concentration and transduces the signal downstream. The SOS pathway is depicted in Fig. 24.3. A loss of function mutation that reduces the Ca^{2+}-binding capacity of SOS3 (*sos 3–1*) renders the mutant hypersensitive to salt (see Zhu, 2002). *SOS2* (AtCIPK24) encodes a novel serine/threonine protein kinase known as the CBL–interacting protein kinase (CIPK). *SOS3* activates SOS2 protein kinase activity in a calcium–dependent manner (Mahajan *et al.*, 2006).

The first target of the SOS3–SOS2 pathway was identified by genetic analysis of the *sos1* mutant of *Arabidopsis*. SOS1 is a Na^+/H^+ antiporter, and the *sos1* mutant was hypersensitive to salt and showed an impaired osmotic/ ionic balance. Genetic analysis confirmed that SOS3, SOS2, and SOS1 function in a common pathway of salt tolerance (see Mahajan and Tuteja, 2005; Zhang *et al.*, 2004; Zhu, 2002). The SOS3–SOS2 kinase complex was found to phosphorylate SOS1 directly. The SOS pathway is depicted in Figs. 24.2 and 24.3. The SOS pathway also seems to have other branches, which help remove excess Na^+ ions out of the cell, thereby maintaining cellular ion homeostasis. In *Arabidopsis*, Na^+ entry into root cells during salt stress appears to be mediated by AtHKTI, a low–affinity Na^+ transporter, which blocks the entry of Na^+ (Fig. 24.3) (see Mahajan and Tuteja, 2005; Zhang *et al.*, 2004; Zhu, 2002). SOS2 also interacts and activates the NHX, resulting in the sequestration of excess Na^+ ions and pushing it into vacuoles (Fig. 24.3), thereby contributing further to Na^+ ion homeostasis. Some other calcium–binding proteins, such as calnexin and calmodulin, also sense the increased level of calcium and can interact and activate the NHX. CAX1 has been identified as an additional target for SOS2 activity, reinstating cytosolic Ca^{2+} homeostasis (Fig. 24.3).

This reflects that the components of SOS pathway may cross talk and interact with other branching components to maintain cellular ion homeostasis, which helps in salinity tolerance.

5. ABSCISIC ACID AND TRANSCRIPTION FACTORS IN SALINITY STRESS TOLERANCE

Abscisic acid is a phytohormone that regulates plant growth and development and also plays an important role in the response of the plant to abiotic stresses, including salinity stress (reviewed in Chinnusamy *et al.*, 2004; Mahajan and Tuteja, 2005; Zhu, 2002). The role of ABA in salinity stress was confirmed by a study of Dr. J. K. Zhu's group, where they showed that ABA-deficient mutants performed poorly under salinity stress (Xiong *et al.*, 2001). The ABA level is known to induce under stress conditions, which is mainly because of the induction of genes for enzymes responsible for ABA biosynthesis. ABA is synthesized from β-carotene through several

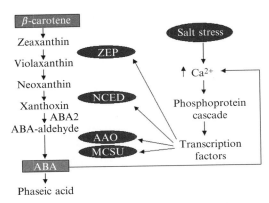

Figure 24.4 ABA biosynthesis pathway and its regulation by osmotic stress. ABA is synthesized from β-carotene via the oxidative cleavage of neoxanthin and conversion of xanthoxin to ABA via ABA-aldehyde. Stresses, including salinity stress, stimulate ABA biosynthesis and accumulation by activating genes involved in the ABA biosynthetic pathway, which itself could be mediated by a calcium-dependent phosphorylation cascade. ABA can also upregulate the expression of ABA biosynthetic genes via calcium signaling pathways (see Xiong *et al.*, 2002; Zhu, 2002). ZEP, zeaxanthin epoxidase; NCED, 9-*cis*-epoxycarotenoid dioxygenase; AAO, ABA-aldehyde oxidase; MCSU, molybdenum cofactor sulfurase. (See color insert.)

enzymatic steps (Fig. 24.4). The salinity stress-induced activation of many ABA biosynthetic genes, such as zeaxanthin oxidase, 9-*cis*-epoxycarotenoid dioxygenase, ABA-aldehyde oxidase, and molybdenum cofactor sulfurase, appear to be regulated through a calcium-dependent phosphorylation pathway (Fig. 24.4) (reviewed by Chinnusamy *et al.*, 2004; Xiong *et al.*, 2002; Zhu, 2002). The accumulation of ABA can also stimulate the expression of ABA biosynthetic genes through a calcium-signaling pathway and can also activate the ABA catabolic enzymes to degrade the ABA.

The induction of osmotic stress responsive genes imposed by salinity is transmitted through either ABA-dependent or ABA-independent pathways, although some others are only partially ABA dependent (Shinozaki *et al.*, 1997). However, the components involved in these pathways often cross talk through calcium in stress signaling pathways.

The transcript accumulation of the *RD29A* gene is reported to be regulated in both ABA-dependent and ABA-independent manners (Yamaguchi-Shinozaki *et al.*, 1993). The proline accumulation in plants can be mediated by both ABA-dependent and ABA-independent signaling pathways (see Zhu, 2002). The salinity stress-induced upregulation of the transcript of pea DNA helicase 45 (*PDH45*) followed the ABA-dependent pathway (Sanan-Mishra *et al.*, 2005), whereas the CBL and CIPK from pea followed the ABA-independent pathway (Mahajan *et al.*, 2006b). The role of calcium in ABA-dependent induction of the *P5CS* gene during salinity stress has been reported by Kinight *et al.* (1997). It is also suggested that phospholipase D, along with ABA and calcium, acts as a negative regulator of proline biosynthesis in Arabidopsis (Thiery *et al.*, 2004). The signaling mechanism behind the activation of these genes is not well known, but the transcriptional activation of a few stress-induced genes represented by RD29A is known to some extent (see Xiong *et al.*, 2002). It is known that the expression of *RD29A, RD22, COR15A, COR47,* and *P5CS* genes was reduced in the *los5* mutant (Xiong *et al.*, 2001). Overall, the ABA-dependent pathways are involved essentially in osmotic stress gene expression.

The transcriptional regulatory network of *cis*-acting elements and transcription factors involved in ABA and salinity stress responsive gene expression is depicted in Fig. 24.5. Promoters of the stress-induced genes contain *cis*-regulatory elements such as DRE/CRT, ABRE, MYC recognition sequence (MYCRS), and MYB recognition sequence (MYBRS), which are regulated by various upstream transcriptional factors (Fig. 24.5) (see Mahajan *et al.*, 2005; Zhu, 2002). The ABA-dependent salinity stress signaling activates basic leucine zipper transcription factors called AREB, which bind to ABRE element to induce the stress responsive gene (*RD29A*). Transcription factors such as DREB2A and DREB2B transactivate the DRE *cis* element of osmotic stress genes and thereby are involved in maintaining the osmotic equilibrium of the cell (see Mahajan and Tuteja, 2005). Some genes, such as *RAD22*, lack the typical CRT/DRE elements

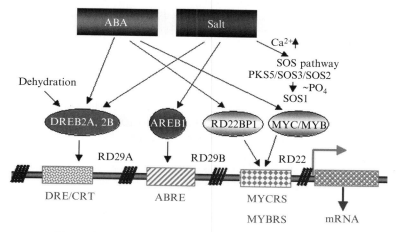

Figure 24.5 Transcriptional regulatory network of *cis*-acting elements and ABA-dependent transcription factors involved in salinity stress gene expression. Osmotic stress signaling generated via salinity stress seems to be mediated by transcription factors such as DREB2A/DREB2B, AREB1, and MYC/MYB transcription activators, which interact with DRE/CRT, ABRE, and MYCRS/MYBRS elements in the promotion of stress genes, respectively. AtMYC2 and AtMYB2 act cooperatively to activate the expression of ABA-inducible genes such as RD22. Transcription factor-binding sites are represented as rectangles at the bottom of the figure, with the representative promoters. Salinity works mainly through the SOS pathway, reinstating cellular ionic equilibrium. (See color insert.)

in their promoter, suggesting their regulation by some other mechanism. The MYC/MYB transcription factors RD22BP1 and AtMYB2 could bind MYCRS and MYBRS elements, respectively, and help in activation of the *RD22* gene (Fig. 24.5). Overall, these transcription factors may also cross talk with each other for their maximal response to stress tolerance.

6. MITOGEN-ACTIVATED PROTEIN KINASES AND SALINITY STRESS

Mitogen-activated protein kinases are a specific class of plant serine/threonine protein kinases that play a central role in the transduction of various extracellular and intracellular signals, including stress signals. These generally function as a cascade where MAPK is phosphorylated and activated by MAPK kinase (MAPKK), which itself is activated by MAPKK kinase (MAPKKK). All three of these kinases are interlinked together and are also called extracellular receptor kinases. The roles of several plant MAP kinases in response to salinity stress have been reported (reviewed by Sanan-Mishra *et al.*, 2006; Zhang *et al.*, 2006b). A transient increase in transcript for genes encoding the AtMPK1, AtMAPKK, AtMKK2, AtMEKK1 (MAPKKK), AtMPK3 (MAPK), AtMPK4, and AtMPK6 levels was observed after stimulating *Arabidopsis* plants with salinity stress. AtMPK4 and AtMPK6 were found to be phosphorylated by AtMKK2. A 46-kDa MsK7 or salt stress–inducible MAP kinase (SIMK) and its upstream activator kinase SIMKK from alfalfa were reported to be activated by salinity. It has been shown that tobacco protoplasts exposed to salinity and osmotic stresses showed enhancement of a 48-kDa kinase, the salicylic acid–induced protein kinase (see Sanan-Mishra *et al.*, 2006; Zhang *et al.*, 2006b). A new MAP kinase gene (*CbMAPK3)* has been isolated from the *Chorispora bungeana* plant, and its transcript level was upregulated in response to salinity and cold stresses (Zhang *et al.*, 2006a).

Collectively, these reports of salinity-induced MAP kinases suggest that a MAP kinase signaling cascade helps in mediating salinity stress tolerance in plants.

7. GLYCINE BETAINE AND PROLINE IN SALINITY STRESS

Plants initiate some defensive machinery in order to cope with stress; one of them is associated with changes in metabolites. Glycine betaine (*N, N,N*-trimethylglycine betaine) and proline are two major osmoprotectant osmolytes, which are synthesized by many plants (but not all) in response to stress, including salinity stress, and thereby help in maintaining the osmotic status of the cell to ameliorate the abiotic stress effect (reviewed by

Chinnusamy *et al.*, 2005; Vinocur and Altman, 2005). In plants where GB is not produced, the transgenics with overexpressing GB synthesizing genes resulted in the production of enough amount of GB, which lead plants to tolerate stresses, including salinity stress (see Rhodes and Hanson, 1993). GB is synthesized from choline by the action of choline monooxygenase and betaine aldehyde dehydrogenase enzymes. Overexpression of the genes encoding betaine aldehyde decarboxylase from halophyte *Suaeda liaotungensis* improved the salinity tolerance in tobacco plants. The choline dehydrogenase gene (*codA*) from *Arthrobacter globiformis* helped salinity tolerance in rice (see Vinocur and Altman, 2005). Overexpression of the *N*-methyl transferase gene in cyanobacteria and Arabidopsis resulted in accumulation of GB in higher levels and improved salinity tolerance (see Mahajan and Tuteja, 2005). It was also reported that foliar application of GB exogenously to low- or nonaccumulating plants helped in improving the growth of plants in the salinity stress condition as reported in the case of *Oryza sativa* (Harinasut *et al.*, 1996) and *Zea mays* (Yang and Lu, 2005). It was also observed that GB-treated plants under salinity stress had significantly decreased Na^+ and increased K^+ concentrations in shoots compared with untreated plants. Hence, GB may also help in salinity tolerance through its role in signal transduction and ion homeostasis.

The amino acid proline accumulates (normally in cytosol) under stress and is correlated with osmotic adjustment to improve plant salinity tolerance (see Vinocur and Altman, 2005). Proline also plays roles in scavenging free radicals, stabilizing subcellular structures, and buffering cellular redox potential under stresses. The salinity stress responsive genes, whose promoters contain proline responsive elements (PRE, ACTCAT), are also known to be induced by proline (see Chinnusamy *et al.*, 2005). In higher plants, proline is synthesized by glutamic acid by the actions of two enzymes, pyrroline-5–carboxylate synthetase (P5CS) and pyrroline-5–carboxylate reductase (P5CR) (Delauney and Verma, 1993). Overexpression of the *P5CS* gene in transgenic tobacco resulted in increased production of proline and salinity/drought tolerance (Kishor *et al.*, 1995). The exogenous application of proline also provided osmoprotection and facilitated the growth of salinity-stressed plants. Proline can also protect cell membranes from salinity-induced oxidative stress by upregulating activities of various antioxidants (Yan *et al.*, 2000).

8. REACTIVE OXYGEN SPECIES IN SALINITY STRESS

Reactive oxygen species produced in response to oxidative stress can cause permanent damage to the cellular apparatus. Reactive oxygen intermediates (ROI) typically result from the excitation of O_2 to form singlet oxygen ($O_2{}^1$) or the transfer of one, two, or three electrons to O_2 to form superoxide radical (O_2^-), hydrogen peroxide (H_2O_2), or a hydroxyl radical (OH^-), respectively. The enhanced production of ROIs during stresses can pose a threat to plants because they are unable to detoxify effectively by the ROI scavenging machinery. The unquenched ROIs react spontaneously with organic molecules and cause membrane lipid peroxidation, protein oxidation, enzyme inhibition, and DNA and RNA damage (see Vinocur and Altman, 2005). Oxidative stress arises under environmental stresses, including salinity stress, and may exceed the scavenging capacity of the natural defense system of the plant. The major ROI-scavenging mechanisms of plants include superoxide dismutase, ascorbate peroxidase, catalase, and GSH reductase, which help in the deactivation of active oxygen species in multiple redox reactions, thereby contributing to the protective system against oxidative stress. The ROS scavengers can increase the plant resistance to salinity stress. Overexpression of the aldehyde dehydrogenase gene in Arabidopsis has been reported to confer salinity tolerance. The aldehyde dehydrogense catalyzes the oxidation of toxic aldehydes, which accumulate as a result of side reactions of ROS with lipids and proteins. The enhancement of stress tolerance in transgenic tobacco plants has been shown by overexpressing *Chlamydomonas* glutathion peroxidase in chloroplast or cytosol (see Vinocur and Altman, 2005).

9. DEAD-Box Helicases in Salinity Stress Tolerance

Because abiotic stresses affect the cellular gene expression machinery, it is possible that molecules involved in nucleic acid metabolism, including helicases, might be involved in stress signaling. Several genes, including genes for helicases, are known to be expressed under the influence of various abiotic stresses, including salinity (reviewed in Owttrim, 2006; Vashisht and Tuteja, 2006). Helicases are ubiquitous enzymes that catalyze the unwinding of energetically stable duplex DNA (DNA helicases) or duplex RNA secondary structures (RNA helicases) (Tuteja and Tuteja, 2004). Most helicases are members of the DEAD-box protein superfamily that play essential roles in basic cellular processes regulating plant growth and development, such as DNA replication, repair, recombination, transcription, ribosome biogenesis, and translation initiation. It seems therefore that the DEAD-box helicase might also be playing an important role in stabilizing growth in plants under stress conditions by regulating some stress–induced pathways. Because RNA molecules are more prone to forming stable nonfunctional secondary structures, their proper functioning requires RNA chaperones. DEAD-box RNA helicases are the best candidates for RNA chaperones because these proteins can use energy derived from ATP hydrolysis to actively disrupt misfolded RNA structures so that correct folding can occur (see Vashisht and Tuteja, 2006). There are now few reports on the upregulation of DEAD-box helicases in response to abiotic stresses. The first stress–induced helicase gene (*CsdA*) was reported from *Escherichia coli*, and later a putative DEAD-box RNA helicase from sorghum plant was observed to be induced under salinity stress (see Owttrim, 2006).

The transcript of the first biochemically active plant DNA helicase (pea DNA helicase 45, PDH45) has been reported to be induced in pea seedling in response to high salinity stress; when this gene was transformed to tobacco, it provided the salinity tolerance (Sanan-Mishra *et al.*, 2005). This response was specific to Na^+ ions stress because treatment with Li^+ did not induce the transcript. The *PDH45* transcript was also upregulated in response to other abiotic stresses (dehydration, wounding, and low temperature), which suggested that the transcript increase could be because of water stress resulting from salinity- and manitol-induced desiccation. The *PDH45* transcript was observed to be induced by ABA, which suggested that the stress effect might be mediated through ABA-mediated pathways. The exact mechanism of PDH45-mediated tolerance of salinity stress is not understood. This protein may act at translational level or may associate with a DNA multisubunit protein complex to alter gene expression (Sanan-Mishra *et al.*, 2005). A unique bipolar helicase that contains both 3′-5′ and 5′-3′ directional helicase activities has been shown to increase its transcript level in both shoot and root in response to salinity and cold stresses (Vashisht *et al.*, 2005).

The exact mechanism of helicase-mediated tolerance of stress is not currently understood. There could be two possible sites of action for the helicases: (i) at the level of transcription or translation to enhance or stabilize protein synthesis or (ii) in an association with DNA multisubunit protein complexes to alter gene expression. It is evident that mRNA and protein synthesis are very sensitive to stress, so factors involved in transcription and translation are potential targets of salt toxicity in plants. The possible mechanism of helicase action during stress is depicted in Fig. 24.6. In response to stress the extra secondary structures could be formed in the 5′-untranslated region in mRNA of many essential genes, which could be inhibitory for translation. These inhibitory secondary structures need to resolve in order to have active translation, as otherwise these RNA will act as nonfunctional RNA where protein synthesis cannot proceed. The stress-induced RNA helicase(s) recognized these nonfunctional RNAs and unwound to resolve the secondary structures, which permit the translation initiation to proceed (Fig. 24.6) (see Vashisht and Tuteja, 2006). Overall, these stress–induced helicases help in recovering the functions of the genes for stress adaptation, which were stopped previously because of the negative impact of the stress.

Stress-regulated DEAD-box helicases can also exert their influence by many other mechanisms. For example, the RNA helicase activity of DEAD-box proteins could facilitate transcription by altering the structure

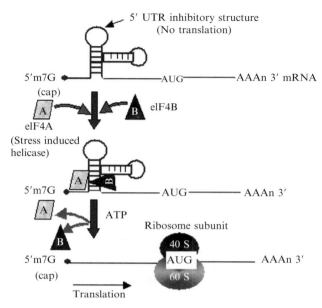

Figure 24.6 Possible mechanism of stress tolerance by a helicase. Eukaryotic initiation factor 4A (eIF4A) is a prototypic member of the DEAD-box RNA helicase family. Abiotic stresses enhance formation of the inhibitory secondary structure at the 5′ UTR of mRNA. This protein is responsible for removal of the secondary structure of the mRNA. eIF4A, along with eIF4B, binds to 5′ UTR and unwinds the inhibitory secondary structure in an ATP-dependent manner. This facilitates the binding of ribosome. After this ribosome scans for the start codon (AUG) and protein synthesis begins normally, which was inhibited due to negative impact of the stress. (See color insert.)

of nascent RNA, a process that can stimulate reinitiation and/or elongation. Helicases can also alter their localization within the cell under stress conditions, which enable helicases to perform other roles in different subcellular compartments. Stress-induced helicases can also get phosphorylated, which provides an opportunity to link their activities with stress sensing signal transduction phosphorylation cascades (see Owttrim, 2006).

 ## 10. CROSS-TOLERANCES BETWEEN STRESSES

Because plants can be exposed to more than one stress at the same time, they have developed mechanisms to cope with different stresses at a time. This suggests that different stress-signaling pathways are also cross talking with each other, which can lead plants to cross-tolerance (reviewed in Chinnusamy et al., 2004; Genoud and Metraux, 1999; Ma et al., 2006). Cross-tolerance is defined as a biological phenomenon by the virtue of which a plant, which is resistant to one stress, is able to develop tolerance to another form of stress. For example, endophyte–infected grasses show increased tolerance to abiotic stresses also (Malinowski and Belesky, 2000). Salinity stress in tomato also causes the accumulation of proteinase inhibitors and the activation of wound–related genes. The CBL1 was known to be regulated differentially by salinity, drought, and cold responses in Arabidopsis, whereas CDPKs mediate cross talk between different signaling pathways, leading to cross-tolerance. Capiati *et al.* (2006) reported that wounding increases salinity tolerance in tomato plants, suggesting cross talk between these stresses. It has been shown that transgenic tobacco plants overexpressing the genes encoding glyoxalase pathway enzymes (*GlyI* and *GlyII*) not only have better tolerance to high salinity stress, but also grow and set viable seeds in zinc-spiked soils, suggesting cross-tolerance between these stresses (Singla-Pareek *et al.*, 2006).

Many stress–induced genes in Arabidopsis have been reported to overlap for salinity, drought, and cold stresses. In general the common consequences of salinity, drought, and cold stresses are osmotic or oxidative stress, suggesting further the possibility of cross talk among these stresses. The various probable candidates involved in cross talk among stresses leading to cross-tolerance include stress sensors, histidine kinases, receptor-like protein kinases, calcium, calcium sensors, calcium-binding proteins, CDPKs, MAPK cascades, ABA, and transcription factors (see Chinnusamy *et al.*, 2004). The exact mechanism of cross-tolerance still needs to be studied.

Ma *et al.* (2006) suggested that large-scale microarray data could be used to recognize the cross talk between different signaling pathways. In Arabidopsis the transcript profile of various genes under salinity and other stresses has been made available publicly through several databases, such as TAIR, NASC, and Genevestigator (see Ma *et al.*, 2006). The genes exclusively upregulated by salinity stress are listed in Ma *et al.* (2006). The various Web sites for further study are as follow.

1. Affymetrix microarray data: http://www.arabidopsis.org/info/expression/ATGeneExpression.jsp
2. Abiotic transcript profile data (Affimatrix microarray data): http://www.weigeworld.org/resources/microarray/ATGeneExpress/
3. Glass microarray data: http://ag.arizona.edu/microarra/; the resulting GPR files can be analyzed by TIGR-TM4 (http://www.tmr.org/)

 ## 11. Future Prospective

Overall, high salinity stress signaling is an important area with respect to an increase in crop yield under suboptimal conditions. The involvement of salinity stress responsive genes in various metabolic processes to enhance the stress tolerance in plants might have general implications. The mechanism of high salinity tolerance is just beginning to be understood. The overall progress of research on salinity stress responsive genes and their products reflects their central role in plant growth and development under stress conditions. Much effort is still required to uncover in detail each product of genes induced by salinity stress and their interacting partners to understand the complexity of the high salinity stress signal transduction pathways. The role of endogenous siRNA in regulating the salinity tolerance in Arabidopsis is also being studied (Borsani et al., 2005), which will be further helpful in our better understanding of the mechanism of salinity stress. Determination of the upstream receptors or sensors that monitor the stimuli, as well as the downstream effectors that regulate the responses, is essential, which will also expedite our understanding of salinity stress signaling mechanisms in plants. The pyramiding of salinity stress-induced genes of either the same pathway or different pathways can help enhance the salinity tolerance. Improvements in salinity tolerance also result from close interactions among molecular biologists, geneticists, biotechnologists, and physiologists and benefit from feedback from plant breeders and agronomists (Munns et al., 2006). A novel role of pea G-protein in high salinity stress tolerance has been observed (unpublished data). A few reports of stress-induced DNA and RNA helicases suggested that salinity stress affects the stability of nucleic acid base pairing. Therefore, the exploitation of salinity stress responsive genes of new pathways, including DNA/RNA metabolism and G-proteins signaling pathways, will be useful in elucidating the less known stress signaling networks and will also be helpful for engineering salinity-tolerant crop plants.

ACKNOWLEDGMENTS

I thank Dr. Renu Tuteja for critical reading and corrections on the chapter and Dr. Shilpi Mahajan and Miss Shikha Misra for help in preparation of the illustrations. This work was partially supported by grants from the Department of Biotechnology, Department of Science and Technology, and Defence Research and Development Organisation, Government of India. I apologize if some references were not cited due to space constraints.

REFERENCES

Borsani, O., Zhu, J., Verslues, P. E., Sunkar, R., and Zhu, J.-K. (2005). Endogenous siRNAs derived from a pair of natural cis-antisense transcripts regulate salt tolerance in Arabidopsis. Cell 123, 1279–1291.

Capiati, D. A., Pais, S. M., and Tellez-Inon, M. T. (2006). Wounding increases salt tolerance in tomato plants: evidence on the participation of calmodulin-like activities in cross-tolerance signaling. J. Exp. Bot. 57, 2391–2400.

Chinnusamy, V., Jagendorf, A., and Zhu, J.-K. (2005). Understanding and improving salt tolerance in plants. *Crop Sci.* **45,** 437–448.

Chinnusamy, V., Schumaker, K., and Zhu, J.-K. (2004). Molecular genetic perspectives on cross-talk and specificity in abiotic stress signaling in plants. *J. Exp. Bot.* **55,** 225–236.

Colmer, T. D., Flowers, T. J., and Munns, R. (2006). Use of wild relatives to improve salt tolerance in wheat. *J. Exp. Bot.* **57,** 1059–1078.

Delauney, A. J., and Verma, D. P. S. (1993). Proline biosynthesis and osmoregulation in plants. *Plant J.* **4,** 215–223.

Desai, M. K., Mishra, R. N., Verma, D., Nair, S., Sopory, S. K., and Reddy, M. K. (2006). Structural and functional analysis of a salt stress inducible gene encoding voltage dependent anion channel (VDAC) from pearl millet (*Pannisetum glaucum*). *Plant Physiol. Biochem.* **44,** 483–493.

Flowers, T. J. (2004). Improving crop salt tolerance. *J. Exp. Bot.* **55,** 307–319.

Genoud, T., and Metraux, J. P. (1999). Crosstalk in plant cell signaling: Structure and function of genetic network. *Trends Plant Sci.* **4,** 503–507.

Harinasut, P., Tsutsui, K., Takabe, T., Nomura, M., and Kishitani, S (1996). Exogenous glycine beatin accumulation and increaded salt tolerance in rice seedlings. *Biosci. Biotechnol. Biochem.* **60,** 366–368.

Hu, Y., and Schmidhalter, U. (2005). Drought and salinity: A comparison of their effects on mineral nutrition of plants. *J. Plant Nutr. Soil Sci.* **168,** 541–549.

Kinight, H., Trewavas, A. J., and Knight, M. R. (1997). Calcium signaling in *Arabidopsis thaliana* responding to drought and salinity. *Plant J.* **12,** 1067–1078.

Kishor, P. B. K., Hong, Z., Miao, G. H., Hu, C. A. A., and Verma, D. P. S. (1995). Overexpression of [delta]-pyrroline-5-carboxylate synthetase increases proline production and confers osmotolerance in transgenic plants. *Plant Physiol.* **108,** 1387–1394.

Ma, S., Gong, Q., and Bohnert, H. J. (2006). Dissecting salt stress pathways. *J. Exp. Bot.* **57,** 1097–1107.

Mahajan, S., Sopoy, S. K., and Tuteja, N. (2006a). CBL-CIPK paradigm: Role in calcium and stress signaling in plants. *Proc. Indian Natn. Sci. Acad.* **72,** 63–78.

Mahajan, S., Sopoy, S. K., and Tuteja, N. (2006b). Cloning and characterization of CBL-CIPK signaling components from a legume (*Pisum sativum*). *FEBS J.* **273,** 907–925.

Mahajan, S., and Tuteja, N. (2005). Cold, salinity and drought stresses: An overview. *Arch. Biochem. Biophys.* **444,** 139–158.

Malinowski, D. P., and Belesky, D. P. (2000). Adaptation of endophyte-infected cool-season grasses to environmental stresses: Mechanisms of drought and mineral stress tolerance. *Crop Sci.* **40,** 923–940.

Munns, R., James, R. A., and Lauchli, A. (2006). Approaches to increasing the salt tolerance of wheat and other cereals. *Exp. J. Bot.* **57,** 1025–1043.

Owttrim, G. W. (2006). RNA helicase and abiotic stress. *Nucleic Acids Res.* **34,** 3220–3230.

Platten, D. J., Cotsaftis, O., Berthomieu, P., Bohnert, H., Davenport, R. J., Fairbairn, D. J., Horie, T., Leigh, R. A., Lin, H.-X., Luan, S., Maser, P., Pantoja, O., Rodriguez-Navarro, A., Schachtman, D. P., Schroeder, J. I., Sentenac, H., Uozumi, N., Very, A. A., Zhu, J.-K., Dennis, E. S., and Tester, M. (2006). Nomenclature for HKT transporters, key determinants of plant salinity tolerance. *Trends Plant Sci.* **11,** 372–374.

Rhodes, D., and Hanson, A. D. (1993). Quaternary ammonium and tertiary sulfonium compounds in higher-plants. *Annu. Rev. Plant Physiol. Plant Mol. Biol.* **44,** 357–384.

Sanan-Mishra, N., Phan, X. H., Sopory, S. K., and Tuteja, N. (2005). Pea DNA helicase 45 overexpression in tobacco confers high salinity tolerance without affecting yield. *Proc. Natl. Acad. Sci. USA* **102,** 509–514.

Sanan-Mishra, N., Tuteja, R., and Tuteja, N. (2006). Signaling through MAP kinase networks in plants. *Arch. Biochem. Biophys.* **452,** 55–68.

Seki, M., Narusaka, M., Ishida, J., Nanjo, T., Fujita, M., Oono, Y., Kamiya, A., Nakajima, M., Enju, A., Sakurai, T., Satou, M., Akiyama, K., Taji, T., Yamaguchi-

Shinozaki, K., Carninci, P., Kawai, J., Hayashizaki, Y., and Shinozaki, K. (2002). Monitering the expression of profiles of 7000 Arabidopsis genes under drought, cold, and high salinity stresses using a full-length cDNA microarray. *Plant J.* **31,** 279–292.

Shinozaki, K., and Yamaguchi-Shinozaki, K. (1997). Gene expression and signal transduction in water-stress response. *Plant Physiol.* **115,** 327–334.

Shoji, T., Suzuki, K., Abe, T., Kaneko, Y., Shi, H., Zhu, J. K., Rus, A., Hasegawa, P. M., and Hashimoto, T. (2006). Salt stress affects cortical microtubule organization and helical growth in Arabidopsis. *Plant Cell Physiol.* **47,** 1158–1168.

Singla-Pareek, S. L., Yadav, S. K., Pareek, A., Reddy, M. K., and Sopory, S. K. (2006). Transgenic tobacco overexpressing glyoxalase pathway enzymes grow and set viable seeds in zinc-spiked soils. *Plant Physiol.* **140,** 613–623.

Thiery, L., Leprince, A., Lefebvre, D., Ghars, M. A., Debabieux, E., and Savoure, A. (2004). Phospholipase D is a negative regulator of proline biosynthesis in *Arabidopsis thaliana*. *J. Biol. Chem.* **279,** 14812–14818.

Tuteja, N., and Tuteja, R. (2004). Prokaryotic and eukaryotic DNA helicases: Essential molecular motor proteins for cellular machinery. *Eur. J. Biochem.* **271,** 1835–1848.

Vashisht, A. A., Pradhan, A., Tuteja, R., and Tuteja, N. (2005). Cold and salinity stress-induced bipolar pea DNA helicase 47 is involved in protein synthesis and stimulated by phosphorylation with proteon kinase C. *Plant J.* **44,** 76–87.

Vashisht, A. A., and Tuteja, N. (2006). Stress responsive DEAD-box helicases: A new pathway to engineer plant stress tolerance. *J. Phytochem. Photobiol.* **84,** 150–160.

Vinocur, B., and Altman, A. (2005). Recent advances in engineering plant tolerance to abiotic stress: Achievements and limitations. *Curr. Opin. Biotech.* **16,** 123–132.

Wu, Y., Lei, D., and Zhu, J. K. (1996). *SOS1*, a genetic locus essential for salt tolerance and potassium acquisition. *Plant Cell* **8,** 617–627.

Xiong, L., Ishitini, M., Lee, H., and Zhu, J.-K. (2001). The Arabidopsis LOS5/ABA3 locus encodes a molybdenum cofactor sulfurase and modulates cold stress and osmotic stress responsive gene expression. *Plant Cell* **13,** 2063–2083.

Xiong, L., Schumaker, K. S., and Zhu, J.-K. (2002). Cell signaling during cold, drought and salt stress. *Plant Cell* **14,** S165–S183.

Yamaguchi-Shinozaki, K., and Shinozaki, K. (1993). Characterization of the expression of a desiccation-responsive *rd29* gene of *Arabidopsis thaliana* and analysis of its promoter in transgenic plants. *Mol. Gen. Genet.* **236,** 331–340.

Yan, H., Gong, L. Z., Zhao, C. Y., and Guo, W. Y. (2000). Effects of exogenous proline on the physiology of soybean plantlets regenerated from embryos *in vitro* and on the ultrastructure of their mitochondria under NaCl stress. *Soybean Sci.* **19,** 314–319.

Yang, X., and Lu, C. (2005). Photosynthesis is improved by exogenous glycine-betain in salt-stressed maize plants. *Physiol. Plant* **124,** 343–352.

Zhang, J. Z., Creelman, R. A., and Zhu, J.-K. (2004). From laboratory to field: Using information from Arabidopsis to engineer salt, cold, and drought tolerance in crops. *Plant Physiol.* **135,** 615–621.

Zhang, T., Liu, Y., Xue, L., Xu, S., Chen, T., Yang, T., Zhang, L., and An, L. (2006a). Molecular cloning and characterization of a novel MAP kinase gene in *Chorispora bungeana*. *Plant Physiol. Biochem.* **44,** 78–84.

Zhang, T., Liu, Y., Yang, T., Zhang, L., Xu, S., Xue, L., and An, L. (2006b). Diverse signals converge at MAPK cascades in plant. *Plant Physiol. Biochem.* **44,** 274–283.

Zhu, J.-K. (2002). Salt and drought stress signal transduction in plants. *Annu. Rev. Plant Biol.* **53,** 247–273.

CHAPTER TWENTY-FIVE

PHENOTYPE OF THE TAURINE TRANSPORTER KNOCKOUT MOUSE

Ulrich Warskulat,* Birgit Heller-Stilb,* Evelyn Oermann,[†]
Karl Zilles,[†,‡] Helmut Haas,[§] Florian Lang,[¶] and Dieter Häussinger*

Contents

Abstract

This chapter reports present knowledge on the properties of mice with disrupted gene coding for the taurine transporter (*taut−/−* mice). Study of those mice unraveled some of the roles of taurine and its membrane transport for the development and maintenance of normal organ functions and morphology. When compared with wild-type controls, *taut−/−* mice have decreased taurine levels in skeletal and heart muscle by about 98%, in brain, kidney, plasma, and retina by 80 to 90%, and in liver by about 70%. *taut−/−* mice exhibit a lower body mass as well as a strongly reduced exercise capacity compared with

* Clinic for Gastroenterology, Hepatology and Infectiology, University of Düsseldorf, Germany
† C.&O. Vogt Institute for Brain Research, University of Düsseldorf, Germany
‡ Institute of Medicine, Research Center Jülich, Germany
§ Department of Neurophysiology, University of Düsseldorf, Germany
¶ Department of Physiology, University of Tübingen, Germany

Methods in Enzymology, Volume 428
ISSN 0076-6879, DOI: 10.1016/S0076-6879(07)28025-5

taut+/− and wild-type mice. Furthermore, *taut*−/− mice show a variety of pathological features, for example, subtle derangement of renal osmoregulation, changes in neuroreceptor expression, and loss of long-term potentiation in the striatum, and they develop clinically relevant age-dependent disorders, for example, visual, auditory, and olfactory dysfunctions, unspecific hepatitis, and liver fibrosis.

Taurine-deficient animal models such as acutely dietary-manipulated foxes and cats, pharmacologically induced taurine-deficient rats, and taurine transporter knockout mouse are powerful tools allowing identification of the mechanisms and complexities of diseases mediated by impaired taurine transport and taurine depletion (Chapman *et al.*, 1993; Heller-Stilb *et al.*, 2002; Huxtable, 1992; Lake, 1993; Moise *et al.*, 1991; Novotny *et al.*, 1991; Pion *et al.*, 1987; Timbrell *et al.*, 1995; Warskulat *et al.*, 2004, 2006b). Taurine, which is the most abundant amino acid in many tissues, is normally found in intracellular concentrations of 10 to 70 mmol/kg in mammalian heart, brain, skeletal muscle, liver, and retina (Chapman *et al.*, 1993; Green *et al.*, 1991; Huxable, 1992; Timbrell *et al.*, 1995). These high taurine levels are maintained by an ubiquitous expression of Na^+-dependent taurine transporter (TAUT) in the plasma membrane (Burg, 1995; Kwon and Handler, 1995; Lang *et al.*, 1998; Liu *et al.*, 1992; Ramamoorthy *et al.*, 1994; Schloss *et al.*, 1994; Smith *et al.*, 1992; Uchida *et al.*, 1992; Vinnakota *et al.*, 1997; Yancey *et al.*, 1975). Taurine is not incorporated into proteins. It is involved in cell volume regulation, neuromodulation, antioxidant defense, protein stabilization, stress responses, and via formation of taurine-chloramine in immunomodulation (Chapman *et al.*, 1993; Green *et al.*, 1991; Huxtable, 1992; Timbrell *et al.*, 1995). On the basis of its functions, taurine may protect cells against various types of injury (Chapman *et al.*, 1993; Green *et al.*, 1991; Huxtable, 1992; Kurz *et al.*, 1998; Park *et al.*, 1995; Stapleton *et al.*, 1998; Timbrell *et al.*, 1995; Welch and Brown, 1996; Wettstein and Häussinger, 1997).

In order to examine the multiple taurine functions, murine models have several intrinsic advantages for *in vivo* research compared to other animal models, including lower cost, maintenance, and rapid reproduction rate. Further, experimental reagents for cellular and molecular studies are widely available for the mouse. In particular, mice can be easily genetically manipulated by making transgene and knockout mice. This chapter focuses on the phenotype of the TAUT-deficient murine model (*taut*−/−; Heller-Stilb *et al.*, 2002), which may help researchers elucidate the diverse roles of taurine in development and maintenance of normal organ functions and morphology.

1. Targeted Disruption of the *TAUT* Gene

An 18-kb *Xho*I genomic clone was isolated from a λ phage library containing mouse strain 129SvJ genomic DNA by screening with a cDNA *taut* probe (Heller–Stilb *et al.*, 2002). To disrupt the *taut* locus, an isogenic targeting vector was designed to delete exon 1 of *taut*. Embryonal stem cells

(ES) cells (derived from 129/SvJ strain) were electroporated with the linearized vector, and colonies underwent positive–negative selection with geneticin and gancyclovir (Mansour *et al.*, 1988). Double-resistant clones were screened for the desired homologous recombination by polymerase chain reaction (PCR). After analysis of PCR-positive clones by Southern blot hybridization, ES clones containing the targeting event were injected into blastocysts of C57BL/6 mice to produce chimeras. Heterozygous (*taut+/−*) mice from ES clones, which transmitted the *taut* mutation through the germ line, were intercrossed to produce wild-type (*taut+/+*), heterozygous (*taut+/−*), and homozygous (*taut−/−*) animals in the F2 progeny.

2. REDUCED TAURINE LEVELS LEAD TO VARIOUS DISEASES IN *TAUT−/−* MICE

The deletion of exon 1 of the *taut* gene leads to a truncated, nonfunctional protein of 450 amino acids (wild-type: 621) as shown by *in vitro* transcription and translation assays and measurement of taurine uptake in fibroblasts from the null mutation (Heller-Stilb *et al.*, 2002).

taut−/− mice exhibit an up to 25% lower body mass compared with *taut+/-* and wild-type mice (Fig. 25.1). Taurine tissue levels are strongly reduced in *taut−/−* mice when compared with wild-type mice (Heller-Stilb *et al.*, 2002; Warskulat *et al.*, 2004, 2006b): *taut−/−* mice exhibit a

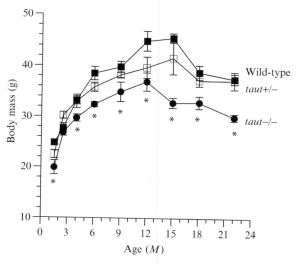

Figure 25.1 *taut−/−* mice exhibit a lower body mass compared with *taut+/−* and wild-type mice. Body masses of male *taut−/−*, *taut+/−* and wild-type mice are shown. Data are expressed as means $\pm$ SEM ($n = 5$–17; $^{*}P < 0.05$ vs wild-type mice).

decrease in taurine levels in skeletal and heart muscle by about 98% and in brain, retina, kidney, liver, and plasma by 70 to 90%. In addition to a reduced fertility (Heller-Stilb *et al.*, 2002), *taut*−/− mice show various pathologies, for example, reduced exercise capacity (Warskulat *et al.*, 2004), changes in neuroreceptor expression (Oermann *et al.*, 2005), and loss of long-term potentiation in the striatum (Sergeeva *et al.*, 2003), and develop loss of vision (Heller-Stilb *et al.*, 2002; Rascher *et al.*, 2004) and hearing (Jiang *et al.*, 2005), olfactory dysfunction (Witt *et al.*, 2003), altered renal osmoregulation (Huang *et al.*, 2006), and liver disease (Warskulat *et al.*, 2006b). Interestingly, circulating erythrocytes from *taut*−/− mice are slightly but significantly more resistant against suicidal cell death following osmotic shock (Lang *et al.*, 2003).

3. REDUCED EXERCISE CAPACITY IN *TAUT*−/− MICE

Taurine was proposed to play a role in the excitation–contraction coupling mechanism of muscle fibers (De Luca *et al.*, 2000). In the myocardium, taurine may have antiarrhythmic properties, positive inotropic effects, and cytoprotective functions (Baum and Weiß, 2001; Chapman *et al.*, 1993; Sawamura *et al.*, 1990; Schaffer and Azuma, 1992). Although *taut*−/− mice have a near complete depletion of taurine in heart and skeletal muscle, these mice show an almost normal heart function (Warskulat *et al.*, 2004). *In vivo* magnetic resonance imaging yielded no evidence of any anatomical alterations of the heart and showed comparable values for diastolic and systolic volumes, as well as cardiac output in wild-type and *taut*−/− mice. More detailed hemodynamic and metabolic measurements on isolated perfused hearts also revealed no significant differences between *taut*-deficient and wild-type hearts.

Because taurine deficiency has been implicated as a potential cause of dilated cardiomyopathy in other animal models (Lake, 1993; Moise *et al.*, 1991; Novotny *et al.*, 1991; Pion *et al.*, 1987), it is surprising that mice with chronically reduced cardiac taurine levels show a largely normal heart function. It should be noted that taurine deficiency induced by acute dietary manipulation results in cardiomyopathy in fox and cat (Moise *et al.*, 1991; Novotny *et al.*, 1991; Pion *et al.*, 1987). Moreover, rats treated with the taurine transport antagonist guanidinoethane sulfonate showed disordered contractile filaments and clear losses of myofibrillar bundles (Lake, 1993). In clinical studies, a correlation between low plasma taurine concentrations and echocardiographic findings representative of dilated cardiomyopathy was reported (Pion *et al.*, 1992).

Ergometric analysis demonstrated that the exercise capacity of *taut*−/− mice is reduced by more than 80% compared to wild-type controls (Fig. 25.2). Analysis of *taut*−/− skeletal muscle revealed electromyographic abnormalities (Warskulat *et al.*, 2004). X-ray studies of the skeleton did

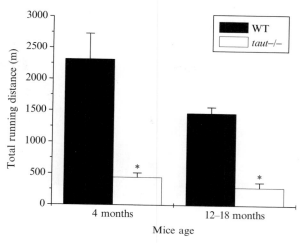

Figure 25.2 Exercise capacity of wild-type and *taut—/—* mice. Three- to 4-month- and 12- to 18-month-old wild-type and *taut—/—* mice were subjected to treadmill exercise. After mice were allowed to equilibrate, treadmill activity was initiated at 4 m/min, 14° inclination, and increased to 8 m/min 10 min later. Young and old wild-type mice ran a total distance of 2316 ± 413 m (294 ± 52 min) and 1464 ± 99 m (188 ± 12 min) to exhaustion, whereas age-matched *taut—/—* mice ran only a total distance of 443 ± 70 m (60 ± 9 min) and 287 ± 83 m (41 ± 10 min), respectively. Data are expressed as means ± SEM ($n = 3–4$ for 3- to 4-month-old mice and $n = 9–12$ for 12- to 18-month-old mice; $^{*}P < 0.05$ vs wild-type mice). From Warskulat *et al.* (2004).

not reveal morphological abnormalities in *taut—/—* mice, indicating that their reduced physical performance is not a result of skeletal malformation. ^{1}H-NMR spectroscopy of *taut—/—* skeletal muscle extracts showed a deficit of more than 10 mM in total organic osmolyte concentration, probably leading to functional defects (Fig. 25.3). In the heart of *taut—/—* mice the lack of taurine is compensated by the upregulation of various organic solutes, resulting in maintenance of a normal osmolyte balance (Fig. 25.3). The reason why heart but not skeletal muscle can compensate for the loss of taurine is presently not known.

For further examination of the taurine-compensative mechanisms in the *taut—/—* heart, age-dependent cardiac gene expression was studied in wild-type and *taut—/—* mice using a mouse-specific DNA microarray that includes 251 genes with relevance for heart function (Warskulat *et al.*, 2006a). α-Actin type 1 mRNA levels were decreased by 70% in the heart of *taut—/—* mice compared to wild-type controls. Interestingly, the hearts of *taut—/—* mice showed a switch in α-actin 1 to 2 expression as confirmed by real-time PCR and Western blot analysis. In addition, mRNA levels of biomarkers for pressure overload and hypertension were upregulated in *taut—/—* hearts, pointing to a stress situation in the heart of *taut—/—* mice even under low stress laboratory conditions.

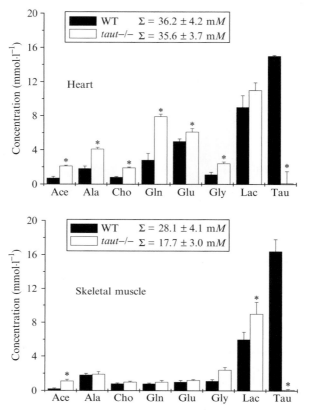

Figure 25.3 Concentration of organic osmolytes in perchloric acid extracts of heart (left) and skeletal muscle (right) of 15-month-old *taut−/−* and wild-type (WT) mice. Ace, acetate; Ala, alanine; Cho, cholines (including choline, phosphocholine, and glycerophosphocholine); Gln, glutamine; Glu, glutamate; Gly, glycine; Lac, lactate; Tau, taurine. Data are expressed as means ± SD ($n = 3$, $^*P < 0.05$ vs wild-type mice). From Warskulat *et al.* (2004).

4. PATHOPHYSIOLOGICAL CHANGES IN BRAIN, RETINA, OLFACTORY BULB, AND INNER EAR OF *TAUT−/−* MICE

4.1. Deficit in taurine-evoked synaptic enhancement and changes in neuroreceptor expression in *taut−/−* mice

Taurine binds to glycine and $GABA_A$ receptors and may act either as a classical neurotransmitter or as a modulator of neuronal excitability (Bernardi, 1985; Bureau and Olsen, 1991; del Olmo *et al.*, 2000; Haas and Hösli, 1973; Hussy *et al.*, 1997; Pow *et al.*, 2002; Sergeeva and Haas, 2001).

Taurine evokes a long-lasting enhancement (LLE$_{TAU}$) of synaptic transmission in hippocampal and corticostriatal slices (Chepkova *et al.*, 2002; Sergeeva *et al.*, 2003). *taut*$-/-$ mice display a low ability to develop LLE$_{TAU}$ in the striatum, but not in the hippocampus. The taurine accumulation triggering long-term enhancement of synaptic responses in the two studied brain structures is provided by different transport systems: striatal LLE$_{TAU}$ needs activation of the taurine (not GABA) transporter, whereas GABA (not taurine) transporters are involved in hippocampal LLE$_{TAU}$. The different mechanisms of taurine-induced synaptic plasticity may reflect the different vulnerabilities of these brain regions under pathological conditions such as brain edema and hepatic encephalopathy that are accompanied by osmotic changes, volume regulation by taurine efflux, and loss of taurine from cells (Butterworth, 2001; Saransaari and Oja, 1991; Scheller *et al.*, 2000).

Hyperammonemia impairs long-term potentiation, a cellular model of learning and memory, evoked by high-frequency stimulation in the hippocampus. Taurine and the mitochondrial enhancer L-carnitine (but not several antioxidants) prevent this impairment (Chepkova *et al.*, 2006).

The balance between excitation and inhibition in the central nervous system (CNS) is regulated not only at the neurotransmitter level, but also at the level of the respective receptors. Changes in neuroreceptor expression and binding are present under pathological conditions such as hepatic encephalopathy (Butterworth, 2001; Saransaari and Oja, 1991; Scheller *et al.*, 2000). Such changes were also found in *taut*$-/-$ mice (Oermann *et al.*, 2005) using quantitative *in vitro* receptor autoradiography, *taut*$-/-$ mice show significantly higher GABA$_A$ receptor densities in the molecular layer of the hippocampal dentate gyrus and in the cerebellum compared to wild-type animals. The densities of kainate receptors are significantly higher in the caudate-putamen, the CA1, and hilus regions of the hippocampus and in the cerebellum of *taut*$-/-$ animals. Significantly increased AMPA receptor densities were also found in the caudate-putamen and the cerebellum. It is exceptional that the cerebellum of *taut*$-/-$ mice showed enhanced densities of all three mentioned receptor types.

4.2. Apoptosis leads to retinal degeneration in *taut*$-/-$ mice

The most prominent morphological feature of young *taut*$-/-$ mice is a severe and progressive retinal degeneration (Fig. 25.4). At 2 weeks postnatal age, the usual age of eye opening, an approximately equal number of rows of photoreceptor nuclei in *taut*$-/-$ mice and wild-type mice is still apparent. Inner and outer photoreceptor segments in *taut*$-/-$ mice are clearly distinguishable but are already smaller than in the controls. Retinal function is observable at this stage as determined by recording of electroretinogram (ERG), but amplitudes are reduced to about one-third of those of wild-type controls (Heller-Stilb *et al.*, 2002). One week later, ERG amplitudes

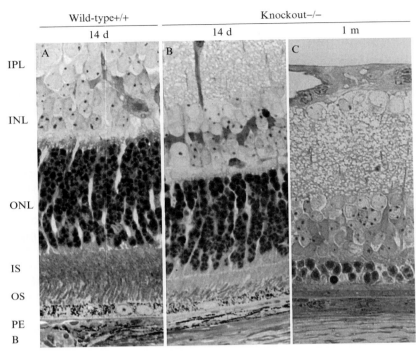

Figure 25.4 Retinal morphology of wild-type and *taut−/−* mice. (Left) Retina of wild-type (*taut+/+*) mice at postnatal day 14. Retinas of *taut−/−* mice at postnatal day 14 (center) and 1 month (right) (magnification ×450). Adapted from Heller-Stilb *et al.* (2002). IPL, inner plexiform layer; INL, inner nuclear layer; ONL, outer nuclear layer (photoreceptor nuclei); IS/OS, inner/outer segments of photoreceptors; PE, pigment epithelium; B, membrane of Bruch. (See color insert.)

dropped to 10 to 15% of normal. At 1 month of age, the number of photoreceptor cell nuclei is reduced markedly in *taut−/−* mice and the retina reveals advanced stages of photoreceptor cell degeneration (Fig. 25.4), which is associated with a complete loss of retinal function (Heller-Stilb *et al.*, 2002). Similar to previous observations in taurine-deficient cats (Pasantes-Morales *et al.*, 1986), light deprivation slows but does not prevent the loss of photoreceptors in *taut−/−* mice (Rascher *et al.*, 2004).

Apoptosis underlies the rapid loss of photoreceptor cells in *taut−/−* mice (Heller-Stilb *et al.*, 2002). Apoptosis is characterized by cell shrinkage, nuclear condensation, DNA fragmentation, and apoptotic body formation and has been viewed as the "final common pathway" in photoreceptor degeneration (Chang *et al.*, 1993; Petrosian *et al.*, 1996; Portera-Cailliau *et al.*, 1994; Remé *et al.*, 1998). Most of the retinal taurine is located in the photoreceptor cells (Huxtable, 1992; Lake and Verdone-Smith, 1989;

Pow *et al.*, 2002), and it was shown previously that decreased taurine levels lead to photoreceptor cell degeneration and retinal malfunction in cats and rats (Hayes and Carey, 1975; Lake and Verdone-Smith, 1989; Lombardini, 1999; Pasantes-Morales *et al.*, 1986; Schmidt *et al.*, 1976; Wright *et al.*, 1987). It has been discussed that an absence of taurine as a radical scavenger may accelerate apoptosis (Obrosova *et al.*, 2001). Thus, reduced scavenging of free radicals may play a major role in the retinal degeneration of *taut−/−* mice. Alternatively, decreased taurine levels may cause retinal degeneration in *taut−/−* mice by impairment of retinoid transport, as taurine seems to be an essential component in the transport of retinoids between pigment epithelium and retina: taurine conjugates with retinaldehyde to 11-*cis*-tauret, which regenerates rhodopsin in the rod outer segments (Petrosian and Haroutounian, 1998, 2000; Petrosian *et al.*, 1996). In this regard, rhodopsin deficiency induces retinal degeneration in rhodopsin gene knockout mice (Humphries *et al.*, 1997).

Apoptosis plays a major role in various human and animal retinal dystrophies, including human retinitis pigmentosa (RP) (Lombardini, 1999; Wong, 1994). RP is a genetically heterogeneous group of hereditary retinal dystrophies (Arshinoff *et al.*, 1981; Pasantes-Morales *et al.*, 2002). It is characterized by the progressive degeneration and loss of retinal cells and represents, with an incidence of about 1:4000, the most common cause of inherited blindness in the adult. The amino acid profiles of RP patients have been found to be abnormal (van Soest *et al.*, 1999). Taurine is only one of these amino acids, but it may be of major importance in a subgroup of patients only (Gal *et al.*, 1997). No mutated genes affecting taurine transport or metabolism have been identified so far in human RP. However, reduced taurine uptake and decreased plasma taurine levels in some RP patients have been described (Benini *et al.*, 1988; Lombardini, 1999; Uma *et al.*, 1983). Thus, *taut* may be another candidate gene whose mutation may underlie RP in a subgroup of patients.

4.3. Olfactory dysfunction of *taut−/−* mice

The high taurine content in the olfactory system has been the subject of various studies (Belluzzi *et al.*, 2004; Chaput *et al.*, 2004; Kamisaki *et al.*, 1996; Kratskin *et al.*, 2000). Olfactory receptor neurons (ORN) are unique in the CNS for their ability to regenerate, especially after injury (Farbman *et al.*, 1988; Schwob, 2002). The normal olfactory epithelium is a pseudo-stratified arrangement of two types of basal (progenitor) cells, olfactory receptor neurons, and large supporting cells.

In contrast to the taurine-deficient retina, no major degeneration is seen in the olfactory epithelium (OE) of *taut−/−* mice. The overall structure of OE and olfactory bulb in *taut−/−* mice appears to be normal. However, immunohistochemical and electrophysiological studies in *taut−/−* mice suggest an altered cellular stress response and changed cell dynamics in the OE (Witt *et al.*, 2003). In particular, an increased number of immature ORN is observed in *taut−/−* animals. The significantly higher proliferative rate in *taut−/−* mice may be one way to compensate for the taurine deficiency. Because of the loss of protective capacity of taurine against oxidative stress (Huxtable, 1992) or excitotoxicity (El Idrissi and Trenkner, 1999), there is a need for a faster replacement of ORN. Furthermore, increased proliferative activity seems to be restricted to immature, basally located cells. However, many proliferating cells apparently do not become mature, either being held in an immature state or differentiating into the direction of supporting cells. Degenerative processes often imply a higher apoptotic rate (Cowan and Roskams, 2002; Magrassi and Graziadei, 1995; Mahalik, 1996). Indeed, *taut−/−* mice with higher proliferative activity also show more apoptotic cells (Witt *et al.*, 2003).

Analysis of electro-olfactograms revealed a decreased responsiveness to olfactory (H_2S) and trigeminal stimuli (CO_2) in *taut−/−* mice compared to control mice (Witt *et al.*, 2003). It is suggested that impaired chemosensitivity could be a general feature of taurine deficiency.

4.4. Age-dependent loss of hearing in *taut−/−* mice

taut−/− mice also show enhanced degeneration in the inner ear and auditory nerve (Jiang *et al.*, 2005). The accelerated losses of outer and inner hair cells (IHC) are accompanied by 20-dB higher auditory brain stem response thresholds in 6-month-old *taut−/−* mice compared to wild-type animals. Peripheral auditory nerve fibers are present in the habenula perforata except in regions of the cochlea associated with missing IHC and spiral ganglion neurons (SGN). Loss of SGN is greater in *taut−/−* mice than in wild-type mice. Interestingly, the central region of auditory nerve fibers in the modiolus shows significantly more degeneration in *taut−/−* than in wild-type mice because most of the glial cells in this region are missing in *taut−/−* mice, but are present in wild-type mice. Thus, the phenotype of *taut−/−* mice suggests that loss of the taurine transporter in glial cells may exacerbate degeneration of auditory nerve fibers in addition to accelerating age-related hair cell loss.

5. Renal Function in *TAUT−/−* Mice

Taurine is accumulated in renal medullary cells during dehydration or infusion of hypertonic saline (Garcia-Perez and Burg, 1991; Nakanishi *et al.*, 1992), an effect paralleled by upregulation of TAUT expression in the outer medulla (Bitoun *et al.*, 2001). Taurine transporter gene knockout leads to a significant renal loss of taurine and subsequent hypotaurinemia in mice (Huang *et al.*, 2006). An increased plasma urea concentration is paralleled by a similar enhanced cellular urea content in *taut*−/− mice (Huang *et al.*, 2006; Lang *et al.*, 2003). Urea is taken up via members of the urea transporter family UT (Hediger *et al.*, 1996). An increase in urea transport can be due to enhanced release of vasopressin, which is known to stimulate urea transport (Shayakul and Hediger, 2004). Compared to control animals, taurine-depleted rats show a higher plasma vasopressin level, which leads to the production of hyperosmolar urine (Mozaffari *et al.*, 2001). Thus, increased vasopressin levels and urea transport may maintain urinary concentrating ability despite the lack of taurine.

The ability of *taut*−/− mice to concentrate or dilute urine upon water deprivation or single water load appears to be normal (Huang *et al.*, 2006). Repeating the oral water load in *taut*−/− mice immediately after normalization of urine flow rate results in delayed diuresis and a higher urinary vasopressin/creatinine ratio, which is a surrogate parameter for daily mean plasma vasopressin concentration (Akizuki *et al.*, 2001; Yao *et al.*, 2004). Water deprivation for 36 h leads to similar antidiuresis and increases of urinary osmolality in *taut*−/− and wild-type mice but to higher plasma osmolality, K$^+$, and urea concentrations in *taut*−/− mice (Huang *et al.*, 2006). These findings point to subtle alterations of renal functions and fluid homeostasis at the hypothalamic level. A persistent urinary concentration and evidence for sustained hyperactivity of the vasopressin system following free access to water after a period of dehydration point to an involvement of taurine in mechanisms required to suppress vasopressin release after normalization of plasma osmolarity. Such a role of taurine may also explain the finding that the diuretic response to repeated water loading but not to repeated vasopressin V$_2$ receptor blockade is delayed in the absence of the taurine transporter.

6. Chronic Liver Disease in *taut*−/− Mice

Taurine was identified as an important organic osmolyte in hepatic parenchymal and nonparenchymal cells with impact for immune functions such as phagocytosis, prostanoid, and cytokine formation (Peters–Regehr *et al.*, 1999; Warskulat *et al.*, 1997; Weik *et al.*, 1998). Plasma taurine levels are decreased in liver cirrhosis (Weisdorf *et al.*, 1987; Yamamoto, 1996).

In contrast to other tissues that exhibit severe taurine depletion, such as skeletal and heart muscle from *taut*−/− mice (Heller-Stilb *et al.*, 2002; Warskulat *et al.*, 2004), liver taurine levels are decreased by only about 70% in adult *taut*−/− mice (Warskulat *et al.*, 2006b). Hepatic taurine depletion in *taut*−/− mice is in part compensated osmotically by increased levels of other organic solutes (Warskulat *et al.*, 2006b). Only mildly lowered taurine levels were found in *taut*+/- liver.

When assessed electron microscopically by quantitative analysis of taurine immunogold labeling, the taurine content of liver parenchymal cells from *taut*−/− mice is decreased by only about 30% (Warskulat *et al.*, 2006b). However, Kupffer and sinusoidal endothelial cells exhibit a decrease by more than 80%, indicating severe taurine depletion in nonparenchymal liver cells from *taut*−/− mice.

taut−/− mice develop moderate unspecific hepatitis and liver fibrosis (Fig. 25.5). The process of liver injury appears to progress slowly and major liver pathology is seen beyond the age of 1 year in *taut*−/− mice (Warskulat *et al.*, 2006b). Whereas during the first 6 months of life no significant signs of liver injury are detectable, about 80% of the 1.5- to 2-year-old *taut*−/− mice present with liver inflammation and fibrosis, compared to only 20% of

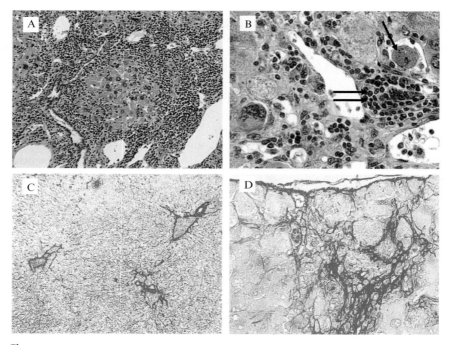

Figure 25.5 Light microscopy of liver sections from wild-type and *taut*−/− mice. *taut*−/− mice develop variable degrees of hepatitis and fibrosis. (A and B) Chronic hepatitis with presence of apoptotic bodies (arrow) and mixed leukocytic infiltrates (double arrow) in a 19-month-old *taut*−/− mouse (×100, ×400). Liver tissue sections were stained with hematoxylin and eosin. (C and D) Liver fibrosis in *taut*−/− mouse. Liver tissue sections were stained with Sirius red. (C) Seventeen-month-old wild-type; (D) 15-month-old *taut*−/− mouse (magnification ×100). Adapted from Warskulat *et al.* (2006b). (See color insert.)

wild-type mice. Interestingly, also 50% of heterozygous *taut+/-* mice develop liver disease at an age beyond 1 year. Thus, it appears that both homo- and heterozygous *taut* knockout mice are prone to unspecific hepatitis and liver fibrosis, but the process may evolve slower in *taut+/-* mice than in *taut-/-* mice.

Hepatocyte destruction and apoptosis are hallmarks of liver pathology in *taut-/-* mice (Warskulat *et al.*, 2006b). The progressive loss of hepatocytes is expected to stimulate hepatocyte proliferation. In *taut-/-* mice, this regenerative process apparently involves oval cell proliferation, which is found abundantly in livers from *taut-/-* mice with chronic hepatitis beyond 1 year of age.

Activation of the CD95 system was found in liver tissue of *taut-/-* mice: increased epidermal growth factor (EGFR) activation, EGFR/CD95 complex formation, and high levels of tyrosine-phosphorylated CD95 with death-inducing complex recruitment (Warskulat *et al.*, 2006b). These events were shown to be required for CD95 membrane trafficking, CD95 activation, and apoptosis induction in hepatocytes (Reinehr *et al.*, 2003a,b, 2004a,b). Activation of the CD95 system is also found in livers from *taut+/-* mice, however, to a lesser degree than in *taut-/-* mice (Warskulat *et al.*, 2006b). No increase in tyrosine-nitrated CD95 is found in livers from *taut*-deficient animals, suggesting that no antiapoptotic nitration of CD95 occurred (Reinehr *et al.*, 2004b). Hydrophobic bile acids are known to trigger hepatocyte apoptosis through activation of the CD95 system (Graf *et al.*, 2002). However, the bile acid conjugation pattern is normal in *taut-/-* mice (Warskulat *et al.*, 2006b). Thus, an increased bile acid-induced activation of the CD95 system in livers from *taut-/-* mice seems unlikely. Activation of the hepatic CD95 system in *taut+/-* and *taut-/-* mice is accompanied by an increased number of apoptotic cells, as shown by staining for activated caspase 3 (Warskulat *et al.*, 2006b). Exposure of young (10 weeks old) wild-type and *taut-/-* mice *in vivo* to an acute oral dose of ethanol (6 g/kg body weight) leads to a stronger activation of the CD95 system in *taut-/-* liver, suggesting that livers from *taut*-deficient mice are more susceptible to liver damage by ethanol (Warskulat *et al.*, 2006b).

In summary, several mechanisms may contribute to activation of the CD95 system in *taut-/-* and *taut+/-* mice: (i) Taurine depletion may lead to hepatocyte shrinkage, which was shown to activate the CD95 system (Reinehr and Häussinger, 2006; Reinehr *et al.*, 2003a,b, 2004a,b); (ii) taurine acts as an antioxidant and taurine deficiency may augment the formation of reactive oxygen species (ROS), which are known upstream triggers of CD95 activation (Medan *et al.*, 2005); (iii) TAUT deficiency results in mitochondrial abnormalities (Warskulat *et al.*, 2006b), which may further augment ROS formation; and (iv) increased tumor necrosis factor-α levels in the plasma of *taut*-deficient mice (Warskulat *et al.*, 2006b) point to an upregulation of proapoptotic cytokines following osmotic stress.

7. AVAILABLE ORGANIC OSMOLYTE TRANSPORTER TRANSGENE AND KNOCKOUT MICE

The phenotype of the *taut*−/− mouse suggests that disorders in osmolyte (shown here for taurine) transport or osmolyte deficiency may predispose to organ damage. Interestingly, liver disease also develops in heterozygous TAUT knockout (*taut*+/−) mice, which display near-normal taurine plasma levels (Warskulat *et al.*, 2006b). However, defective taurine transport in heterozygotes may be unmasked in these animals upon challenges of taurine homeostasis during stress conditions.

It will be of interest to assess the phenotype of other organic osmolyte-deficient or nontransporting mouse models. Only a few organic osmolyte transporter transgene and knockout mice have been available until now. Newborn sodium/*myo*-inositol cotransporter (SMIT) knockout mice, which have an 84% reduction in whole body *myo*-inositol, die in contrast to *taut*−/− mice (Heller-Stilb *et al.*, 2002) shortly after birth as a consequence of hypoventilation (Berry *et al.*, 2003). *myo*-Inositol acts like taurine as an osmolyte in the brain, but SMIT is also essential for the development and function of the peripheral nerves in mice (Chau *et al.*, 2005). Heterozygous SMIT knockout mice exhibit no neurobehavioral disturbances (Shaldubina *et al.*, 2007). Likewise, young heterozygous TAUT knockout mice show no retinal and olfactory malfunctions in contrast to the null mutation (Heller-Stilb *et al.*, 2002; Witt *et al.*, 2003). However, overexpression of the *smit* gene in mouse lens led to an osmotic cataract (Cammarata *et al.*, 1999; Jiang *et al.*, 2000). Mouse models for betaine depletion or betaine transporter knockout are not yet published; however, transgenic mice have been generated that contain in their genome 2.4 kb of the 5′-flanking region of the canine betaine transporter subunit 1 gene (BGT-1) in front of a reporter gene (Kaneko *et al.*, 1997). It was shown in these mice that this promoter region mediates an increase in transcription of reporter gene in response to hyperosmolarity in the renal medulla.

Conditional gene knockout mice have not yet been used for studies on osmolyte transporters. Conditional gene knockout murine models have several intrinsic advantages for *in vivo* research compared to conventional gene knockouts, including tissue-specific and age-dependent gene knockout examinations, which may reduce lethality often seen in conventional gene knockout. Thus, further studies in conditional osmolyte transporter gene knockout mice are of great interest.

REFERENCES

Akizuki, N., Uchida, S., Sasaki, S., and Marumo, F. (2001). Impaired solute accumulation in inner medulla of Clcnk1−/− mice kidney. *Am. J. Physiol. Renal Physiol.* **280,** F79–F87.
Arshinoff, S., McCulloch, J. C., Macrae, W., Stein, A. N., and Maliss, E. B. (1981). Amino acids in retinitis pigmentosa. *Br. J. Ophthal.* **65,** 626–630.

Baum, M., and Weiβ, M. (2001). The influence of a taurine containing drink on cardiac parameters before and after exercise measured by echocardiography. *Amino Acids* **20**, 75–82.

Belluzzi, O., Puopolo, M., Benedusi, M., and Kratskin, I. (2004). Selective neuroinhibitory effects of taurine in slices of rat main olfactory bulb. *Neuroscience* **124**, 929–944.

Benini, G., Marconini, C., Wirth, A., and Montagnoli, G. (1988). Is taurine a marker for retinitis pigmentosa? *Ophthalmologia* **197**, 130–135.

Bernardi, N. (1985). On the role of taurine in the cerebellar cortex: A reappraisal. *Acta Physiol. Pharmacol. Latinoam.* **35**, 153–164.

Berry, G. T., Wu, S., Buccafusca, R., Ren, J., Gonzales, L. W., Ballard, P. L., Golden, J. A., Stevens, M. J., and Geer, J. J. (2003). Loss of murine Na+/myo–inositol cotransporter leads to brain myo–inositol depletion and central apnea. *J. Biol. Chem.* **278**, 18297–18302.

Bitoun, M., Levillain, O., and Tappaz, M. (2001). Gene expression of the taurine transporter and taurine biosynthetic enzymes in rat kidney after antidiuresis and salt loading. *Pflüg. Arch.* **442**, 87–95.

Bureau, M. H., and Olsen, R. W. (1991). Taurine acts on a subclass of GABA$_A$ receptors in mammalian brain *in vitro*. *Eur. J. Pharmacol.* **207**, 9–16.

Burg, M. H. (1995). Molecular basis of osmotic regulation. *Am. J. Physiol.* **268**, F983–F996.

Butterworth, R. F. (2001). Neurotransmitter dysfunction in hepatic encephalopathy: New approaches and new findings. *Metab. Brain Dis.* **16**, 55–65.

Cammarata, P. R., Zhou, C., Chen, G., Singh, I., Reeves, R. E., Kuszak, J. R., and Robinson, M. L. (1999). A transgenic animal model of osmotic cataract. 1. Overexpression of bovine Na$^+$/myo–inositol cotransporter in lens fibers. *Invest. Ophthlmol. Vis. Sci.* **40**, 1727–1737.

Chang, G. Q., Hao, Y., and Wong, F. (1993). Apoptosis: Final common pathway of photoreceptor death in rd, rds, and rhodopsin mutant mice. *Neuron* **11**, 595–605.

Chapman, R. A., Suleiman, M. S., and Earm, Y. E. (1993). Taurine and the heart. *Cardiovasc. Res.* **27**, 358–363.

Chaput, M. A., Palouzier-Paulignan, B., Delaleu, J. C., and Duchamp-Viret, P. (2004). Taurine action on mitral cell activity in the frog olfactory bulb *in vivo*. *Chem. Senses* **29**, 83–91.

Chau, J. F., Lee, M. K., Law, J. W., Chung, S. K., and Chung, S. S. (2005). Sodium/myo–inositol cotransporter-1 is essential for the development and function of the peripheral nerves. *FASEB J.* **19**, 1887–1889.

Chepkova, A. N., Doreulee, N., Yanovsky, Y., Mukhopadhyay, D., Haas, H. L., and Sergeeva, O. A. (2002). Long-lasting enhancement of corticostriatal neurotransmission by taurine. *Eur. J. Neurosci.* **16**, 1523–1530.

Chepkova, A. N., Sergeeva, O. A., and Haas, H. L. (2006). Taurine rescues hippocampal long-term potentiation from ammonia-induced impairment. *Neurobiol. Dis.* **23**, 512–521.

Cowan, C. M., and Roskams, A. J. (2002). Apoptosis in the mature and developing olfactory neuroepithelium. *Microsc. Res. Tech.* **58**, 204–215.

del Olmo, N., Bustamante, J., del Rio, R. M., and Solis, J. M. (2000). Taurine activates GABA(A) but not GABA(B) receptors in rat hippocampal CA1 area. *Brain Res.* **864**, 298–307.

De Luca, A., Pierno, S., Tricarico, D., Desaphy, J. F., Liantonio, A., Barbieri, M., Camerino, C., Montanari, L., and Camerino, D. C. (2000). Taurine and skeletal muscle ion channels. *Adv. Exp. Med. Biol.* **483**, 45–56.

El Idrissi, A., and Trenkner, E. (1999). Growth factors and taurine protect against excitotoxicity by stabilizing calcium homeostasis and energy metabolism. *J. Neurosci.* **19**, 9459–9468.

Farbman, A. I., Brunjes, P. C., Rentfro, L., Michas, J., and Ritz, S. (1988). The effect of unilateral naris occlusion on cell dynamics in the developing rat olfactory epithelium. *J. Neurosci.* **8**, 3290–3295.

Gal, A., Apfelstedt-Sylla, E., Janecke, A. R., and Zrenner, E. (1997). Rhodopsin mutations in inherited retinal dystrophies and dysfunctions. *Prog. Retinal Eye Res.* **16**, 51–79.

Garcia-Perez, A., and Burg, M. B. (1991). Renal medullary organic osmolytes. *Physiol. Rev.* **71**, 1081–1115.

Graf, D., Kurz, A. K., Reinehr, R., Fischer, R., Kircheis, G., and Häussinger, D. (2002). Prevention of bile acid-induced apoptosis by betaine in rat liver. *Hepatology* **36**, 829–839.

Green, T. R., Fellman, J. H., Eicher, A. L., and Pratt, K. L. (1991). Antioxidant role and subcellular localization of hypotaurine and taurine in human neutrophils. *Biochim. Biophys. Acta* **1073**, 91–97.

Haas, H. L., and Hösli, L. (1973). The depression of brain stem neurones by taurine and its interaction with strychnine and bicuculline. *Brain Res.* **52**, 399–402.

Hayes, K. C., and Carey, R. E. (1975). Retinal degeneration associated with taurine deficiency in the cat. *Science* **188**, 949–950.

Hediger, M. A., Smith, C. P., You, G., Lee, W. S., Kanai, Y., and Shayakul, C. (1996). Structure, regulation and physiological roles of urea transporters. *Kidney Int.* **49**, 1615–1623.

Heller-Stilb, B., van Royen, C., Rascher, K., Hartwig, H. G., Huth, A., Seeliger, M. W., Warskulat, U., and Häussinger, D. (2002). Disruption of the taurine transporter gene *(taut)* leads to retinal degeneration in mice. *FASEB J.* **16**, 231–233.

Huang, D. Y., Boini, K. M., Lang, P. A., Grahammer, F., Duszenko, M., Heller-Stilb, B., Warskulat, U., Häussinger, D., Lang, F., and Vallon, V. (2006). Impaired ability to increase water excretion in mice lacking the taurine transporter gene TAUT. *Eur. J. Physiol.* **451**, 668–677.

Humphries, M. M., Rancourt, D., Farrar, G. J., Kenna, P., Hazel, M., Bush, R. A., Sieving, P. A., Sheils, D. M., McNally, N., Creighton, P., Erven, A., Boros, A., *et al.* (1997). Retinopathy induced in mice by targeted disruption of the rhodopsin gene. *Nat. Genet.* **15**, 115–117.

Hussy, N., Deleuze, C., Pantaloni, A., Desarmenien, M. G., and Moos, F. (1997). Agonist action of taurine on glycine receptors in rat supraoptic magnocellular neurones: Possible role in osmoregulation. *J. Physiol.* **502**, 609–621.

Huxtable, R. J. (1992). Physiological actions of taurine. *Physiol. Rev.* **72**, 101–163.

Jiang, H., Ding, D., Müller, M., Pfister, M., Warskulat, U., Häussinger, D., and Salvi, R. (2005). Central auditory nerve degeneration in knockout mice lacking the taurine transporter. Meeting of the Association for Research in Otolaryngology, Daytona. [Abstract]

Jiang, Z., Chung, S. K., Zhou, C., Cammarata, P. R., and Chung, S. S. (2000). Over-expression of Na$^+$-dependent myo-inositol transporter in mouse lens led to congenital cataract. *Invest. Ophthlmol. Vis. Sci.* **41**, 1467–1472.

Kamisaki, Y., Wada, K., Nakamoto, K., and Itoh, T. (1996). Effects of taurine on GABA release from synaptosomes of rat olfactory bulb. *Amino Acids* **10**, 49–57.

Kaneko, T., Takenaka, M., Okabe, M., Yoshimura, Y., Yamauchi, A., Horio, M., Kwon, H. M., Handler, J. S., and Lmai, E. (1997). Osmolarity in renal medulla of transgenic mice regulates transcription via 5′-flanking region of canine BGT1 gene. *Am. J. Physiol.* **272**, F610–F616.

Kratskin, I. L., Rio, J. P., Kenigfest, N. B., Doty, R. L., and Reperant, J. (2000). A light and electron microscopic study of taurine-like immunoreactivity in the main olfactory bulb of frogs. *J. Chem. Neuroanat.* **18**, 87–101.

Kurz, A. K., Schliess, F., and Häussinger, D. (1998). Osmotic regulation of the heat shock response in primary hepatocytes. *Hepatology* **28**, 774–781.

Kwon, M. H., and Handler, J. S. (1995). Cell volume regulated transporters of compatible osmolytes. *Curr. Biol.* **7**, 465–471.

Lake, N. (1993). Loss of cardiac myofibrils: Mechanism of contractile deficits induced by taurine deficiency. *Am. J. Physiol.* **264,** H1323–H1326.

Lake, N., and Verdone-Smith, C. (1989). Immuno-cytochemical localization of taurine in the mammalian retina. *Curr. Eye Res.* **8,** 163–173.

Lang, F., Busch, G. L., Ritter, M., Völkl, H., Waldegger, S., Gulbins, E., and Häussinger, D. (1998). Functional significance of cell volume regulatory mechanisms. *Physiol. Rev.* **78,** 247–306.

Lang, P. A., Warskulat, U., Heller-Stilb, B., Huang, D. Y., Grenz, A., Myssina, S., Duszenko, M., Lang, F., Häussinger, D., Vallon, V., and Wieder, T. (2003). Blunted apoptosis of erythrocytes from taurine transporter deficient mice. *Cell. Physiol. Biochem.* **13,** 337–346.

Liu, Q. R., Lopez-Corcuera, B., Nelson, H., Mandian, S., and Nelson, N. (1992). Cloning and expression of a cDNA encoding the transporter of taurine and β-alanine in mouse brain. *Proc. Natl. Acad. Sci. USA* **89,** 12145–12149.

Lombardini, J. B. (1999). Taurine: Retinal function. *Brain Res. Rev.* **16,** 151–169.

Magrassi, L., and Graziadei, P. P. (1995). Cell death in the olfactory epithelium. *Anat. Embryol. (Berl.)* **192,** 77–87.

Mahalik, T. J. (1996). Apparent apoptotic cell death in the olfactory epithelium of adult rodents: Death occurs at different developmental stages. *J. Comp. Neurol.* **372,** 457–464.

Mansour, S. L., Thomas, K. R., and Capecchi, M. R. (1988). Disruption of the proto-oncogene int-2 in mouse embryo-derived stem cells: A general strategy for targeting mutations to non-selectable genes. *Nature (London)* **336,** 348–352.

Medan, D., Wang, L., Toledo, D., Lu, B., Stehlik, C., Jiang, B. H., Shi, X., and Rojanasakul, Y. (2005). Regulation of Fas (CD95)-induced apoptotic and necrotic cell death by reactive oxygen species in macrophages. *J. Cell Physiol.* **203,** 78–84.

Moise, N. S., Pacioretty, L. M., Kallfelz, F. A., Stipanuk, M. H., King, J. M., and Gilmour, R. F., Jr. (1991). Dietary taurine deficency dilated cardiomyopathy in the fox. *Am. Heart J.* **121,** 541–547.

Mozaffari, M. S., and Schaffer, D. (2001). Taurine modulates arginine vasopressin-mediated regulation of renal function. *J. Cardiovasc. Pharmacol.* **37,** 742–750.

Nakanishi, T., Uyama, O., Yamada, T., and Sugita, M. (1992). Sustained metabolic alkalosis associated with development of the milk-alkali syndrome. *Nephron* **60,** 251.

Novotny, M. J., Hogan, P. M., Paley, D. M., and Adams, H. R. (1991). Systolic and diastolic dysfunction of the left ventricle induced by dietary taurine deficiency in cats. *Am. J. Physiol.* **261,** H121–H127.

Obrosova, I. G., Fathallah, L., and Stevens, M. J. (2001). Taurine counteracts oxidative stress and nerve growth factor deficit in early experimental diabetic neuropathy. *Exp. Neurol.* **172,** 211–219.

Oermann, E., Warskulat, U., Heller-Stilb, B., Häussinger, D., and Zilles, K. (2005). Taurine-transporter gene knockout-induced changes in $GABA_A$, kainate and AMPA but not NMDA receptor binding in mouse brain. *Anat. Embryol.* **210,** 363–372.

Park, E., Schuller-Levis, G., and Quinn, M. R. (1995). Taurine chloramine inhibits production of nitric oxide and TNF-μ in activated RAW 264.7 cells by mechanisms that involve transcriptional and translational events. *J. Immunol.* **154,** 4778–4784.

Pasantes-Morales, H., Dominguez, L., Compomanes, M. A., and Pacheco, P. (1986). Retinal degeneration induced by taurine deficiency in light-deprived cats. *Exp. Eye Res.* **43,** 55–60.

Pasantes-Morales, H., Quiroz, H., and Quesada, O. (2002). Treatment with taurine, diltia-zem, and vitamin E retards the progressive visual field reduction in retinitis pigmentosa: A 3-year follow-up study. *Metab. Brain Dis.* **17,** 183–197.

Peters-Regehr, T., Bode, J. G., Kubitz, R., and Häussinger, D. (1999). Organic osmolyte transport in quiescent and activated rat hepatic stellate cells (Ito cells). *Hepatology* **29,** 173–180.

Petrosian, A. M., and Haroutounian, J. E. (1998). The role of taurine in osmotic, mechanical, and chemical protection of the retinal rod outer segments. *Adv. Exp. Med. Biol.* **442,** 407–413.

Petrosian, A. M., and Haroutounian, J. E. (2000). Taurine as a universal carrier of lipid soluble vitamins: A hypothesis. *Amino Acids* **19,** 409–421.

Petrosian, A. M., Haroutounian, J. E., and Zueva, L. V. (1996). Tauret: A taurine-related endogenous substance in the retina and its role in vision. *Adv. Exp. Med. Biol.* **403,** 333–342.

Pion, P. D., Kittleson, M. D., Rogers, Q. R., and Morris, J. G. (1987). Myocardial failure in cats associated with low plasma taurine: A reversible cardiomyopathy. *Science* **237,** 764–768.

Pion, P. D., Kittleson, M. D., Thomas, W. P., Skiles, M. L., and Rogers, Q. R. (1992). Clinical findings in cats with dilated cardiomyopathy and relationship of findings to taurine deficiency. *J. Am. Vet. Med. Assoc.* **201,** 167–274.

Portera-Cailliau, C., Sung, C. H., Nathans, J., and Adler, R. (1994). Apoptotic photoreceptor cell death in mouse models of retinitis pigmentosa. *Proc. Natl. Acad. Sci. USA* **91,** 974–978.

Pow, D. V., Sullivan, R., Reye, P., and Hermanussen, S. (2002). Localization of taurine transporters, taurine, and (3)H taurine accumulation in the rat retina, pituitary, and brain. *GLIA* **37,** 153–168.

Ramamoorthy, S., Leibach, F. H., Mahesh, V. B., Han, H., Yang-Feng, T., Blakely, R. D., and Ganapathy, V. (1994). Functional characterization and chromosomal localization of a cloned taurine transporter from human placenta. *Biochem. J.* **300,** 893–900.

Rascher, K., Servos, G., Berthold, G., Hartwig, H., Warskulat, U., Heller-Stilb, B., and Häussinger, D. (2004). Light deprivation slows but does not prevent the loss of photoreceptors in taurine transporter knockout mice. *Vision Res.* **44,** 2091–2100.

Reinehr, R., Becker, S., Höngen, A., and Häussinger, D. (2004a). The Scr family kinase Yes triggers hyperosmotic activation of the epidermal growth factor receptor and CD95. *J. Biol. Chem.* **279,** 23977–23987.

Reinehr, R., Görg, B., Höngen, A., and Häussinger, D. (2004b). CD95-tyrosine nitration inhibits hyperosmotic and CD95 ligand-induced CD95 activation in rat hepatocytes. *J. Biol. Chem.* **279,** 10364–10373.

Reinehr, R., Graf, D., and Häussinger, D. (2003a). Bile salt-induced hepatocyte apoptosis involves epidermal growth factor receptor-dependent CD95 tyrosine phosphorylation. *Gastroenterology* **125,** 839–853.

Reinehr, R., and Häussinger, D. (2006). Hyperosmotic activation of the CD95 death receptor system. *Acta Physiol. (Oxf.)* **187,** 199–203.

Reinehr, R., Schliess, F., and Häussinger, D. (2003b). Hyperosmolarity and CD95L trigger CD95/EGFR association and tyrosine phosphorylation of CD95 as prerequisites for CD95 membrane trafficking and DISC formation. *FASEB J.* **17,** 731–733.

Remé, C. E., Grimm, C., Hafezi, F., Marti, A., and Wenzel, A. (1998). Apoptotic cell death in retinal degenerations. *Prog. Retinal Eye Res.* **17,** 443–464.

Saransaari, P., and Oja, S. S. (1991). Excitatory amino acids evoke taurine release from cerebral cortex slices from adult and developing mice. *Neuroscience* **45,** 451–459.

Sawamura, A., Sada, H., Azuma, J., Kishimoto, S., and Sperelakis, N. (1990). Taurine modulates ion influx through cardiac Ca^{2+} channels. *Cell. Calcium* **11,** 251–259.

Schaffer, S. W., and Azuma, J. (1992). Myocardial physiological effects of taurine and their significance. *In* "Taurine: Nutritional Value and Mechanisms of Action" (J. B. Lombardini, S. W. Schaffer, and J. Azuma, eds.), pp. 105–120. Plenum Press, New York.

Scheller, D., Szathmary, S., Kolb, J., and Tegtmeier, F. (2000). Observations on the relationship between the extracellular changes of taurine and glutamate during cortical spreading depression, during ischemia, and within the area surrounding a thrombotic infarct. *Amino Acids* **19,** 571–583.

Schloss, P., Puchel, A. W., and Betz, H. (1994). Neurotransmitter transporters: New members of known families. *Curr. Opin. Cell Biol.* **4,** 595–599.

Schmidt, S. Y., Berson, E. L., and Hayes, K. C. (1976). Retinal degeneration in the taurine-deficient cat. *Trans. Am. Acad. Ophthalmol. Otolaryngol.* **81,** OP687–OP693.

Schwob, J. E. (2002). Neural regeneration and the peripheral olfactory system. *Anat. Rec.* **269,** 33–49.

Sergeeva, O. A., Chepkova, A. N., Doreulee, N., Eriksson, K. S., Poelchen, W., Mönnighoff, I., Heller-Stilb, B., Warskulat, U., Häussinger, D., and Haas, H. L. (2003). Taurine-induced long-lasting enhancement of synaptic transmission in mice: Role of transporters. *J. Physiol.* **550,** 911–919.

Sergeeva, O. A., and Haas, H. L. (2001). Expression and function of glycine receptors in striatal cholinergic interneurons from rat and mouse. *Neuroscience* **104,** 1043–1055.

Shaldubina, A., Buccafusca, R., Johanson, R. A., Agam, G., Belmaker, R. H., Berry, G. T., and Bersudsky, Y. (2007). Behavioural phenotyping of sodium-myo-inositol cotransporter heterozygous knockout mice with reduced brain inositol. *Genes Brain Behav.* **6,** 253–259.

Shayakul, C., and Hediger, M. A. (2004). The SLC14.gene family of urea transporters. *Pflüg. Arch.* **447,** 603–609.

Smith, K. E., Borden, L. A., Wang, C. D., Hartig, P. R., Branchek, T. A., and Weinshank, R. L. (1992). Cloning and expression of a high affinity taurine transporter from rat brain. *Mol. Pharmacol.* **42,** 563–569.

Stapleton, P. P., O'Flaherty, L., Redmond, P., and Bouchier-Hayes, D. J. (1998). Host defense: A role for the amino acid taurine? *J. Parent. Ent. Nutr.* **22,** 42–48.

Timbrell, J. A., Saebra, V., and Waterfield, C. J. (1995). The *in vivo* and *in vitro* protective properties of taurine. *Gen. Pharmacol.* **26,** 453–462.

Uchida, S., Kwon, H. M., Yamauchi, A., Preston, A. S., Marumo, F., and Handler, J. S. (1992). Molecular cloning of the cDNA for an MDCK cell Na$^+$- and Cl$^-$-dependent taurine transporter that is regulated by hypertonicity. *Proc. Natl. Acad. Sci. USA* **89,** 8230–8234.

Uma, S. M., Satapathy, M., and Sitaramayya, A. (1983). Decreased plasma taurine levels in retinitis pigmentosa. *Biochem. Med.* **30,** 49–52.

van Soest, S., Westerveld, A., De Jong, P., Bleeker-Wagemakers, E., and Bergen, A. (1999). Retinitis pigmentosa: Defined from a molecular point of view. *Surv. Ophthalmol.* **43,** 321–334.

Vinnakota, S., Qian, X., Egal, H., Sarthy, V., and Sarkar, H. K. (1997). Molecular characterization and *in situ* localization of a mouse retinal taurine transporter. *J. Neurochem.* **69,** 2238–2250.

Warskulat, U., Andree, B., Lüsebrink, J., Köhrer, K., and Häussinger, D. (2006a). Switch from actin alpha 1 to alpha 2 expression and up-regulation of biomarkers for pressure overload and cardiac hypertrophy in taurine-deficient mouse heart. *Biol. Chem.* **387,** 1449–1454.

Warskulat, U., Borsch, E., Reinehr, R., Heller-Stilb, B., Mönnighoff, I., Buchczyk, D., Donner, M., Flögel, U., Kappert, G., Soboll, S., Beer, S., Pfeffer, K., *et al.* (2006b). Chronic liver disease is triggered by taurine transporter knockout in the mouse. *FASEB J.* **20,** 574–576.

Warskulat, U., Flögel, U., Jacoby, C., Hartwig, H. G., Thewissen, M., Merx, M. W., Molojavyi, A., Heller-Stilb, B., Schrader, J., and Häussinger, D. (2004). Taurine transporter knockout depletes muscle taurine levels and results in severe skeletal muscle impairment but leaves cardiac function uncompromised. *FASEB J.* **18,** 577–579.

Warskulat, U., Zhang, F., and Häussinger, D. (1997). Taurine is an osmolyte in rat liver macrophages (Kupffer cells). *J. Hepatol.* **26,** 1340–1347.

Weik, C., Warskulat, U., Bode, J. G., Peters-Regehr, T., and Häussinger, D. (1998). Compatible organic osmolytes in rat liver sinusoidal endothelial cells. *Hepatology* **27,** 569–575.

Weisdorf, S. A., Freese, D. K., Fath, J. J., Tsai, M. Y., and Cerra, F. B. (1987). Amino acid abnormalities in infants with extrahepatic biliary atresia and cirrhosis. *J. Pediatr. Gastroenterol. Nutr.* **6,** 860–864.

Welch, W. J., and Brown, R. (1996). Influence of molecular and chemical chaperones on protein folding. *Cell Stress Chap.* **1,** 109–115.

Wettstein, M., and Häussinger, D. (1997). Cytoprotection by the osmolytes betaine and taurine in ischemia-reoxygenation injury in the perfused rat liver. *Hepatology* **26,** 1560–1566.

Witt, M., Roth, C., Warskulat, U., Bidmon, H. J., Häussinger, D., and Hummel, T. (2003). Increased proliferation and chaperone activity in the olfactory mucosa of taurine transporter knockout mice. *Chem. Senses* **28,** A108–A109.

Wong, P. (1994). Apoptosis, retinitis pigmentosa and degeneration. *Biochem. Cell. Biol.* **72,** 489–498.

Wright, C. E., Tallan, H. H., Wright, T. R., Gillam, B., Gaull, G. E., and Sturman, J. A. (1987). Taurine availability and function in neurogenetic retinopathies. *Adv. Exp. Med. Biol.* **217,** 101–112.

Yamamoto, S. (1996). Plasma taurine in liver cirrhosis with painful muscle cramps. *Adv. Exp. Med. Biol.* **403,** 597–600.

Yancey, P. H., Clark, M. E., Hand, S. C., Bowlus, R. D., and Somero, G. N. (1975). Living with water stress: Evolution of osmolyte systems. *Science* **217,** 1214–1222.

Yao, L., Huang, D. Y., Pfaff, I. L., Nie, X., Leitges, M., and Vallon, V. (2004). Evidence for a role of protein kinase C–alpha in urine concentration. *Am. J. Physiol. Renal Physiol.* **287,** F299–F304.

CHAPTER TWENTY-SIX

MOLECULAR BASIS OF OSMOLYTE EFFECTS ON PROTEIN AND METABOLITES

Jörg Rösgen

Contents

Abstract

Osmolytes can have strong effects on biochemical reactions, such as protein folding or protein–ligand interaction. These effects are mediated through solvation—the nonspecific interaction between the solution components.

Department of Biochemistry and Molecular Biology, University of Texas Medical Branch, Galveston, Texas

Methods in Enzymology, Volume 428
ISSN 0076-6879, DOI: 10.1016/S0076-6879(07)28026-7

459

Therefore, understanding the impact of osmolytes on cellular biochemistry requires an understanding of the underlying solvation processes. This chapter discusses the thermodynamic effects of osmolytes on proteins and small organic molecules in terms of the solvation of these molecules, as derived from Kirkwood–Buff theory. This approach allows experimental determination of solvation properties from thermodynamic data. Knowledge of solvation at this level provides insight into the observed behavior of proteins and small molecules in osmolyte solution on a microscopic level. As examples, we provide solvation effects on protein folding, ligand binding, and osmolyte thermodynamics.

1. INTRODUCTION

1.1. Importance of chemical activities for life

It was recognized two centuries ago that chemical activities of biochemical compounds are crucial properties that determine much of the processes of life (Ostwald, 1893). This is because chemical activities determine both equilibrium and kinetic behaviors of all chemical reactions (Scatchard, 1921). Unfortunately, the term "chemical activity" has no strict analog in common language, but "effective concentration" is a term that roughly captures the gist of chemical activity. Every compound has some effective concentration (chemical activity), even if it is not actually participating in a chemical reaction.

The natural logarithm of chemical activities a—called chemical potentials $\mu = \mu^0 + RT \ln a$—is the older concept (Gibbs, 1876/1878) and frequently occurs throughout thermodynamics. Chemical potentials μ are commonly subdivided into three contributions: (1) a concentration-independent offset μ^0 (the standard chemical potential), (2) a directly concentration-dependent term $RT \ln c$, and (3) a correction term $RT \ln \gamma$. The activity coefficient γ converts the concentration c into chemical activity $a = c \cdot \gamma$.

This chapter investigates how biochemical compounds mutually alter their chemical activities, how this impacts biochemical reactions, and what is the molecular basis of this behavior in terms of the enrichment or depletion of some compounds around others.

1.2. Solvation vs binding

Solvation and binding are two modes in which small molecules can interact with macromolecules. The major distinction between solvation and binding is the strength of the interaction, the affinity. High affinity normally means that the interaction is specific. Dissociation constants for specific binding are millimolar or smaller, whereas apparent "dissociation constants" of solvation by osmolytes are in the molar range, and values as high as 30 M have been reported for GdnHCl (Tanford, 1968). In view of such weak interaction, it is important to consider explicitly the water molecules, which are competing for interaction with the surface of the protein (Schellman, 1994). This chapter uses the Kirkwood–Buff approach (Kirkwood and Buff, 1951), which captures both specific binding and solvation and permits assessment of the hydration and osmolyte solvation of the protein from experimental data. In this approach the thermodynamics of the solution is derived from the structure of the solution, and therefore it is important to define first what the term "structure of the solution" means in this context.

1.3. Solvation, solution structure, and water structure

There is no doubt that the addition of osmolytes somehow alters "water structure." However, there are many ways to define water structure, for example, in terms of numbers of hydrogen bonds and their length, or the average density of water molecules at various distances from other water molecules. Each of these various definitions has its own purpose and is useful in different contexts (see also Chapter 22). In the context of the thermodynamics of osmolyte action on proteins, the proper definition of the structure of the solution is in terms of pair-distribution functions (Kirkwood and Buff, 1951), as shown in Fig. 26.1A. This definition is similar to the one X-ray crystallography uses to define crystal structures, namely angle-dependent distance distributions. In solution all molecules diffuse and tumble, and thus the angle dependence is averaged out. Therefore, the structure of the solution is given by angle-averaged distance distributions, which are also called pair-distribution functions.

The most basic feature of such pair-distribution functions is the inability of molecules to overlap. This leads to a region of exclusion (ordinate value of 0 in Fig. 26.1A), which is limited by the large peaks of the first solvation shell. For simplicity, we display protein P, osmolyte O, and water W as spheres. The center of the protein is at the origin of the coordinate system (Fig. 26.1A), and a water (small sphere) and an osmolyte molecule (intermediate sphere) are drawn at the position of closest approach to the protein. Note that the region of mutual exclusion around a protein is larger for osmolytes than for water because the naturally occurring organic osmolytes are always larger than water. At large distances between molecules the pair correlations decay to bulk density, which is represented by a value of unity (right side of Fig. 26.1A). Solvation peaks occur between these two regions of exclusion and of bulk density. These peaks depend on the specific type of molecules involved and will therefore be different for different osmolytes. Incidentally, the solvation peaks seen in Fig. 26.1A do not necessarily indicate attractive interactions, since such peaks are also caused by plain packing effects in crowded solutions (Hansen and McDonald, 1986). Rather, attractive interaction will modulate height and position of the peaks.

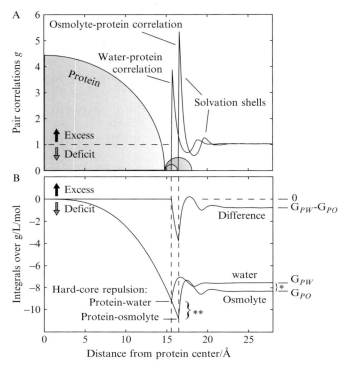

Figure 26.1 Schematic figure of protein solvation by water and osmolyte. (A) Pair correlations between molecules approximated by spheres. For facilitating discussion of the
effects of steric exclusion, the molecular species are drawn into the figure as spheres.
The protein ($\approx$18 kDa) is represented by the big sphere and water (18 D) and osmolyte
($\sim$140 D) by the small and intermediate spheres, respectively. Water and osmolyte are
drawn at the point of closest approach to the protein surface. Lines represent water–
protein and osmolyte–protein correlations as a function of distance from the center of
mass of the protein. Because curves are normalized to bulk density, values larger/smaller
than unity indicate an excess/deficit, respectively, of water and osmolyte. (B) Preferential interaction derived from integration of the pair correlations. The figure shows integral over pair correlations from A ($\int(g-1)dV = \int(g-1)4\pi r^2 dr$). The overall solvation
G_{PO} and G_{PW} is given by the complete integration (see bottom and right side of B). The
preferential interaction is also shown (see top and right of B, also seen at the bottom
right marked by an asterisk). Taking into account only plain hard-core repulsion at the
protein surface would lead to an overestimate of the preferential interaction (indicated
by a double asterisk).

1.4. Overall solvation derived from pair correlations

The pair-distribution function g is normalized with respect to bulk solution
density, and so subtracting unity ($g-1$) gives the deviation from bulk
density. Therefore, integration of $g-1$ yields the overall excess or deficit
of one kind of molecule around another. It is this overall correlation that

determines the thermodynamic response of the solution components to a change in solution composition (Kirkwood and Buff, 1951).

Two examples for this, protein–water solvation and protein–osmolyte solvation, are shown in Fig. 26.1B. Integrating the protein–water pair-distributions g_{PW} results in strongly negative values, which are only partially compensated by the large solvation peaks. As distance increases toward the region of bulk density, the integral tails off into the overall correlation G_{PW}. This overall correlation is negative because protein and water molecules do not overlap, that is, a volume element occupied by protein excludes water molecules. The same holds for the overall correlation between protein and osmolyte G_{PO}. The difference between G_{PW} and G_{PO} is a measure of the preferential interaction of the protein with either water or osmolyte.

Note that in Fig. 26.1B $G_{PW} - G_{PO}$ is smaller than the difference between the hard-core repulsion values at the protein surface. This is because of the solvation peaks, which occur even in the absence of attractive interaction because of packing effects at the protein surface as noted earlier. Excluded volumes calculated without taking solvation peaks into account (Schellman, 2003) have therefore to be taken with a grain of salt (Schurr et al., 2005).

1.5. Experimental data useful for deriving thermodynamic solvation

Kirkwood–Buff analysis of the thermodynamic solvation in solution is possible in forward and inverse direction. The original Kirkwood–Buff theory starts with knowledge of the atomic coordinates, from which thermodynamics is derived (Kirkwood and Buff, 1951). However, it is also possible to derive solvation information (Kirkwood–Buff integrals G_{ij}) from thermodynamic data. This procedure is called "inverse Kirkwood–Buff theory" (Ben-Naim, 1977).

Several thermodynamic properties are useful for a calculation of G_{ij}. Most of these properties are slopes with regard to molecular concentrations. The partial volume $\bar{v}_i$ of each component can be obtained from density measurements as a function of the concentration of all solution components. The second set of properties is given by the slopes of chemical activities ("effective concentrations"). In an n-component system there are n derivatives for the activity of each component, that is, a total of n^2 derivatives. Measuring all of these is neither practical nor necessary, and this chapter discusses strategies to reduce the number of needed derivatives, for example, by useful approximations. In a system with a small number of components it is typically sufficient to determine the chemical activity of water as a function of the concentrations of other components. This can be done by measuring the water vapor pressure. Other useful methods used to determine chemical activities include osmotic methods, ion selective electrodes, diffusion measurements, centrifugation, or solubility measurements.

The last thermodynamic property of interest for a Kirkwood–Buff analysis is compressibility. For liquids, however, the compressibility is normally small enough to be insignificant for practical purposes in the Kirkwood–Buff analysis.

1.6. Useful approximations

Frequently, protein concentrations are low in experiments that measure the impact of a high concentration of osmolyte on protein. In such cases in which the concentration of one component is very low, it can be useful to approximate many properties of the solution by the known properties of a solution that does not contain the low-concentration component. It is therefore of interest to look at the chemical activities and the partial volumes in solutions of osmolytes that do not contain protein.

The chemical activity of a large number of naturally occurring organic osmolytes is well captured by a simple statistical-thermodynamic approach (Rösgen *et al.*, 2004a,b). The chemical activities of about half of the osmolytes were found to be well described by the first-order equation

$$a_{O,1st} = \frac{c_O}{1 - c_O/c_1},\tag{26.1}$$

which was derived using terms up to first order in the partition function. Figure 26.2 shows the activity coefficient $\gamma_{O,1st} = a_{O,1st}/c_O$ for several osmolytes. The abscissa is normalized by c_1 to demonstrate the uniformity in behavior. The other osmolytes require the second-order equation

$$a_{O,2nd} = \frac{g_2/2}{2 - c_O/c_2}\left[-(1 - c_O/c_1) + \sqrt{(1 - c_O/c_1)^2 + \frac{4c_O}{g_2}(2 - c_O/c_2)}\right],$$

$$\tag{26.2}$$

where c_O is the osmolyte molarity, and the constants c_1, c_2, and g_2 are given in Table 26.1 for several osmolytes. For the Kirkwood–Buff analysis we need the slope of the chemical activity with respect to c_O. This is given to first and second orders by

$$\left(\frac{\partial \ln a_{O,1st}}{\partial \ln c_O}\right)_{p,T} = \frac{1}{1 - c_O/c_1}\tag{26.3}$$

and

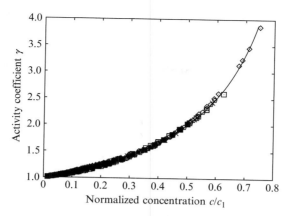

Figure 26.2 Uniform behavior of first-order activity coefficient. Activity coefficient data for 12 osmolytes were fitted to the first-order activity coefficient equation $1/(1 - c/c_1)$ (Rösgen *et al.*, 2004a, 2005). Experimental data are replotted with the concentration normalized by c_1 (symbols). The line represents the general first-order activity coefficient equation $1/(1 - c/c_1)$. As explained under case 1, this uniform first-order behavior is because of a constant difference in hydration and self-solvation of the osmolytes. The osmolytes reported are erythritol, mannitol, sorbitol, alanine, sarcosine, raffinose, maltose, xylose, glucose, mannose, fucose, and rhamnose.

$$\left(\frac{\partial \ln a_{O,2nd}}{\partial \ln c_O}\right)_{p,T} = \left(-\frac{1}{1 + 2a_{O,2nd}/g_2} + \frac{2 + a_{O,2nd}/c_1}{1 + a_{O,2nd}/c_1 + a_{O,2nd}^2/(c_2 g_2)}\right)^{-1},$$

$$(26.4)$$

respectively (Rösgen *et al.*, 2005).

Data on solution physical properties are often given in terms of *molal* concentrations m_i (moles per kg of water) because they allow for high precision and accuracy of data. However, *molar* concentrations c_i (moles per liter of solution) are more popular in biochemical and biophysical applications. Also, molar concentrations are more useful for data interpretation (Rösgen *et al.*, 2004a, 2005). Conversion from molality m_i of compound type i to its molarity c_i is straightforward if the density of the solution ρ is known:

$$c_i = \frac{m_i \rho}{1 + \sum_k m_k M_k / 1000}.$$

$$(26.5)$$

M_k is the molar mass of compound number k, and the factor of 1000 comes from the conversion of kilograms (in m_k) to grams (in M_k).

Table 26.1 Parameters for the calculation of osmolyte chemical activities [Eqs. (26.1) and (26.2)], their slopes [Eqs. (26.3) and (26.4)], and concentration scale conversion [Eq. (26.6)] (Rösgen *et al.*, 2005)

	g_2 mol/liter	c_1 mol/liter	$2c_2$ mol/liter	c_{max} mol/liter	Valid up to mol/liter
Xylose	–	7.6	–	10.16	2.6
Glucose	–	6.28	–	8.670	4
Fucose	–	4.89	–	9.05	1.9
Glactose	790	8.27	c_{max}	8.99	2.5
Rhamnose	–	4.4	–	9.07	1.2
Mannose	–	7.04	–	8.54	3.5
Maltose	–	3.135	–	4.27	1.8
Raffinose	–	1.523	–	2.46	0.22
Sucrose	70.4	2.466	c_{max}	4.617	2.6
Glycerol	19	4.8	c_{max}	13.69	7.1
Mannitol	–	7.35	–	8.173	1.1
meso- Erythritol	–	9.3	–	11.88	3.8
Sorbitol	–	6.475	–	8.17	4.8
Urea	21.6	20.3	c_{max}	22.03	10.1
Glycine	3.765	3.260	c_{max}	21.41	2.8
Alanine	–	14.40	–	16.07	1.7
Proline	120.5	5.38	c_{max}	12.52	4.5
Sarcosine	–	8.68	–	16.29	5.1
Glycine betaine	16.88	1.97	c_{max}	10.72	3.4

A convenient approximation that makes possible the same conversion without the necessity of knowing the solution density has been provided by Rösgen *et al.* (2004a). The molarity c_O and molality m_O of organic osmolytes without a net charge can easily be converted into one another by the following equations

$$c_O \approx \frac{m_O \rho_W}{1 + m_O \rho_W / c_{max}}, \quad m_O \approx \frac{c_O / \rho_W}{1 - c_O / c_{max}}, \quad (26.6)$$

where ρ_W is the density of plain water (0.99705 kg/liter at 25°) and c_{max} is calculated from the density of pure osmolyte (see Table 26.1). In analogy to Eq. (26.5) we might extend Eq. (26.6) to

$$c_O \approx \frac{m_O \rho_W}{1 + \sum_k m_k \rho_W / c_{max, k}}. \quad (26.7)$$

This equation is expected to be valid if the partition function is linear in the molar concentrations of all components (Rösgen et al., 2004a). However, this has not yet been tested experimentally for aqueous solutions containing more than a single concentrated additive.

 ## 2. GENERAL KIRKWOOD–BUFF EQUATIONS

The Kirkwood–Buff equations discussed here may serve two different purposes. On the one hand, solvation information can be calculated from experimental data. This is done most conveniently in the molal concentration scale because experimental data on solution physical properties should normally be collected in terms of molal concentrations for reasons of accuracy and precision. On the other hand, basic principles may be derived from data. For osmolytes, this works best in the molar concentration scale, as demonstrated elsewhere (Rösgen et al., 2005) and discussed later.

2.1. Working in the molal scale

The general Kirkwood–Buff expressions in the molal scale are very straightforward, although they can become lengthy if the number of components in solution is large. Because these equations contain cofactors of determinants of matrices, we will review their use shortly. Given a matrix A [see Eq. (26.8)] and its determinant $|A|$, the cofactor of $|A|$ with respect to element A_{ik} is indicated by the symbol $|A|_{ik}$ and can be calculated as the derivative of the determinant $|A|$ with respect to element A_{ik}. The following example is for two dimensions

$$A = \begin{pmatrix} A_{11} & A_{12} \\ A_{21} & A_{22} \end{pmatrix} \quad |A|_{12} = \frac{\partial}{\partial A_{12}} |A| = \frac{\partial}{\partial A_{12}} (A_{11}A_{22} - A_{12}A_{21}) = -A_{21} \quad (26.8)$$

The overall correlation between components i and k (the Kirkwood–Buff integral G_{ik}) can then be calculated by (Kirkwood and Buff, 1951)

$$G_{ik} = \frac{|A|_{ik}}{|A|} \frac{1}{c_i c_k} - \frac{\delta_{ik}}{c_i}, \quad (26.9)$$

where δ_{ik} is the Kronecker delta ($\delta_{ik} = 1$ for $i = k$, and $\delta_{ik} = 0$ otherwise). The elements A_{ik} of matrix A are

$$A_{ik} = V \left(\frac{\partial(\mu_i/RT)}{\partial N_k} \right)_{T,p} + \frac{\bar{v}_i \bar{v}_k}{\kappa RT} = a_{ik} + \frac{\bar{v}_i \bar{v}_k}{\kappa RT}, \qquad (26.10)$$

where κ is the compressibility, R is the gas constant, and T is the absolute temperature in K. Matrix a is defined as

$$a_{ik} = V \left(\frac{\partial(\mu_i/RT)}{\partial N_k} \right)_{T,p} = \frac{1000}{c_W M_W} \left(\frac{\partial(\mu_i/RT)}{\partial m_k} \right)_{T,p}, \qquad (26.11)$$

where c_W is the molarity of water in the sample and M_W is the molecular weight of water. The first derivative term is most convenient for the elimination of parameters described later, whereas the second derivative term is the one needed for evaluating experimental data, which are normally given in terms of molalities m_k.

2.2. Elimination of parameters based on the given system

For an n-component system there are n^2 different a_{ik}. More than half of all a_{ik} for each system can be directly calculated from the other a_{ik}, leaving $n(n-1)/2$ of the a_{ik} to be measured. First, the a_{ik} are symmetric ($a_{ik} = a_{ki}$), which eliminates $n(n-1)/2$ of the a_{ik}. An additional n of the a_{ik} can be eliminated using the Gibbs–Duhem relation $\sum_i N_i \cdot a_{ik} = 0$. Generally, the availability of experimental methods will dictate which of the a_{ik} should be retained. For instance, existence of water vapor pressure data for the given system will make a_{iW} the best choice for being retained. However, a_{WW} should always be eliminated because of the definition of the molality, which renders a derivative with respect to m_W undefined. Furthermore, note that some of the a_{ik} can be approximated by $a_{i\,i} = 1/m_i$. This is the case for dilute components because the dependence of their chemical potential on their own concentration m_i then has the form $\mu_i = \mu_i^\circ + RT \ln(m_i)$.

Take, for example, a dilute protein that is dissolved in a concentrated aqueous osmolyte solution, which contains a dilute buffer component in addition. This is a four-component system ($n = 4$) with $n^2 = 16$ different a_{ik}. Out of these $n(n-1)/2 = 6$ are redundant, and $n = 4$ can be eliminated through the Gibbs–Duhem relation. The a_{PP} and a_{BB} of the dilute components can be equated to $1/m_P$ and $1/m_B$, respectively, and given that protein and buffer do not interact, we have $a_{PB} = 0$ in addition. This leaves $16 - 6 - 4 - 3 = 3$ of the a_{ik} to be measured.

2.3. Working in the molar scale

Expressing the overall solvation (Kirkwood–Buff integrals) in terms of molarities c_i requires a slightly different approach than seen earlier for molalities m_i. In the molal scale we expressed the G_{ik} in terms of a matrix

A of experimental properties. In the molar scale we do the opposite and express experimental properties in terms of a matrix B of Kirkwood–Buff integrals. We start with the derivative of the chemical potential, which is (Kirkwood and Buff, 1951)

$$\left(\frac{\partial \mu_k}{\partial N_i}\right)_{T,V} = RT \frac{|B|_{ik}}{V|B|}. \tag{26.12}$$

The matrix elements are

$$B_{ik} = c_i(\delta_{ik} + c_k G_{ik}). \tag{26.13}$$

Transformation from constant volume to constant pressure results in

$$\left(\frac{\partial \mu_k}{\partial N_i}\right)_{T,p} = RT \frac{|B|_{ik}}{V|B|} - \frac{\bar{v}_i \bar{v}_k}{\kappa V}, \tag{26.14}$$

where the partial volumes are given by

$$\bar{v}_i = \frac{\sum_k c_k |B|_{ik}}{\sum_{rl} c_k c_l |B|_{kl}}, \tag{26.15}$$

and the compressibility is

$$\kappa RT = \frac{|B|}{\sum_{rl} c_k c_l |B|_{kl}}. \tag{26.16}$$

Because all of these equations contain determinants and cofactors, their handling can become quite tedious. This is obvious, seeing that the number of terms in a determinant increases more than exponentially with the number of rows and columns. An elegant way of getting around this problem, at least for the volumetric properties $\bar{v}_i$ and κ, was originally introduced by Ben-Naim (1988) and has been generalized by Smith (Chitra and Smith, 2002; Smith, 2006):

$$\bar{v}_k = \kappa RT - \sum_i \bar{v}_i c_i G_{ik} \approx -\sum_i \bar{v}_i c_i G_{ik}. \tag{26.17}$$

Generally, the compressibility can be neglected for aqueous solutions, as done on the right-hand side of the equation.

The use of Eqs. (26.14) through (26.17) is best understood by applying them, and thus we proceed now to simple examples. We focus on the expressions in the molar scale because this facilitates interpretation of data.

3. CASE 1: OSMOLYTE SOLVATION

As a first example we discuss solvation in a two-component system containing water and osmolyte. The solvation of osmolyte molecules themselves is of direct importance to their effects on proteins, as we shall see when proceeding to the other cases, which involve protein. Our current example is divided into two parts. The first part serves to demonstrate the detailed pathway for developing the equations, and some readers may wish to skip this part. The second part discusses the meaning of the equations for osmolyte solvation and osmolyte behavior.

3.1. Derivation of equations

For calculations in a two-component system we need a 2×2 B matrix [see Eq. (26.13)]

$$B = \begin{pmatrix} c_W(1 + c_W G_{WW}) & c_O c_W G_{OW} \\ c_O c_W G_{OW} & c_O(1 + c_O G_{OO}) \end{pmatrix}. \qquad (26.18)$$

Its determinant is

$$|B| = c_W(1 + c_W G_{WW})c_O(1 + c_O G_{OO}) - (c_O c_W G_{OW})^2, \qquad (26.19)$$

and the cofactors are

$$|B|_{WW} = c_O(1 + c_O G_{OO}), \qquad (26.20)$$
$$|B|_{OO} = c_W(1 + c_W G_{WW}), \qquad (26.21)$$

and

$$|B|_{OW} = |B|_{WO} = -c_O c_W G_{OW}. \qquad (26.22)$$

These are used in Eq. (26.15) to calculate the partial volumes of the osmolyte

$$\bar{v}_O = \frac{c_W|B|_{WO} + c_O|B|_{OO}}{c_Oc_O|B|_{OO} + 2c_Wc_O|B|_{WO} + c_Wc_W|B|_{WW}}$$
$$= \frac{1 + c_W(G_{WW} - G_{WO})}{c_W + c_O + c_Oc_W(G_{WW} + G_{OO} - 2G_{WO})} \tag{26.23}$$

and the water

$$\bar{v}_W = \frac{1 + c_O(G_{OO} - G_{WO})}{c_W + c_O + c_Oc_W(G_{WW} + G_{OO} - 2G_{WO})}. \tag{26.24}$$

The compressibility

$$\kappa = \frac{c_W(1 + c_WG_{WW})c_O(1 + c_OG_{OO}) - (c_Oc_WG_{OW})^2}{c_W + c_O + c_Oc_W(G_{WW} + G_{OO} - 2G_{WO})} \tag{26.25}$$

is calculated analogously from Eq. (26.16). The equation for the slope of the chemical activity [Eq. (26.14)] is formulated in terms of particle numbers rather than molar concentrations. As a result, we have to do a base transform

$$\left(\frac{\partial}{\partial N_O}\right)_{T,p} = \left(\frac{\partial c_O}{\partial N_O}\right)_{T,p} \left(\frac{\partial}{\partial c_O}\right)_{T,p}$$
$$= \left(\frac{1}{V} - \frac{N_O\bar{v}_O}{V^2}\right) \left(\frac{\partial}{\partial c_O}\right)_{T,p} = \frac{c_W\bar{v}_W}{V} \left(\frac{\partial}{\partial c_O}\right)_{T,p}, \tag{26.26}$$

where we used $c_O = N_O/V$ and $c_O\bar{v}_O = 1 - c_W\bar{v}_W$ (the volume fractions and add up to unity). Using Eq. (26.26), Eq. (26.14) becomes

$$\left(\frac{\partial \ln a_O}{\partial c_O}\right)_{T,p} = \left(\frac{\partial(\mu_O/RT)}{\partial c_2}\right)_{T,p} = \frac{1}{c_W\bar{v}_W} \left(\frac{|B|_{OO}}{|B|} - \frac{\bar{v}_O\bar{v}_O}{RT\kappa}\right), \tag{26.27}$$

and we get after a straightforward (but lengthy) calculation

$$\left(\frac{\partial \ln a_O}{\partial c_O}\right) = \frac{1}{c_O} + \frac{G_{WO} - G_{OO}}{1 - (G_{WO} - G_{OO})c_O}. \tag{26.28}$$

This is the equation we will primarily use, along with the volumetric equations

$$\bar{v}_O = RT\kappa - \bar{v}_O c_O G_{OO} - \bar{v}_W c_W G_{WO} \qquad (26.29)$$

and

$$\bar{v}_W = RT\kappa - \bar{v}_O c_O G_{WO} - \bar{v}_W c_W G_{WW} \qquad (26.30)$$

derived from Eq. (26.17).

3.2. Application of equations

Evaluation of the solvation properties in osmolyte solution requires knowledge of the chemical activities of osmolytes. In uncovering the basis for the observed behavior of osmolytes in solution we first compare the experimental behavior of most protecting osmolytes [represented by Eq. (26.3) along with Table 26.1] with the result of Kirkwood–Buff theory [Eq. (26.28)]. Equation (26.3) can be reformulated to yield

$$\left(\frac{\partial \ln a_{O,1st}}{\partial c_0}\right) = \frac{1}{c_O} + \frac{V_{app}}{1 - V_{app} c_O}, \qquad (26.31)$$

where the apparent hydrated volume of the osmolyte $V_{app} = 1/c_1$ replaces the constant c_1. Comparison with Eq. (26.28) shows that the apparent hydrated volume equals the difference between osmolyte hydration G_{WO} and osmolyte self-solvation G_{OO}. This means that most osmolytes have a concentration-independent difference $G_{WO} - G_{OO}$, as pointed out elsewhere (Rösgen et al., 2005). Consequently, any change in preference of the osmolyte for water is balanced by the same change in preference for other osmolyte molecules, and vice versa. Such concentration-independent solvation behavior might be a prerequisite for a compound to be useful as an osmolyte (Rösgen et al., 2005). Compounds such as alcohols, which strongly change their solvation preferences as their concentration increases, are normally not used as osmolytes in living organisms.

Next, we derive expressions for G_{WO}, G_{WW}, and G_{OO}. These three unknowns are contained in Eqs. (26.28) through (26.30). Solving these equations results in (Rösgen et al., 2007)

$$G_{WO} = RT\kappa - \frac{\bar{v}_O}{a_{OO}} \qquad (26.32)$$

for osmolyte hydration (which equals solvation of water by osmolyte for symmetry reasons),

$$G_{OO} = RT\kappa - \frac{\bar{v}_O}{a_{OO}} + \frac{1 - a_{OO}}{a_{OO}c_O} \qquad (26.33)$$

for osmolyte self-solvation, and

$$G_{WW} = RT\kappa - \frac{1}{c_W} + \frac{\bar{v}_O}{a_{OO}} \cdot \frac{\bar{v}_O c_O}{1 - \bar{v}_O c_O} \qquad (26.34)$$

for water self-hydration. The term $RT\kappa$ is of the order of 1 ml/mol for aqueous solutions (Lide, 2004) and can therefore be neglected relative to the other contributions, which are shown in Fig. 26.3. Solvation in urea solution (black lines) is very similar for all three kinds of water and osmolyte self-solvation, and their mutual interaction. The small difference among G_{WW}, G_{WO}, and G_{OO} results in the nearly ideal behavior of urea solutions (Rösgen et al., 2004a, 2005, 2007). This can be seen from Eq. (26.28), in which only the ideal term $1/c_O$ remains if the solvation differences are zero. Sorbitol (gray lines in Fig. 26.3) has a much larger spread in its solvation properties. G_{WW}, G_{WO}, and G_{OO} are distinctly different. However, G_{WO} and G_{OO} are nearly parallel, which leads to the first-order behavior mentioned earlier.

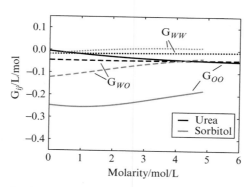

Figure 26.3 Solvation in aqueous osmolyte solutions. Urea (black lines) and sorbitol (gray lines) have quite different solvation behaviors. Urea self-solvation G_{OO} (continuous line), hydration G_{WO} (dashed line), and water self-hydration G_{WW} (dotted line) are all close to 0 liter/mol. Therefore, the solvation term in Eq. (26.28) vanishes and only the ideal term $1/c_O$ is left, and the activity coefficient is nearly unity over the whole range of solubility (Rösgen et al., 2004a, 2005). Sorbitol shows a much larger spread in solvation and therefore behaves nonideally.

4. CASE 2: M VALUES

Protein stability is well known to respond to the presence of osmolytes. The denaturing osmolyte urea unfolds proteins (Ramsden, 1902; Spiro, 1900), whereas other osmolytes can force proteins to fold (Baskakov and Bolen, 1998). Such effects are the consequence of a difference in preferential interaction of the native and denatured state with osmolyte and water (Casassa and Eisenberg, 1964). It was believed for a long time that it was impossible to determine the effects of water and osmolyte separately in a model-independent manner, and much effort has been devoted to separating the effects of water and osmolyte by implementing model assumptions (Eisenberg, 1994; Parsegian *et al.*, 1995; Schellman, 1994; Smith, 2004; Timasheff, 1998; Zhang *et al.*, 1996). Such approaches can be very useful and reasonable, yet they have the drawback of yielding results that depend on the choice of particular models. Kirkwood–Buff theory, however, enables us to derive both the denaturational change in hydration and the change in osmolyte solvation in a model–independent fashion, directly from experimental data (Rösgen *et al.*, 2005, 2007; Shimizu, 2004; Shimizu and Smith, 2004).

4.1. Derivation of equations

We consider an unfolding equilibrium:

$$N \leftrightarrow D \quad K = \frac{[D]}{[N]} \tag{26.35}$$

The response of K to the addition of osmolyte is given by the derivative $(\partial \ln K / \partial c_O)$, which is normally written in terms of the so-called m value $m = -RT(\partial \ln K / \partial c_O)$ (Greene and Pace, 1974). To connect the m value with Kirkwood–Buff theory, we must express K in terms of chemical potentials. This is done through the equilibrium condition

$$\begin{aligned} \mu_N = \mu_D &\Leftrightarrow \mu_N^0 + RT\ln[N] + RT\ln[\gamma_N] \\ &= \mu_D^0 + RT\ln[D] + RT\ln[\gamma_D]. \end{aligned} \tag{26.36}$$

Rearranging this equation yields

$$RT\ln K = RT\ln\frac{[D]}{[N]} = -RT\ln\frac{\gamma_D}{\gamma_N} - \Delta\mu^0. \tag{26.37}$$

In the derivatives the protein concentration is kept constant, therefore the relation

$$\left(\frac{\partial(\mu_N^0 + RT\ln\gamma_N)}{\partial c_O}\right) = \left(\frac{\partial\mu_N}{\partial c_O}\right) \tag{26.38}$$

holds for the native and an analogous expression for the denatured state. The slope of K with osmolyte concentration is then

$$-\left(\frac{\partial\ln K}{\partial c_O}\right) = \left(\frac{\partial(\mu_D - \mu_N)/RT}{\partial c_O}\right). \tag{26.39}$$

This relation can be used directly with Eq. (26.14) to calculate the m value. Because of the determinants the calculation is tedious. However, this calculation can be automated using tools such as the program Mathematica. Note that the base transform from N_2 to c_2 in Eq. (26.26) does not apply to the case of a three–component system. Changing from (N_2, N_3) to (c_2, c_3) we need the following transform

$$\left(\frac{\partial}{\partial c_O}\right)_{c_P} = \frac{V}{\bar{v}_W c_W}\left[(1 - \bar{v}_P c_P)\left(\frac{\partial}{\partial N_O}\right)_{N_P} + \bar{v}_O c_P\left(\frac{\partial}{\partial N_P}\right)_{N_O}\right]. \tag{26.40}$$

Because of space limitation we refrain from giving a derivation of this transform. As a simplification of the equations we restrict the result to the case of dilute protein ($c_P \to 0$). The more complex equation for arbitrary protein concentration can be found elsewhere (Smith, 2006). We proceed by applying Eqs. (26.14) and (26.40) first to μ_D

$$\left(\frac{\partial\mu_D/RT}{\partial c_O}\right) = \frac{G_{PW} - G_{PO}}{1 - (G_{WO} - G_{OO})c_O}. \tag{26.41}$$

Then the analogous equation for μ_N is subtracted to yield

$$\frac{m}{RT} = \frac{\Delta(G_{PW} - G_{PO})}{1 - (G_{WO} - G_{OO})c_O}. \tag{26.42}$$

The difference Δ is between denatured and native states, that is, the difference in Eq. (26.42) is $\Delta(G_{PW} - G_{PO}) = (G_{PW}^{denat.} - G_{PO}^{denat.}) - (G_{PW}^{native} - G_{PO}^{native})$. Also, we use $\Delta(G_{PO}) = G_{PO}^{denat.} - G_{PO}^{native}$ and $\Delta(G_{PW}) = G_{PW}^{denat.} - G_{PW}^{native}$, given later.

4.2. Interpretation

As mentioned previously, this very simple equation for the m value [Eq. (26.42)] is valid for the case of dilute protein in osmolyte solution. The denominator of the solvation expression for the m value [Eq. (26.42)] is independent of the protein and just contains properties of the osmolyte. This can be seen from a comparison of Eq. (26.42) with a modified version of Eq. (26.28)

$$\frac{1}{1 - (G_{WO} - G_{OO})c_O} = \left(\frac{\partial \ln a_O}{\partial \ln c_0}\right). \qquad (26.43)$$

The numerator in Eq. (26.42), however, contains the balance between protein hydration G_{PW} and protein–osmolyte solvation G_{PO}, that is, preferential interaction.

Using Eq. (26.43), we can solve Eq. (26.42) for the change in solvation preference upon unfolding

$$\Delta(G_{PW} - G_{PO}) = \frac{m}{RT} \bigg/ \left(\frac{\partial \ln a_O}{\partial \ln c_O}\right). \qquad (26.44)$$

The quantities on the right-hand side of this equation can be measured, and thus $\Delta(G_{PW} - G_{PO})$ can be calculated. Moreover, general principles can be derived from this equation. The m value is established to be constant for urea (Courtenay et al., 2000; Ferreon and Bolen, 2004; Greene and Pace, 1974; Makhatadze, 1999; Santoro and Bolen, 1988; Timasheff and Xie, 2003), and there is some evidence that this also holds for protecting osmolytes (Felitsky and Record, 2004; Holthauzen and Bolen, 2007; Mello and Barrick, 2003). Equation (26.44) therefore indicates that the relative variation in $\Delta(G_{PW} - G_{PO})$ is independent of the specific protein. Only the amplitude of this variation (given by the m value) depends on the protein type. The magnitude of the m value in turn is dominated by the interaction between osmolyte and protein backbone (Auton and Bolen, 2005; Bolen and Baskakov, 2001), the most numerous chemical group in proteins.

We discussed already in case 1 the simple self-solvation and hydration of osmolyte. The findings in the current case 2 add another example for the very predictable solvation behavior of osmolytes, which is most likely a necessity for a compound to be useful as an osmolyte. Osmolytes should reasonably be expected to have an effect on proteins that is as uniform as possible. Otherwise, if osmolytes impacted each protein differently, they would wreak havoc on the cells they are supposed to protect.

Examples for $\Delta(G_{PW} - G_{PO})$ of ribonuclease T1 in the presence of urea and sorbitol are given in Fig. 26.4A (Rösgen et al., 2007). Positive and negative numbers in this plot correspond to stabilizing solvation and destabilizing solvation, respectively. The change in solvation preference upon

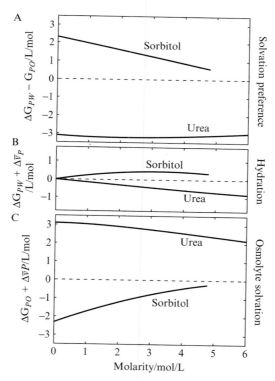

Figure 26.4 Change in protein solvation upon unfolding in osmolyte solution. Curves were calculated from Eqs. (26.46) and (26.47) using m values for RNase T1 (Auton and Bolen, 2005) and published data on osmolyte solutions (Rösgen et al., 2004a, 2005). See also Table 26.1 for osmolyte data. (A) The change in solvation preference upon unfolding depends strongly on sorbitol concentration, but is essentially independent of urea concentration. (B) The overall hydration of RNase T1 changes little upon unfolding. (C) The change in osmolyte solvation of RNase T1 upon unfolding is large and depends strongly on sorbitol concentration, but less so on urea concentration.

unfolding $\Delta(G_{PW} - G_{PO})$ is essentially insensitive to the concentration of the denaturant urea. In the case of the protecting osmolyte sorbitol, however, $\Delta(G_{PW} - G_{PO})$ strongly depends on concentration. This strong concentration dependence of the solvation preference of the protein is shared by most protecting osmolytes (Rösgen et al., 2005).

To this point, the hydration and osmolyte solvation contributions in $\Delta(G_{PW} - G_{PO})$ are not yet separated. This can be done in a straightforward manner by first applying Eq. (26.17) to the case of protein folding under conditions of low protein concentration ($c_P \to 0$)

$$\Delta \bar{v}_P = RT\kappa - \bar{v}_O c_O \Delta(G_{PO}) - \bar{v}_W c_W \Delta(G_{PW}). \qquad (26.45)$$

Equations (26.44) and (26.45) give us two equations with two unknowns that can be directly solved to yield

$$\Delta(G_{PO}) = -\Delta \bar{v}_P - (1 - \bar{v}_O c_O) \frac{m/RT}{(\partial \ln a_O / \partial \ln c_O)} \qquad (26.46)$$

(using $\bar{v}_O c_O + \bar{v}_W c_W = 1$) and

$$\Delta(G_{PW}) = -\Delta \bar{v}_P + \bar{v}_O c_O \frac{m/RT}{(\partial \ln a_O / \partial \ln c_O)}. \qquad (26.47)$$

If the protein volume change upon unfolding $\Delta \bar{v}_P$ is not known, it may be set to zero. This is because volume changes upon protein unfolding are normally small, in the order of 2% or less of the protein volume (Chalikian, 2003; Lin et al., 2002; Rösgen and Hinz, 2000; Royer, 2002).

Figure 26.4B and C show two examples for hydration and osmolyte solvation change upon unfolding of ribonuclease T1. Note that $\Delta \bar{v}_P$ was not taken into account, and thus we have to keep in mind that solvation data might need an offset of ±0.2 liter/mol or less. Sorbitol acts as a stabilizer of RNase T1 with respect to both hydration ΔG_{PW} and osmolyte solvation ΔG_{PO}. Urea, in contrast, acts as a denaturant through both ΔG_{PW} and ΔG_{PO}. The only strongly concentration-dependent one among these four solvation curves is protein–sorbitol solvation ΔG_{PO} (Fig. 26.4C). The difference between native and denatured state hydration ΔG_{PW} is not very sensitive to the concentrations of either urea or sorbitol. Also, the change in urea solvation upon unfolding does not change much with urea concentration.

At low sorbitol concentration ΔG_{PW} approaches zero, and at the solubility limit ΔG_{PO} comes close to zero. Therefore, $\Delta \bar{v}_P$ could potentially bring either ΔG_{PW} or ΔG_{PO} into the destabilizing regime at those concentrations. The same holds in the case of urea, where ΔG_{PW} could switch over to the stabilizing regime at low concentration.

5. CASE 3: SPECIFIC BINDING OF AN OSMOLYTE

Osmolytes shift not only conformational equilibria but they also affect binding reactions. This section discusses the effect of an osmolyte on its own binding to a protein. Specific binding is normally viewed as a binding reaction with a specific stoichiometry and involves at least two protein species (bound and free state) and one ligand species (free ligand). Alternatively, specific binding could be considered a solvation phenomenon that involves two proteins states that merely differ in their solvation characteristics. In this point of view, the n-fold liganded state is not considered to be "liganded." Rather it is seen to have a "solvation preference" for one of the solution components (the ligand), and this preferential solvation is worth the number n of bound ligands. Both of these approaches are discussed, using as an example an osmolyte taking the two roles of being a specifically binding ligand on the one hand and a highly concentrated cosolute on the other hand.

5.1. Specific binding as a solvation phenomenon

We are dealing with a specific binding reaction of an osmolyte to a protein, with stoichiometry n:

$$N + nO \leftrightarrow NO_n \quad K = \frac{[NO_n]}{[N] \cdot c_O}. \quad (26.48)$$

As long as we consider just the protein and ignore (or are not aware of) the participation of the ligand, we can just copy Eq. (26.42) to obtain the m value for this reaction

$$-\left(\frac{\partial \ln(c_{NO_n}/c_N)}{\partial \ln c_O}\right) = \frac{m}{RT} = \frac{\Delta(G_{PW} - G_{PO})c_O}{1 - (G_{WO} - G_{OO})c_O}. \quad (26.49)$$

In that case, however, the binding of the ligand is included in the preferential interaction $\Delta(G_{PW} - G_{PO})$. To unravel these contributions, we first have a quick look at preferential interaction. There are many kinds of preferential interaction parameters (Anderson *et al.*, 2002; Shulgin and Ruckenstein, 2006). The one that is useful for our case is (Schurr *et al.*, 2005)

$$\Gamma_{O(P\to0)} = -(G_{PW} - G_{PO})c_O = -\left(\bar{X}_W \frac{c_O}{c_W} - \bar{X}_O\right), \quad (26.50)$$

where the excess or deficit in number of waters and osmolyte around the protein is $\bar{X}_W = G_{PW}c_W$ and $\bar{X}_O = G_{PO}c_O$, respectively. To distinguish between general solvation of the protein and events at the binding site, we reserve the symbol m for general solvation and the symbols $\bar{X}_W$ and $\bar{X}_O$ for the binding site. Equation (26.49) then becomes

$$-\left(\frac{\partial\ln(c_{NO_n}/c_N)}{\partial\ln c_O}\right) = \frac{m'}{RT}c_O + \left(\Delta\bar{X}_W \frac{c_O}{c_W} - \Delta\bar{X}_O\right)\cdot\left(\frac{\partial\ln a_O}{\partial\ln c_O}\right).$$

$$(26.51)$$

We use prime $'$ to distinguish the m' value in this equation from the m value in Eq. (26.49), which does not consider the specific binding separately. $\Delta\bar{X}_O$ equals the stoichiometry n. All terms containing the factor c_O in this equation vanish in the region of dilute osmolyte, and $\partial\ln a_O/\partial\ln c_O$ becomes 1. The only term that is significant under such conditions is the $\Delta\bar{X}_O$ term. Thus, the term $\Delta\bar{X}_W$ should also be counted as solvation and we get

$$-\left(\frac{\partial\ln(c_{NO_n}/c_N)}{\partial\ln c_O}\right) = \frac{m''}{RT}c_O - \Delta\bar{X}_O\cdot\left(\frac{\partial\ln a_O}{\partial\ln c_O}\right). \quad (26.52)$$

The solvation term m''/RT is insignificant at low c_O, and thus the binding equilibrium reacts to the presence of osmolyte solely by specific binding. At molar concentrations of c_O the solvation term becomes significant and will therefore influence the binding equilibrium.

5.2. Specific binding explicitly taken into account

Under normal circumstances the specific binding will be known, and then it is more straightforward to take it into account directly. Along the same line of reasoning as for Eqs. (26.42) and (26.49) we can formulate the dependence of the equilibrium constant K on osmolyte concentration. The only difference with respect to Eq. (26.49) is that we have to also consider γ_O:

$$
-\left(\frac{\partial \ln K}{\partial \ln c_O}\right) = \left(\frac{\partial \ln[\gamma_{NO_n}/(\gamma_N \gamma_O^{\Delta \bar{X}_O})]}{\partial \ln c_O}\right)
$$
$$
= \left(\frac{\partial \ln(a_{NO_n}/a_N)}{\partial \ln c_O}\right) - \Delta \bar{X}_O \left(\frac{\partial \ln \gamma_O}{\partial \ln c_O}\right). \qquad (26.53)
$$

The derivative $\partial \ln(a_{NO_n}/a_N)/\partial \ln c_O$ could be written as an m value or (equivalently) as a solvation expression similar to Eq. (26.49). For the sake of simplicity we use the m value notation:

$$
-\left(\frac{\partial \ln K}{\partial \ln c_O}\right) = \frac{m}{RT} c_O - \Delta \bar{X}_O \cdot \left(\frac{\partial \ln \gamma_O}{\partial \ln c_O}\right). \qquad (26.54)
$$

Note that the specific binding itself only impacts the pK, if the activity coefficient of the osmolyte γ_O differs significantly from ideal behavior ($\gamma_O = 1$). According to Eq. (26.28) such deviation is based on the solvation properties of the osmolyte, and the deviation from ideality occurs at high c_O if osmolyte hydration G_{WO} and self-solvation G_{OO} differ.

We use binding of glucose to a protein as an example. Equation (26.54) can be easily integrated to yield the shift in ligand dissociation pK as a function of osmolyte concentration. Inserting the known activity coefficient of glucose (see Table 26.1) we obtain

$$
\Delta pK = \frac{m}{RT\ln 10} c_O + \Delta \bar{X}_O \frac{\ln(1 - c_O/c_1)}{\ln 10}. \qquad (26.55)
$$

Figure 26.5 shows this pK shift as a continuous line. Solvation of the free osmolyte [second term in Eq. (26.55)] leads to a significant deviation of the pK shift from the one that merely takes into account plain protein solvation [m-value term in Eq. (26.55)]. It is thus important to consider solvation of the free osmolyte ligand rather than focusing attention on nonspecific solvation alone.

Equation (26.54) itself does not contain any explicit solvation term. However, the m value, as well as the activity coefficient term, contains solvation terms implicitly, as given by Eqs. (26.49) and (26.28), respectively. The logical sequence in analyzing protein and osmolyte solvation is to first determine thermodynamic parameters experimentally, such as the m value, and then equate them to their solvation expressions for analyzing the solvation properties.

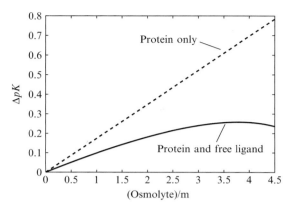

Figure 26.5 Shift in osmolyte dissociation pK due to solvation effects. Both solvation of the protein and solvation of the unbound osmolyte ligand have an important effect on the pK. The dashed line indicates the effect of protein solvation alone on the pK [m-value term in Eq. (26.55)]. Additionally taking into account the solvation of the free osmolyte [second term in Eq. (26.55)] has a drastic impact on the pK. The calculation is based on an m value of 1 kJ/molM and the activity coefficient of glucose (see Table 26.1).

6. CASE 4: SPECIFIC BINDING OF A DILUTE LIGAND

It is known that the osmolytes urea and glycerol affect the activity coefficient of ATP (Sinha and Kundu, 1998), and we provided some evidence that CMP is also affected by urea (Ferreon *et al.*, 2007). This demonstrates that in addition to osmolyte–protein solvation, we must also consider osmolyte–ligand solvation. It is thus of importance to investigate the four-component system water–protein–ligand–osmolyte.

The dependence of protein chemical activity on osmolyte concentration can be calculated in a straightforward manner from Eq. (26.14) with the help of appropriate software, such as Mathematica. The resulting expression is bulky, but can be simplified by considering only cases of dilute protein and ligand ($c_P \rightarrow 0$, $c_L \rightarrow 0$). In calculating this limit, all terms containing ligand binding $\bar{X}_L = G_{PL} c_L$ have to be retained. In order to do the calculations, we again need to do a base transform, which is

$$
\left(\frac{\partial}{\partial c_O} \right)_{c_P} = \frac{V}{\bar{v}_W c_W} \left[(\bar{v}_W c_W + \bar{v}_O c_O) \left(\frac{\partial}{\partial N_O} \right)_{N_P, N_L} \right.
$$
$$
\left. + \bar{v}_O c_P \left(\frac{\partial}{\partial N_P} \right)_{N_O, N_L} + \bar{v}_O c_L \left(\frac{\partial}{\partial N_L} \right)_{N_O, N_P} \right]. \tag{26.56}
$$

The overall result is then

$$\left(\frac{\partial \ln a_P}{\partial c_O}\right) = \frac{G_{PW} - G_{PO} + (G_{LO} - G_{LW})G_{PL}c_L}{(1 - G_{PL}^2 c_P c_L)[1 - (G_{WO} - G_{OO})c_O]}. \qquad (26.57)$$

We first note that the denominator contains a protein- and ligand-dependent term in addition to the osmolyte solvation term (in square brackets) that also occurs in the three-component case [Eq. (26.49)]. For calculating the m value for ligand binding, however, we do not have to consider ligand binding explicitly. Rather, we have to calculate the slope of the chemical activity of the liganded and unliganded state separately. This results in

$$-\left(\frac{\partial \ln K}{\partial c_O}\right) = \left(\frac{\partial \ln(a_{P_B}/a_{P_U} a_L)}{\partial c_O}\right) = \frac{\Delta(G_{PW} - G_{PO}) - (G_{LW} - G_{LO})}{1 - (G_{WO} - G_{OO})c_O}.$$

$$(26.58)$$

There are six solvation parameters (Kirkwood–Buff integrals) in Eq. (26.58) that have to be determined. This task is best divided into several subtasks. Solvation in the bulk osmolyte solution (G_{WO} and G_{OO}) can be derived from data on two-component solutions of aqueous osmolyte, such as vapor pressure data. For many osmolytes, G_{WO} and G_{OO} are calculated readily for many osmolytes from Eqs. (26.32) and (26.33) combined with Eq. (26.3) or (26.4) using the parameters given in Table 26.1. The solvation of free ligand by water and osmolyte (G_{LW} and G_{LO}) can then be determined using the proper form of Eq. (26.17)

$$\bar{v}_L = RT\kappa - \bar{v}_O c_O G_{LO} - \bar{v}_W c_W G_{LW} \qquad (26.59)$$

along with either solubility (Sinha and Kundu, 1998) or vapor pressure data (Rafflenbeul *et al.*, 1973; Uedaira, 1972) as a measure for the slope of a_L with c_O. Finally, the change in hydration and osmolyte solvation of the protein upon binding (ΔG_{PW} and ΔG_{PO}) has to be determined from the measured m value [Eq. (26.58)] and volumetric Eq. (26.45).

ACKNOWLEDGMENTS

Part of this work was funded by NIH (R01GM049760). The author thanks Wayne Bolen and Luis Holthauzen for critically reading and discussing the manuscript for this chapter.

REFERENCES

Anderson, C. F., Courtenay, E. S., and Record, M. T. (2002). Thermodynamic expressions relating different types of preferential interaction coefficients in solutions containing two solute components. *J. Phys. Chem.* B **106**, 418–433.

Auton, M., and Bolen, D. W. (2005). Predicting the energetics of osmolyte-induced protein folding/unfolding. *Proc. Natl. Acad. Sci. USA* **102**, 15065–15068.

Baskakov, I., and Bolen, D. W. (1998). Forcing thermodynamically unfolded proteins to fold. *J. Biol. Chem.* **273**, 4831–4834.

Ben-Naim, A. (1977). Inversion of Kirkwood-Buff theory of solutions: Application to water-ethanol system. *J. Chem. Phys.* **67**, 4884–4890.

Ben-Naim, A. (1988). Theory of preferential solvation of nonelectrolytes. *Cell Biophys.* **12**, 255–269.

Bolen, D. W., and Baskakov, I. V. (2001). The osmophobic effect: Natural selection of a thermodynamic force in protein folding. *J. Mol. Biol.* **310**, 955–963.

Casassa, E. F., and Eisenberg, H. (1964). Thermodynamic analysis of multicomponent solutions. *Adv. Protein Chem.* **19**, 287–395.

Chalikian, T. V. (2003). Volumetric properties of proteins. *Annu. Rev. Biophys. Biomol. Struct.* **32**, 207–235.

Chitra, R., and Smith, P. E. (2002). Molecular association in solution: A Kirkwood-Buff analysis of sodium chloride, ammonium sulfate, guanidinium chloride, urea, and 2,2, 2-trifluoroethanol in water. *J. Phys. Chem.* B **106**, 1491–1500.

Courtenay, E. S., Capp, M. W., Saecker, R. M., and Record, M. T., Jr. (2000). Thermodynamic analysis of interactions between denaturants and protein surface exposed on unfolding: Interpretation of urea and guanidinium chloride m-values and their correlation with changes in accessible surface area (ASA) using preferential interaction coefficients and the local-bulk domain model. *Proteins Suppl.* **4**, 72–85.

Eisenberg, H. (1994). Protein and nucleic-acid hydration and cosolvent interactions: Establishment of reliable base-line values at high cosolvent concentrations. *Biophys. Chem.* **53**, 57–68.

Felitsky, D. J., and Record, M. T., Jr. (2004). Application of the local-bulk partitioning and competitive binding models to interpret preferential interactions of glycine betaine and urea with protein surface. *Biochemistry* **43**, 9276–9288.

Ferreon, A. C., and Bolen, D. W. (2004). Thermodynamics of denaturant-induced unfolding of a protein that exhibits variable two-state denaturation. *Biochemistry* **43**, 13357–13369.

Ferreon, A. C., Ferreon, J. C., Bolen, D. W., and Rösgen, J. (2007). Protein phase diagrams. II. Nonideal behavior of biochemical reactions in the presence of osmolytes. *Biophys. J.* **92**, 245–256.

Gibbs, J. W. (1876/1878). On the equilibrium of heterogeneous substances. *Trans. Connect. Acad.* **3**, 108–248, 343–542.

Greene, R. F., Jr., and Pace, C. N. (1974). Urea and guanidine hydrochloride denaturation of ribonuclease, lysozyme, alpha-chymotrypsin, and beta-lactoglobulin. *J. Biol. Chem.* **249**, 5388–5393.

Hansen, J. P., and McDonald, I. R. (1986). "Theory of Simple Liquids." Academic Press, London.

Holthauzen, L. M., and Bolen, D. W. (2007). Mixed osmolytes: The degree to which one osmolyte affects the protein stabilizing ability of another. *Protein Sci.* **16**, 293–298.

Kirkwood, J. G., and Buff, F. P. (1951). The statistical mechanical theory of solutions. I. *J. Chem. Phys.* **19**, 774–777.

Lide, D. R. (2004). "CRC Handbook of Chemistry and Physics." CRC Press, Boca Raton, FL.

Lin, L. N., Brandts, J. F., Brandts, J. M., and Plotnikov, V. (2002). Determination of the volumetric properties of proteins and other solutes using pressure perturbation calorimetry. *Anal. Biochem.* **302,** 144–160.

Makhatadze, G. I. (1999). Thermodynamics of protein interactions with urea and guanidinium hydrochloride. *J. Phys. Chem. B* **103,** 4781–4785.

Mello, C. C., and Barrick, D. (2003). Measuring the stability of partly folded proteins using TMAO. *Protein Sci.* **12,** 1522–1529.

Ostwald, W. (1893). On chemical energy. *J. Am. Chem. Soc.* **15,** 421–430.

Parsegian, V. A., Rand, R. P., and Rau, D. C. (1995). Macromolecules and water: Probing with osmotic stress. *Methods Enzymol.* **259,** 43–94.

Rafflenbeul, L., Pang, W.-M., Schönert, H., and Haberle, K. (1973). Zur Thermodynamik der hydrophoben Wechselwirkung; die Systeme Wasser+Glycin+Harnstoff und Wasser+Alanin+Harnstoff bei 25 C. *Zeitschr. Naturforsch.* **28,** 533–554.

Ramsden, W. (1902). Some new properties of urea. *J. Physiol.* **28,** 23–26.

Rösgen, J., and Hinz, H. J. (2000). Response functions of proteins. *Biophys. Chem.* **83,** 61–71.

Rösgen, J., Pettitt, B. M., and Bolen, D. W. (2004a). Uncovering the basis for nonideal behavior of biological molecules. *Biochemistry* **43,** 14472–14484.

Rösgen, J., Pettitt, B. M., and Bolen, D. W. (2005). Protein folding, stability, and solvation structure in osmolyte solutions. *Biophys. J.* **89,** 2988–2997.

Rösgen, J., Pettitt, B. M., and Bolen, D. W. (2007). An analysis of the molecular origin of osmolyte-dependent protein stability. *Protein Sci.* **16,** 733–743.

Rösgen, J., Pettitt, B. M., Perkyns, J., and Bolen, D. W. (2004b). Statistical thermodynamic approach to the chemical activities in two-component solutions. *J. Phys. Chem. B* **108,** 2048–2055.

Royer, C. A. (2002). Revisiting volume changes in pressure-induced protein unfolding. *Biochim. Biophys. Acta* **1595,** 201–209.

Santoro, M. M., and Bolen, D. W. (1988). Unfolding free energy changes determined by the linear extrapolation method. 1. Unfolding of phenylmethanesulfonyl alpha-chymotrypsin using different denaturants. *Biochemistry* **27,** 8063–8068.

Scatchard, G. (1921). The speed of reaction in concentrated solutions and the mechanism of the inversion of sucrose. *J. Am. Chem. Soc.* **43,** 2387–2406.

Schellman, J. A. (1994). The thermodynamics of solvent exchange. *Biopolymers* **34,** 1015–1026.

Schellman, J. A. (2003). Protein stability in mixed solvents: A balance of contact interaction and excluded volume. *Biophys. J.* **85,** 108–125.

Schurr, J. M., Rangel, D. P., and Aragon, S. R. (2005). A contribution to the theory of preferential interaction coefficients. *Biophys. J.* **89,** 2258–2276.

Shimizu, S. (2004). Estimation of excess solvation numbers of water and cosolvents from preferential interaction and volumetric experiments. *J. Chem. Phys.* **120,** 4989–4990.

Shimizu, S., and Smith, D. J. (2004). Preferential hydration and the exclusion of cosolvents from protein surfaces. *J. Chem. Phys.* **121,** 1148–1154.

Shulgin, I. L., and Ruckenstein, E. (2006). A protein molecule in a mixed solvent: The preferential binding parameter via the Kirkwood-Buff theory. *Biophys. J.* **90,** 704–707.

Sinha, R., and Kundu, K. K. (1998). Transfer Gibbs energies of ATP in aqueous mixtures of non-ionic glycerol and urea and ionic NaNO3. *Indian J. Chem. Sect. a Inorg. Bio-Inorgan. Phys. Theor. Anal. Chem.* **37,** 789–794.

Smith, P. E. (2004). Local chemical potential equalization model for cosolvent effects on biomolecular equilibria. *J. Phys. Chem. B* **108,** 16271–16278.

Smith, P. E. (2006). Chemical potential derivatives and preferential interaction parameters in biological systems from kirkwood-buff theory. *Biophys. J.* **91,** 849–856.

Spiro, K. (1900). Ueber die Beeinflussung der Eiweisscoagulation durch stricksstoffhaltige Substanzen. *Zeitschr. Physiol. Chem.* **30,** 182–199.

Tanford, C. (1968). Protein denaturation. *Adv. Protein Chem.* **23,** 121–282.

Timasheff, S. N. (1998). Control of protein stability and reactions by weakly interacting cosolvents: The simplicity of the complicated. *Adv. Protein Chem.* **51,** 355–432.

Timasheff, S. N., and Xie, G. (2003). Preferential interactions of urea with lysozyme and their linkage to protein denaturation. *Biophys. Chem.* **105,** 421–448.

Uedaira, H. (1972). Activity coefficients for the system glycylglycine-urea-water. *Bull. Chem. Soc. Jpn* **45,** 3068–3072.

Zhang, W., Capp, M. W., Bond, J. P., Anderson, C. F., and Record, M. T., Jr. (1996). Thermodynamic characterization of interactions of native bovine serum albumin with highly excluded (glycine betaine) and moderately accumulated (urea) solutes by a novel application of vapor pressure osmometry. *Biochemistry* **35,** 10506–10516.

CHAPTER TWENTY-SEVEN

METHODS OF CHANGING BIOPOLYMER VOLUME FRACTION AND CYTOPLASMIC SOLUTE CONCENTRATIONS FOR *IN VIVO* BIOPHYSICAL STUDIES

Michael C. Konopka,* James C. Weisshaar,* *and* M. Thomas Record, Jr.*,†

Contents

Abstract

In vitro changes in polymer volume fraction (macromolecular crowding) and changes in solute or salt concentration typically have large effects on protein and nucleic acid processes (e.g., folding, binding, assembly, precipitation, crystallization). However, the large changes in these concentration variables, which occur *in vivo* as part of cellular responses to osmotic stress, appear to have much less dramatic effects on cellular biopolymer processes. Methods of changing intracellular concentrations by varying the extracellular osmolality or the concentration of a permeable solute or by titrating cells with an impermeable solute (plasmolysis) under conditions where an active response is suppressed are reviewed. The first *in vivo* biophysical studies of protein folding and protein diffusion performed as a function of these variables are also discussed.

* Department of Chemistry, University of Wisconsin–Madison, Madison, Wisconsin
† Department of Biochemistry, University of Wisconsin–Madison, Madison, Wisconsin

Methods in Enzymology, Volume 428
ISSN 0076-6879, DOI: 10.1016/S0076-6879(07)28027-9

1. Introduction

Most biochemical studies are performed in dilute solutions, which differ greatly from the cell cytoplasm both in total concentration of biopolymers and in the nature and concentration of salt ions and other solutes. Relatively few quantitative *in vivo* studies and *in vivo–in vitro* comparisons of biopolymer processes (diffusion, transport, folding, assembly, binding, catalysis) exist. These are of particular importance because of the large effects of changes in solution conditions on many biopolymer processes *in vitro*, the large differences between typical *in vitro* and *in vivo* solution conditions, and the variability of *in vivo* solution conditions in some, if not all, cells.

In vivo, the total biopolymer concentration (volume fraction) and concentrations of osmotically significant ions and solutes can be large (Cayley and Record, 2003; Cayley *et al.*, 1991, 1992, 2000). These concentrations vary over wide ranges and in predictable ways in response to environmental osmotic stress. Exponentially growing *Escherichia coli* responds to increases in growth osmolality by increasing cytoplasmic amounts of K^+, glutamate$^-$, trehalose, and (when available) osmoprotectants such as proline and/or glycine betaine. These increases in amounts of cytoplasmic solutes are insufficient to cause retention of as much cytoplasmic water as at the optimal growth osmolality (0.28 Osm); consequently, the total concentration of biopolymers (and presumably that of other nonosmoregulated solutes) also increases significantly with osmolality of growth. At high osmolality of growth, the biopolymer volume fraction approaches 0.3, and solute concentrations can exceed 1 molal. Higher biopolymer volume fractions (up to 0.5, where all remaining cytoplasmic water appears to be water of hydration) and higher solute concentrations can be achieved by "plasmolysis" titrations of cell suspensions with membrane-impermeable solutes. High osmolality plasmolysis conditions greatly reduce the diffusion coefficient of cytoplasmic green fluorescent protein (GFP) in fluorescence recovery after photobleaching (FRAP) experiments (Konopka *et al.*, 2006). Cytoplasmic concentrations of membrane-permeable (small, uncharged) solutes such as urea and glycerol vary with changes in extracellular concentration, without affecting concentrations of biopolymers or other solutes. In particular, urea concentrations of up to 3 *M* have been used to denature marginally stable proteins in the cytoplasm of *E. coli* without affecting viability (Ghaemmaghami and Oas, 2001; Ignatova and Gierasch, 2004).

2. Varying Biopolymer Volume Fraction and Cytoplasmic Solute Concentrations Osmotically

Escherichia coli is the only cell for which there is quantitative data on water, osmolyte, and biopolymer amounts as a function of osmolality and growth condition. *E. coli* exhibits both active and passive responses to osmotic upshifts (see reviews, Csonka and Epstein, 1996; Csonka and Hanson, 1991; Wood, 1999), which result in large changes in the amount of cytoplasmic water, the concentrations and volume fraction of cytoplasmic biopolymers, and the concentrations of cytoplasmic and periplasmic solutes.

By incompletely understood mechanisms of osmosensing and osmoregulation (Wood, 1999), exponentially growing cells stressed by an increase in growth osmolality adapt by increasing the amounts of various intracellular osmolytes. Accumulation of these osmotically active solutes by transport or biosynthesis increases the amount of intracellular water. Because water is membrane permeable and osmolytes are not, the amount of water in a cell compartment and the volume of that compartment at a specified external osmolality (and turgor pressure, if applicable) are determined entirely by the amounts of various osmolytes in that compartment. Contributions to osmotic nonideality, arising from interactions with themselves, other osmolytes, or biopolymers, as well as from excluded volume effects, must be taken into account. Uptake of cytoplasmic solutes and water allows the cell to resume exponential growth. The growth rate in minimal medium as a function of osmolality and the presence of osmoprotectants increase linearly with the amount of free cytoplasmic water, extrapolating to zero at high osmolality conditions where the cytoplasm cannot accumulate sufficient solutes to retain any free water (Cayley and Record, 2003; Record et al., 1998). The observation of this general correlation between growth rate and amount of cytoplasmic water led to the proposal that the effect of biopolymer volume fraction (crowding and/or confinement) on diffusion of key cytoplasmic biopolymers might be a determinant of the growth rate of osmotically stressed E. coli. In vivo measurements of diffusion coefficients of a model protein (GFP) as a function of cytoplasmic biopolymer volume fraction are described in the last section of this chapter.

Osmolytes used by many types of cells have been identified (Yancey et al., 1982). Cytoplasmic and periplasmic E. coli osmolytes and other osmotically regulated E. coli solutes are listed in Table 27.1. Common organic osmolytes include sugars (trehalose in E. coli, sucrose in some algae and plants), other polyols (glycerol in animals and mannitol in some algae and plants), amino acids and their derivatives [in E. coli glutamate$^-$ and, when available, proline, ectoine ((S)-2-methyl-1,4,5,6-tetrahydropyrimidine-4-carboxylic acid), glycine betaine (GB; N,N,N-trimethyl glycine)], and trimethyl amine oxide (in many marine animals). K$^+$ is the only inorganic osmolyte. E. coli, growing at high osmolality in MOPS-buffered minimal glucose medium (MBM) (Neidhardt et al., 1974), actively accumulates cytoplasmic potassium glutamate and KMOPS by transport of K$^+$ (Epstein and Schultz, 1965) and MOPS$^-$ (Cayley et al., 1989) and synthesis of glutamate (Dinnbier et al., 1988; McLaggan et al., 1994). The cytoplasmic concentration of K$^+$ increases in response to osmotic stress not only because of increases in the amount and concentration of cytoplasmic K$^+$ salts, but also because the nucleic acid concentration increases with increasing osmolality of growth (Cayley et al., 1991). The nucleic acid concentration (also the

Table 27.1 Common osmolytes and concentrations

		Osmolyte	Transporter or precursor[a]	In vivo concentration in MBM[b,c]
Cytoplasmic	Transported	K^+ (1.32 mM external)	trkA (C) kdp (O)	0.22 M at 0.1 Osm; 0.81 M at 1.0 Osm
		Glycine betaine (1 mM external)	proU (O) and proP (O)	0.52 M at 1.0 Osm + GB 1.67; M at 2.2 Osm + GB
		Proline (1 mM external)	proU (O) and proP (O)	0.48 M at 1.0 Osm + pro
		MOPS$^-$ (40 mM external)		<0.01 M at 0.1 Osm; 0.21 M at 1.0 Osm
		Choline[d]	betT	
		Ectoine	proP (O) and proU (O)	
	Biosynthetic	Glutamate$^-$	Glutamine	0.03 M at 0.1 Osm; 0.19 M at 1.0 Osm
		Trehalose	Glucose	<0.01 M at 0.1 Osm; 0.31 M at 1.0 Osm
		Putrescine^{2+}	Ornithine	0.05 M at 0.1 Osm; <0.01 M at 1.0 Osm
Periplasmic	Biosynthetic	Membrane–derived oligosaccharides (MDOs)	UPD–glucose	0.05 M at 0.1 Osm; <0.005 M at 0.8 Osm
	Freely diffusing	External solutes < about 1 kDa		

[a] Unless noted otherwise, Wood (1999) and references therein.

[b] MBM, MOPS-buffered minimal glucose medium (Neidhardt et al., 1974).

[c] Amounts of cytoplasmic and periplasmic water and of cytoplasmic osmolytes are from Cayley et al. (1989, 1991, 1992, 2000) and Cayley and Record (2003). Amounts of cytoplasmic putrescine are from Munro et al. (1972). Amounts of periplasmic MDOs are from Kennedy (1982) and Lacroix et al. (1989), as analyzed by Cayley et al. (2000).

[d] Converted to glycine betaine by choline dehydrogenase and betaine aldehyde dehydrogenase (Landfald and Strom, 1986).

biopolymer volume fraction) increases with growth osmolality because the amount of water per unit mass of biopolymer in growing cells decreases with increasing osmolality, as discussed later. In *E. coli*, cytoplasmic K^+ is both an osmolyte and an unbound (but locally accumulated) nucleic acid counterion. The high cytoplasmic concentration of negatively charged nucleic acid phosphates (primarily in ribosomes) is compensated in large part by locally accumulated but unbound K^+, as well as by site- or diffusely bound Mg^{2+}, putrescine^{2+}, and other oligocations (Cayley *et al.*, 1991).

Escherichia coli responds passively to an increase in external osmolality by an efflux of cytoplasmic water and a reduction in turgor pressure (Cayley and Record, 2003; Cayley *et al.*, 1991, 1992, 2000). Because the cytoplasmic amounts of solutes and biopolymers per cell are unchanged in this passive response, the result is an increase in their cytoplasmic concentrations. Cells plasmolyzed in this manner are incapable of resuming growth without making osmoregulated changes in amounts of various cytoplasmic solutes in order to take up additional water (Wood, 1999).

Immediately after an osmotic upshift, exponentially growing *E. coli* undergoes the passive response described earlier (loss of some cytoplasmic water and turgor) and subsequently responds actively during an extended lag phase before resuming exponential growth (Csonka and Epstein, 1996; Csonka and Hanson, 1991; Wood, 1999). In this active response, some osmolytes are transported and others are synthesized (Table 27.1) in order to increase the amount of osmotically significant solutes. As a result, the amount of cytoplasmic water increases and the biopolymer volume fraction and the concentrations of most (nonosmolyte) cytoplasmic solutes decrease. Figure 27.1 summarizes amounts of cytoplasmic water, osmotically regulated solutes, and nucleic acid phosphates for growth at optimal osmolality (0.28 Osm; A) and for passive and active responses to an osmotic upshift of cells grown at 0.1 Osm (B) to 1.0 Osm $+/-$ glycine betaine (C–E). Figure 27.2 plots biopolymer volume fractions (A and B) and K^+ and GB concentrations (C–E) as functions of osmolality of growth and of plasmolysis for different growth conditions.

Taken together, Figs. 27.1 and 27.2 illustrate the range of possible cytoplasmic solution conditions encountered by all covalent and noncovalent cell processes occurring in the growth of osmotically stressed *E. coli*. *In vitro*, most noncovalent interactions of proteins or nucleic acids, as well as many enzyme-catalyzed covalent reactions, would be affected greatly by changes of this magnitude in solute or salt concentrations (e.g., potassium glutamate [Ha *et al.*, 1992; Leirmo *et al.*, 1987] and GB [Hong *et al.*, 2005; Kontur *et al.*, 2006]) or biopolymer volume fraction (Ellis, 2001; Luby-Phelps, 2000; Minton, 2001; Zimmerman and Minton, 1993).

Figure 27.1A illustrates amounts of free and bound cytoplasmic water (left axis) and of osmoregulated cytoplasmic solutes and nucleic acids (right axis) in *E. coli* K-12 (MG1655) growing in minimal medium at the osmolality where

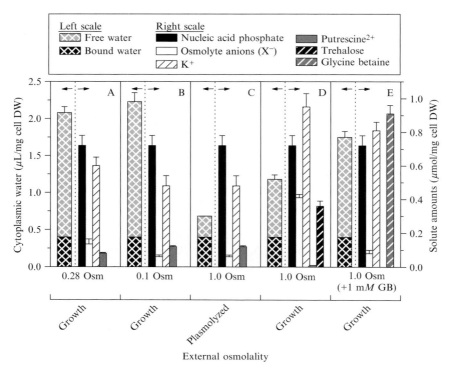

Figure 27.1 Amounts of cytoplasmic water, nucleic acid phosphate, and most significant, osmotically regulated cytoplasmic solutes in *E. coli* MG1655 grown in MOPS-buffered minimal glucose medium and harvested during exponential growth at various
external osmolalities. Cytoplasmic amounts of bound and free water determined by
plasmolysis titrations (see text) are expressed in units of microliters per milligram cell
dry weight on the left axis scale. Cytoplasmic amounts of nucleic acid phosphate and
solutes (glutamate$^-$, osmolyte anions X$^-$, K$^+$, putrescine^{2+}, trehalose, and glycine betaine [GB]) are expressed in units of micromoles per milligram cell dry weight on the
right axis. From left to right, the osmotic conditions are (A) growth at 0.28 Osm, (B)
growth at 0.1 Osm, (C) plasmolysis to 1.0 Osm from 0.10 Osm growth, (D) growth at
1.0 Osm, and (E) growth at 1.0 Osm with 1 mM glycine betaine. Putrescine was not
determined at 1.0 Osm with glycine betaine. Adapted from Record *et al.* (1998).

growth rate is maximum (0.28 Osm). Bound water (determined by plasmolysis titration [Cayley *et al.*, 1991] and interpreted as water of hydration) is about
20% of total cytoplasmic water. The molar amount of K$^+$, the principal
cytoplasmic solute, greatly exceeds that of all small cytoplasmic anions (mostly
glutamate$^-$); the large excess of K$^+$ over small anions is a clear demonstration
that the cytoplasm is a concentrated polyelectrolyte solution of nucleic acid
polyanions (>0.25 M in phosphate groups; cf. Fig. 27.1), whose negative
charge is compensated in large part by unbound K$^+$ ions.

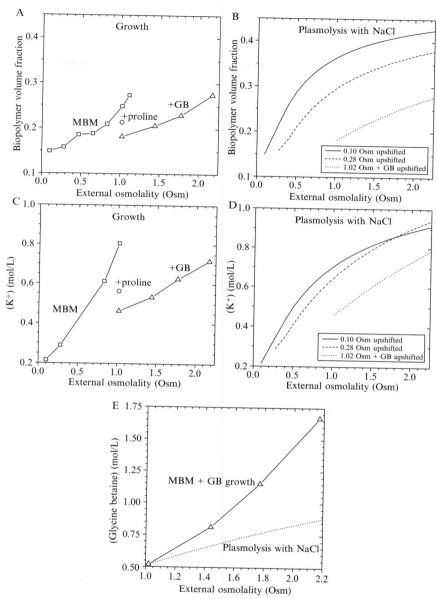

Figure 27.2 Biopolymer volume fraction and concentrations of cytoplasmic K^+ and glycine betaine (GB) are plotted vs external osmolality for both growth and plasmolysis conditions. (A) Biopolymer volume fraction vs external osmolality of growth in MBM (squares), MBM + proline (circle), and MBM + GB (triangles). (B) Biopolymer volume fraction vs external osmolality of plasmolysis of cells grown in 0.10 Osm MBM (solid line), 0.28 Osm MBM (dashed line), and 1.02 Osm MBM + GB (dotted line). *(cont.)*

Figures 27.1B through 27.1E summarize the changes that occur in amounts of cytoplasmic water and solutes when cells grown at low (suboptimal) osmolality (0.1 Osm; B) are plasmolyzed to 1.0 Osm under nongrowing conditions (C) and after adaptation and resumption of exponential growth at 1.0 Osm in the absence (D) and presence (E) of an osmoprotectant (GB). Plasmolysis from 0.1 to 1.0 Osm reduces the amount of water in the cytoplasm greatly without affecting amounts of cytoplasmic solutes, thereby increasing all solute concentrations. Before adaptation, the majority of cytoplasmic water is bound. Because the amounts of cytoplasmic biopolymers and solutes do not change in plasmolysis, their concentrations increase greatly in a plasmolysis titration, as shown in Fig. 27.2B for the biopolymer volume fraction (increasing from 0.15 at 0.1 Osm growth to 0.36 at 1.0 Osm) and Fig. 27.2D for cytoplasmic K^+ (increasing from 0.21 M at 0.1 Osm to 0.70 M at 1.0 Osm).

To resume growth at 1.0 Osm in a MOPS-buffered minimal medium in the absence of added osmoprotectants (Fig. 27.1D), K^+ (Epstein and Schultz, 1965) and $MOPS^-$ (Cayley et al., 1989) are transported and glutamate$^-$ and trehalose are synthesized (Dinnbier et al., 1988; McLaggan et al., 1994), increasing the amount of cytoplasmic water somewhat and reducing the biopolymer volume fraction. The amount of free water in these growing cells (Fig. 27.1D) is nearly three times as large as in plasmolyzed cells at 1.0 Osm (Fig. 27.1C), but only half as large as in cells growing at low osmolality before plasmolysis (Fig. 27.1B). With 1 mM glycine betaine in the growth medium at 1.0 Osm (Fig. 27.1E), cells accumulate GB as the primary solute instead of trehalose and potassium salts. The total molar amount of cytoplasmic osmolytes in cells growing at 1.0 Osm $+/-$ GB is similar (Cayley and Record, 2003; Cayley et al., 1992), but GB is so much more effective an osmolyte than those it replaces that the amount of cytoplasmic water increases significantly to a level approaching that of the original low osmolality cells. This may be explained by the observation that GB greatly prefers to interact with water rather than with itself or with anionic groups of biopolymers or other solutes (Felitsky et al., 2004) so that the osmolality of a concentrated GB-biopolymer solution greatly exceeds the ideal mixing value.

Although the amount of free water in the cytoplasm may be reduced to very low levels by plasmolysis or growth at high osmolality, the activity of cytoplasmic water is not similarly reduced (Cayley and Record, 2003).

(C and D) Cytoplasmic K^+ concentrations (in moles per liter of cytoplasmic water) are plotted vs osmolality of (C) growth or (D) plasmolysis for the same conditions as A and B. (E) Cytoplasmic glycine betaine concentration is plotted vs osmolality of growth (triangles) and of plasmolysis (dotted line) of cells grown in 1.02 Osm MBM + GB.

The turgor pressure of *E. coli* is quite small at high osmolality and even at low osmolality does not exceed 4 to 5 atm so the activity of cytoplasmic water never differs greatly from that of the growth medium (Cayley *et al.*, 2000), even when the amount of free cytoplasmic water is very small. This implies that the activity coefficient of cytoplasmic water exceeds unity at high osmolality.

These varied responses of *E. coli* to osmotic stress provide many ways to investigate how changes in biopolymer volume fraction (macromolecular crowding, confinement) and in cytoplasmic osmolyte concentrations affect biopolymer processes, including diffusion (Konopka *et al.*, 2006), folding, and aggregation (Ignatova and Gierasch, 2004, 2006) of proteins in the *E. coli* cytoplasm to complement similar *in vitro* work (Minton, 2005; Muramatsu and Minton, 1988; Ping *et al.*, 2004; Tokuriki *et al.*, 2004). Because both biopolymer volume fraction and osmolyte concentrations change with osmolality of growth or of plasmolysis, it is important to investigate different conditions to separate the effects of these variables. For example, Fig. 27.2A reveals that exponentially growing cells in MBM (0.83 Osm), MBM + proline (1.02 Osm), and MBM + GB (1.44 Osm) all exhibit a biopolymer volume fraction of 0.21. Figure 27.2B shows that this biopolymer volume fraction (0.21) can also be achieved by plasmolysis of cells grown in 0.10 Osm MBM (upshifted to 0.28 Osm) or 0.28 Osm MBM (upshifted to 0.5 Osm). All five of these conditions have the same biopolymer volume fraction, but very different cytoplasmic osmolyte compositions and concentrations and different osmolalities. Figure 27.2C and D indicate some choices of osmolality (and biopolymer volume fraction) with the same cytoplasmic K^+ concentration. Figure 27.2E summarizes more limited data for cytoplasmic glycine betaine.

A strong three-way correlation among cytoplasmic K^+ concentration, amount of cytoplasmic water (or biopolymer volume fraction), and growth rate has been observed for osmotically stressed, exponentially growing *E. coli*. This motivates the proposal that compensation among these variables might explain why the large *in vitro* effects of crowding and K^+ concentration on protein–nucleic acid interactions are not simply manifested *in vivo* (Cayley *et al.*, 1991; Richey *et al.*, 1987; Zimmerman and Harrison, 1987; Zimmerman and Trach, 1988, 1991).

3. BIOPHYSICAL STUDIES *IN VIVO* USING PERMEABLE SOLUTES: PROTEIN UNFOLDING BY UREA TITRATION

While biological membranes are impermeable to ions and large solutes, a few small solutes are membrane permeable. For example, the cytoplasmic membrane of *E. coli* is permeable to glycerol, urea, and dimethyl sulfoxide.

Cytoplasmic concentrations of permeable solutes can be increased without affecting other cytoplasmic solute or biopolymer concentrations. Because urea is found to drive unfolding of marginally stable proteins at concentrations that do not affect cell viability, urea has been of particular interest.

Two studies have quantitatively compared the stability of a protein *in vitro* and *in vivo* using urea titration. Ghaemmaghami and Oas (2001) compared the stability of the folded state of a monomeric N-terminal domain of λ-repressor ($\lambda \star_{6-85}$) in the *E. coli* cytoplasm and in a cell lysate. Figure 27.3A shows the *in vitro* urea denaturation curve of the purified $\lambda \star_{6-85}$ repressor protein obtained by circular dichroism (CD) at 222 nm. Application of a two-state model yielded an *in vitro* stability in the absence of urea of $\Delta G^{\circ} = 6.16 \pm 0.1$ kcal-mol^{-1}. Urea titrations *in vivo* are possible because *E. coli* remains viable for at least 30 min in up to 3 M urea, as shown by Fig. 27.3B. In place of CD, inapplicable *in vivo*, hydrogen/deuterium (H/D) exchange of amide protons, which are water inaccessible in the folded state, detected by matrix-assisted laser desorption/ionization mass spectrometry (SUPREX) was used to monitor unfolding during urea titrations.

For cells grown in LB medium at moderate osmolality, where concentrations of cytoplasmic osmolytes and the biopolymer volume fraction are not high, the stability of cytoplasmic $\lambda \star_{6-85}$ repressor protein to urea denaturation was observed by SUPREX to be the same as in a cell lysate. Stabilities (6.7 ± 0.1 and 6.7 ± 0.2 kcal-mol^{-1}, respectively) determined by H/D exchange differed by approximately 0.5 kcal-mol^{-1} from those determined by CD. For cells resuspended in LB + 1.2 M NaCl and equilibrated for 2 h (Fig. 27.3C, triangles), where the removal of much cytoplasmic water increases the biopolymer volume fraction and all solute concentrations, no unfolding of cytoplasmic $\lambda \star_{6-85}$ repressor protein could be detected by SUPREX, even at 3 M urea (Fig. 27.3C). Presumably the stabilization of the protein to urea-induced unfolding is attributable to the increases in macromolecular crowding and cytoplasmic solute concentrations. Unfortunately, no quantitative data for amounts of cytoplasmic water, biopolymers, and osmolytes are available for these growth conditions.

Ignatova and Gierasch (2004, 2006) investigated protein unfolding *in vivo* using a fluorescence assay. Incorporation of a tetra-cysteine motif into cellular retinoic acid-binding protein I (CRABP I) allowed them to specifically label this protein with fluorescein arsenical hairpin binder (FlAsH) dye. FlAsH only fluoresces strongly when complexed with a tetra-cysteine motif, making it a specific probe for the protein of interest (Griffin *et al.*, 1998). FlAsH fluorescence increases in the unfolded state, making it possible to differentiate between folded and unfolded proteins. Fluorescence intensity is then used to monitor the fraction of protein unfolded in urea denaturation curves. For cells grown in LB at moderate osmolality, the stability of CRABP I was the same *in vivo* and *in vitro* (Ignatova and Gierasch, 2004).

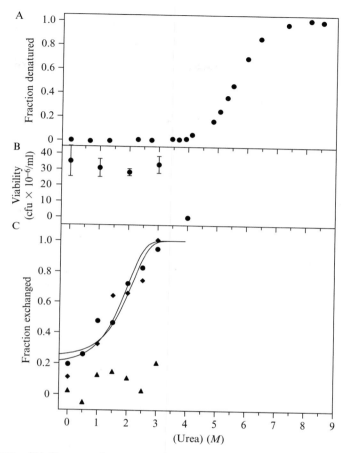

Figure 27.3 (A) Denaturation curve of purified λ repressor domain ($\lambda\star_{6-85}$) using circular dichroism. (B) Viability of *E. coli* after a 30-min exposure to the indicated urea concentration. (C) Stability measurements of $\lambda\star_{6-85}$ using SUPREX measurements of the amide proton exchange. *In vitro* measurements of cell lysates at 15° (circles) and *in vivo* measurements from viable *E. coli* growing in normal LB (diamonds) and equilibrated in LB + 1.2 *M* NaCl for 2 h (triangles). Reprinted with permission from Macmillan Publishers Ltd. (Ghaemmaghami and Oas, 2001).

Subsequent *in vivo* experiments with FlAsH-labeled wild-type and mutant CRABP I by the same authors focused on *in vivo* differences in stability and in aggregation of this protein in cells grown at low osmolality and osmotically stressed in the presence of the osmoprotectant proline (Ignatova and Gierasch, 2006). Stabilities of this protein were compared using FlAsH-monitored urea titrations. For a mutant (P39A) aggregation-prone CRABP I (see Fig. 27.4), present in cells grown at low osmolality in the absence of proline (closed symbols), the stability (ΔG^{o}) was found to be

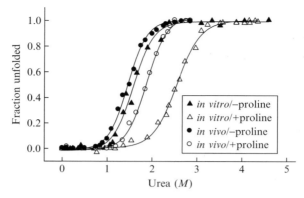

Figure 27.4 *In vitro* (triangles) and *in vivo* (circles) fluorescence measurements of unfolding of P39A CRABP mutant in the presence (open symbols) and absence (solid symbols) of 0.5 *M* proline. *In vitro* experiments were performed by monitoring unfolding by tryptophan fluorescence of protein equilibrated at different urea concentrations at 37°. *In vivo* measurements were made on P39A tetra-Cys CRABP-expressing cells prelabeled with FlAsH–EDT$_2$. Some cells were treated with salt to stimulate proline uptake. Two hours following induction, various final urea concentrations were used and incubated for 30 min prior to fluorescence measurements of FlAsH. Reproduced with permission from Ignatova and Gierasch (2006). Copyright © 2006 National Academy of Sciences, USA

similar *in vitro* (triangle; 3.60 ± 0.10 kcal-mol^{-1}) and *in vivo* (circles; 3.2 ± 0.1 kcal-mol^{-1}). For cells grown at the same osmolality and osmotically upshifted in the presence of proline (open symbol), the stability of this protein increased *in vivo* (4.0 ± 0.12 kcal-mol^{-1}), very similar to that observed *in vitro* in 0.5 *M* proline (4.95 ± 0.15 kcal-mol^{-1}). Because amounts of cytoplasmic water, proline, other osmolytes, and biopolymers are not known for these growth conditions, the roles of proline and of crowding or other osmolytes on stability and aggregation of CRABP I cannot be dissected.

4. EFFECTS OF BIOPOLYMER VOLUME FRACTION AND/ OR OTHER *IN VIVO* FACTORS ON PROTEIN DIFFUSION IN THE *E. COLI* CYTOPLASM

Large effects of high concentrations of flexible and globular biopolymers (volume fractions up to 0.25) on the diffusion of proteins are observed *in vitro*. These effects are interpreted by macromolecular crowding models such as scaled particle theory (SPT) (Han and Herzfeld, 1993; Muramatsu and Minton, 1988). Table 27.2 summarizes several studies comparing the mean diffusion coefficient $< D >$ of the tracer molecule in a protein background with a volume fraction of 0.15 or higher to the diffusion coefficient

Table 27.2 Selected *in vitro* and *in vivo* crowding effects on diffusion data

	Tracer particle	Background particles	Tracer/crowder size (kDa)	D_0 in dilute solution ($\mu m^2\text{-}s^{-1}$)	$\langle D \rangle$ in crowded solution ($\mu m^2\text{-}s^{-1}$)	ϕ	$D_0/\langle D \rangle$
Selected in vitro	Hemoglobin[a]	Hemoglobin	65/–	72 ± 5.8	30 ± 2.4	0.15	2.4
	Ovalbumin[a]	Ovalbumin	43.5/–	60	7 ± 0.6	0.25	10
	Lumbricus Hb[a]	Lumbricus Hb	3700/–	13 ± 1[b]	6.0 ± 0.5	0.15	2.2
	Cyanmetmyoglobin[c]	BSA	17/70	102 ± 5	41	0.15	2.5
	HbCO[c]	BSA	65/70	69 ± 2.7	27	0.15	2.6
	FITC-BSA[c]	BSA	70/70	61 ± 2.2	20	0.15	3.0
	FITC-aldolase[c]	BSA	150/70	41 ± 2.6	11	0.15	3.9
In vivo — growing cells	GFP[d]	E. coli cytoplasm	27/cyto	87 ± 3[e]	7.7 ± 2.5		11
	MBP-GFP[d]	E. coli cytoplasm	72/cyto		2.5 ± 0.6		
	his-tag β-gal-GFP[d]	E. coli cytoplasm	464/cyto		≈0		
	GFP (0.24 Osm LB)[f]	E. coli cytoplasm	27/cyto	87 ± 3[e]	6.1 ± 2.4		14
plasmolyzed cells	GFP (0.24→0.63 Osm)[f]	E. coli cytoplasm	27/cyto	87 ± 3[e]	0.94 ± 0.55		93
	GFP (0.24→0.68 Osm)[f]	E. coli cytoplasm	27/cyto	87 ± 3[e]	0.32 ± 0.50		270
	GFP (0.24→0.94 Osm)[f]	E. coli cytoplasm	27/cyto	87 ± 3[e]	0.014 ± 0.021		6200

[a] Data from Gros (1978) and references therein assuming a specific volume of 0.75 ml/g.

[b] Obtained by linearly extrapolating to zero concentration.

[c] Data from Muramatsu and Minton (1988) using their assumption of a specific volume of 0.75 ml/g.

[d] Diffusion data in live *E. coli* from (Elowitz et al., 1999). ϕ is unknown for these growth conditions.

[e] GFP diffusion data in dilute solution from Terry, B. R. Matthews, E. K., and Haseloff, J. (1995). Molecular characterisation of recombinant green fluorescent protein by fluorescence correlation microscopy. *Biochem. Biophys. Res. Commun.* **217**, 21–27.

[f] Diffusion data in crowded solution from cells grown in LB from Konopka et al. (2006).

Notes: Lumbricus Hb, *Lumbricus terrestris* hemoglobin; HbCO, human carbonmonoxyhemoglobin; BSA, bovine serum albumin; FITC, fluorescein isothiocyanate label; MBP, maltose–binding protein; ϕ, biopolymer volume fraction.

in buffer D_0. For *in vitro* studies at $< \phi > = 0.15$, $D_0/< D >$ lies in the narrow range 2 to 4.

Elowitz *et al.* (1999) measured the diffusion coefficient of GFP and several GFP–fusion proteins in the cytoplasm of live *E. coli* by FRAP. They found $D_0/< D > = 11$ for GFP. The biopolymer volume fraction has not been measured for their growth conditions, although an estimate of ≈ 0.15 to 0.2 appears reasonable. Hence, the reduction in diffusion coefficient of GFP in the *E. coli* cytoplasm from that obtained in dilute solution is greater than that expected from crowding alone (Table 27.2). Other studies of protein diffusion in the *E. coli* cytoplasm used fluorescence correlation spectroscopy (Cluzel *et al.*, 2000) and single particle tracking of large mRNA molecules (Golding and Cox, 2004, 2006); protein diffusion in the *E. coli* periplasm has been monitored by FRAP (Mullineaux *et al.*, 2006).

Using GFP as a test protein, Konopka *et al.* (2006) investigated the effect on protein diffusion in the *E. coli* cytoplasm by removing cytoplasmic water and increasing both biopolymer volume fraction and cytoplasmic solute concentrations. For these experiments, cells were grown in LB medium at moderate osmolality and plasmolyzed with NaCl. Figure 27.5 illustrates the dramatic reduction in $< D >$ (between two and three orders of magnitude for $\approx 50\%$ reduction in cytoplasmic volume) and the increase in dispersion in D among cells as external osmolality is increased and the volume of cytoplasmic water is reduced. SPT can explain the roughly exponential decrease in $< D >$ at high osmolalities, but not the drop in $< D >$ from D_0 at low osmolality. The increase of dispersion in D may be because of cell-to-cell heterogeneity, as discussed elsewhere (Konopka *et al.*, 2006).

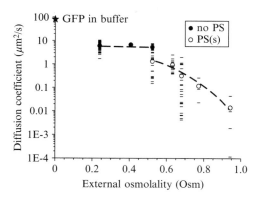

Figure 27.5 Semilog plot of green fluorescent protein (GFP) diffusion coefficient vs external osmolality of exponential growing cells harvested from 0.24 Osm LB and upshifted to the total osmolality shown. Closed circles represent the arithmetic mean of cells without plasmolysis spaces (PS). Open circles are the arithmetic mean of cells with one or more PS. Dashes represent individual measurements. The star indicates the diffusion coefficient of GFP in buffer. From Konopka *et al.* (2006).

The aforementioned decrease in $< D >$ for GFP in *E. coli* is more dramatic than the sixfold decrease in $< D >$ found for a small dye molecule in the cytoplasm of Swiss 3T3 fibroblasts when shrunk a factor of 3 by osmotic upshift observed by Kao *et al.* (1993). The difference may be a result of the smaller probe molecule or that crowding/confinement effects are not as significant in eukaryotic cell due to compartmentalization by organelles.

Currently, we are investigating the diffusion of GFP as functions of osmolality of growth and plasmolysis for the *E. coli* strain and conditions quantified in Figs. 27.1 and 27.2. In particular, different conditions yielding the same biopolymer volume fraction are being compared. Preliminary results (M. Konopka, T. Record, and J. Weisshaar, unpublished) indicate that significant differences in $< D >$ for GFP exist for growing cells and for plasmolyzed cells with the same biopolymer volume fraction, which raise interesting questions as to how cells growing at high osmolality without much cytoplasmic water and with a high biopolymer volume fraction might "reorganize" their cytoplasm to increase protein diffusion.

5. SUMMARY

Because differences in biopolymer volume fraction and solute concentrations between typical *in vitro* and *in vivo* solution conditions are large, it is surprising that only a few quantitative *in vivo* studies and comparison of *in vivo–in vitro* results have been performed to date. The methods described in this chapter provide the tools for making quantitative measurements *in vivo* while varying cytoplasmic volume fraction and solute concentrations. Such experiments hold the promise of bridging between solution studies of biochemical processes and the much more complex environment of living cells.

ACKNOWLEDGMENTS

Research from the authors' laboratories cited here and the preparation of this review were supported by NIH GM47022 (MTR) and NSF CHE-0452375 (JCW), as well as the University of Wisconsin-Madison Graduate School and Department of Chemistry (JCW).

REFERENCES

Cayley, D. S., Guttman, H. J., and Record, M. T., Jr. (2000). Biophysical characterization of changes in amounts and activity of *Escherichia coli* cell and compartment water and turgor pressure in response to osmotic stress. *Biophys. J.* **78**, 1748–1764.

Cayley, S., Lewis, B. A., Guttman, H. J., and Record, M. T., Jr. (1991). Characterization of the cytoplasm of *Escherichia coli* K–12 as a function of external osmolarity: Implications for protein-DNA interactions *in vivo*. *J. Mol. Biol.* **222**, 281–300.

Cayley, S., Lewis, B. A., and Record, M. T., Jr. (1992). Origins of the osmopr-otective properties of betaine and proline in *Escherichia coli* K-12. *J. Bacteriol.* **174,** 1586–1595.

Cayley, S., and Record, M. T., Jr. (2003). Roles of cytoplasmic osmolytes, water, and crowding in the response of *Escherichia coli* to osmotic stress: Biophysical basis of osmo-protection by glycine betaine. *Biochemistry* **42,** 12596–12609.

Cayley, S., Record, M. T., Jr., and Lewis, B. A. (1989). Accumulation of 3-(N-morpholino) propanesulfonate by osmotically stressed *Escherichia coli* K-12. *J. Bacteriol.* **171,** 3597–3602.

Cluzel, P., Surette, M., and Leibler, S. (2000). An ultrasensitive bacterial motor revealed by monitoring signaling proteins in single cells. *Science* **287,** 1652–1655.

Csonka, L. N., and Epstein, W. (1996). In "*Escherichia coli* and *Salmonella typhimurium*: Cellular and Molecular Biology" (F. C. Neidhardt, ed.), pp. 1210–1233. ASM Press, Washington, DC.

Csonka, L. N., and Hanson, A. D. (1991). Prokaryotic osmoregulation: Genetics and physiology. *Annu. Rev. Microbiol.* **45,** 569–606.

Dinnbier, U., Limpinsel, E., Schmid, R., and Bakker, E. P. (1988). Transient accumulation of potassium glutamate and its replacement by trehalose during adaptation of growing cells of *Escherichia coli* K-12 to elevated sodium chloride concentrations. *Arch. Microbiol.* **150,** 348–357.

Ellis, R. J. (2001). Macromolecular crowding: Obvious but underappreciated. *Trends Bio-chem. Sci.* **26,** 597–604.

Elowitz, M. B., Surette, M. G., Wolf, P. E., Stock, J. B., and Leibler, S. (1999). Protein mobility in the cytoplasm of *Escherichia coli*. *J. Bacteriol.* **181,** 197–203.

Epstein, W., and Schultz, S. G. (1965). Cation transport in *Escherichia coli*. V. Regulation of cation content. *J. Gen. Physiol.* **49,** 221–234.

Felitsky, D. J., Cannon, J. G., Capp, M. W., Hong, J., Van Wynsberghe, A. W., Anderson, C. F., and Record, M. T. (2004). The exclusion of glycine betaine from anionic biopolymer surface: Why glycine betaine is an effective osmoprotectant but also a compatible solute. *Biochemistry* **43,** 14732–14743.

Ghaemmaghami, S., and Oas, T. G. (2001). Quantitative protein stability measurement *in vivo*. *Nat. Struct. Biol.* **8,** 879–882.

Golding, I., and Cox, E. C. (2004). RNA dynamics in live *Escherichia coli* cells. *Proc. Natl. Acad. Sci. USA* **101,** 11310–11315.

Golding, I., and Cox, E. C. (2006). Physical nature of bacterial cytoplasm. *Phys. Rev. Lett.* **96,** 098102.

Griffin, B. A., Adams, S. R., and Tsien, R. Y. (1998). Specific covalent labeling of recombinant protein molecules inside live cells. *Science* **281,** 269–272.

Gros, G. (1978). Concentration dependence of the self-diffusion of human and *Lumbricus terrestris* hemoglobin. *Biophys. J.* **22,** 453–468.

Ha, J. H., Capp, M. W., Hohenwalter, M. D., Baskerville, M., and Record, M. T., Jr. (1992). Thermodynamic stoichiometries of participation of water, cations and anions in specific and non-specific binding of lac repressor to DNA: Possible thermodynamic origins of the "glutamate effect" on protein–DNA interactions. *J. Mol. Biol.* **228,** 252–264.

Han, J., and Herzfeld, J. (1993). Macromolecular diffusion in crowded solutions. *Biophys. J.* **65,** 1155–1161.

Hong, J., Capp, M. W., Saecker, R. M., and Record, M. T., Jr. (2005). Use of urea and glycine betaine to quantify coupled folding and probe the burial of DNA phosphates in lac repressor-lac operator binding. *Biochemistry* **44,** 16896–16911.

Ignatova, Z., and Gierasch, L. M. (2004). Monitoring protein stability and aggregation *in vivo* by real-time fluorescent labeling. *Proc. Natl. Acad. Sci. USA* **101,** 523–528.

Ignatova, Z., and Gierasch, L. M. (2006). Inhibition of protein aggregation *in vitro* and *in vivo* by a natural osmoprotectant. *Proc. Natl. Acad. Sci. USA* **103,** 13357–13361.

Kao, H. P., Abney, J. R., and Verkman, A. S. (1993). Determinants of the translational mobility of a small solute in cell cytoplasm. *J. Cell Biol.* **120,** 175–184.

Kennedy, E. P. (1982). Osmotic regulation and the biosynthesis of membrane-derived oligosaccharides in *Escherichia coli. Proc. Natl. Acad. Sci. USA* **79,** 1092–1095.

Konopka, M. C., Shkel, I. A., Cayley, S., Record, M. T., and Weisshaar, J. C. (2006). Crowding and confinement effects on protein diffusion *in vivo. J. Bacteriol.* **188,** 6115–6123.

Kontur, W. S., Saecker, R. M., Davis, C. A., Capp, M. W., and Record, M. T., Jr. (2006). Solute probes of conformational changes in open complex (RPo) formation by *Escherichia coli* RNA polymerase at the lambdaPR promoter: Evidence for unmasking of the active site in the isomerization step and for large-scale coupled folding in the subsequent conversion to RPo. *Biochemistry* **45,** 2161–2177.

Lacroix, J. M., Tempete, M., Menichi, B., and Bohin, J. P. (1989). Molecular cloning and expression of a locus (mdoA) implicated in the biosynthesis of membrane-derived oligosaccharides in *Escherichia coli. Mol. Microbiol.* **3,** 1173–1182.

Landfald, B., and Strom, A. R. (1986). Choline-glycine betaine pathway confers a high level of osmotic tolerance in *Escherichia coli. J. Bacteriol.* **165,** 849–855.

Leirmo, S., Harrison, C., Cayley, D. S., Burgess, R. R., and Record, M. T., Jr. (1987). Replacement of potassium chloride by potassium glutamate dramatically enhances protein-DNA interactions *in vitro. Biochemistry* **26,** 2095–2101.

Luby-Phelps, K. (2000). Cytoarchitecture and physical properties of cytoplasm: Volume, viscosity, diffusion, intracellular surface area. *Int. Rev. Cytol.* **192,** 189–221.

McLaggan, D., Naprstek, J., Buurman, E. T., and Epstein, W. (1994). Interdependence of K+ and glutamate accumulation during osmotic adaptation of *Escherichia coli. J. Biol. Chem.* **269,** 1911–1917.

Minton, A. P. (2001). The influence of macromolecular crowding and macromolecular confinement on biochemical reactions in physiological media. *J. Biol. Chem.* **276,** 10577–10580.

Minton, A. P. (2005). Influence of macromolecular crowding upon the stability and state of association of proteins: Predictions and observations. *J. Pharm. Sci.* **94,** 1668–1675.

Mullineaux, C. W., Nenninger, A., Ray, N., and Robinson, C. (2006). Diffusion of green fluorescent protein in three cell environments in *Escherichia coli. J. Bacteriol.* **188,** 3442–3448.

Munro, G. F., Hercules, K., Morgan, J., and Sauerbier, W. (1972). Dependence of the putrescine content of *Escherichia coli* on the osmotic strength of the medium. *J. Biol. Chem.* **247,** 1272–1280.

Muramatsu, N., and Minton, A. P. (1988). Tracer diffusion of globular proteins in concentrated protein solutions. *Proc. Natl. Acad. Sci. USA* **85,** 2984–2988.

Neidhardt, F. C., Bloch, P. L., and Smith, D. F. (1974). Culture medium for enterobacteria. *J. Bacteriol.* **119,** 736–747.

Ping, G., Yuan, J. M., Sun, Z., and Wei, Y. (2004). Studies of effects of macromolecular crowding and confinement on protein folding and protein stability. *J. Mol. Recogn.* **17,** 433–440.

Record, M. T., Jr., Courtenay, E. S., Cayley, D. S., and Guttman, H. J. (1998). Responses of *E. coli* to osmotic stress: Large changes in amounts of cytoplasmic solutes and water. *Trends Biochem. Sci.* **23,** 143–148.

Richey, B., Cayley, D. S., Mossing, M. C., Kolka, C., Anderson, C. F., Farrar, T. C., and Record, M. T., Jr. (1987). Variability of the intracellular ionic environment of *Escherichia coli*: Differences between *in vitro* and *in vivo* effects of ion concentrations on protein-DNA interactions and gene expression. *J. Biol. Chem.* **262,** 7157–7164.

Terry, B. R., Matthews, E. K., and Haseloff, J. (1995). Molecular characterisation of recombinant green fluorescent protein by fluorescence correlation microscopy. *Biochem. Biophys. Res. Commun.* **217,** 21–27.

Tokuriki, N., Kinjo, M., Negi, S., Hoshino, M., Goto, Y., Urabe, I., and Yomo, T. (2004). Protein folding by the effects of macromolecular crowding. *Protein Sci.* **13,** 125–133.

Wood, J. M. (1999). Osmosensing by bacteria: Signals and membrane-based sensors. *Microbiol. Mol. Biol. Rev.* **63,** 230–262.

Yancey, P. H., Clark, M. E., Hand, S. C., Bowlus, R. D., and Somero, G. N. (1982). Living with water stress: Evolution of osmolyte systems. *Science* **217,** 1214–1222.

Zimmerman, S. B., and Harrison, B. (1987). Macromolecular crowding increases binding of DNA polymerase to DNA: An adaptive effect. *Proc. Natl. Acad. Sci. USA* **84,** 1871–1875.

Zimmerman, S. B., and Minton, A. P. (1993). Macromolecular crowding: Biochemical, biophysical, and physiological consequences. *Annu. Rev. Biophys. Biomol. Struct.* **22,** 27–65.

Zimmerman, S. B., and Trach, S. O. (1988). Macromolecular crowding extends the range of conditions under which DNA polymerase is functional. *Biochim. Biophys. Acta* **949,** 297–304.

Zimmerman, S. B., and Trach, S. O. (1991). Estimation of macromolecule concentrations and excluded volume effects for the cytoplasm of *Escherichia coli. J. Mol. Biol.* **222,** 599–620.

CHAPTER TWENTY-EIGHT

CHARACTERIZATION OF PLANT AQUAPORINS

Ralf Kaldenhoff,* Adam Bertl,* Beate Otto,*
Menachem Moshelion,† *and* Norbert Uehlein*

Contents

Abstract

Plants have been reported to contain a large set of aquaporins (38 for *Arabidopsis*), which has been divided into four subfamilies on the basis of similarities in their amino acid sequences. They belong to the large superfamily of major intrinsic proteins (MIP), which was the basis for the nomenclature PIP, TIP, and NIP, also indicating the subcellular localization plasma membrane, tonoplast, and nodule of the respective founding member. The fourth subfamily of small and basic intrinsic proteins is not well characterized so far. The increasing number of reports dealing with various aspects of plant aquaporins is starting to advance our understanding of aquaporin biology in plants. Fundamental questions include: what is the basic function of the different plant aquaporins, what is their primary substrate, and what is the consequence of function/ malfunction of a particular aquaporin for the overall function of the plant?

* Institute of Botany, Applied Plant Sciences, Darmstadt University of Technology, Darmstadt, Germany,
† The Robert H. Smith Institute of Plant Sciences and Genetics in Agriculture, Faculty of Agricultural, Food & Environmental Quality Science, The Hebrew University of Jerusalem, Rehovot, Israel

Methods in Enzymology, Volume 428
ISSN 0076-6879, DOI: 10.1016/S0076-6879(07)28028-0

Biochemical and biophysical techniques can be employed to get information on the basic functional characteristics of plant aquaporins. An impressive set of techniques has been used to study aquaporin function on molecular, subcellular, and cellular levels in plants, as well as in heterologous expression systems. The physiological role of aquaporins in plants is much less well understood, but reports unraveling the physiological role of aquaporins, mainly employing genetic techniques and functional measurement on the whole plant level, are emerging. The goal of this chapter is to give an overview on the applied methods, together with some exemplary findings.

1. MOLECULAR FUNCTION OF PLANT AQUAPORINS

In comparison to other living organisms, plants have a large number of genes that code for aquaporins. Angiosperm plants possess about 35 different aquaporin genes. Regarding the current nomenclature and based on sequence similarities the proteins can be divided into four different subfamilies. Two of these group allocations relate to localization, that is, aquaporins preferentially inserted into the plasma membrane are named plasma membrane intrinsic proteins (PIP) and those that integrate into the tonoplast are tonoplast intrinsic proteins (TIP). NIP aquaporins were initially identified in nodulated roots of soybean where they are located in the peribacteroid membrane, which is part of the plasma membrane (Niemietz and Tyerman, 2000). The smallest and least analyzed group is formed by the small and basic intrinsic proteins (SIPs), which were shown to be integrated into the membrane of the endoplasmic reticulum (Ishikawa *et al.*, 2005). In the past few years analysis of plant aquaporins *in planta* has provided more insight into function and localization, indicating that the current aquaporin classification cannot be generally accepted anymore (Barkla *et al.*, 1999). Reports have emerged showing that PIP aquaporins are also located in organellar membranes, as well as TIP aquaporins also being localized in the plasma membrane. The large number of different aquaporin genes in plants is still enigmatic. It could indicate the contribution of different aquaporins in fine-tuning of water transport during changing environmental conditions and developmental stages or different substrate properties of different aquaporin isoforms.

Since the late 1990s functional analyses have revealed a set of possible substrates for aquaporin-facilitated membrane transport in plants besides water. These are, for example, glycerol, urea, NH_3/NH_4^+, CO_2, silicon, methylammonium, formamide, malate, glycine (Kaldenhoff and Fischer, 2006), and hydrogen peroxide (Bienert *et al.*, 2006). Functional analysis of plant aquaporins is often accomplished using easy-to-handle heterologous expression systems such as *Xenopus* oocytes or yeast. Obtained data give first insight into the permeability properties of single aquaporins, but do just give a hint for the function *in planta*. The physiological role of single aquaporins and their transport specificity can only be determined correctly in the system where the respective aquaporins are expressed naturally: the homologous plant system with its cellular network and other aquaporins. The plant-specific pattern of post-transcriptional modification and regulation, which may not exist in the heterologous systems, is also important in identifying the function of a specific plant aquaporin. Comparing normal control plants with plants having a modified aquaporin expression at the cellular and whole plant level provides a suitable experimental strategy to unravel the role of plant aquaporins in plants. In doing so, it is of minor importance if transgenic plants or aquaporin mutants are employed and even inhibitors could be useful, with certain restrictions.

1.1. Aquaporin function in plant membranes

A potent way to understand aquaporin function *in planta* and transport properties on organ and subcellular levels is to analyze the permeability of different types of membranes. Different aquaporin-dependent membrane permeabilities can be expected for the plasma membrane of root and leaf cells. Special techniques exist to isolate those membranes from different plant tissues. After purification membrane vesicles can be subjected to permeability studies using stopped-flow spectrometry (see later).

1.1.1. Isolation of plasma membranes

Following the procedure of Kjellbom and Larsson (1984), the first step of preparing plasma membranes is the isolation of crude membrane fractions, the microsomes. Starting material can be leaf or root tissue. One hundred grams fresh weight is homogenized three times for 30 s in 300 ml 0.33 M sucrose, 50 mM HEPES/KOH, pH 7.5, 5 mM EDTA, 5 mM dithiothreitol (DTT), 5 mM ascorbic acid, 0.5 mM phenylmethylsulfonyl fluoride, 0.2% bovine serum albumin (BSA), 0.2% casein (boiled for 10 min), and 0.6% polyvinylpolypyrrolidone (to remove phenolic compounds) using a kitchen homogenizer. The homogenate is filtered through three layers of miracloth, and the bulk of chloroplasts and mitochondria is sedimented by centrifugation at 5000g for 10 min. The membranes in the supernatant are collected at 100,000g for 1 h. This pellet is resuspended to a total volume of 10 ml in 0.33 M sucrose, 5 mM HEPES/KOH, pH 7.5, 5 mM KCl, 1 mM DTT, and 0.1 mM EDTA. The two-phase partitioning system (Lundborg *et al.*, 1981; Widell and Larsson, 1981) is widely used to isolate plasma membranes from microsomes. Using this method, membrane particles are separated according to their surface properties (Albertsson, 1971) rather than according to their size and density. Thus, preparations of membrane vesicles with uniform surface properties and thereby uniform sidedness can be obtained. An aliquot (9 g) of the microsomal suspension is added to 27 g of the phase mixture to give a total phase system of 36 g with a final composition of 6.2% (w/w) Dextran T 500, 6.2% (w/w) polyethylene glycol (PEG) 3350, 0.33 M sucrose, 5 mM potassium phosphate buffer, pH 7.8, 5 mM KCl, 1 mM DTT, and 0.1 mM EDTA. The phase system is mixed by several inversions and centrifuged in a swinging bucket rotor for 10 min at 1500g. The upper phase is re-extracted three times with a fresh lower phase. The plasma membrane enriched final upper phase is diluted fivefold with buffer and centrifuged at 100,000g for 1 h. The pellet is resuspended in 3 mM HEPES/KOH, pH 7.5, 0.25 M sucrose, 50 mM KCl, and 1 mM DTT.

1.1.2. Isolation of chloroplast envelope membranes

To separate and purify chloroplast envelope membranes, intact chloroplasts have to be isolated from leaves. A fractionation of starch granules and intact and broken chloroplasts from homogenized leaves is achieved by centrifugation on a two-step Percoll gradient in an isoosmotic buffer solution (Bock, 1998). For isolation and separation of envelope membranes the chloroplasts are incubated in a hypertonic buffer solution at 0° to promote shrinkage of the chloroplasts and separation of the different envelope membranes. To disrupt these chloroplasts they are subjected to a freeze/thaw cycle and subsequently homogenized using a Potter homogenizer. For the following accumulation of envelope membranes different protocols exist. Purification of inner and outer envelope membranes can be achieved by flotation centrifugation as described by Cline *et al.* (1981) or by ultracentrifugation on a three-step sucrose gradient (1.0, 0.8, and 0.46 M in 10 mM Tricine, 2 mM EDTA, pH 7.5; Keegstra and Yousif, 1986). The two resulting fractions differ in their constituent polypeptides and exhibit different densities. The light and heavy membrane fractions have been tentatively identified as the outer and inner envelope membranes, respectively (Cline *et al.*, 1981).

1.1.3. Isolation of peribacteroid membranes

During interaction between plant and rhizobia a highly specialized type of membrane is produced emanating from the plant plasma membrane, which is the peribacteroid membrane (PBM) surrounding the symbiosome. This membrane contains transport systems for substance exchange between both symbiotic partners. Among others the aquaporin Nodulin 26 (NOD26) was shown to be localized there, and the permeability of PBM-derived membrane vesicles was analyzed (Niemietz and Tyerman, 2000). The isolation and purification of peribacteroid membrane vesicles have been described by Christiansen *et al.* (1995).

1.2. Analysis of plant aquaporins in heterologous and artificial expression systems

1.2.1. *Xenopus* oocytes

Swelling assay Molecular identification and functional characterization of plant aquaporins have relied on expression in heterologous systems. Oocytes from *Xenopus laevis* have proven a convenient system for injecting heterologous genetic material and characterizing the translational product functionally. For functional analysis, the genetic information coding for the heterologous protein can be injected as RNA into the interface of the animal and vegetal pole or as cDNA directly into the nucleus. The latter requires some skills in manipulation; however, DNA is in general more stable and easier to handle than mRNA. Good starting points for plant aquaporin expression are quantities of 50 ng of RNA or 10 ng of DNA in volumes of about 50 and 10 nl, respectively, but the amount of RNA or cDNA that needs to be injected will be determined greatly on how well the plant aquaporin expresses in oocytes.

The function of plant proteins in transmembrane water transport was first shown for an abundant vacuolar protein of about 27 kDa, which was named TIP, based on its subcellular location and sequence homology to members of the large family of MIP. Functional analysis was achieved by expressing the protein in *Xenopus* oocytes and comparing water transport of oocytes injected with water (negative control) and those injected with γ-TIP cRNA from *Arabidopsis thaliana* (Maurel *et al.*, 1993). The water transport activity of *Xenopus* oocytes is given as the osmotic permeability coefficient P_f, which can be calculated from the initial rate of oocyte volume increase $d(V/V_0)/dt$ in response to a sudden hypotonic challenge (Zhang *et al.*, 1990). Oocyte volume changes are usually monitored by means of video microscopy, allowing the osmotic permeability coefficient to be calculated by

$$ P_f = V_0[d(V/V_0)/dt]/[S_0 \times V_w(Osm_{in} - Osm_{out})] \qquad (28.1) $$

with the initial oocyte volume (V_0), the initial oocyte surface (S_0), and the molar volume of water (V_w).

Xenopus oocytes display rather low intrinsic water permeability (P_f in the range of 5-10×10^{-4} cm/s; Maurel *et al.*, 1993) and consequently a slow increase in cell volume of <2% per minute, if challenged by a hypoosmotic shift of about 100 to 150 mOsmol/kg. Injection of γ-TIP cRNA from *A. thaliana* into *Xenopus* oocytes gave rise to a roughly 10-fold increase in osmotic water permeability of the eggs (P_f in the range of 1.5×10^{-2} cm/s), resulting in rapid swelling with an increase in cell volume of about 15 to 20% per minute and rapid bursting (within 5 min) when bathed in hypotonic solutions (Maurel *et al.*, 1993). Osmotic water permeabilities of

Xenopus oocytes have been reported to increase by about 10- to 50-fold upon expression of plant aquaporins (Ciavatta *et al.*, 2001; Holm *et al.*, 2005; Secchi *et al.*, 2006; Siefritz *et al.*, 2001), but some plant aquaporins show no or only minor water permeability when expressed in oocytes (Secci *et al.*, 2006; Siefritz *et al.*, 2001; Temmei *et al.*, 2005; Wallace and Roberts, 2005). This raises the question about the physiological function of an aquaporin that does not conduct water. A PIP1-type aquaporin from *Mimosa pudica*, MpPIP1;1, exhibited no water permeability in *Xenopus* oocytes, but selectively increased water permeability of MpPIP2;1 in a phosphorylation-dependent manner (Temmei *et al.*, 2005). This may hint to modulation of water transport activity via direct interaction of different aquaporin species.

Water-impermeable plant aquaporins were also found to function in facilitating the transport of other substrates, such as gases (CO_2, NH_3), ions (NH_4^+), or small uncharged substrate (glycerol, H_2O_2) as revealed by other experimental approaches applied to *Xenopus* oocytes.

Electrical charge movement   A key question concerning aquaporin function is the selectivity of these proteins; in particular, do aquaporins conduct ions in addition to the water molecules? This question can be addressed by measuring the movement of electrical charges associated with the function of the aquaporins. The large diameter (>1 mm) of *Xenopus* oocytes and the relatively low intrinsic ionic conductivity (Dascal, 1987) make them especially suitable for electrophysiological recording, that is, analysis of the transport of electrical charges. Direct analysis of electrical charge translocation (electrical current) is best analyzed via two electrode voltage clamp. Two microelectrodes are inserted into the oocyte, one of which is used to monitor the actual membrane voltage and clamp the membrane to a suitable voltage. The activity of charge translocating proteins in the oocyte membrane would change the set voltage, which is compensated for by feeding electrical current into the cell via the second microelectrode. The current necessary to maintain that voltage can be recorded and is directly related to the current flowing through the charge translocating membrane protein.

Two electrode voltage clamp experiments in oocytes expressing plant aquaporins (γ-TIP) revealed that these aquaporins are in general not permeable for ions (Maurel *et al.*, 1993). This general notion is, however, still under debate, as some exceptions for this rule have been reported, in particular from animal systems. Electrophysiological analyses in oocytes revealed a finite permeability of AQP6 for anions, mainly NO_3^- (Ikeda *et al.*, 2002; Liu *et al.*, 2005; Yasui *et al.*, 1999). In oocytes expressing the wheat aquaporin TaTIP2;1, Holm *et al.* (2005) reported TaTIP2;1-mediated ionic currents under voltage clamp conditions and fast depolarization of the plasma membrane upon abrupt isoosmotic replacement of Na^+ by NH_4^+, which was also associated with oocyte swelling. Because both ionic current and depolarization increased with alkaline going extracellular pH in the range of pH 7 to pH 8, where the concentration of NH_3 increases roughly 10-fold and NH_4^+ remains virtually unchanged, a NH_3-gated NH_4^+ transport mechanism has been discussed for TaTIP2;1 (Holm et al., 2005).

Intra/extracellular pH   In addition to water, aquaporin homologues transport alternative substrates, such as glycerol, NH_3, or CO_2 (reviewed in Kaldenhoff, 2006). Using functional complementation of a yeast mutant deficient in ammonium uptake (31019b: *mep1-3*Δtriple deletion; Marini et al., 1997), three aquaporins from wheat (TaTIP2;1; TaTIP2;2; and TaTIP2;3) were identified that restored the ability of the yeast mutant to grow on low NH_4Cl (2 mM) as the sole nitrogen source. Growth of TIP2-transformed yeast strains was improved at more alkaline pH, suggesting that NH_3 is the substrate taken up by the yeast via the plant aquaporins. In order to analyze the transport activity of these aquaporins, TaTIP2;1 was expressed in *Xenopus* oocytes and extracellular pH from 20 oocytes was monitored in a weakly buffered solution. Addition of 20 mM NH_4Cl resulted in an acidification of the medium, which was significantly faster in the TaTIP2;1 expressing oocytes than in the water-injected controls. This clearly indicated that NH_3, rather than NH_4^+, is the substrate transported by TaTIP2;1 (Jahn et al., 2004).

Liquid ion exchanger-based microelectrodes (Felle and Bertl, 1986) were used to monitor intracellular pH changes in *Xenopus* oocytes associated with the uptake of CO_2. Provided that the transport process is the limiting step, CO_2 permeability can be determined from the initial rate in intracellular pH change following perfusion of the recording chamber with CO_2-aerated buffer. To fulfill this requirement, the oocytes were injected with carbonic anhydrase, an enzyme that catalyzes the conversion of CO_2 to HCO_3^-. CO_2 permeability in oocytes expressing the plant aquaporin NtAQP1 was found to be 45% higher than that of oocytes injected with the carbonic anhydrase alone (Uehlein et al., 2003).

1.2.2. Yeast

Yeasts have increasingly been used for expression and functional characterization of plant membrane transport proteins. The bakers' yeast *Saccharomyces cerevisiae* has proven especially useful for cloning plant membrane transport proteins via functional complementation of yeast mutants, resulting in cloning of the first plant ion channels (Anderson et al., 1992; Sentenac et al., 1992), sugar transporters (Riesmeier et al., 1992; Sauer and Stadler, 1993), amino acids transporters (Frommer et al., 1993; Hsu et al., 1993), and others. A diverse set of experimental approaches was used for functional characterization of these transport proteins in yeast, that is, electrophysiology (Bertl et al., 1995), growth assays (Nakamura et al., 1998; Riesmeier et al., 1992), or uptake of radiolabeled substrate (Frommer et al., 1993; Riesmeier et al., 1992).

Heterologous aquaporins were functionally expressed in yeasts, *Pichia pastoris* (Daniels *et al.*, 2006), *Schizosaccharomyces pombe* (Tanghe *et al.*, 2005), and mainly in *S. cerevisiae* (Bienert *et al.*, 2006; Loque *et al.*, 2005; Pettersson *et al.*, 2006). *S. cerevisiae* has proven especially useful for functional analysis of heterologously expressed aquaporins because of the low intrinsic water permeability of the plasma membrane. The *S. cerevisiae* genome harbors four genes, which encode MIP-like proteins, two aquaporins, *AQY1* and *AQY2*, and two aquaglyceroporins, *FPS1* and *Yfl05*.

Swelling assay   Water transport in yeasts was studied by monitoring the kinetics of bursting of osmotically challenged protoplasts as a decrease in optical density (OD_{600}), which was influenced by the presence of heterologous aquaporins, but not aquaglyceroporins (Pettersson *et al.*, 2006). Upon expression in a osmosensitive yeast mutant (*fps1Δ*), the cauliflower aquaporin BobTIP1;1 was shown to be localized in the vacuolar membrane and complement a defect in osmoregulation of the *fps1Δ* mutant. Changes in cell volume adjustments associated with the function of BobTIP1;1 were determined via an image analysis system connected to an inverted light microscope (Prudent *et al.*, 2005).

In order to characterize aquaporin function in a more quantitative way, stopped-flow spectrometry is the method of choice. This can be applied to membrane vesicles, protoplasts, or intact yeast cells. Membrane vesicles or protoplasts are subjected to an osmotic challenge by rapidly mixing the vesicle/protoplast suspension with an equal volume of hypo-/hypertonic buffer. Light of 436 nm is directed to the observation chamber (8-μl volume) by means of an optical fiber and the change in scattered light is monitored at a 90° angle. An increase in vesicle/protoplast volume indicating water influx gives rise to a decrease in the intensity of the scattered light, which follows an exponential relaxation. This signal can be described by an exponential function, yielding a time constant τ, which allows for determining the osmotic water permeability coefficient (P_f) of the aquaporin expressing membrane. Post-Golgi vesicles (about 100 nm diameter) isolated from a temperature-sensitive yeast mutant, displaying a defect in the secretory pathway (*sec6-4*), have been shown to exhibit osmotic water permeability coefficients in the range of 1.3×10^{-3} cm/s, which was increased to 4.2×10^{-3} cm/s upon expression of a heterologous aquaporin, specifically the human *CHIP28*, demonstrating the suitability of both the yeast secretory vesicles and the stopped-flow technique for quantitative analysis of aquaporin-mediated water transport (Laize *et al.*, 1995). A faster, easier, and more economic system for studying water transport via stopped-flow spectrometry is yeast protoplasts. Using this system, it was shown that yeast strain Σ1278b, containing the two functional aquaporins *AQY1-1* and *AQY2-1* (Bonhivers *et al.*, 1998), exhibits endogenous aquaporin activity, giving rise to an osmotic water permeability coefficient at 10° of

1.7×10^{-2} cm/s, which compares to 1.8×10^{-3} cm/s for the control strain FL100, containing the nonfunctional *AQY1-2* allel (Laize *et al.*, 2000). In contrast to these findings, deletion of both aquaporin genes, *AQY1* and *AQY2*, in a Σ1278b–derived strain (10560-6B) had only a minor effect on the osmotic water permeability of the yeast mutant, that is, no effect on hypertonic challenge and a 50% decrease in P_f upon hypotonic shift, but only at extremely low temperature (7°) (Soveral *et al.*, 2006). Overexpression of AQY1 or AQY2 in the double deletion background resulted in a 9- and 5-fold increased water permeability of the yeast protoplasts at 7° (Soveral *et al.*, 2006). Compared to this, plant aquaporins seem to have a significantly higher water permeability, as expression of TaTIP1;1 was reported to increase the osmotic water permeability (given as an increase in the rate constant $k = 1/\tau$ determined from fitting exponential functions to light-scattering data) of yeast protoplasts 25-fold (Bienert *et al.*, 2006). High water permeability can be found in TaTIP2;2 as well, which increases P_f of yeast protoplasts upon overexpression about 200-fold at 10° (A. Bertl and R. Kaldenhoff, unpublished experiments).

An alternative approach for studying water transport in yeast protoplasts takes advantage of concentration-dependent self-quenching properties of fluorescent dyes. For that, yeast protoplasts are loaded for 10 min at 30° with the nonfluorescent precursor carboxyfluorescein diacetate (1 m*M* in isotonic solution), which is cleaved by endogenous esterases to form the fluorescent dye carboxyfluorescein. Changes in fluorescence intensity resulting from osmotically induced volume changes can be monitored by the same stopped-flow apparatus as described earlier. Swelling of protoplasts after a hypotonic challenge would lead to dilution of the fluorescent dye within the protoplasts and to an increase in fluorescence intensity. The time course of the fluorescence signal can again be described by an exponential function, and the time constant τ can be used to calculate the osmotic water permeability coefficient P_f as described for the light-scattering experiments. The osmotic water permeability coefficients determined via concentration-dependent fluorescence quenching were in good agreement with values determined from the corresponding light-scattering experiments, demonstrating the suitability of this technique for quantitative analysis of aquaporin-mediated water transport in yeast (Soveral *et al.*, 2006).

However, it should be noted that the fluorescent dyes carboxyfluorescein and fluorescein leak out of the cells, which is especially true at higher temperatures (>10°), as the efflux of the dyes (shown for fluorescein) apparently involves energy-dependent transport processes via the ATP-binding cassette transporter such as the yeast Pdr12 (Holyoak *et al.*, 1999).

Intracellular pH   The fluorescence intensity of the dye fluorescein and its derivatives is not only concentration dependent, thus reporting osmotic-induced volume changes, it also depends strongly on the pH.

The latter property has been used to analyze aquaporin function in the yeast mutant 31019b (*mep1-3Δ*) expressing TaTIP2;2 aquaporins (A. Bertl and R. Kaldenhoff, unpublished data) by means of stopped-flow spectrometry. For that, cells are loaded with the fluorescent dye by incubation at 30° for 10 to 20 min in buffer containing 50 mM HEPES/NaOH, pH 7.0, 5 mM 2-deoxyglucose, and 50 μM fluorescein diacetate. Fluorescence changes induced by the application of 25 mM NH$_4$Cl (mixing yeast cells suspended in 50 mM NaCl with an equal volume of 50 mM NH$_4$Cl) are monitored behind a 515-nm cutoff filter (excitation wavelength 480 nm) in a stopped-flow apparatus. Exposure of the cells to NH$_4$Cl resulted in an increase in fluorescence intensity, indicating NH$_4$Cl-induced cytosolic alkalinization. The initial rate of the observed change in fluorescence intensity (alkalinization) was sensitive to external pH, increasing with more alkaline pH in the range of pH 6 to pH 9. Because the concentration of NH$_3$ increases 670-fold over this range of external pH, while the concentration of NH$_4{}^+$ decreases by 30%, it is quite obvious that NH$_3$ transport across the yeast plasma membrane must account for the observed fluorescence changes (alkalinization). At pH 8 and 10°, we observed a 66-fold increase in the rate of alkalinization and therefore in the permeability of the yeast plasma membrane upon overexpression of TaTIP2;2.

1.3. Single cell analysis

1.3.1. Determination of P_f

One of the approaches in quantifying the permeability of a plant cell to water consists of isolating protoplasts and monitoring the initial rate of change of their volume on an osmotic challenge. The osmotic water permeability (termed P_f or P_{os}) can be deduced from the initial rate of volume relaxation (Verkman, 2000; Zhang *et al.*, 1990), that is, from the slope dV/dt of the linear phase (first few seconds) of the volume versus the time plot, using Eqs. (28.2) and (28.3). This determination is based on the premise that the rate of bath solution exchange is instantaneous, such that the external concentration, C_{out}, attained its final value immediately and that the internal concentration, C_{in}, remained at its initial value:

$$dV/dt = P_{f_i} \cdot S_{0_i} \cdot V_w \cdot (C_{in} - C_{out}) \qquad (28.2)$$

where S_{0i} is the initial surface area of the cell and V_W is the partial molar volume of water (18 cm$^3 \cdot$mol^{-1}). P_{f_i} is extracted by rearrangement:

$$P_{f_i} = (dV/dt)/[S_{o_i} \cdot V_w \cdot (C_{in} - C_{out})] \qquad (28.3)$$

There are at least two problems with the aforementioned premise: (a) even when an instantaneous change of solution is possible, a systematic error is introduced, causing an underestimate of P_f because, already during the initial phase of protoplast swelling, the volume, the surface area, and the internal concentration of solutes do not remain constant and (b) instantaneous bath perfusion has technical, and physiological, limitations: unlike animal cells, isolated plant cell protoplasts do not stick well to the chamber bottom and defeat attempts of rapid (by no means instantaneous) solution flushes. To overcome these difficulties, a numerical approach must be adopted. This approach consists of a simple experimental procedure, followed by an off-line curve fitting, which yields relatively accurate P_f values over a large span of water permeability values. This analysis is performed with the "P_fFit program" (http://departments.agri.huji.ac.il/plantscience/staff-eng/moran-pffit.html). The equations underlying the various models of cell swelling and incorporating the corrections based on the rate of bath perfusion are detailed in Moshelion et al. (2004). In addition to calculation of the initial P_f, that is, the P_f at the onset of cell volume change, the "P_fFit program" charts the time course of P_f, taking advantage of 15-s-long (and longer) records of volume changes rather than of only the few initial seconds. This approach increased considerably the overall throughput of the experiment and analysis, as well as their reliability. Moreover, based on specific hypotheses and model choices, this approach enabled revealing and describing quantitatively a dynamic change in P_f of protoplasts during the osmotic challenge.

1.3.2. Protoplast isolation
The following protocol is suitable for protoplast isolation of plant parenchymatic and meristematic tissues. (a) Chop the tissue in cell wall digestion solution with a final concentration of 1% cellulase (CEL, Worthington, NJ), 0.02% pectolyase Y23 (Karlan, California), 0.05% BSA (Sigma), and 0.025% polyvinylpyrrolidone K30 (Fluka) in isotonic solution. (b) Place the chopped cells in the digestion solution on a rotary shaker (150 rpm) for 45 min at 30°. (c) Stop the enzymatic reaction by washing the cells with 5 ml of isotonic solution through a nylon filter (20-μm pore size). (d) Collect the cells accumulated on the filter and enhance protoplast release from the cell walls; immediately transfer the filter (upside down) into fresh isotonic solution and incubate further on a shaker (20 min, 150 rpm, 30°). (e) After filtration through a second nylon filter (50-μm pore size), collect the protoplasts into a test tube.

1.3.3. Experimental protoplast solutions
The isotonic solution (10 mM KCl, 1 mM CaCl$_2$, 8 mM MES, pH 5.75) should be adjusted with sorbitol to an osmolarity equal to that of the tissue (as measured from the freeze/thaw solution extraction).

1.3.4. Bath perfusion

The incubation chamber consists of a Lucite (Plexiglas) plate (2-mm-thick microscopic sample glass size), with a longitudinal groove in the middle of the plate and a cut-through slit within the groove. The elongated simple shape of the chamber groove minimizes turbulence. Once started, bath perfusion should be maintained at the same pace. To obtain quantitative information on the time course of bath solution exchange, the absorbance of transmitted light should be monitored by distribution of an indicator dye such as xylene cyanol (ICN Biomedicals, Irvine, CA; 1 mg/50 ml in the incoming solution).

1.3.5. Immobilizing the cells

The bottom of the chamber consists of a coverslip, sealed to the Lucite plate with silicon grease (Merck, Darmstadt, Germany) for the duration of the experiment. To augment the stickiness of the globular protoplasts to the bath bottom, spread a 0.5-μl drop of 1% protamine sulfate (Sigma; diluted in water, kept in aliquots at $-20°$ and mixed well before use) on the glass bottom for 2 min (or until completely dry) and then rinse three to four times ($\approx$1 ml) with isotonic solution. Discard the remaining solution and replace with a drop of the protoplast-containing solution, allowing the protoplasts to settle down for 5 to 10 min. Then fill up the whole groove with isotonic solution.

1.3.6. Image acquisition

The analysis relies on a reasonably well-focused cell contour. To improve visibility, place another coverslip on top of the chamber before perfusion starts, with its edges sealed with silicon grease to the surface of the Lucite plate. The first several images recorded in the fast-flowing isotonic solution constitute a baseline. The solution should then be switched to a hypo- or hypertonic one, while simultaneously acquiring images at a rate of 1 Hz.

1.3.7. Image analysis

The aim of the analysis is to extract the P_f of the membrane by curve fitting of an experimental obtained time course of cell volume changes with a simulated one. Simulation requires reconstruction of the time course of bath perfusion from measurements of density of an indicator dye.

1.3.8. Establishing the time course of cell volume changes

The image sequence should be loaded into the ImageJ software environment (http://rsb.info.nih.gov/ImageJ) and processed with a plug-in designed to ease the tracking of protoplasts ("protoplast analyzer" program; Volkov *et al.*, 2007). The protoplast area is determined, following automated detection and vectorization of their edges. Calibration and standardization of the conversion procedure to maintain strictly the exact cell size and verifying the size estimate using similarly processed images of 15-μm–diameter beads (Molecular Probes, Eugene, OR) and images of a microscope calibration slide (10 μm/div). The cells cross-sectional areas within the contours are calculated using the aforementioned method and converted to cell volumes based on geometry of a sphere.

2. AQUAPORIN FUNCTION IN PLANTS

2.1. Expression analysis

2.1.1. Northern analysis and *in situ* polymerase chain reaction (PCR)

Analysis of plant aquaporin expression is hampered by the high number of aquaporin genes in many plant species. A screen of about 470,000 expressed sequence tags from 215 maize cDNA libraries revealed 31 nucleotide sequences encoding 30 different aquaporin homologous proteins (Chaumont et al., 2001). By computational analysis of genomic sequence data from the model plant A. thaliana (Ward, 2001) (http://www.arabidopsis.org/browse/genefamily/Aquaporins.jsp), it was concluded that 35 genes could potentially code for aquaporin homologues. Because some of these not only belong to the same subfamily but share a high degree in sequence similarity in the coding region, a cross-hybridization with closely related sequences cannot be excluded. This becomes relevant for Northern, Southern, and PCR approaches if specific aquaporins rather than a complete subfamily are supposed to be tested. Fortunately, most aquaporin mRNAs contain untranslated regions large enough to synthesize gene-specific probes for hybridization or a PCR primer that ensures amplification of a single cDNA species. Taking these considerations into account, standard techniques for Northern or real-time PCR are applicable to analyze the mRNA expression of a specific aquaporin. Because the policy of the *Arabidopsis* community is to share experimental data, expression analysis for specific genes such as aquaporins is also available online, for example, at https://www.genevestigator.ethz.ch/at/, a page that was initiated and is run by the Gruissem Laboratory, ETH Zurich, Switzerland (Zimmermann et al., 2005) and provides information collected from many microarray expression analyses. However, in some instances, the experimental approach should be checked carefully to avoid misleading data interpretation from possible cross-reaction with highly homologous sequences.

2.1.2. Western blot

For a first analysis of aquaporin proteins, 5 to 10 μg protein per lane can be separated on a 12% polyacrylamide gel (SDS-PAGE according to Laemmli, 1970). Because aquaporins tend to aggregate easily, it is important not to heat the samples. Denaturing of the proteins is obtained by incubation in loading buffer for 30 min at room temperature. Protein transfer should be carried out in a tank transfer system (Amersham) filled with 10 mM 3-(cyclohexylamino)-1-propanesulfonic acid, 10% methanol, pH 11 (adjusted with NaOH), to a nitrocellulose membrane for 2 h at 80 V. Semidry blotting is less effective. To verify the transfer, membrane-bound proteins should be detected by a reversible, colloidal silver stain. Subsequently, the membrane has to be incubated in a solution containing 2% sodium-citrate, 0.8% $FeSO_4$-heptahydrate, and 0.2% $AgNO_3$ for 10 min under vigorous agitation followed by an extensive wash with demineralized water. After documentation, the membrane is destained with 15 mM potassium hexacyanoferrat(III) and 50 mM sodium thiosulfate. Immunodetection is performed directly after the destaining procedure. The first step is blocking with fat-free milk powder in phosphate-buffered saline (PBS)/0.5% Tween 20, followed by incubation with the antibodies. Detection is done via a chemiluminescent substrate.

2.1.3. *In situ* immunostaining

In situ immunostaining works fine following the procedure given by Ludevid *et al.* (1992) and Tautz and Pfeifle (1989) with slight modifications (M. Bennet, personal communication). To identify false-positive signals produced by the DIG antibody, tissues should be treated with DIG antibody alone. Plant material is sampled by cutting tissue into small pieces with a razor blade. Fixation of samples is done in a solution containing PBS, 6 mM EGTA, and 6% formaldehyde for 25 min at room temperature. If leaf material is used, pigmentation of the tissue can be removed by several washings in ethanol. Antibody incubation (aquaporin-specific antibody and a phosphatase-conjugated secondary antibody) and additional washing steps according to standard protocols follow. During the last step the solution should be supplemented with Levamisol, an inhibitor of lysosomal phosphatases. Detection is carried out in a small volume of a NBT/BCIP dilution in the last washing buffer for several hours. Visualization and documentation can be done under microscope or binocular.

2.2. Modification of aquaporin expression

2.2.1. Antisense

The antisense technique can be used for targeted disruption of messenger RNAs. When antisense versions of specific mRNAs are expressed, double-stranded RNA molecules with the intrinsic, native mRNA form and translation of the targeted gene are blocked. The procedure was applied successfully to block the activity of aquaporin genes (Kaldenhoff *et al.*, 1998; Siefritz *et al.*, 2002). However, as the constructs used for establishing an antisense system often extend over the whole coding sequence, there is the possibility that the antisense effect also affects the expression of closely related aquaporin isoforms. If the aquaporin under investigation has a profound role in plant physiology, it can be anticipated that plant morphology and/or physiology adjusts to the nonoptimal conditions. Accordingly, compensation effects, such as an increase of the root/shoot mass ratio or reduction of transpiration, were observed (Kaldenhoff *et al.*, 1998; Martre *et al.*, 2002). Aharon *et al.* (2003) detected a decrease in root/shoot mass ratio when they overexpressed *Arabidopsis* PIP1b in tobacco, indicating that root volume or surface is regulated in correlation to the root hydraulic conductance, which also correlates positively with the expression level of aquaporins.

Martre *et al.* (2002) produced double-antisense *Arabidopsis* plants with reduced expression of PIP1 and PIP2 aquaporins by sexual crossing of the respective antisense lines. Compared to controls, the osmotic hydraulic conductivity of isolated root and leaf protoplasts was reduced 5- to 30-fold. The double antisense plants had a 3-fold decrease in root hydraulic conductivity expressed on the basis of root dry mass, but a compensating 2.5-fold increase in the ratio of root/leaf dry mass. As a result, the hydraulic conductance of the whole plant was nearly unchanged (Martre *et al.*, 2002).

2.2.2. RNAi

The use of short interfering RNA fragments to induce formation of double-stranded RNA affects the interference with the gene of interest and appears to have a more specific effect than the antisense technique if the appropriate region of the mRNA is chosen as the target. Bots *et al.* (2005b) applied the RNAi technique on tobacco to suppress the expression of aquaporins. They used a 450-bp fragment specific to the PIP2 family of aquaporins cloned into the vector pGSA1165 (*Arabidopsis* Biological Resource Center [ABRC], Accession No. CD3-450), which in turn is derived from pCAMBIA 1200 (Cambia, Canberra, Australia). The fragment was inserted correctly and in an inverted orientation separated by a chalcone synthase intron, thus building a 450-bp RNA double strand with both strands being connected by the intron loop. Gene constructs encoding intron-spliced inverted repeat RNAs with a hairpin structure can induce post-transcriptional gene silencing with an efficiency of almost 100% (Smith *et al.*, 2000).

In contrast to Bots *et al.* (2005b), whose goal was to suppress the expression of a complete subfamily of aquaporins, Ma *et al.* (2004) intended to use the same technique to induce RNA interference against TIP1;1 in *Arabidopsis*. They used a partial CDS (575 bp) to build their RNAi construct, which, as shown by real-time PCR, also interfered with TIP1;2 expression, as both genes have an overall identity of 81% and a 29 nucleotide stretch of absolute identity. Similarly, a reduction in TIP1;3, which shows a stretch of 19 bp identical to the RNAi segment used, cannot completely be excluded, although its reduction, as seen in microarrays, was less pronounced than that observed for TIP1;2 (Ma *et al.*, 2004).

Overall, RNA interference might not be entirely gene specific, but the specificity depends on the sequence used to silence the gene and its sequence identity to closely homologous genes. When inducing an RNA interference specific to a certain aquaporin gene, compensation of the RNAi effect by close homologues is possible and should be taken into consideration.

2.2.3. Overexpression

The literature provides many examples of overexpression of endogenous and heterologous aquaporin gene *in planta*. To overexpress an aquaporin in plants, the coding sequence of the gene of interest is introduced into the plant genome under the control of a specific promoter. In general, there are three different types of promoters: (1) plant-derived promoters responding to stress signals or activating plant defense, (2) chemically inducible promoters, and (3) constitutive promoters.

Use of the constitutive 35S promoter from cauliflower mosaic virus for overexpression of aquaporins in plants has been described elsewhere (Aharon *et al.*, 2003; Katsuhara *et al.*, 2003; Reisen *et al.*, 2003; Yu *et al.*, 2005). However, under the control of a constitutive promoter, overexpression starts in the young seedling and persists throughout plant life, and thus may influence the expression pattern of other aquaporins during development of the plant.

In contrast, chemical gene induction is a useful tool used to increase the expression of selected aquaporins at any time during plant development. These systems, which use chemically inducible promoters responding to signals that are normally not recognized by plants, such as tetracyclin or ethanol, have been described as systems suitable for overexpression of aquaporins in plants (Gatz, 1997; Uehlein *et al.*, 2003).

Lian *et al.* (2004) used the stress-inducible SWPA2 promoter for homologous overexpression of the aquaporin RWC3 in rice. Compared to the wild-type plant, transgenic rice exposed to water stress exhibited higher root osmotic hydraulic conductivity (L_p), leaf water potential, and relative cumulative transpiration after 10 h of polyethylene glycol treatment. These results indicated that RWC3 probably played a role in drought avoidance in rice.

2.2.4. Reverse genetics

There are many examples in the literature of how plant aquaporin function can be assessed using inhibitors, antisense, or RNAi techniques. No doubt that treatment of a complex organism such as a higher plant with inhibitors of aquaporin function, such as heavy metals, could have unforeseen side effects. Furthermore, the use of transgenes to reduce aquaporin expression might downregulate not only the target gene, but also closely related genes. Particularly in plants where complete genome sequence data are not yet available and thus it is not known if closely related aquaporins exist, this is a point of precariousness. In both cases it is difficult to completely exclude side effects unless the effect on other aquaporins respectively their expression is analyzed. Reverse genetics, in the sense that the aquaporin gene or even just the cDNA is known and the physiological function in the plant is to be characterized, provides tools where single aquaporin gene knockouts can be analyzed without side effects on membranes or on gene expression of other similar genes. In many organisms, homologous recombination is the method of choice if specific genes are subjected to mutagenesis; however, in higher plants it occurs infrequently and random integration predominates. Only in lower plants such as the moss *Physcomitrella patens* does homologous recombination take place at a frequency that allows targeted gene knockout (Hohe *et al.*, 2004). A database search is available at http://www.cosmoss. org/cnt/. New, just recently developed techniques used to increase homologous recombination in higher plants by coexpression of protein factors exist (Wright *et al.*, 2005), but are not yet systematically employed for a general approach. Currently, T-DNA or transposon-tagged lines can be used in routine processes to screen for insertions in any desired gene and also in aquaporin genes. With this regard, *Arabidopsis* provides the most advanced support (The *Arabidopsis* Genome Initiative, 2000), for example, by a www supported search in T-DNA collections called "T-DNA express" at http:// signal.salk.edu/cgi-bin/tdnaexpress?TDNA=S1, which is run and maintained by the Salk Institute Genomic Analysis Laboratory. For any particular gene name, annotation, description, or sequence the location of the T-DNA insertion can be identified and a corresponding line can be ordered from one of the stock centers. The entry "aquaporin" in the field function produces 10 hits for different aquaporin genes, with several lines per genomic location that can be ordered from *Arabidopsis* Biological Resource Center (ABRC, http://www.biosci.ohio-state.edu/~plantbio/Facilities/ abrc/abrchome.htm), RIKEN Bioresource Center (http://www.brc.riken. jp/lab/epd/Eng/index.shtml), or Nottingham *Arabidopsis* Stock Center (NASC, http://arabidopsis.info/). Similar approaches have been developed for other important plants, such as rice at the OryGenesDB at http:// orygenesdb.cirad.fr/index.htm, maize at the Maize GDB, or tomato at the Tomato Genetics Resource Center (http://tgrc.ucdavis.edu/).

2.3. Hydraulic conductivity of plant organs and single cells

2.3.1. High-pressure flowmeter (HPFM)

An easy to handle system used to measure the hydraulic conductivity of intact root or shoot systems or leaves was described by Tyree *et al.* (1995). The authors developed a high-pressure flowmeter working in principle as an inverted "Scholander bomb." To measure a complete root system, shoots are decapitated and roots are connected to the device via a rubber gasket compression fitting. The system is pressurized and thus water is perfused from the root stalk to the soil. Care has to be taken that the HPFM is free of air bubbles and that the compression fitting is not overtightened, as this will severely influence the measurement. The system contains two pressure transducers sensing the pressure before and after switchable plastic capillary tubings with known hydraulic conductivity. The plant object is connected in series after the second pressure transducer. In this way water flows from the first sensor through the capillary tube to the second sensor and from there through the plant material. Assuming that the water flow through the reference capillary equals the flow through the plant object, the hydraulic conductivity of the plant material equals the hydraulic conductivity of the capillary at a given pressure difference. The advantage of using the HPFM is that the plant tissue is not compressed by the applied pressure, as can be the case with the "Scholander bomb."

The experimenter can choose between two recording modes: transient and steady rate. When doing transient measurements the perfusion pressure is increased linearly, whereas flow readings are recorded in specified time intervals. The rate of water flow into the root will exceed the rate of water passage through the root system. At the beginning of pressure application the root may be dehydrated to a certain extent and thus some water will remain in the root to rehydrate the cells. Furthermore, some small air bubbles may be present in the plant material, which will be compressed upon pressure application and will add an additional component to the overall flow. Performing three or more transient measurements in a row on the same object will reduce dehydration and remove small air bubbles. The slope of the calculated flow versus the pressure at the second pressure transducer is a reasonable estimation of the root system. It is recommended to perform transient measurements between 0.25 and 0.5 MPa to reduce the under-estimation of root hydraulic conductance because of compression of air bubbles. The slope of flow versus pressure is related to the root surface area to obtain the specific root hydraulic conductivity.

Steady-state measurements can also be performed with the HPFM. During steady-state measurements the applied pressure is held constant and the flow can be followed over a time range of at least 30 min up to a couple of days. After reaching equilibrium the water flow into the object will equal the water flow out of the object. The flow rate depends on the hydraulic conductivity of the plant material, which changes with the expression level of aquaporins, as reported previously (Siefritz *et al.*, 2002). Analysis of hydraulic conductance over a period of days indicated a correlation of diurnal changes in aquaporin expression and specific hydraulic conductance (Siefritz *et al.*, 2002; Uehlein *et al.*, 2006).

2.3.2. Pressure probe

The cell pressure probe as constructed by Hüsken *et al.* (1978) is a micro-manometer used to continuously measure and manipulate the hydrostatic pressure of single plant cells from suspension cultures, but also *in situ*. It can be used to analyze the hydraulic architecture of tissues, cell growth and movement, and the responses of tissues to changes in water status. Manipulating the cell pressure and measuring the pressure relaxation allow calculating the hydraulic conductivity (L_p) of cell membranes.

Using this technique, Volkov et al. (2007) showed an increased hydraulic conductance of growing versus nongrowing barley leaf tissues. In growing tissues an increased aquaporin activity can be expected, but a direct contribution of aquaporins evoking the measured differences in L_p was not proven.

Characterization of an *Arabidopsis* PIP2;2 knockout line obtained by T-DNA insertion revealed a 27 to 28% reduction in the cell hydraulic conductivity Lp (Javot et al., 2003), indicating that this single isoform was of major importance for water transport in root cortical cells. Despite this huge change in single cell hydraulic conductivity, growth and development of the PIP2;2 knockout plants were not different from wild-type plants.

2.4. Development

2.4.1. Germination and cell elongation

Water uptake is one of the initial steps during seed germination. Subsequently, the radicle and embryo axis elongate (Bewley, 1997). Because the initial steps and cell elongation require high rates of water flow into the cells, a role of plant aquaporins during germination was anticipated (Maurel, 1997). Accordingly, expression of specific aquaporins was observed in seeds of *Arabidopsis* (Johnson et al., 1990) and other plant species, such as ice plant (*Mesembryanthemum crystallinum*) (Fukuhara and Bohnert, 2000; Miyazaki et al., 1999). A systematic screen for gene expression during seed germination after treatment with the germination hormone gibberellic acid showed an increased expression of aquaporins (Ogawa et al., 2003). A first approach for studying the function of aquaporins during seedling germination was shown by determining radicle emergence rates and testa rupture rates and by observing seedling morphology in the absence and presence of aquaporin inhibitors (Vander Willigen et al., 2006).

Despite the processes related to germination, cell elongation can also be observed in adult plants in so-called elongation regions and growth zones of plant organs. As indicative for a correlation between cell elongation and aquaporin function, a localized expression was used. This was assessed by Northern hybridization with tissue-specific RNA or standard *in situ* hybridization techniques (Chaumont et al., 1998; Kaldenhoff et al., 1995; Siefritz et al., 2001; Werner et al., 2001), as well as aquaporin promotor-reporter gene expression (Kaldenhoff et al., 1995, 1996; Mizutani et al., 2006; Siefritz et al., 2004). Most plants do not grow with a constant rate, but show times of increased and reduced growth during a 24-h day. During the so-called epinastic leaf movement, upper and lower sides of the leaf and petiole grow with different velocities, causing a change in leaf position relative to the shoot axis (Otto and Kaldenhoff, 2000). Cell growth is initiated by local cell wall loosening (Cosgrove, 1996) and osmotic driving forces, which are activated by proton pump-powered fluxes of ions (Moran *et al.*, 1996; Philippar *et al.*, 1999) and water following the osmotic gradient. Using antisense plants it was demonstrated that specific aquaporins contribute to the epinastic leave movement, and their physiological role could be monitored by measuring the leaf ankle with regard to the shoot axis in aquaporin antisense plants and comparing these to controls (Siefritz *et al.*, 2004). Aquaporins were also found to be expressed in flower organs (Bots *et al.*, 2005a). Effects of RNAi-triggered repression of a PIP2 aquaporin in these flower organs (anthers) were analyzed by observation of retarded organ development in aquaporin RNAi plants, by determination of flower bud rehydration kinetics and *in vivo* nuclear magnetic resonance relaxometry. Using this technique, individual organs (anthers) can be analyzed; it is possible to measure the water status in great detail in living tissue over a longer time period, and the discrimination of water in different compartments of the anther can be observed (Bots *et al.*, 2005b).

2.5. Water and salt stress

The impact of aquaporins' contribution to whole plant water status can be analyzed by comparing plants with modified aquaporin expression under beneficial growth conditions and abiotic stress such as water or salt stress. The capacity to transport water within and between plant organs plays a major role not only under beneficial growth conditions, but also under conditions of water deficiency.

Siefritz *et al.* (2002) induced water stress of NtAQP1 antisense tobacco plants by drought or application of PEG. Drought produced a relatively lower stress effect, reducing the soil water potential from -0.01 to -0.07 MPa. A more severe water stress was achieved by irrigating the plants with a PEG solution with an osmotic potential of -0.35 MPa (138.4 g/liter PEG 6000). Because PEG is not root permeable it simulates soil desiccation. In contrast to the low stress treatment, where no visible plant reactions were observed, the PEG treatment induced wilting of the plants with impaired expression of NtAQP1, starting about 2 h after onset of treatment. This finding showed a contribution of NtAQP1 to water stress avoidance.

Aharon *et al.* (2003) obtained different results when they applied both drought and salt stress to tobacco plants overexpressing *Arabidopsis* PIP1b. To initiate the drought stress, the irrigation of nearly 3-week-old plants was stopped. Transgenic plants started wilting 20 days after stopping irrigation, which was 6 days earlier than the wild-type control plants. For salt stress conditions, 4-week-old plants were treated with tap water or tap water containing 90 mM NaCl. Upon irrigation with tap water, AthPIP1b overexpressing tobacco plants reached a significantly higher fresh weight. When treated with 90 mM NaCl for a period of 40 days the plant fresh weight was reduced substantially and no difference between transgenic and control plants was observed.

2.6. Photosynthesis

The resistance to diffusion of CO_2 from the atmosphere to the site of carboxylation in the chloroplast stroma is finite and limits leaf photosynthesis. In order to get there, CO_2 coming from the atmosphere has to cross different barriers, among them the plasma membrane of the mesophyll cells and at least two chloroplast envelope membranes. The presence of CO_2 conducting aquaporins in these membrane resistances should directly affect the photosynthetic capacity of the plant. An influence of aquaporin expression on photosynthetic performance of plants was shown (Aharon *et al.*, 2003; Flexas *et al.*, 2006; Hanba *et al.*, 2004; Uehlein *et al.*, 2003).

There are two different ways to evaluate the photosynthetic performance of a plant. Measuring the CO_2 and water vapor exchange rates of leaves includes both light-dependent and light-independent reactions or performing chlorophyll fluorescence analyses photosynthetic capacity of the plant can be estimated via activity of photosystem II.

2.6.1. Gas exchange

Several instruments used to measure leaf or whole plant gas exchange, which is commonly uptake of CO_2 and release of water vapor through the stomata, are available, for example, from LI-COR Biosciences (http://www.licor.com/), Heinz Walz GmbH (http://www.walz.com/), and PP Systems (http://www.ppsystems.com/). An air flow containing a defined concentration of CO_2 and water vapor is moved across a plant leaf enclosed in an assimilation chamber. The cores of these photosynthesis systems are differential infrared gas analyzers measuring the difference in CO_2 and water concentration between the air stream coming from the chamber containing the leaf and another air stream that does not have contact with the leaf. After the gas analyzer the gas mixture leaves the system. This configuration is called open mode.

Analyzing plant gas exchange, Uehlein *et al.* (2003) showed a direct correlation between the expression of NtAQP1 and the photosynthetic performance of tobacco plants. Net photosynthesis of tobacco plants overexpressing NtAQP1 was increased about 40%, whereas upon reduced expression the photosynthesis rate was reduced roughly 40%. Aharon *et al.* (2003) overexpressed the *Arabidopsis* PIP1b in tobacco and showed an induction of net photosynthesis of about 30 to 40% depending on the overexpression line. Hanba *et al.* (2004), who overexpressed HvPIP2;1 in rice, also were able to show an increase in photosynthesis. To date this is the only example pointing at CO_2 conductivity of a PIP2 aquaporin.

2.6.2. Chlorophyll fluorescence

Measuring chlorophyll fluorescence is a useful tool to analyze photosynthetic energy conversion. Chlorophyll fluorimeters can be obtained, for example, from Hansatech Instruments Ltd. (http://www.hansatech-instruments.com/) or Heinz Walz GmbH (http://www.walz.com/).

There is an inverse correlation between chlorophyll fluorescence and photosynthetic efficiency. Light energy absorbed by a photosystem can proceed into different directions: heat dissipation, photochemistry, and fluorescence emission. Measuring the emission of fluorescence light allows an estimation of the proportion of energy flowing to photochemistry.

Aharon *et al.* (2003) performed chlorophyll fluorescence measurements on tobacco plants overexpressing *Arabidopsis* PIP1b. They detected a direct correlation between expression of PIP1b and the maximum quantum efficiency of dark-adapted leaves. The authors showed a higher photochemical quantum efficiency, measured as the maximum chlorophyll photochemical efficiency of photosystem II in dark-adapted leaves (F_v/F_m), which is a measure of photosystem II organization and vitality. Flexas *et al.* (2006) analyzed the response of electron transport rate (ETR) to the substomatal CO_2 concentration (c_i) on tobacco plants with increased and reduced expression of NtAQP1. They showed that at a given c_i ETR was positively correlated with the expression level of NtAQP1, pointing out that NtAQP1 contributes to CO_2 conductivity of mesophyll cells.

ACKNOWLEDGMENT

Financial support by the Deutsche Forschungsgemeinschaft (DFG, Grant KA 1032/15-2) is greatly acknowledged.

REFERENCES

Aharon, R., Shahak, Y., Wininger, S., Bendov, R., Kapulnik, Y., and Galili, G. (2003). Overexpression of a plasma membrane aquaporin in transgenic tobacco improves plant vigor under favorable growth conditions but not under drought or salt stress. *Plant Cell* **15,** 439–447.

Anderson, J. A., Huprikar, S. S., Kochian, L. V., Lucas, W. J., and Gaber, R. F. (1992). Functional expression of a probable *Arabidopsis thaliana* potassium channel in *Saccharomyces cerevisiae*. *Proc. Natl. Acad. Sci. USA* **89,** 3736–3740.

Barkla, B. J., Vera–Estrella, R., Pantoja, O., Kirch, H. H., and Bohnert, H. J. (1999). Aquaporin localization: How valid are the TIP and PIP labels? *Trends Plant Sci.* **4,** 86–88.

Bertl, A., Anderson, J. A., Slayman, C. L., and Gaber, R. F. (1995). Use of *Saccharomyces cerevisiae* for patch–clamp analysis of heterologous membrane proteins: Characterization of Kat1, an inward–rectifying K^+ channel from *Arabidopsis thaliana*, and comparison with endogenous yeast channels and carriers. *Proc. Natl. Acad. Sci. USA* **92,** 2701–2705.

Bewley, J. D. (1997). Seed germination and dormancy. *Plant Cell* **9**, 1055–1066.

Bienert, G. P., Møller, A. L. B., Kristiansen, K. A., Schulz, A., Møller, I. M., Schjoerring, J. K., and Jahn, T. P. (2006). Specific aquaporins facilitate the diffusion of hydrogen peroxide across membranes. *J. Biol. Chem.* **282**, 1183–1192.

Bock, R. (1998). Analysis of RNA editing in plastids. *Methods* **15**, 75–83.

Bonhivers, M., Carbrey, J. M., Gould, S. J., and Agre, P. (1998). Aquaporins in *Saccharomyces*; genetic and functional distinctions between laboratory and wild–type strains. *J. Biol. Chem.* **273**, 27565–27572.

Bots, M., Feron, R., Uehlein, N., Weterings, K., Kaidenhoff, R., and Mariani, T. (2005a). PIP1 and PIP2 aquaporins are differentially expressed during tobacco anther and stigma development. *J. Exp. Bot.* **56**, 113–121.

Bots, M., Vergeldt, F., Wolters–Arts, M., Weterings, K., van As, H., and Mariani, C. (2005b). Aquaporins of the PIP2 class are required for efficient anther dehiscence in tobacco. *Plant Physiol.* **137**, 1049–1056.

Chaumont, F., Barrieu, F., Herman, E. M., and Chrispeels, M. J. (1998). Characterization of a maize tonoplast aquaporin expressed in zones of cell division and elongation. *Plant Physiol.* **117**, 1143–1152.

Chaumont, F., Barrieu, F., Wojcik, E., Chrispeels, M. J., and Jung, R. (2001). Aquaporins constitute a large and highly divergent protein family in maize. *Plant Physiol.* **125**, 1206–1215.

Christiansen, J. H., Rosendahl, L., and Widell, S. (1995). Preparation and characterization of seal inside–out peribacteroid membrane vesicles from *Pisum sativum* L. and *Glycine max*. L. root nodules by aqueous polymer two–phase partitioning. *J. Plant Physiol.* **147**, 175–181.

Ciavatta, T. V., Morillon, R., Pullman, G. S., Chrispeels, M. J., and Cairney, J. (2001). An aquaglyceroporin is abundantly expressed early in the development of the suspensor and the embryo proper of loblolly pine. *Plant Phys.* **127**, 1556–1567.

Cline, K., Andrews, J., Mersey, B., Newcomb, E. H., and Keegstra, K. (1981). Separation and characterization of inner and outer envelope membranes of pea chloroplasts. *Proc. Natl. Acad. Sci. USA* **78**, 3595–3599.

Cosgrove, D. J. (1996). Plant cell enlargement and the action of expansins. *Bioessays* **18**, 533–540.

Daniels, M. J., Wood, M. R., and Yeager, M. (2006). *In vivo* functional assay of a recombinant aquaporin in *Pichia pastoris*. *Appl. Environ. Microbiol.* **72**, 1507–1514.

Dascal, N. (1987). The use of *Xenopus* oocytes for the study of ion channels. *CRC Crit. Rev. Biochem.* **22**, 317–387.

Felle, H., and Bertl, A. (1986). The fabrication of H^+–selective liquid–membrane micro–electrodes for use in plant cells. *J. Exp. Bot.* **37**, 1416–1428.

Flexas, J., Ribas–Carbo, M., Hanson, D. T., Bota, J., Otto, B., Cifre, J., McDowell, N., Medrano, H., and Kaldenhoff, R. (2006). Tobacco aquaporin NtAQP1 is involved in mesophyll conductance to CO_2 *in vivo*. *Plant J.* **48**, 427–439.

Frommer, W. B., Hummel, S., and Riesmeier, J. W. (1993). Expression cloning in yeast of a cDNA encoding a broad specificity amino acid permease from *Arabidopsis thaliana*. *Proc. Natl. Acad. Sci. USA* **90**, 5944–5948.

Fukuhara, T., and Bohnert, H. J. (2000). The expression of a Vp1–like gene and seed dormancy in *Mesembryanthemum crystallinum*. *Genes Genet. Syst.* **75**, 203–209.

Gatz, C. (1997). Chemical control of gene expression. *Annu. Rev. Plant Physiol. Plant Mol. Biol.* **48**, 89–108.

Hanba, Y. T., Shibasaka, M., Hayashi, Y., Hayakawa, T., Kasamo, K., Terashima, I., and Katsuhara, M. (2004). Overexpression of the barley aquaporin HvPIP2;1 increases internal CO_2 conductance and CO_2 assimilation in the leaves of transgenic rice plants. *Plant Cell Physiol.* **45**, 521–529.

Hohe, A., Egener, T., Lucht, J. M., Holtorf, H., Reinhard, C., Schween, G., and Reski, R. (2004). An improved and highly standardised transformation procedure allows efficient production of single and multiple targeted gene–knockouts in a moss, *Physcomitrella patens. Curr. Genet.* **44,** 339–347.

Holm, L. M., Jahn, T. P., Møller, A. L. B., Schjoerring, J. K., Ferri, D., Klaerke, D. A., and Zeuthen, T. (2005). NH_3 and NH_4^+ permeability in aquaporin–expressing *Xenopus* oocytes. *Pflüg. Arch. Eur. J. Physiol.* **450,** 415–428.

Holyoak, C. D., Bracey, D., Piper, P. W., Kuchler, K., and Coote, P. J. (1999). The *Saccharomyces cerevisiae* weak–acid–inducible ABC transporter Pdr12 transports fluorescein and preservative anions from the cytosol by an energy–dependent mechanism. *J. Bacteriol.* **181,** 4644–4652.

Hsu, L. C., Chiou, T. J., Chen, L., and Bush, D. R. (1993). Cloning a plant amino acid transporter by functional complementation of a yeast amino acid transport mutant. *Proc. Natl. Acad. Sci. USA* **90,** 7441–7445.

Hüsken, D., Steudle, E., and Zimmermann, U. (1978). Pressure probe technique for measuring water relations of cells in higher plants. *Plant Physiol.* **61,** 158–163.

Ikeda, M., Beitz, E., Kozono, D., Guggino, W. B., Agre, P., and Yasui, M. (2002). Characterization of aquaporin–6 as a nitrate channel in mammalian cells: Requirement of pore–lining residue threonine 63. *J. Biol. Chem.* **277,** 39873–39879.

Ishikawa, F., Suga, S., Uemura, T., Sato, M. H., and Maeshima, M. (2005). Novel type aquaporin SIPs are mainly localized to the ER membrane and show cell–specific expression in *Arabidopsis thaliana. FEBS Lett.* **579,** 5814–5820.

Jahn, T. P., Møller, A. L., Zeuthen, T., Holm, L. M., Klaerke, D. A., Mohsin, B., Kuhlbrandt, W., and Schjoerring, J. K. (2004). Aquaporin homologues in plants and mammals transport ammonia. *FEBS Lett.* **574,** 31–36.

Javot, H., Lauvergeat, V., Santoni, V., Martin–Laurent, F., Guclu, J., Vinh, J., Heyes, J., Franck, K. I., Schaffner, A. R., Bouchez, D., and Maurel, C. (2003). Role of a single aquaporin isoform in root water uptake. *Plant Cell* **15,** 509–522.

Johnson, K. D., Hofte, H., and Chrispeels, M. J. (1990). An intrinsic tonoplast protein of protein storage vacuoles in seeds is structurally related to a bacterial solute transporter (GIpF). *Plant Cell* **2,** 525–532.

Kaldenhoff, R. (2006). Besides water: Functions of plant membrane intrinsic proteins and aquaporins. *Progr. Bot.* **67,** 206–218.

Kaldenhoff, R., and Fischer, M. (2006). Functional aquaporin diversity in plants. *Biochim. Biophys. Acta* **1758,** 1134–1141.

Kaldenhoff, R., Grote, K., Zhu, J. J., and Zimmermann, U. (1998). Significance of plasmalemma aquaporins for water–transport in *Arabidopsis thaliana. Plant J.* **14,** 121–128.

Kaldenhoff, R., Kolling, A., Meyers, J., Karmann, U., Ruppel, G., and Richter, G. (1995). The blue light–responsive AthH2 gene of *Arabidopsis thaliana* is primarily expressed in expanding as well as in differentiating cells and encodes a putative channel protein of the plasmalemma. *Plant J.* **7,** 87–95.

Kaldenhoff, R., Kolling, A., and Richter, G. (1996). Regulation of the *Arabidopsis thaliana* aquaporin gene AthH2 (PIP1b). *J. Photochem. Photobiol. B* **36,** 351–354.

Katsuhara, M., Koshio, K., Shibasaka, M., Hayashi, Y., Hayakawa, T., and Kasamo, K. (2003). Over–expression of a barley aquaporin increased the shoot/root ratio and raised salt sensitivity in transgenic rice plants. *Plant Cell Physiol.* **44,** 1378–1383.

Keegstra, K., and Yousif, A. E. (1986). Isolation and characterization of chloroplast envelope membranes. *Methods Enzymol.* **118,** 316–325.

Kjellbom, P., and Larsson, C. (1984). Preparation and polypeptide composition of chlorophyll–free plasma–membranes from leaves of light–grown spinach and barley. *Physiol. Plant.* **62,** 501–509.

Laemmli, U. K. (1970). Cleavage of structural proteins during the assembly of the head of bacteriophage T4. *Nature* **227**, 680–685.

Laize, V., Rousselet, G., Verbavatz, J. M., Berthonaud, V., Gobin, R., Roudier, N., Abrami, L., Ripoche, P., and Tacnet, F. (1995). Functional expression of the human CHIP28 water channel in a yeast secretory mutant. *FEBS Lett.* **373**, 269–274.

Laize, V., Tacnet, F., Ripoche, P., and Hohmann, S. (2000). Polymorphism of *Saccharomyces cerevisiae* aquaporins. *Yeast* **16**, 897–903.

Lian, H. L., Yu, X., Ye, Q., Ding, X., Kitagawa, Y., Kwak, S. S., Su, W. A., and Tang, Z. C. (2004). The role of aquaporin RWC3 in drought avoidance in rice. *Plant Cell Physiol.* **45**, 481–489.

Liu, K., Kozono, D., Kato, Y., Agre, P., Hazama, A., and Yasui, M. (2005). Conversion of aquaporin 6 from an anion channel to a water–selective channel by a single amino acid substitution. *Proc. Natl. Acad. Sci. USA* **10**, 2192–2197.

Loque, D., Ludewig, U., Yuan, L., and von Wiren, N. (2005). Tonoplast intrinsic proteins AtTIP2;1 and AtTIP2;3 facilitate NH_3 transport into the vacuole. *Plant Physiol.* **137**, 671–680.

Ludevid, D., Hofte, H., Himelblau, E., and Chrispeels, M. J. (1992). The expression pattern of the tonoplast intrinsic protein gamma–TIP in *Arabidopsis thaliana* is correlated with cell enlargement. *Plant Physiol.* **100**, 1633–1639.

Lundborg, T., Widell, S., and Larsson, C. (1981). Distribution od ATPases in wheat root membranes separated by phase partition. *Physiol. Plant.* **52**, 89–95.

Ma, S., Quist, T. M., Ulanov, A., Joly, R., and Bohnert, H. J. (2004). Loss of TIP1;1 aquaporin in *Arabidopsis* leads to cell and plant death. *Plant J.* **40**, 845–859.

Marini, A. M., Soussi–Boudekou, S., Vissers, S., and Andre, B. (1997). A family of ammonium transporters in *Saccharomyces cerevisiae*. *Mol. Cell. Biol.* **17**, 4282–4293.

Martre, P., Morillon, R., Barrieu, F., North, G. B., Nobel, P. S., and Chrispeels, M. J. (2002). Plasma membrane aquaporins play a significant role during recovery from water deficit. *Plant Physiol.* **130**, 2101–2110.

Maurel, C. (1997). Aquaporins and water permeability of plant membranes. *Annu. Rev. Plant Physiol. Plant Mol. Biol.* **48**, 399–429.

Maurel, C., Reizer, J., Schroeder, J. I., and Chrispeels, M. J. (1993). The vacuolar membrane protein γ–TIP creates water specific channels in *Xenopus* oocytes. *EMBO J.* **12**, 2241–2247.

Miyazaki, S., Koga, R., Bohnert, H. J., and Fukuhara, T. (1999). Tissue– and environmental response–specific expression of 10 PP2C transcripts in *Mesembryanthemum crystallinum*. *Mol. Gen. Genet.* **261**, 307–316.

Mizutani, M., Watanabe, S., Nakagawa, T., and Maeshima, M. (2006). Aquaporin NIP2;1 is mainly localized to the ER membrane and shows root–specific accumulation in *Arabidopsis thaliana*. *Plant Cell Physiol.* **47**, 1420–1426.

Moran, N., Yueh, Y. G., and Crain, R. G. (1996). Signal transduction and cell volume regulation in plant leaflet movements. *News Physiol. Sci.* **11**, 108–114.

Moshelion, M., Moran, N., and Chaumont, F. (2004). Dynamic changes in the osmotic water permeability of protoplast plasma membrane. *Plant Physiol.* **135**, 2301–2317.

Nakamura, R. L., and Gaber, R. F. (1998). Studying ion channels using yeast genetics. *Methods Enzymol.* **293**, 89–104.

Niemietz, C. M., and Tyerman, S. D. (2000). Channel–mediated permeation of ammonia gas through the peribacteroid membrane of soybean nodules. *FEBS Lett.* **465**, 110–114.

Ogawa, M., Hanada, A., Yamauchi, Y., Kuwahara, A., Kamiya, Y., and Yamaguchi, S. (2003). Gibberellin biosynthesis and response during *Arabidopsis* seed germination. *Plant Cell* **15**, 1591–1604.

Otto, B., and Kaldenhoff, R. (2000). Cell–specific expression of the mercury–insensitive plasma–membrane aquaporin NtAQP1 from *Nicotiana tabacum*. *Planta* **211**, 167–172.

Pettersson, N., Hagstrom, J., Bill, R. M., and Hohmann, S. (2006). Expression of heterologous aquaporins for functional analysis in *Saccharomyces cerevisiae*. *Curr. Genet.* **50,** 247–255.

Philippar, K., Fuchs, I., Luthen, H., Hoth, S., Bauer, C. S., Haga, K., Thiel, G., Ljung, K., Sandberg, G., Bottger, M., Becker, D., and Hedrich, R. (1999). Auxin–induced K(+) channel expression represents an essential step in coleoptile growth and gravitropism. *Proc. Natl. Acad. Sci. USA* **96,** 12186–12191.

Prudent, S., Marty, F., and Charbonnier, M. (2005). The yeast osmosensitive mutant fps1Δ transformed by the cauliflower BobTIP1;1 aquaporin withstand a hypo–osmotic shock. *FEBS Lett.* **579,** 3872–3880.

Reisen, D., Leborgne–Castel, N., Ozalp, C., Chaumont, F., and Marty, F. (2003). Expression of a cauliflower tonoplast aquaporin tagged with GFP in tobacco suspension cells correlates with an increase in cell size. *Plant Mol. Biol.* **52,** 387–400.

Riesmeier, J. W., Willmitzer, L., and Frommer, W. B. (1992). Isolation and characterization of a sucrose carrier cDNA from spinach by functional expression in yeast. *EMBO J.* **11,** 4705–4713.

Sauer, N., and Stadler, R. (1993). A sink–specific H+/monosaccharide co–transporter from *Nicotiana tabacum*: Cloning and heterologous expression in baker's yeast. *Plant J.* **4,** 601–610.

Secchi, F., Lovisolo, C., Uehlein, N., Kaldenhoff, R., and Schubert, A. (2006). Isolation and functional characterization of three aquaporins from olive (*Olea europaea* L.). *Planta* **225,** 381–392.

Sentenac, H., Bonneaud, N., Minet, M., Lacroute, F., Salmon, J. M., Gaymard, F., and Grignon, C. (1992). Cloning and expression in yeast of a plant potassium ion transport system. *Science* **256,** 663–665.

Siefritz, F., Biela, A., Eckert, M., Otto, B., Uehlein, N., and Kaldenhoff, R. (2001). The tobacco plasma membrane aquaporin NtAQP1. *J. Exp. Bot.* **52,** 1953–1957.

Siefritz, F., Otto, B., Bienert, G. P., van der Krol, A., and Kaldenhoff, R. (2004). The plasma membrane aquaporin NtAQP1 is a key component of the leaf unfolding mechanism in tobacco. *Plant J.* **37,** 147–155.

Siefritz, F., Tyree, M. T., Lovisolo, C., Schubert, A., and Kaldenhoff, R. (2002). PIP1 plasma membrane aquaporins in tobacco: From cellular effects to function in plants. *Plant Cell.* **14,** 869–876.

Smith, N. A., Singh, S. P., Wang, M. B., Stoutjesdijk, P. A., Green, A. G., and Waterhouse, P. M. (2000). Total silencing by intron–spliced hairpin RNAs. *Nature* **407,** 319–320.

Soveral, G., Veiga, A., Loureiro–Dias, M. C., Tanghe, A., Van Dijck, P., and Moura, T. F. (2006). Water channels are important for osmotic adjustments of yeast cells at low temperature. *Microbiology* **152,** 1515–1521.

Tanghe, A., Kayingo, G., Prior, B. A., Thevelein, J. M., and Van Dijck, P. (2005). Heterologous aquaporin (*AQY2–1*) expression strongly enhances freeze tolerance of *Schizosaccharomyces pombe*. *J. Mol. Microbiol. Biotechnol.* **9,** 52–56.

Tautz, D., and Pfeifle, C. (1989). A non–radioactive *in situ* hybridization method for the localization of specific RNAs in *Drosophila* embryos reveals translational control of the segmentation gene hunchback. *Chromosoma* **98,** 81–85.

Temmei, Y., Uchida, S., Hoshino, D., Kanzawa, N., Kuwahara, M., Sasaki, S., and Tsuchiya, T. (2005). Water channel activities of *Mimosa pudica* plasma membrane intrinsic proteins are regulated by direct interaction and phosphorylation. *FEBS Lett.* **579,** 4417–4422.

The *Arabidopsis* Genome Initiative (2000). Analysis of the genome sequence of the flowering plant *Arabidopsis thaliana*. *Nature* **408,** 796–815.

Tyree, M. T., Patino, S., Bennink, J., and Alexander, J. (1995). Dynamic measurements of root hydraulic conductance using a high–pressure flowmeter in the laboratory and field. *J. Exp. Bot.* **46,** 83–94.

Uehlein, N., and Kaldenhoff, R. (2006). Aquaporins and biological rhythm. *Biol. Rhythm Res.* **37,** 315–322.

Uehlein, N., Lovisolo, C., Siefritz, F., and Kaldenhoff, R. (2003). The tobacco aquaporin NtAQP1 is a membrane CO_2 pore with physiological functions. *Nature* **425,** 734–737.

Vander Willigen, C., Postaire, O., Tournaire–Roux, C., Boursiac, Y., and Maurel, C. (2006). Expression and inhibition of aquaporins in germinating *Arabidopsis* seeds. *Plant Cell Physiol.* **47,** 1241–1250.

Verkman, A. S. (2000). Water permeability measurement in living cells and complex tissues. *J. Membr. Biol.* **173,** 73–87.

Volkov, V., Hachez, C., Moshelion, M., Draye, X., Chaumont, F., and Fricke, W. (2007). Water permeability differs between growing and non–growing barley leaf tissues. *J. Exp. Bot.* **58,** 377–390.

Wallace, I. S., and Roberts, D. M. (2005). Distinct transport selectivity of two structural subclasses of the Nodulin–like intrinsic protein family of plant aquaglyceroporin channels. *Biochemistry* **44,** 16826–16834.

Ward, J. M. (2001). Identification of novel families of membrane proteins from the model plant *Arabidopsis thaliana. Bioinformatics* **17,** 560–563.

Werner, M., Uehlein, N., Proksch, P., and Kaldenhoff, R. (2001). Characterization of two tomato aquaporins and expression during the incompatible interaction of tomato with the plant parasite. *Cuscuta reflexa. Planta* **213,** 550–555.

Widell, S., and Larsson, C. (1981). Separation of presumptive plasma–membranes from mitochondria by partition in an aqueous polymer 2–phase system. *Physiol. Plant.* **51,** 368–374.

Wright, D. A., Townsend, J. A., Winfrey, R. J., Irwin, P. A., Rajagopal, J., Lonosky, P. M., Hall, B. D., Jondle, M. D., and Voytas, D. F. (2005). High–frequency homologous recombination in plants mediated by zinc–finger nucleases. *Plant J.* **44,** 693–705.

Yasui, M., Hazama, A., Kwon, T. H., Nielsen, S., Guggino, W. B., and Agre, P. (1999). Rapid gating and anion permeability of an intracellular aquaporin. *Nature* **402,** 184–187.

Yu, Q. J., Hu, Y. L., Li, J. F., Wu, Q., and Lin, Z. P. (2005). Sense and antisense expression of plasma membrane aquaporin BnPIP1 from *Brassica napus* in tobacco and its effect on plant drought resistance. *Plant Sci.* **169,** 647–656.

Zhang, R., Logee, K., and Verkman, A. (1990). Expression of mRNA coding for kidney and red cell water channels in *Xenopus* oocytes. *J. Biol. Chem.* **265,** 15375–15378.

Zimmermann, P., Hennig, L., and Gruissem, W. (2005). Gene–expression analysis and network discovery using Genevestigator. *Trends Plant Sci.* **10,** 407–409.

Author Index

G

Subject Index

A

Abscisic acid
  osmotic regulation, 488–492, 494–495
  salinity stress tolerance role, 427–429
  stress response pathways, 111, 422
Acid sphingomyelinase, hepatocyte osmotic stress
    effects on activity, 148–151
Actin cytoskeleton
  biochemical analysis during cell volume
      changes, 230–231
  DNase I inhibition assay for quantification,
      233–235
  dynamics assessment, 232–233
  functional overview, 228
  immunoblot analysis of detergent-insoluble
      pellets and supernatants, 236–237
  morphological changes during cell volume
      changes, 229–230
  rhodamine–phalloidin fluorescence
      measurement of filamentous actin,
      235–236
  signaling pathways linking reorganization and
      cell volume regulation, 231–232
Adipocyte, see Glucose transport
Anhydrosignaling
  anhydrobiotic engineering of mammalian cells,
      270, 272
  cell viability test, 273
  gene expression quantification, 275
  mammalian cell culture and desiccation,
      272–273
  overview, 271–272
  Western blot, 273–275
Annexin V, Cell Lab Quant SC staining in
    apoptotic volume decrease, 169–170,
    173, 175
Antisense oligonucleotides, aquaporin
    knockdown in plants, 518–519
Apoptotic volume decrease
  CD95 hyperosmotic activation in hepatocytes
    acid sphingomyelinase activity, 148–151
    chloride channel upregulation, 151
    epidermal growth factor receptor
        interactions, 152–153
    inhibition studies, 156–157
    ion exchange effects, 148
    oligomerization, 153–156
    pH changes, 148–151

  protein kinase C activation, 151–152
  translocation to plasma membrane, 148
  Yes activation, 152
  definition, 162–163
  flow cytometry analysis
    Cell Lab Quant SC
      annexin V staining, 169–170, 173, 175
      caspase activity detection, 170–171,
          175, 177
      cell preparation, 169
      cell size analysis, 169–170
      data acquisition, 172
      data interpretation, 172, 174–175, 177,
          179–180
      fluorescence calibration, 166–167
      instrument start-up and calibration, 166
      nucleic acid staining, 168, 173
      overview, 165–166
      plasma membrane potential
          determination, 171–172, 177,
          179–180
      volume calibration, 167
    classical assays, 163, 165
  overview, 146
Aquaporins, plants
  development studies, 523–524
  expression analysis
    immunostaining, 518
    Northern blot, 517
    polymerase chain reaction, 517
    Western blot, 517–518
  functional overview, 506–507
  hydraulic conductivity studies
    high-pressure flowmeter, 521–522
    pressure probe, 522–523
  knockdown
    antisense oligonucleotides, 518–519
    RNA interference, 519–520
  membrane isolation for function studies
    chloroplast membrane, 508
    peribactoid membrane, 508
    plasma membrane, 507–508
  overexpression in plants, 520
  photosynthesis assays
    chlorophyll fluorescence, 526
    gas exchange, 525
  reverse genetics studies, 520–521
  single cell analysis
    bath perfusion, 516

571

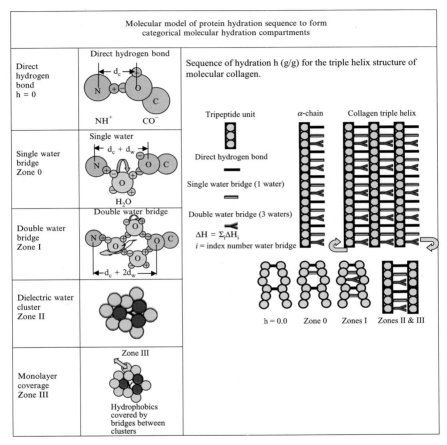

Gary D. Fullerton and Ivan L. Cameron, Figure 1.2 This conceptual diagram shows the sequential formation of single water bridges (zone 0), double water bridges (zone I), dielectric water clusters (zone II), and monolayer coverage (zone III) with completed fourfold hydrogen bonding of all bound waters in the dielectric clusters. Water molecules in each zone have differences in free energy, melting temperature, osmosis, and other properties of biological significance.

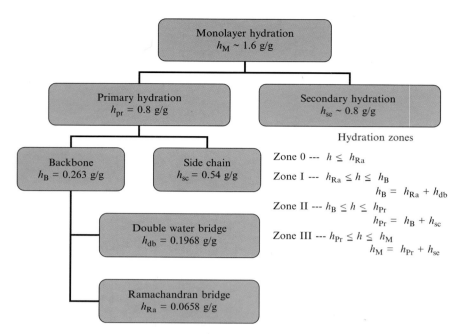

Gary D. Fullerton and Ivan L. Cameron, Figure 1.3 There are a total of seven water hydration compartments predicted by the molecular hydration model as shown here for collagen. Of the seven compartments, four capacities for zones 0, I, II, and III, h_{Ra}, h_B, h_{Pr}, and h_M, can be measured directly, whereas the other three are calculated from the measured results. Reproduced with permission from Fullerton and Rahal (2007).

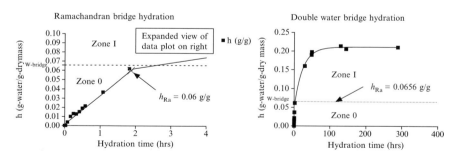

Gary D. Fullerton and Ivan L. Cameron, Figure 1.6 The rehydration of collagen from the "absolute" dry state in a wet atmosphere of 35% relative humidity at 22° is described by two zones with boundary between zone 0 (Ramachandran bridges) and zone I (double water bridges) as defined in Fig. 1.3. Rehydration of the single water bridges occurs in less than 2 h as shown on the left, whereas accumulation of water with a rehydration time of several days to equilibrium is shown on the right. The nonlinear best fit to the two-zone data set governing gives $h_{Ra} = 0.06$ g/g at the intersection shown by the "X" at the intersection that is compared with the theoretical value $h_{Ra} = 0.0656$ shown by the dotted line or a measurement error of approximately 2%. Reproduced with permission from Cameron et al. (2007).

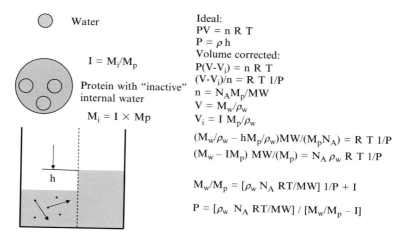

Osmotically inactive volume
(water)

Water

$I = M_i/M_p$

Protein with "inactive"
internal water

$M_i = I \times M_p$

Ideal:
$PV = n\,R\,T$
$P = \rho\,h$
Volume corrected:
$P(V-V_i) = n\,R\,T$
$(V-V_i)/n = R\,T\,1/P$
$n = N_A M_p/MW$
$V = M_w/\rho_w$
$V_i = I\,M_p/\rho_w$

$(M_w/\rho_w - hM_p/\rho_w)MW/(M_p N_A) = R\,T\,1/P$
$(M_w - IM_p)\,MW/(M_p) = N_A\,\rho_w\,R\,T\,1/P$

$M_w/M_p = [\rho_w\,N_A\,RT/MW]\,1/P + I$

$P = [\rho_w\,N_A\,RT/MW] / [M_w/M_p - I]$

Gary D. Fullerton and Ivan L. Cameron, Figure 1.10   Derivation of the volume-corrected expression for osmotic pressure that accounts for encapsulated water fraction $h_{enc} = I$, including water bridges, dielectric water clusters, and internal bulk water. As shown in Figs. 1.11 and 1.12, the encapsulated water fraction is variable but generally in the range from 1.4 to 4.0 g/g for most proteins under native mammalian conditions (Fullerton et al., 1993).

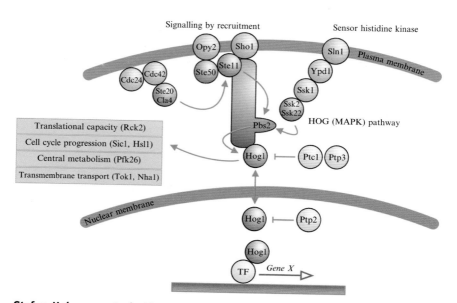

Stefan Hohmann et al., Figure 2.1   The yeast HOG pathway signaling system and overview of response mechanisms. The pathway consists of two branches: the Sho1 branch (Cdc42/24: G protein, Ste20: PAK protein kinase, Opy2, Sho1: membrane protein scaffold, Ste11/Ste50: MAPKKK, Pbs2: MAPKK) and the Sln1 branch (with Sln1-Ypd1-Ssk1: sensing histidine kinase phosphorelay system, Ssk2/22: MAPKKK), Hog1: MAPK, Ptc1, Ptp2/3: protein phosphatases, TF: transcription factor.

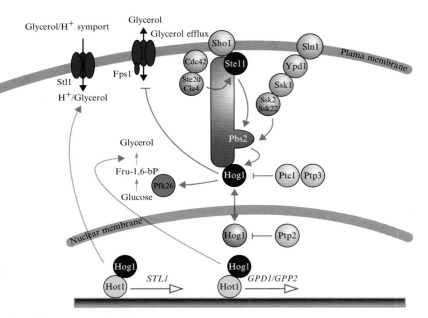

Stefan Hohmann *et al.*, Figure 2.2 The HOG pathway controls water and glycerol homeostasis. At the level of gene expression the capacity for glycerol uptake and production are increased by Hog1. Hog1 also appears to stimulate glycolytic flux by activating Pfk26 to produce fructose–2,6–bisphosphate. Finally, Hog1 may participate in the control of the glycerol efflux channel Fps1.

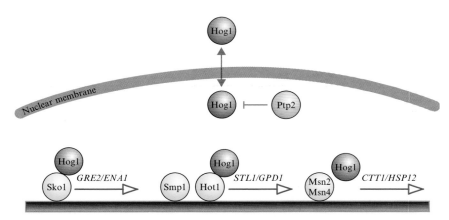

Stefan Hohmann *et al.*, Figure 2.3 Hog1 and control of gene expression. Hog1 interacts with different transcription factors in the nucleus, such as Sko1, Hot1, and probably also Smp1, to control different sets of genes (some typical examples are shown). It appears that Hog1 thereby itself is the activating factor.

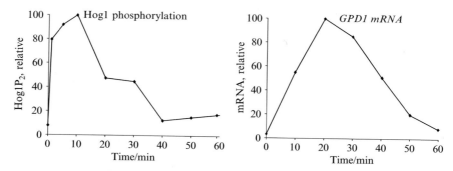

Stefan Hohmann *et al.*, Figure 2.4 Osmotic responses are transient. An example of quantification of the level of phospho-Hog1 and the mRNA of the Hog1-target gene *GPD1* is shown. Cells were treated with 0.5 *M* NaCl at time 0.

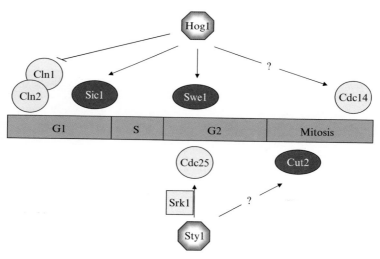

J. Clotet and F. Posas, Figure 4.3 The osmocheckpoint modulates cell cycle transitions through SAPK activation. SAPKs from different yeasts act on several factors that control cell cycle progression positively (clear circles) or negatively (dark ellipses). Question marks indicate that the exact mechanism of action is not known. This figure suggests that SAPKs are able to protect cells from osmoinsults in any phase of the cell cycle.

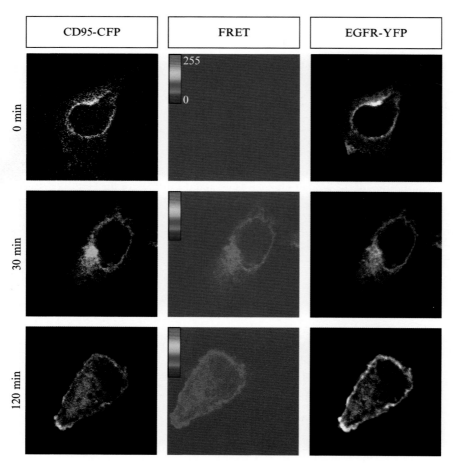

Roland Reinehr and Dieter Häussinger, Figure 8.1 Hyperosmolarity-induced CD95–CFP/EGFR-YFP association in Huh7 cells. Huh7 hepatoma cells were cotransfected with CD95-CFP and EGFR-YFP and then exposed to hyperosmolarity (405 mOsm/liter) for 0, 30, and 120 min. FRET pictures were taken according to Eberle *et al.* (2005) and then normalized with respect to FRET efficiencies (scale indicates FRET efficiency from blue/0 to red/255). Hyperosmolarity induced an intracellular FRET signal within 30 min, indicating an intracellular CD95–CFP/EGFR-YFP association, and within 120 min a membrane translocation of the protein complex, as indicated by a FRET signal at the plasma membrane.

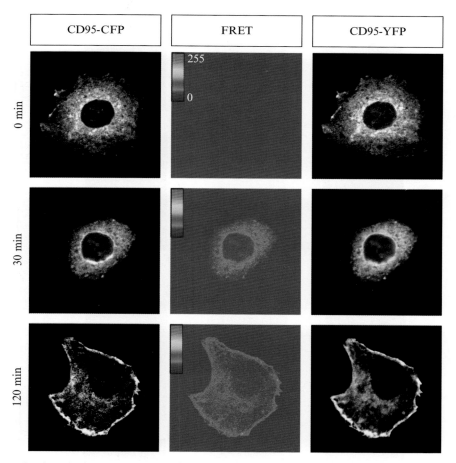

Roland Reinehr and Dieter Häussinger, Figure 8.2 Hyperosmolarity-induced CD95–CFP/CD95–YFP oligomerization in Huh7 cells. Huh7 hepatoma cells were cotransfected with CD95–CFP and CD95–YFP and then exposed to hyperosmolarity (405 mOsm/liter) for 0, 30, and 120 min. FRET pictures were taken according to Eberle *et al.* (2007) and then normalized with respect to FRET efficiencies (scale indicates FRET efficiency from blue/0 to red/255). Hyperosmolarity induced an intracellular FRET signal within 30 min, indicating an intracellular CD95–CFP/CD95–YFP association, and within 120 min a FRET signal at the plasma membrane, suggestive for a translocation of the former formed protein complex.

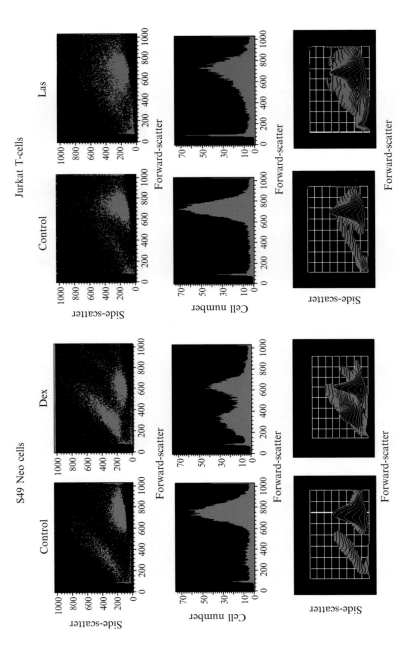

Carl D. Bortner et al., Figure 9.1 Changes in the light-scattering properties of dexamethasone (Dex)–treated S49 Neo- and FasL–treated Jurkat T cells. The observation of normal and shrunken apoptotic cells by flow cytometry is based on changes in the light-scattering properties of the cells. The degree of this change in light scatter is both cell type and stimulus specific. The shrunken population of apoptotic cells can be clearly observed by flow cytometry for S49 Neo cells treated with Dex for 48 h. In both a forward-scatter versus side-scatter dot plot and a forward-scatter histogram, two distinct populations are present comparing Dex-treated cells to control cells. However, similar examination of Jurkat T cells treated with Fas ligand for 4 h shows that the normal and shrunken populations of these cells cannot be clearly distinguished.

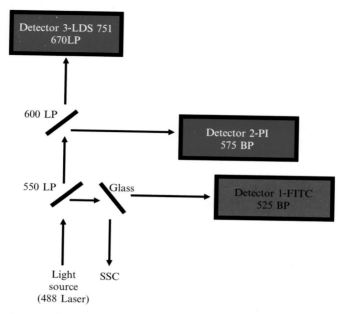

Carl D. Bortner *et al.*, Figure 9.2 Optical setup for the Cell Lab Quanta SC. The Cell Lab Quanta SC is equipped with a 488-nm argon laser, three photomultiplier tubes (PMTs) for fluorescence analysis, and a side-scatter channel. Our optical setup uses a series of long-pass filters to split the emitted fluorescent signal to detector 1 (525 nm BP; FITC), detector 2 (575 nm BP; PI), and detector 3 (670LP; LDS 751).

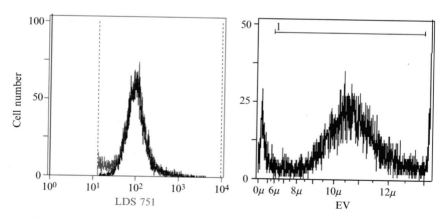

Carl D. Bortner *et al.*, Figure 9.3 Initial gating strategy for examining AVD during apoptosis. LDS 751, a cell-permeant nucleic acid dye, was added to all the samples. For our optical setup, LDS 751 is detected in the FL-3 channel. The addition of LDS 751 allows for a better signal-to-noise ratio, as well as an unbiased analysis of the electronic volume of the cells. Additionally, a gate was used on the electronic volume histogram to eliminate debris and clumps of cells.

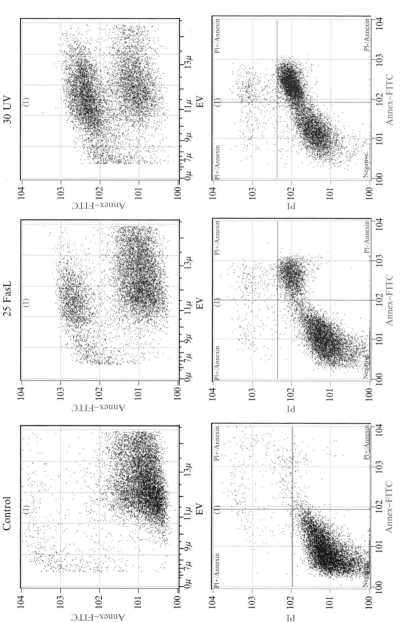

Carl D. Bortner et al., Figure 9.4   Analysis of Annexin V–FITC and AVD during apoptosis. Jurkat T cells were treated with 25 ng of Fas ligand or 30 mJ/cm² of UV for 3 h to induce apoptosis. These samples were then stained with Annexin–FITC and PI and examined on the Cell Lab Quanta SC. A distinct population of Annexin–FITC–positive stained cells can be observed on an electronic volume versus an Annexin–FITC dot plot for the treated samples, indicating externalization of the phosphatidylserine (PS). Examination of these cells on an Annexin–FITC versus PI dot plot shows that the externalization of PS occurs prior to the loss of membrane integrity, indicative of apoptosis. Quadrants set on the Annexin–FITC versus PI dot plots were used to determine changes in cell size based on the electronic volume parameter.

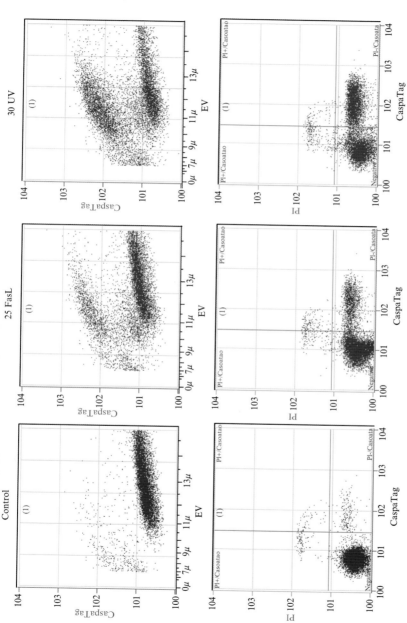

Carl D. Bortner et al., Figure 9.5 Analysis of CaspaTag and AVD during apoptosis. Jurkat T cells were treated with 25 ng of Fas ligand or 30 mJ/cm² of UV for 3 h to induce apoptosis. These samples were then stained with CaspaTag reagent for an additional hour prior to the addition of PI and examination on the Cell Lab Quanta SC. A distinct population of CaspaTag positive-stained cells can be observed on an electronic volume versus CaspaTag dot plot for treated samples, indicating caspase activity. Examination of these cells on an CaspaTag versus PI dot plot shows that the caspase activity occurs prior to the loss of membrane integrity. Quadrants set on the CaspaTag versus PI dot plot were used to determine changes in cell size based on the electronic volume parameter.

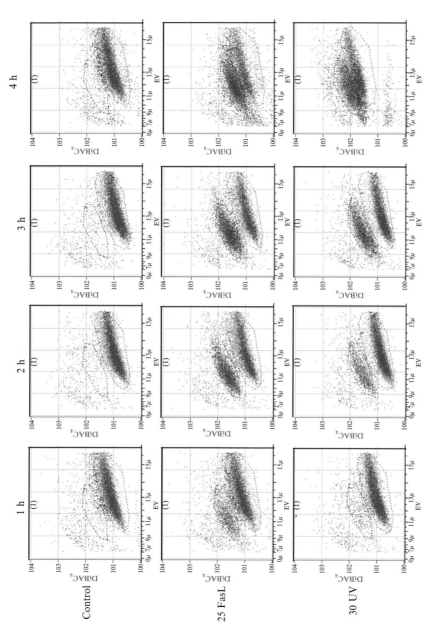

Carl D. Bortner et al., Figure 9.6 Analysis of DiBAC₄ fluorescence and AVD during apoptosis. Jurkat T cells were treated with 25 ng/ml FasL or 30 mJ/cm² UV to induce apoptosis. DiBAC₄ was used to determine changes in the plasma membrane potential over a period of 4 h. Dot plots of electronic volume (EV) versus DiBAC₄ show a time-dependent increase in DiBAC₄ fluorescence for the treated cells (green population). Gates were set on these plots to determine changes in cell size and volume for each population of cells.

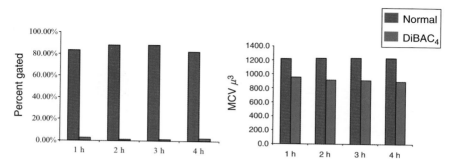

Carl D. Bortner *et al.*, Figure 9.7 Percentage gated and mean cell volume (MCV) for control DiBAC$_4$-stained cells over time. No significant change in the percentage of cells gated or the MCV was observed for the normal (red) population of cells or the small percentage of DiBAC$_4$-positive (green) cells over time, indicating that DiBAC$_4$ is not toxic to the cells.

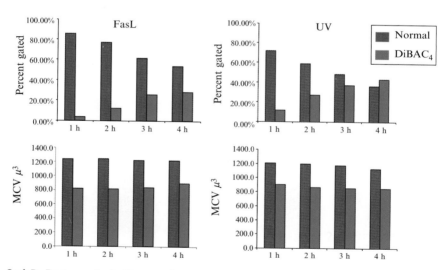

Carl D. Bortner *et al.*, Figure 9.8 Percentage gated and mean cell volume (MCV) for FasL- and UV-treated DiBAC$_4$-stained cells over time. Under each apoptotic condition, an increase in the number of DiBAC$_4$-positive cells, along with a simultaneous decrease in the number of normal (DiBAC$_4$ negative) cells, was observed over a period of 4 h. In contrast, no change in MCV was observed for either FasL- or UV-treated normal or DiBAC$_4$-positive populations of cells, suggesting that the change in MCV remains constant over this period of time.

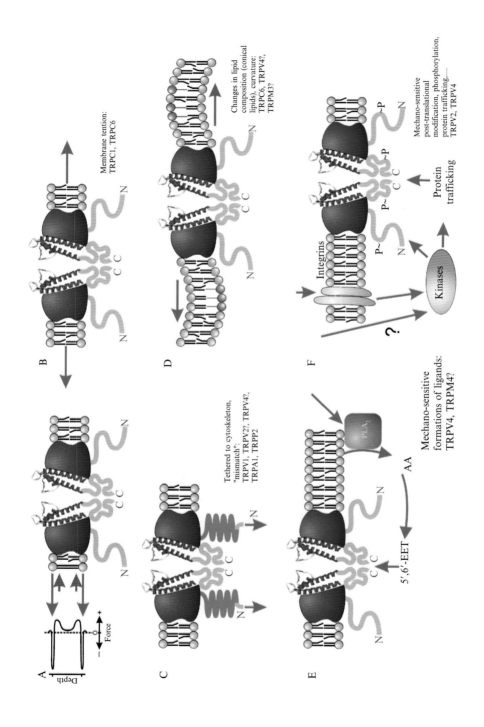

A

Depth

Force + − 0

B

Membrane tention:
TRPC1, TRPC6

N

C C

N

C

Tethered to cytoskeleton,
"mismatch":
TRPV1, TRPV2?, TRPV4?,
TRPA1, TRPP2

N

C C

N

D

Changes in lipid
composition (conical
lipids), curvature:
TRPC6, TRPV4?,
TRPM3?

N

C C

E

Mechano-sensitive
formations of ligands:
TRPV4, TRPM4?

PLA₂

AA

5',6'-EET

N

C

F

Integrins

Protein
trafficking

Mechano-sensitive
post-translational
modification, phosphorylation,
protein trafficking......
TRPV2, TRPV4

Kinases

P~

P~ ~P

N C C ~P

N

?

Stine Falsig Pedersen and Bernd Nilius, Figure 10.1 Some possible mechanisms of mechano- or osmosensing by TRP channels. (A) Any channel in a plasma membrane is exposed to lateral forces from the bilayer (see energy profile at the left-hand side), which create positive or negative forces (lateral pressure indicated by the arrows) acting on the channel. In equilibrium, the conformational energy of the channel matches this energy profile. Any changes in this equilibrium may cause modulation of channel activity, e.g., gating. (B) Channel gating by changes in bilayer tension, which will alter the tension profile shown in A (referred to in the text as mechanism 1). (C) Channel gating by tethering to cytoskeletal elements, which are reorganized by mechanical or osmotic stress (mechanism 2). (D) Channel gating by changes in membrane curvature (mechanism 3). (E) Channel gating by an intracellular messenger that can be produced by an enzyme acting as a mechano/osmo/volume sensor (e.g., volume-sensitive activation of PLA_2, resulting in the generation of arachidonic acid, which either itself or in the form of its downstream metabolic products gates the channel (mechanism 4). (F) Channel gating as a consequence of regulation by signaling events induced by mechanical or osmotic stress can induce signaling events, e.g., via integrins or kinases, which in turn regulate both the activity of channels resident in the plasma membrane and the membrane insertion/retrieval of channels (mechanism 5). See text for details.

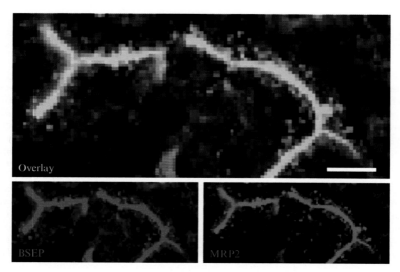

Ralf Kubitz and Dieter Häussinger, Figure 18.2 Transporter distribution in human liver. Cryofixed human liver (from surgically resected liver tissue) was stained for the bile salt export pump BSEP (red) and the bilirubin transporter MRP2 (green). Most immunoreactivity is found in a tubular configuration of the canaliculi, and some is found in vesicular structures within the cells. The two transporter proteins are distributed unequally within these vesicles. Bar: 10 μm.

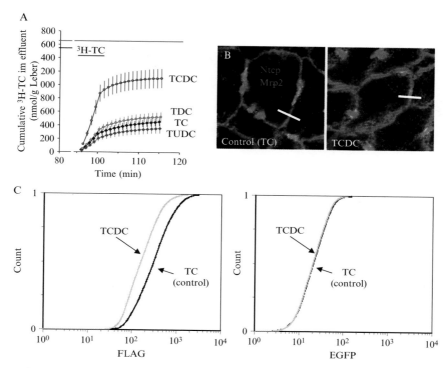

Ralf Kubitz and Dieter Häussinger, Figure 18.3 Bile salt-dependent modulation of bile salt uptake. (A) Rat livers were perfused with 25 μmol/liter of bile salts [taurocholate (TC), taurochenodeoxycholate (TCDC), taurodeoxycholate (TDC), or tauroursodeoxycholate (TUDC)] for 30 min. After a washout period of 20 min, a 5-min pulse of radiolabeled taurocholate ([³H]TC) was administered. Recovery of radioactivity in the effluent revealed that TCDC but not the other bile salts reduced net uptake of [³H]TC. (B) TCDC induced a retrieval of Ntcp, as shown by immunofluorescence of rat liver slices after perfusion with 25 μmol/liter TCDC + 100 μmol/liter TC compared to 25 μmol/liter TC + 100 μmol/liter TC (control). (C) HepG2 cells were transfected with double-tagged Ntcp. An extracellular FLAG tag was used to quantify Ntcp at the cell membrane by an anti-FLAG antibody, while intracellular EGFP was used to measure total Ntcp. As compared to TC (control, black line), incubation with TCDC (100 μmol/liter; 1 h; gray line) induced a shift of FLAG fluorescence toward lower values, indicating a reduction of membrane-bound Ntcp. Total Ntcp was not changed, as shown by constant EGFP fluorescence under both conditions.

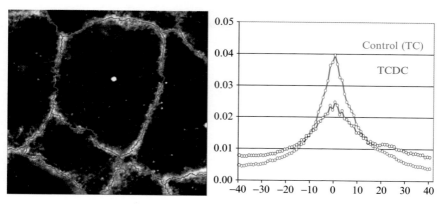

Ralf Kubitz and Dieter Häussinger, Figure 18.4 Quantification of transporter distribution by fluorescence densitometry. (A) Rat livers were perfused with 25 μmol/liter of TC (control) or TCDC. After immunostaining with an anti–Ntcp antibody, fluorescence pictures were recorded by confocal scanning microscopy (as shown in Fig. 18.3B). Pictures were analyzed by a semiautomatic software: single cells were identified by their border strips. Thereafter, distribution of fluorescence intensity perpendicular to all straight stretches of single cells was measured and averaged as shown on the right side. In TCDC-treated livers, fluorescence profiles of Ntcp were flattened as compared to TC (control)-treated livers.

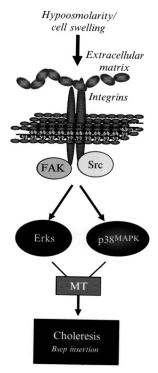

Ralf Kubitz and Dieter Häussinger, Figure 18.5 Cell swelling-induced signaling pathway. Cell swelling triggers integrin attachment to the extracellular matrix (ECM) and integrin activation. Focal adhesion kinases (FAK) and src kinases are thereby activated. Further signal transduction involves the dual activation of Erk-type and p38-type MAP kinases. Eventually, transporter bearing vesicles are inserted into the target membrane in a microtubule (MT)-dependent manner, along with an increase of transporter activity. A similar mechanism is activated nonosmotically by tauroursodeoxycholate and triggers choleresis.

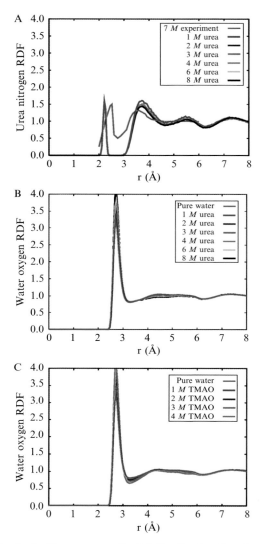

David A. C. Beck *et al.*, Figure 22.2 Radial distribution functions as a function of increasing cosolvent concentration. (A) Nitrogen-centered radial distribution function for urea from simulations at 1, 2, 3, 4, 6, and 8 *M* urea and experiment (Turner *et al.*, 1991); (B) water oxygen radial distribution function for urea at 1, 2, 3, 4, 6, and 8 *M*; and (C) water oxygen radial distribution function for TMAO at 1, 2, 3, and 4 *M*.

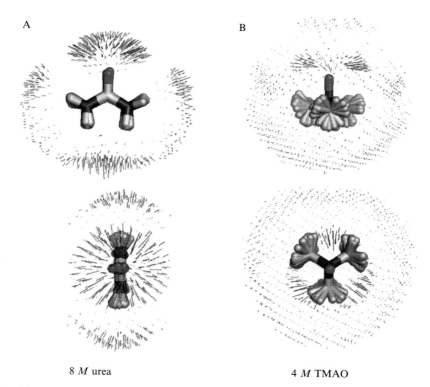

A

B

8 *M* urea

4 *M* TMAO

David A. C. Beck *et al.*, Figure 22.4 Hydration site population and orientation from cosolvent simulations. (A) Urea and (B) TMAO. The population color scale increases from lowest (red) to highest (blue). Only the top 15% populated hydration sites are displayed; i.e., only sites with a $G_{XO}(r)$ above 1.8 pairs are shown.

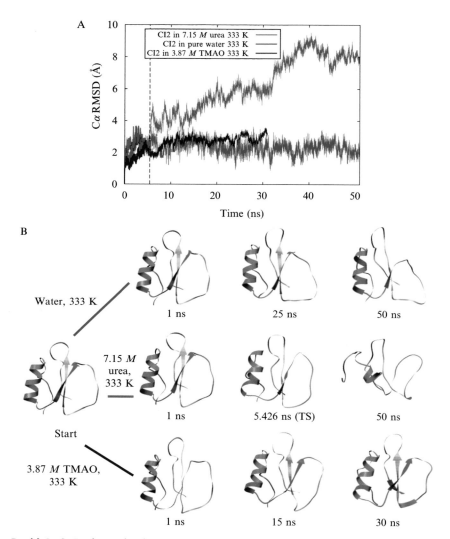

David A. C. Beck *et al.*, Figure 22.5 Cα RMSD as a function of time and snapshots from simulations of CI2 in pure water, 7.15 *M* urea, and 3.87 *M* TMAO. (A) The Cα RMSD of simulation structures to the crystal structure are plotted in green for CI2 in pure water control, red for CI2 in 7.15 *M* urea, and blue for CI2 in 3.87 *M* TMAO. Simulations of CI2 in pure water and 3.87 *M* TMAO at 333 K are stable and well folded; however, in the presence of 7.15 *M* urea, the protein begins to unfold, resulting in the large Cα RMSD. (B) Snapshots of CI2 from the pure water, 7.15 *M* urea, and 3.87 *M* TMAO simulations. The protein is shown as ribbons. In pure water and TMAO, CI2 maintains its native hydrophobic core and overall topology over the course of the simulations. In urea, at the beginning of the simulation (1 ns) CI2 is well structured and waters are excluded from the core. However, at the unfolding TS, which occurs at 5.429 ns, water begins to enter the core of the protein and the active site loop has changed conformation. After this time, urea molecules begin to join and replace waters that have penetrated and disrupted the core (50 ns).

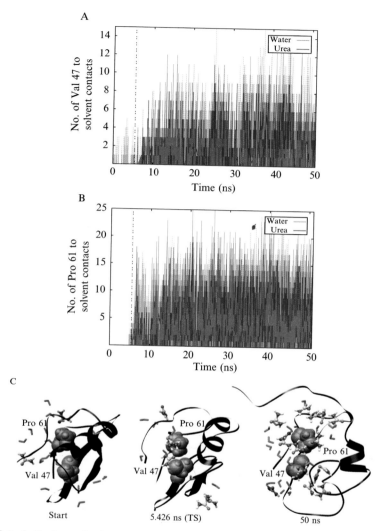

David A. C. Beck *et al.*, Figure 22.6 Water and urea contacts to CI2 core residues as a function of simulation time accompanied by snapshots of unfolding. Simulation of CI2 in 7.15 *M* urea: (A) The number of heavy atom contacts between CI2's Val 47 and water (cyan) and urea (red) as the MD simulation progresses. (B) The number of heavy atom contacts between CI2's Pro 61 and water (cyan) and urea (red) vs simulation time. (C) Snapshots of CI2 showing hydration waters and urea within 5.0 Å of Val 47 and Pro 61. The protein is shown as ribbons with two of the hydrophobic residues of the core shown in space filling. At the beginning of the simulation, there are no urea to Val 47 or Pro 61 contacts, although several water contacts can be observed (a and c). After water attacks the core at the TS (vertical bar in a and b at 5.426 ns), urea molecules are observed entering into the hydrophobic nucleus. By the end of the simulation (50 ns) the protein is denatured and the core is solvated by urea.

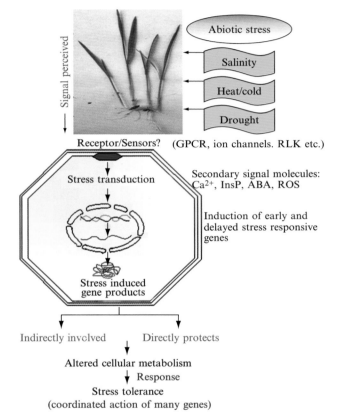

Narendra Tuteja, Figure 24.1 Generic pathway for plant response to stress. The extracellular stress signal is first perceived by the membrane receptors and then activates a large and complex signaling cascade intracellularly, including the generation of secondary signal molecules. The signal cascade results in the expression of multiple stress responsive genes, the products of which can provide the stress tolerance directly or indirectly. Overall, the stress response could be a coordinated action of many genes, which may cross talk with each others. GPCR, G-protein-coupled receptor; RLK, receptor-like kinase; InsP, inositol phosphate; ABA, abscisic acid; ROS, reactive oxygen species.

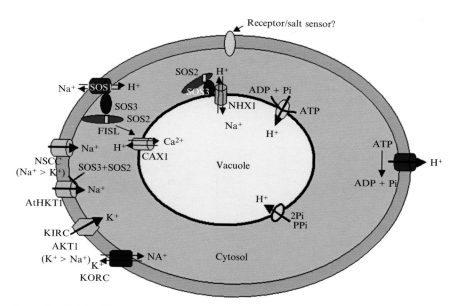

Narendra Tuteja, Figure 24.2 Regulation of ion homeostasis by various ion pumps. The salinity stress signal is perceived by a receptor or salt sensor present at the plasma membrane of the cell. This signal is responsible for activating various ion pumps present at plasma and vacuolar membranes. This signal also activates the SOS pathway, the components of which help in regulating some of these pumps. The various pumps/channels are the K^+ inward-rectifying channel (KIRC), histidine kinase transporter (HKT), non-specific cation channels (NSCC), K^+ outward-rectifying channel (KORC), Na^+/H^+ antiporters (SOS1), vacuolar Na^+/H^+ exchanger (NHX), and H^+/Ca^+ antiporter (CAX1). Na^+ extrusion from plant cells is powered by the electrochemical gradient generated by H^+-ATPases, which permits the Na^+/H^+ antiporters to couple the passive movement of H^+ inside along the electrochemical gradient and extrusion of Na^+ out of the cytosol. The stress signal sensed by SOS3 activates SOS2, which activates SOS1 (for details, see Fig. 24.5 and text).

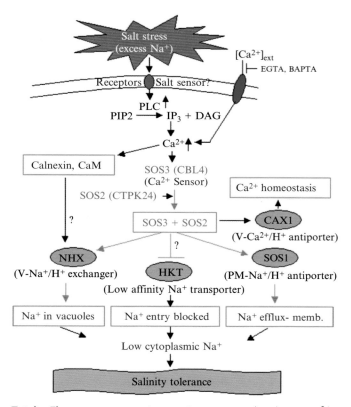

Narendra Tuteja, Figure 24.3 Regulation of ion (e.g., Na^+, K^+, and Ca^{2+}) homeostasis by SOS and related pathways in relation to salinity stress tolerance. High salinity (Na^+) stress initiates a calcium signal that activates the SOS pathway. The signal first activates phospholipase C (PLC), which hydrolyses phosphatidylinositol bisphosphate (PIP2) to generate inositol trisphosphate (IP_3), and diacylglycerol (DAG) resulted in an increased level of Ca^{2+} ions. This induces a cytosolic Ca^{2+} ion sensed by a calcium sensor such as SOS3, which interacts with the SOS2 protein kinase. This SOS3–SOS2 protein kinase complex phosphorylates SOS1, a Na^+/H^+ antiporter, resulting in an efflux of excess Na^+ ions. The SOS3–SOS2 complex interacts with and influences other salt-mediated pathways, resulting in ionic homeostasis. This complex inhibits HKT1 activity (a low-affinity Na^+ transporter), thus restricting Na^+ entry into the cytosol. SOS2 also interacts and activates the vacuolar Na^+/H^+ exchanger (NHX), resulting in the sequestration of excess Na^+ ions, further contributing to Na^+ ion homeostasis. Calnexin and calmodulin (CaM) or other calcium-binding proteins can also interact and activate the NHX or other transporters. The H^+/Ca^{2+} antiporter (CAX1) has been identified as an additional target for SOS2 activity reinstating cytosolic Ca^{2+} homeostasis.

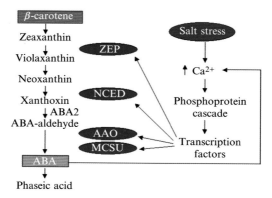

Narendra Tuteja, Figure 24.4 ABA biosynthesis pathway and its regulation by osmotic stress. ABA is synthesized from β-carotene via the oxidative cleavage of neoxanthin and conversion of xanthoxin to ABA via ABA-aldehyde. Stresses, including salinity stress, stimulate ABA biosynthesis and accumulation by activating genes involved in the ABA biosynthetic pathway, which itself could be mediated by a calcium-dependent phosphorylation cascade. ABA can also upregulate the expression of ABA biosynthetic genes via calcium signaling pathways (see Xiong *et al.*, 2002; Zhu, 2002). ZEP, zeaxanthin epoxidase; NCED, 9-*cis*-epoxycarotenoid dioxygenase; AAO, ABA-aldehyde oxidase; MCSU, molybdenum cofactor sulfurase.

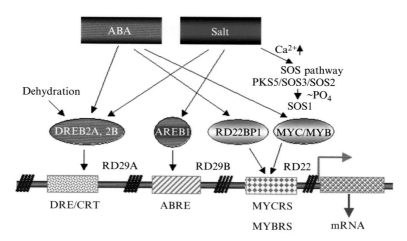

Narendra Tuteja, Figure 24.5 Transcriptional regulatory network of *cis*-acting elements and ABA-dependent transcription factors involved in salinity stress gene expression. Osmotic stress signaling generated via salinity stress seems to be mediated by transcription factors such as DREB2A/DREB2B, AREB1, and MYC/MYB transcription activators, which interact with DRE/CRT, ABRE, and MYCRS/MYBRS elements in the promotion of stress genes, respectively. AtMYC2 and AtMYB2 act cooperatively to activate the expression of ABA-inducible genes such as RD22. Transcription factor-binding sites are represented as rectangles at the bottom of the figure, with the representative promoters. Salinity works mainly through the SOS pathway, reinstating cellular ionic equilibrium.

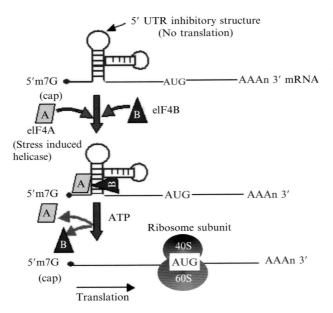

Narendra Tuteja, Figure 24.6 Possible mechanism of stress tolerance by a helicase. Eukaryotic initiation factor 4A (eIF4A) is a prototypic member of the DEAD-box RNA helicase family. Abiotic stresses enhance formation of the inhibitory secondary structure at the 5′ UTR of mRNA. This protein is responsible for removal of the secondary structure of the mRNA. eIF4A, along with eIF4B, binds to 5′ UTR and unwinds the inhibitory secondary structure in an ATP-dependent manner. This facilitates the binding of ribosome. After this ribosome scans for the start codon (AUG) and protein synthesis begins normally, which was inhibited due to negative impact of the stress.

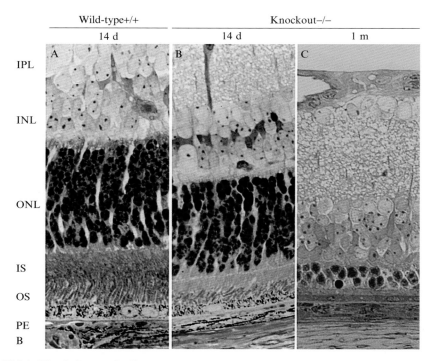

Ulrich Warskulat *et al.*, Figure 25.4   Retinal morphology of wild-type and *taut−/−*
mice. (Left) Retina of wild-type (*taut+/+*) mice at postnatal day 14. Retinas of *taut−/−*
mice at postnatal day 14 (center) and 1 month (right) (magnification ×450). Adapted
from Heller-Stilb *et al.* (2002). IPL, inner plexiform layer; INL, inner nuclear layer;
ONL, outer nuclear layer (photoreceptor nuclei); IS/OS, inner/outer segments of
photoreceptors; PE, pigment epithelium; B, membrane of Bruch.

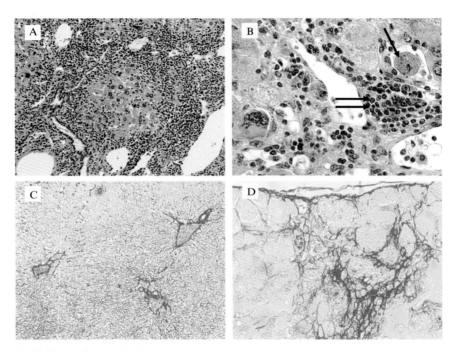

Ulrich Warskulat *et al.*, Figure 25.5   Light microscopy of liver sections from wild-type and *taut−/−* mice. *taut−/−* mice develop variable degrees of hepatitis and fibrosis. (A and B) Chronic hepatitis with presence of apoptotic bodies (arrow) and mixed leukocytic infiltrates (double arrow) in a 19-month-old *taut−/−* mouse (×100, ×400). Liver tissue sections were stained with hematoxylin and eosin. (C and D) Liver fibrosis in *taut−/− * mouse. Liver tissue sections were stained with Sirius red. (C) Seventeen-month-old wild type; (D) 15-month-old *taut−/−* mouse (magnification ×100). Adapted from Warskulat *et al.* (2006b).